MATHS IN ACTION

National 4

R Howat
J McLaughlin
G Meikle
E Mullan
D Murray
E Varrie

Series editor: E Mullan

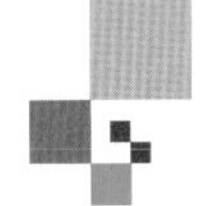

Nelson Thornes

Published in 2013 by:
Nelson Thornes Ltd
Delta Place
27 Bath Road
CHELTENHAM
GL53 7TH
United Kingdom

13 14 15 16 17 / 10 9 8 7 6 5 4 3 2 1

A catalogue record for this book is available from the British Library

ISBN 978 1 4085 1910 3

Cover photograph: Steve Bronstein/Getty
Page make-up by Tech-Set Limited, Gateshead

Printed in China by 1010 Printing International Ltd

Contents

Introduction

This book has been created especially to cater for the needs of the student attempting the National 4 course. It is important that a student sees the relevance of a subject in order to learn it effectively.

With that in mind, wherever possible, real-life scenarios from daily living, the workplace, and a variety of school subjects have been chosen to give questions a context. Inevitably, these contexts do get simplified as other factors not relevant to the occasion are omitted.

It is said that practice makes perfect, and each exercise contains some drill-type questions to facilitate this. There is, however, a larger bank of drill questions available on www.kerboodle.com/mathsinaction. Here you will also find a wealth of differentiated worksheets, assessment support and digital versions of the textbooks.

Each chapter is structured in the same way, and begins with a problem associated with a real-life situation. Although some students may be able to attempt this already, it will prove to be a challenge to most, as they are unlikely to have the necessary skills and experience to solve the problem fully. It should therefore act as an introductory stimulus for bringing into play the strategies for problem solving that students are already familiar with, while perhaps also showing the shortcomings of these for the task in hand. (However, by the time students have worked through the chapter, they should be able to complete the problem successfully, and when the problem is reintroduced at the end of the chapter, they should appreciate how their skills have expanded.)

There follows a feature called 'What you need to know', which contains a few questions that students should be able to tackle using their current knowledge. It is important that students are comfortable with each of these, because the questions exercise the very skills required to get the most from the chapter.

Exposition of the new skills is accompanied by worked examples, which the student is encouraged to mimic in the laying out of their own solutions.

The level of difficulty of questions is indicated in the following ways:

Questions essential to the successful completion of the course are contained in the A exercises. If students can do these, they are on target for success.

A student wishing to take Maths further into National 5, and perhaps Higher, should also be able to handle the strategies needed to complete the B exercises.

A student aiming at Higher should in particular be able to make a good attempt at those B exercise questions that have a green-tinted background.

At the end of each chapter there is a section called 'Preparation for assessment'. If a student cannot do one of these questions they should perhaps revisit the corresponding section in the chapter or ask the teacher for help. This section culminates with a revisit to the introductory question, which by now should be totally accessible. This will act to bring the class together again if students have been working at different levels.

Throughout the book, various icons have been used to identify particular features:

  indicate the most appropriate mode of tackling a problem. As part of the assessment measures students' numerate abilities without the aid of a calculator, it is essential that the parts marked 'non-calculator' are indeed done without a calculator.

 indicates a topic suitable for class discussion. This device is used when, as a topic starts, we are aware that students may have different educational experiences as they advanced through S1 to S3 following the CfE. A class discussion will help to make the necessary knowledge base the same for the whole group.

 indicates a question for which some research or investigative work would enrich the student's experience.

 indicates where a puzzle has been added also for the sake of enrichment. These puzzles can always be taken further.

Finally, a convention is followed throughout the series whereby the decimal point is placed mid-line, e.g. when writing '3 point 14' we would write 3·14 and not 3.14. The latter style is used in spreadsheets, however.

In Britain we often use the 'point on the line' to act as a less conspicuous multiplication sign than '×'. This is especially useful in algebra:

e.g. When $x = 6$ then $3x + 2 = 3.6 + 2 = 18 + 2 = 20$

In conclusion, maths is everywhere, it's relevant, it *is* essential, and it can be enjoyed.

1 Factorising, simplifying and sequences

Before we start...

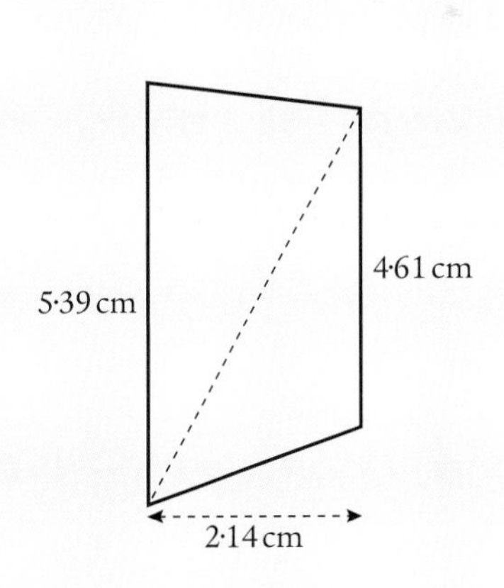

An artist sketching the Charles Rennie Mackintosh fireplace doesn't draw a rectangle but a trapezium to get the perspective correct.

To find its area he draws a diagonal splitting the trapezium into two triangles as shown.

Both triangles have an altitude (height) of 2·14 cm.

His calculation looks like this:

Area $= \frac{1}{2} \times 5{\cdot}39 \times 2{\cdot}14 + \frac{1}{2} \times 4{\cdot}61 \times 2{\cdot}14$

He doesn't have a calculator!

Can you work it out without a calculator ... in two minutes?

And while you're thinking ... can you find a number with exactly 14 factors?

Can you find the smallest such number?

What you need to know

1 Calculate:

a $3 + 4 \times 5$

b $3 \times 5 + 4 \times 5$

c $(3 + 4) \times 5$.

2 Compare your answer to **1b** and **1c** and say what the brackets make happen.

3 **a** Find the prime factors of:

i 105 **ii** 42.

b Hence write down the highest common factor of 105 and 42.

4 Simplify by collecting like terms:

$$3a + 4b + 2a + 6b + 3a^2$$

5 This sum, $5a + 7b$, has two terms. Each term is a product of two factors.

a What is the first term?

b What are the two factors forming the product in the second term?

1.1 Removing brackets (the distributive law)

The **distributive law** will be familiar to you from working with numbers.

For example, suppose you want to multiply 27 by 6.

Instead of thinking, 'six 27s', you would think, 'six 20s plus six 7s'.

You could write 6×27 as (6×20) and (6×7), making the calculation much easier.

So: $6 \times 27 = 6 \times (20 + 7) = (6 \times 20) + (6 \times 7) = 120 + 42 = 162$.

The area of this window frame can be worked out in two ways.

First method

Consider the whole window as one rectangle.

$$\begin{aligned} A &= l \times b \\ &= 80 \times (40 + 60) \\ &= 80 \times 100 \\ &= 8000\text{ cm}^2 \end{aligned}$$

Second method

Consider the window as two smaller rectangles.

The area of the smaller window is:

$$\begin{aligned} A &= l \times b \\ &= 80 \times 40 \\ &= 3200\text{ cm}^2 \end{aligned}$$

The area of the larger window is:

$$\begin{aligned} A &= l \times b \\ &= 80 \times 60 \\ &= 4800\text{ cm}^2 \end{aligned}$$

Total area $= 3200 + 4800 = 8000\text{ cm}^2$

This shows that $80 \times (40 + 60)$ is the same as $\underline{80 \times 40} + \underline{80 \times 60}$.

Now suppose the smaller window was a cm wide, the larger one was b cm wide and the height was c cm.

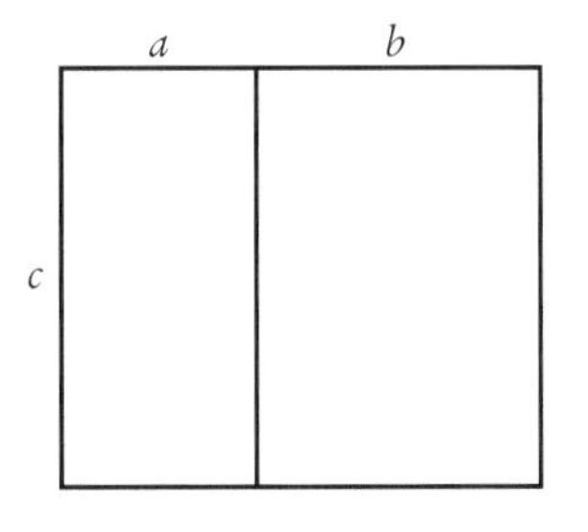

As above, the area can be worked out in two ways.

Treated as one large rectangle:

$$\begin{aligned} \text{area} &= c \times (a + b) \\ &= c(a + b)\text{ cm}^2 \qquad ① \end{aligned}$$

Treated as two rectangles:

area $= \underline{c \times a} + \underline{c \times b}$

$= (ca + cb)\ \text{cm}^2$ ②

Since both these areas are the same it means the expression ① is equal to expression ②.

This means that we can say: $c(a + b) = ca + cb$.

This is the **distributive law** written in symbols. In general we can 'remove brackets' in an expression by multiplying each term inside the brackets by the factor outside. This is called **expanding** the expression.

Example 1

A fencer, when tendering for a job, estimates the cost of materials at £5 per metre and labour at £2 per metre.

Calculate, using two methods, the total cost of putting up 25 metres of fencing.

Method 1: $25 \times 5 + 25 \times 2 = 125 + 50 = 175$

Total cost = £175

Method 2: Total cost for 1 metre = 5 + 2 = £7

Total cost for 25 metres = 25 × £7 = £175

Total cost = £175

This shows that $25 \times (5 + 2) = \underline{25 \times 5} + \underline{25 \times 2}$.

Example 2

When visiting friends, Michael takes 4 bags, each holding 3 mandarins and 2 nectarines.

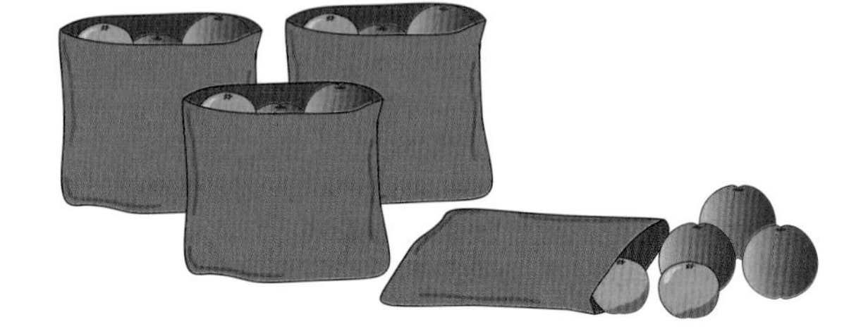

Using m to stand for the weight of a mandarin and n to stand for the weight of a nectarine, write an expression, with and without brackets, for the total weight of the four bags.

$4 \times (3m + 2n) = 4(3m + 2n)$ Four times the total weight of 3 mandarins and 2 nectarines

$= 4 \times 3m + 4 \times 2n$

$= 12m + 8n$

Note that each term in the brackets has been multiplied by the factor outside the brackets.

Example 3

Expand this expression: $3(2x + 5)$.

$3(2x + 5) = 6x + 15$ Each term in the brackets has been multiplied by the factor 3

Example 4

Expand this expression: $-2(3x - 4)$.

$-2(3x - 4)$ can be thought of as $-2(3x + (-4))$.

Expanding the expression, we get:

$\underline{-2 \times 3x} + \underline{-2 \times (-4)} = -6x + 8$.

Exercise 1.1A

1 Expand each expression:

a $2(x + 7)$ b $3(y + 4)$ c $6(d - 5)$

d $8(c + 2)$ e $6(z - 6)$ f $5(h - 1)$

g $3(4 + a)$ h $6(3 + c)$ i $2(9 - t)$

j $7(3 + b)$ k $9(5 - h)$ l $11(6 + k)$.

2 Use the distributive law to express these without brackets:

a $4(2x + 5)$ b $5(6e - 1)$ c $2(9 - 4p)$

d $7(3c + 5)$ e $15(3 + 2b)$ f $8(10h - 1)$

g $5(2c - 7d)$ h $3(5f + 3g)$ i $7(4x - 9y)$

j $9(3m - 2n)$ k $12(5s + 2t)$ l $6(7u + 5v)$.

3 Remembering the rules of multiplying by a negative number, e.g. $-3 \times 4 = -12$ and $-3 \times (-4) = 12$, expand the following:

a $-2(x + 3)$ b $-4(2y + 1)$

c $-5(d - 2)$ d $-7(2c - 3)$

e $-6(z - 1)$ f $-3(3h - 4)$

g $-(k - 1)$ (Remember that this can be thought of as $-1(k - 1)$.)

h $-(4 - 2x)$ i $-(2h + 5)$.

4 At the Riverside Museum in Glasgow they estimate that there will be 160 000 family groups visiting it in a year.

Entry is free, but suppose that on average a family will spend £10 in the gift shop and £15 at the café.

a Estimate, in **two** different ways, the takings in a year from family groups.

b The Maxwells are regular visitors to the museum.

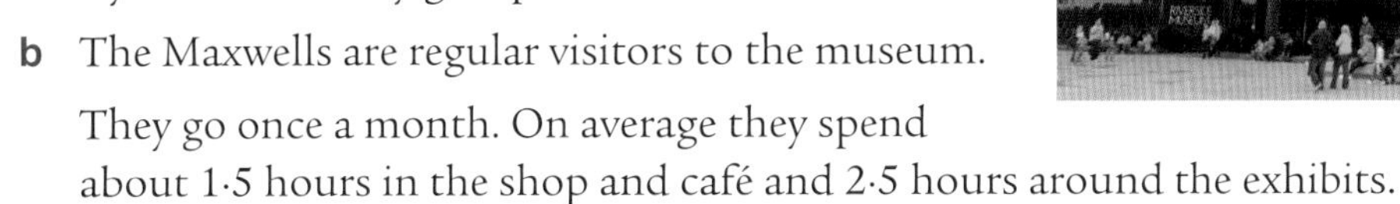

They go once a month. On average they spend about 1·5 hours in the shop and café and 2·5 hours around the exhibits.

Estimate, in **two** different ways, how much time they spend a year at the museum.

c Comment on which method gives you the simpler arithmetic.

Algebraic pictures

It sometimes helps to let numbers look like lines, and products look like areas.

This line might represent x: x

This line might represent 6: 6

Then this represents $x + 6$: x 6

This area would be $6 \times (x + 6)$:

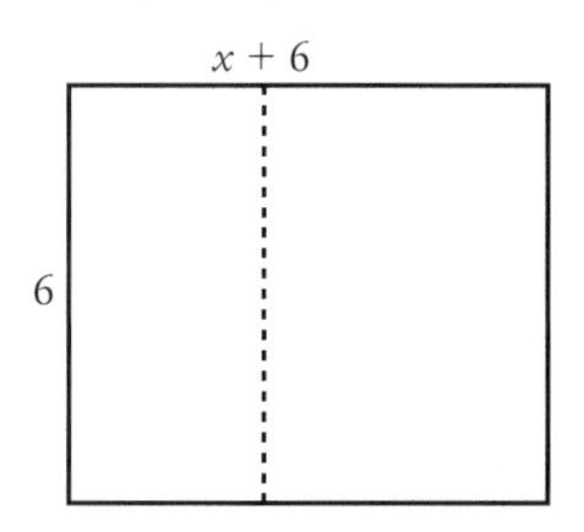

Example 5

a Make an illustration to represent the product $5(a + 4)$.

b Show how it also represents $5a + 5 \times 4$.

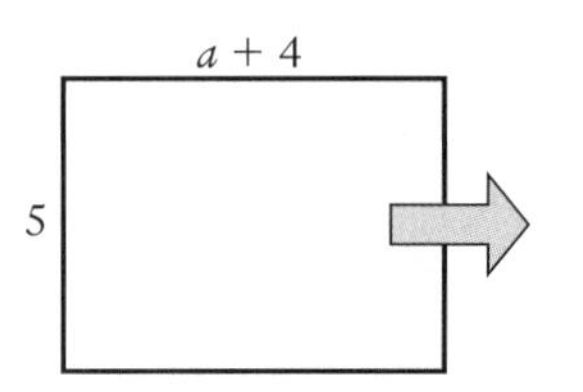

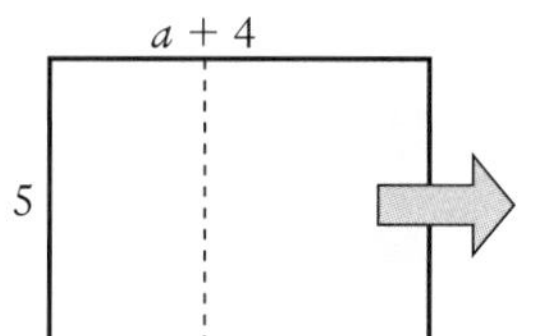

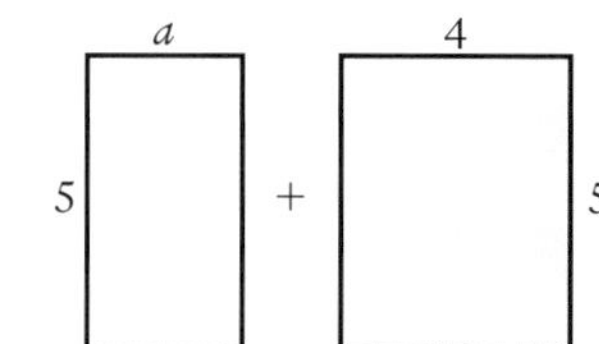

Example 6

Illustrate the product $3(a - 2)$.

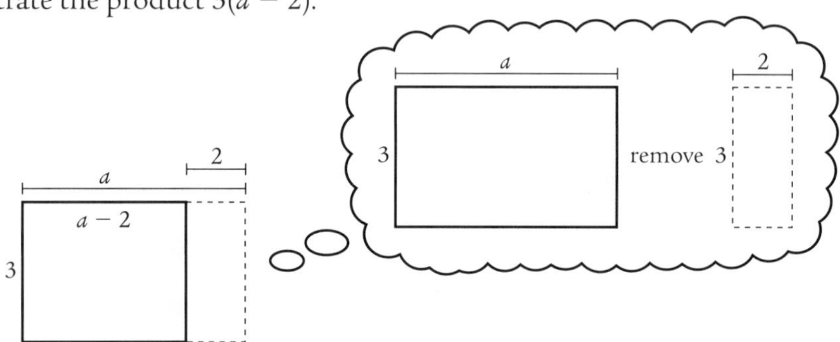

Exercise 1.1B

1 Write an expression for each blue area using brackets, then expand the brackets.

a

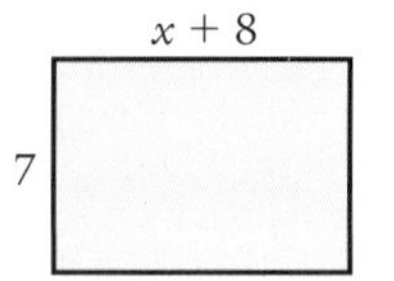

b

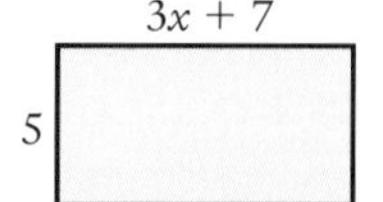

c

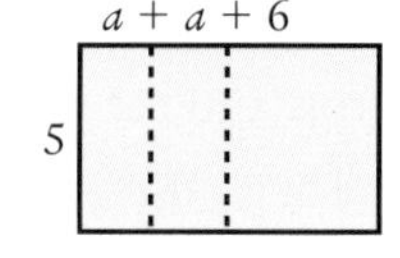

d

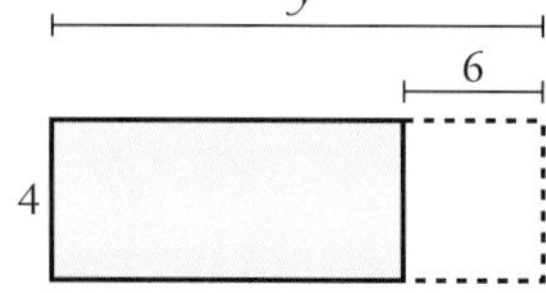

2 **a** Construct an expression for the total weight in the bags, making use of brackets, where a stands for the weight of an apple in grams and b for the weight of a banana in grams.

b Rewrite this expression in an expanded form, without brackets.

3 On the esplanade out in his poem 'Tam O'Shanter'.

The Scots mile was longer than the modern UK mile.

Let m stand for the number of metres in a UK mile and x be the extra needed to make it a Scots mile.

a Write an expression, using brackets, for the number of metres in a Scots mile.

b A league is the distance a person can walk in an hour. It is about 3 Scots miles.

Write down an expression, with the aid of brackets, for the number of metres in a league.

c Expand the brackets and identify which term represents:

i the number of metres in 3 UK miles

ii how much longer 3 Scots miles are.

4 The diagram shows four methane molecules, each made up of 1 carbon atom and 4 hydrogen atoms.

```
    H         H         H         H
    |         |         |         |
H — C — H H — C — H H — C — H H — C — H
    |         |         |         |
    H         H         H         H
```

a Using c to represent the atomic weight of a carbon atom and h the atomic weight of a hydrogen atom, write an expression using brackets for the total atomic weight of the four molecules in the diagram.

b Expand the brackets to find another expression for the total atomic weight.

5 Martin downloaded an application for bird watching. It contained records of 200 birds. For each bird there was memory set aside for data, d kB, memory set aside for a picture, p kB, and memory set aside for the song s kB.

a Write a formula that gives T kB, the **total** memory set aside for the application.

b Remove the brackets to give another formula for T.

c $d = 45, p = 60$ and $s = 25$.

i Calculate T using both formulae.

ii Which formula is the easier to use?

6 Mr Lane the builder has six identical pipes, each of length $4p + 7$ metres.

a Write an expression, using brackets, to show the total length of the six pipes.

b Remove the brackets to give another expression showing the total length.

c A collar is used to protect each joint.

If Mr Lane needs n pipes, he needs $n - 1$ collars.

Every pipe costs £7 and every collar £2.

i Write an expression, with brackets, for the total cost of the collars.

ii Hence, write an expression for the total cost of pipes and collars.

iii Simplify this expression as far as you can.

7 Every month Sarah pays her household bills by direct debit. An amount of £x for rent and three lots of £y for electricity, gas and telephone comes out of Sarah's bank account along with £180 a month for food, £150 for clothes and entertainment and £170 for other expenses.

a Write an expression for the total amount coming out of her bank account every month.

b If Sarah is paid £12 000 a year, find an expression for the amount of money she has left at the end of the year.

c Remove the brackets and simplify the expression if possible.

8 Remembering that $a \times a = a^2$, expand:

a $x(x + 2)$ **b** $e(e - 1)$ **c** $-p(1 - 3p)$

d $c(4c + 3)$ **e** $b(3 + 2b)$ **f** $h(2h - 5)$

g $c(3c - 4d)$ **h** $g(2f + g)$ **i** $x(x - 2y)$

j $n(m - n)$ **k** $-t(2s + 5t)$ **l** $u(4u + uv)$.

9 Remove the brackets from these expressions:

a $5x(x - 8)$ **b** $3a(5a + 3)$ **c** $2m(2m - 7)$

d $4p(3p + 4q + 2)$ (Hint: multiply each term in the bracket by the factor outside.)

e $6a(5b - 6c + 3)$ **f** $-10(5c^2 + 3c + 7)$ **g** $-4(x^2 - 5x - 4)$.

10 A local tourist board makes a one-page leaflet to promote their neighbourhood.

It contains the photo of a sunset.

The printer charges 30p per leaflet.

The photographer charges 10p each time his photo is printed.

Altogether the board orders a run of 5000 leaflets.

a Calculate, in **two** ways, the total cost of the batch of leaflets.

b The printers produce a formula to help customers work out the cost, C pence, of future projects: $C = n(30b + 10i)$, where n is the number of leaflets, b is the number of pages in the leaflet and i is the number of images used.

Find an expression for C that does not use brackets.

1.2 Factorising

Finding factors

We need to be able to spot all the factors in a term.

Example 1

Find all the factors of 72.

First we repeatedly divide by the primes:

We find that $72 = 2^3 \times 3^2$.

2	72
2	36
2	18
3	9
3	3
	1

Then we can make a table, remembering that 1 is a factor of every number:

×		Factors of 2^3			
		1	**2**	$\mathbf{2^2}$	$\mathbf{2^3}$
Factors of 3^2	**1**	1	2	4	8
	3	3	6	12	24
	$\mathbf{3^2}$	9	18	36	72

This shows that 72 has 12 factors:

1, 2, 3, 4, 6, 8, 9, 12, 18, 24, 36 and 72.

Example 2

Find all the factors of $6x^2$.

Imagine the table:

×		Factors of 6			
		1	**2**	**3**	**6**
Factors of x^2	**1**	1	2	3	6
	$\boldsymbol{x}$	x	$2x$	$3x$	$6x$
	$\boldsymbol{x^2}$	x^2	$2x^2$	$3x^2$	$6x^2$

The factors of $6x^2$ are:

1, 2, 3, 6, x, $2x$, $3x$, $6x$, x^2, $2x^2$, $3x^2$ and $6x^2$.

Factorising using common factors

The inverse operation to removing brackets is called **factorising**.

We take a sum of terms and turn it into a product of factors.

Notice that when we got rid of brackets, the terms that resulted shared a common factor, namely the factor that was outside the brackets.

If we could examine the sum of terms and spot the common factors, we could 'take them back outside the brackets'.

We're looking for the highest common factor of the terms.

Example 3

Factorise: $12ab + 8ac$.

The factors of $12ab$ are:

×		Factors of 12					
		1	**2**	**3**	**4**	**6**	**12**
Factors of ab	**1**	1	2	3	4	6	12
	$\boldsymbol{a}$	a	$2a$	$3a$	$\boldsymbol{4a}$	$6a$	$12a$
	$\boldsymbol{b}$	b	$2b$	$3b$	$4b$	$6b$	$12b$
	$\boldsymbol{ab}$	ab	$2ab$	$3ab$	$4ab$	$6ab$	$12ab$

The factors of $8ac$ are:

×		Factors of 8			
		1	**2**	**4**	**8**
Factors of ac	**1**	1	2	4	8
	$\boldsymbol{a}$	a	$2a$	$\boldsymbol{4a}$	$8a$
	$\boldsymbol{c}$	c	$2c$	$4c$	$8c$
	$\boldsymbol{ac}$	ac	$2ac$	$4ac$	$8ac$

The highest common factor (HCF) is $4a$.

So $12ab + 8ac = 4a.3b + 4a.2c$ (Here the dots are being used to signify multiplication)

$\Rightarrow$ $12ab + 8ac = 4a(3b + 2c)$ $4a$ is taken outside the bracket

We can check our answer by mentally removing the brackets again.

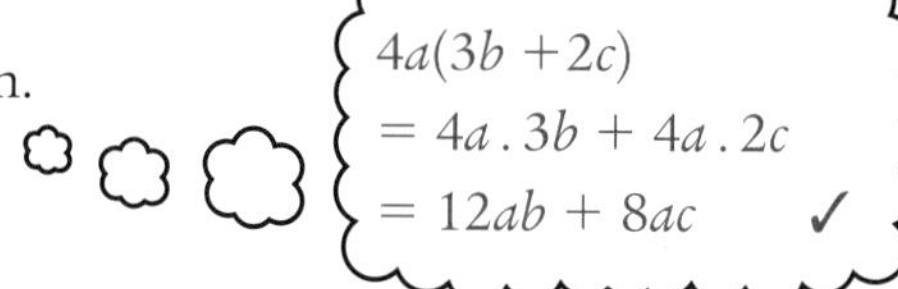

Example 4

Factorise: $35x^2 - 14x$.

By inspection we see that 7 is the HCF of 35 and 14 and that x is the HCF of x and x^2.

So the HCF of the two terms is $7x$.

$$35x^2 - 14x = 7x \,.\, 5x - 7x \,.\, 2 = 7x(5x - 2)$$

Exercise 1.2A

1 Even though the function machine is incomplete, we can deduce that the missing factor is 5.

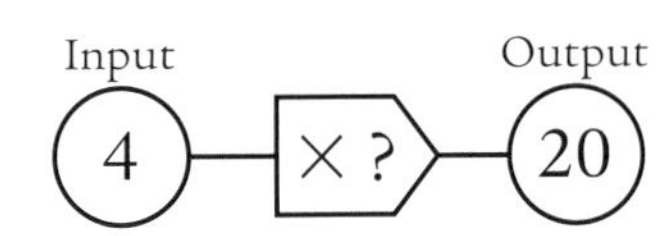

Given the input and the output, identify the missing factor in each case:

	Input	Output	Missing factor
a	3	12	
b	2	$2d$	
c	$5x$	$10x$	
d	$6y$	$6y^2$	
e	$8a$	$32ab$	
f	$7cd$	$21cd^2$	

2 Find all the factors of:

a 15 **b** 28 **c** $16m$

d $34k$ **e** $7x^2$ **f** $9ab$.

3 Find the highest common factor (HCF) of each pair:

a 6 and 8 **b** 10 and 15 **c** 16 and 20

d 7 and 21 **e** $4a$ and 12 **f** 28 and $56b$.

4 Factorise each expression by 'taking out' a common factor:

a $5y + 5$ **b** $7b + 7c$ **c** $9 - 9f$

d $3p - 3q$ **e** $2t - 8$ **f** $6 + 4k$

g $6x + 9xy$ **h** $15d + 12de$ **i** $8e - 10$

j $27v + 18uv$ **k** $35z - 20xyz$ **l** $16a - 24h$

m $10x + 8xy$ **n** $14s + 7st$ **o** $42fg - 35g$

p $55m^2 + 44m$ **q** $36z^2 - 45z$ **r** $32ab - 16b^2$.

5 The density of gold is 19·3 g/cm^3.

That means if you want to find the weight of a quantity of gold you multiply its volume in cubic centimetres by 19·3.

A jeweller had two lots of gold: a gold ring of volume 0·8 cm^3 and a gold chain of volume 9·2 cm^3.

He wrote down the calculation to find the weight:
$19{\cdot}3 \times 0{\cdot}8 + 19{\cdot}3 \times 9{\cdot}2$.

a Factorise the calculation by taking out the common factor.

b What weight of gold did the jeweller have?

6 A group of hikers walks at a steady pace of 2·5 miles per hour for 1 hour and 25 minutes.

After a short break for lunch, they continue for a further 2 hours and 35 minutes at the same pace.

One hiker tries to work out the distance walked using the formula:

distance = speed × time

This gives:

$$\text{distance} = 2{\cdot}5 \times 2\tfrac{35}{60} + 2{\cdot}5 \times 1\tfrac{25}{60}$$

a Factorise the calculation.

b Evaluate it.

7 Take out a common factor first and you'll see how easily you can do these calculations without a calculator:

a $7 \times 34 + 7 \times 66$

b $1{\cdot}2 \times 7{\cdot}9 + 1{\cdot}2 \times 2{\cdot}1$

c $23{\cdot}5 \times 12 - 23{\cdot}5 \times 2$

d $3{\cdot}79 \times 19 - 9 \times 3{\cdot}79$.

8 In Example 1 on page 9 you saw how $72 = 2^3 . 3^2$.

You also saw that 2^3 has 4 factors and that 3^2 has 3 factors, leading to the fact that 72 has $4 \times 3 = 12$ factors.

a **i** Check that $400 = 2^4 . 5^2$.

ii How many factors has 2^4 and 5^2?

iii How many factors has 400?

b **i** What is the value of $3^5 . 5^1$? (5^1 is just 5, but we're emphasising the power.)

ii How many factors does it have? (Don't work them out.)

c **i** Thinking backwards, can you find a number with 14 factors?

ii Can you find the smallest such number?

Example 5

Find all the factors of $8x + 48$.

First factorise the expression:

$8x + 48 = 8(x + 6)$

Now imagine the table:

×		Factors of 8			
		1	**2**	**4**	**8**
Factors of $(x + 6)$	**1**	1	2	4	8
	$\boldsymbol{x + 6}$	$x + 6$	$2(x + 6)$	$4(x + 6)$	$8(x + 6)$

So 1, 2, 4, 8, $(x + 6)$, $2(x + 6)$, $4(x + 6)$, $8(x + 6)$ are the required factors.

Example 6

Pennies are arranged in a rectangular array.

An expression for the total number of coins is $10x + 20$.

This has been found by multiplying an expression for the number of coins in a row by an expression for the number of coins in a column.

Neither expression involves fractions.

List the possible pairs of expressions.

$10x + 20 = 10(x + 2)$

The factors of 10 are 1, 2, 5 and 10; the factors of $(x + 2)$ are 1 and $(x + 2)$.

So all possible pairs that multiply to give $10x + 20$ are:

1 with $10(x + 2)$, 2 with $5(x + 2)$, 5 with $2(x + 2)$, 10 with $(x + 2)$.

Exercise 1.2B

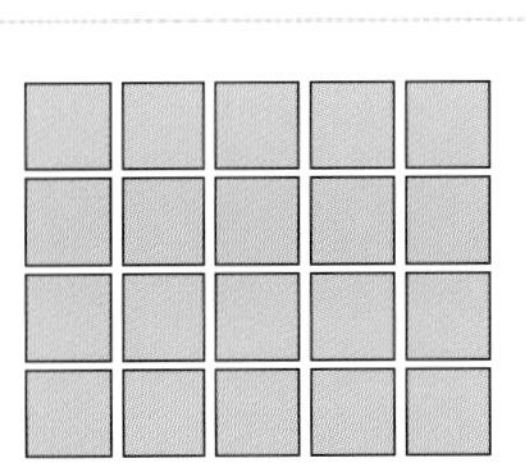

1 An expression for the area of tiling is $(6x + 18)$ units^2.

This has been found by multiplying expressions for its length and breadth, neither of which involves fractions.

- **a** List all factors of $6x + 18$.
- **b** List all possible pairs that multiply to give $6x + 18$.
- **c** Repeat parts **a** and **b** for tiled spaces with area:

  i $(10y - 5)$ units^2 **ii** $(40m + 45)$ units^2 **iii** $(9c + 12cd)$ units^2.

2 A group of friends needs to carry a total of 6 computers and 18 books equally between them.

- **a** Using c to represent the weight of a computer and b for the weight of a book, write an expression to show the total weight the group needs to carry.
- **b** Factorise this expression.
- **c** Write an expression for the weight each person has to carry if there are:

  i 6 friends **ii** 3 friends **iii** 2 friends.

3 A machine makes spaghetti strands with a total length of $300t + 750$ cm.

These are then cut into 150 equal lengths and packaged.

- **a** Factorise $300t + 750$.
- **b** Write an expression to show the length of each strand of spaghetti.

Spaghetti

4 In the Home Economics department they are cooking poultry.

As a basic rule, the cooking time, T minutes, depends on the weight, w kg, being cooked.

$$T = 20 + 40w$$

- **a** **i** A chicken weighs c kg. Express the time to cook the chicken in terms of c.

  ii A turkey is heavier at b kg. Express the time to cook the turkey in terms of b.
- **b** Find an expression in terms of b and c for the difference in the time it takes to cook the chicken and the turkey.
- **c** Factorise this expression and find a formula for D minutes, the difference in cooking times.
- **d** How long before a 2 kg chicken should they start to cook an 8 kg turkey so that they are ready at the same time?

5 A group of friends attends a concert at Hampden Park.

Altogether they bought 6 programmes at £15·50 each, 12 sandwiches costing £3·30 each and 6 drinks each costing £7·90.

- **a** Write an expression for the total spent.
- **b** Factorise this expression to help you calculate the total cost without the aid of a calculator.

6 In the design of cars, manufacturers have always had to worry about the effect of sudden changes of speed on the people in the car. To study this, the engineers use an idea called **momentum**.

Momentum is calculated by multiplying the mass of a person by their velocity in metres/second.

$$\text{Momentum} = mv$$

When a car changes speed over one second, the force, F newtons, acting on the person can be taken as the change in momentum.

- **a** Write down an expression for F when a car goes from v_1 m/s to v_2 m/s in one second.
- **b** Factorise this expression.
- **c** Search the internet for 'safe force that a human can endure' and read about g-forces etc.

7 A lean-to shed is being built. The cost of the wood to build a side is based on the area of the side.

At the tall end, the side is H metres high.
At the short end this is h metres.
The side is w metres wide.

A dotted line divides the side into two triangles.

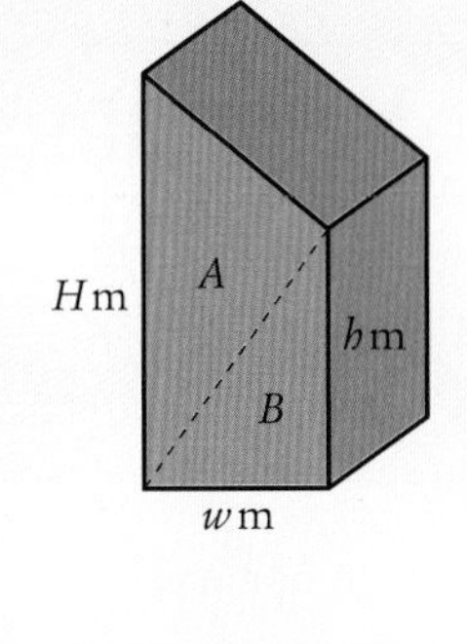

a Write down a formula for:

i area A in terms of H and w

ii area B in terms of h and w.

b Write down a formula for the area of the side of the shed.

c Given that $H = 3{\cdot}3$ m, $h = 2{\cdot}7$ m and $w = 3{\cdot}4$ m, calculate the area of the side of the shed, without the aid of a calculator.

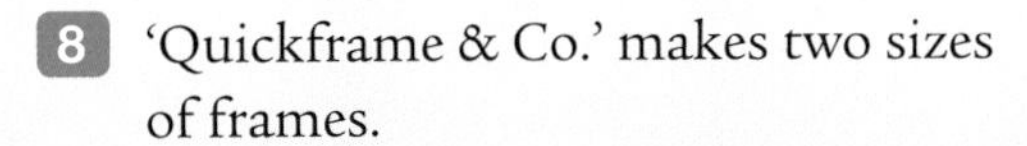

8 'Quickframe & Co.' makes two sizes of frames.

They decide to charge £6 for the smaller frame and twice as much for the larger frame.

They keep a spreadsheet to record how much they charge each customer.

The formula in D2 is '=B2*6+C2*12'.

(Remember that * is computer for ×.)

File | Home | Insert | Page Layout | Formulas | Data | Review | View

D2 | f_x =B2*6+C2*12

Quickframe & Co

	A	B	C	D	E
1	Customer Name	Number of Small Frames	Number of Large Frames	Total Cost	
2	Archie Baxter	3	1	£ 30.00	
3	Cathy Dean	5	2	£ 54.00	
4	Elsa Frier	4	5	£ 84.00	
5	Gordon Howie	1	3	£ 42.00	
6	Ian Johns	2	3	£ 48.00	
7	Kier Lambie	4	6	£ 96.00	
8					
9					
10					

a Can you write this formula using only one *?

b In theory, simplifying this formula wouldn't matter to the computer as it doesn't see one piece of arithmetic as harder than another.

What would be the benefit of simplifying formulae?

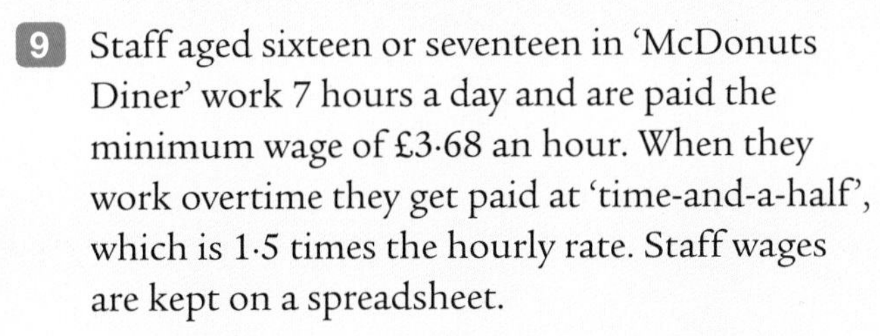

9 Staff aged sixteen or seventeen in 'McDonuts Diner' work 7 hours a day and are paid the minimum wage of £3·68 an hour. When they work overtime they get paid at 'time-and-a-half', which is 1·5 times the hourly rate. Staff wages are kept on a spreadsheet.

The formula in D2 is '= B2*3.68+C2*3.68*1.5'.

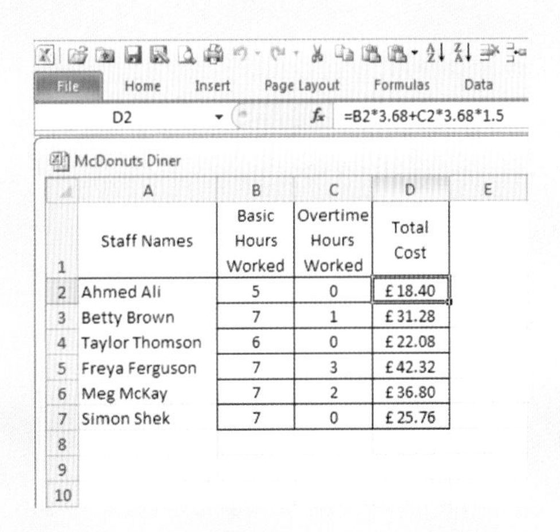

File | Home | Insert | Page Layout | Formulas | Data

D2 | f_x =B2*3.68+C2*3.68*1.5

McDonuts Diner

	A	B	C	D	E
1	Staff Names	Basic Hours Worked	Overtime Hours Worked	Total Cost	
2	Ahmed Ali	5	0	£ 18.40	
3	Betty Brown	7	1	£ 31.28	
4	Taylor Thomson	6	0	£ 22.08	
5	Freya Ferguson	7	3	£ 42.32	
6	Meg McKay	7	2	£ 36.80	
7	Simon Shek	7	0	£ 25.76	
8					
9					
10					

a What formula will be found in D3?

b What other formula could be used in D2? (It will only contain one *.)

c Computers work fast, but they still take time.

Suppose a multiplication took 2 units of time and an addition took only 1 unit of time.

How much time would they save if they used the simplified formula when processing the wages of 500 staff?

1.3 Simplifying expressions

Making things easier

Like terms are terms that are multiples of the same unknown.

For example, a and $7a$, $-6z^2$ and $7z^2$, $4bc$ and $-7bc$ are all pairs of like terms.

Whereas $2v$ and $5w$, $-3z$ and $7z^2$, ac and $-4bc$ are pairs of unlike terms.

Note that $5xy$ and $6yx$ are like terms since the order in which you do a multiplication doesn't matter.

Consider the sum of like terms $6a + 3a$.

Taking out the common factor we get $a(6 + 3) = 9a$.

So we can add, or subtract, like terms simply by writing $6a + 3a = 9a$.

However, $3a + 3ab$, being unlike terms, cannot simply be added ...

... but we can still make use of common factors to simplify:

$$3a + 3ab = 3a(1 + b)$$

We can simplify expressions by collecting like terms.

Example 1

Simplify $3(3 - 2a) + 4(a + 5)$.

$3(3 - 2a) + 4(a + 5) = 9 - 6a + 4a + 20$ First we remove the brackets

$= \underline{9 + 20} \underline{- 6a + 4a}$ Then we collect like terms: note each sign is associated with the term to its right

$= 29 - 2a$

Example 2

A book weighs m kg. A jotter weighs n kg.

Miss Hargreaves has 3 boxes each containing 10 books and 11 jotters.

She puts them on a trolley to move them.

Mr Jaffrey puts 4 boxes each with 7 books and 20 jotters on the trolley.

a Find an expression for the total weight carried by the trolley.

b Express this in its simplest form.

a, b Total weight $= 3(10m + 11n) + 4(7m + 20n)$

$= 30m + 33n + 28m + 80n$

$= \underline{30m + 28m} + \underline{33n + 80n}$

$= 58m + 113n$

The weight carried by the trolley is $(58m + 113n)$ kg.

Exercise 1.3A

1 Simplify:

a $3b + 8b$ b $2xy - xy$ c $4y^2 + 4y$ d $6c - 5bc$ e $2pq + (-pq)$

f $z^2 - z^3$ g $-mn + 7mn$ h $2x^2 + 4x^2$ i $8bc - 2cb$ j $4pq + 7qp$

2 Simplify each expression by collecting like terms.

a $5x + 2x$ b $2d + d$ c $4m + m + 3m$

d $6r + 5r + 7r$ e $3st + 6st$ f $3t^2 + 4t^2$

g $5ab + 2ab + 11ab$ h $6gh + 2hg$ i $3k + 2h - k + h$

3 Collect like terms and simplify the expressions.

a $8x + 3y - 4x - y$ b $12f + g + 3g - f$

c $15z - 2z + 4y - 3z - 4$ d $6r + 5 + 3k + 7r - 2$

e $9g^2 + 3g + 4 + 4g^2 + 5g - 1$ f $5m + 4n - 5m - 4n + 1$

g $7t^2 + t + 1 - 3t^2 + 2t - 1 - 3t^2$ h $5y^3 + 2y^2 - 4y^3 - 4y^2$

i $8w + 4w - 2vw$ j $10cd - 3cd + 2d$

k $18b - 10b + cb$ l $7r + qr - 8r + 1$

m $20pq + 3qp - 13p + 2$

4 Colin is shopping at Costcut supermarket.

He buys 6 cans of soup each weighing t grams, 2 cans of tomatoes each weighing h grams and 5 cans of beans each weighing t grams.

When he picks up the bag, he decides it is too heavy so puts back 3 cans of the same weight.

Write an expression for the total weight of the cans in the bag and simplify it if possible.

5 Multiply out the brackets and then simplify as far as possible.

a $3(x + 4) + x$ b $4(y - 2) + 3$ c $x(3 + x) - x^2$

d $3(2y + 2k) + 4k + 1$ e $3(x + 1) + 4(2x + 5)$ f $5(3x - 4) + 4(2x + 3)$

g $5x + 3(4x - 5) + 1$ h $3z + 4(5 - z) + 10$ i $y + 3(6 + y) - 12$

6 A construction company makes temporary partitions to hide their work.

If they need x partitions, they will need $(x - 1)$ supporting posts, $(3x + 7)$ ties and 2 anchor posts.

Partitions cost £50 each, supporting posts are £5, ties cost £1 each and the anchor posts are £10 each.

a Write down an expression for the total cost of erecting x partitions.

b Simplify this to the point where there are only two terms.

c If the bill comes to £600, why should it be checked?

7 In a magic square, each entry is a different whole number.
Each row, column and main diagonal add up to the same number.

With algebra, they add up to the same expression.

		$2x - 1$
$3x - 2$	$5x - 4$	$7x - 6$

2	7	6
9	5	1
4	3	8

a In this square, what is the magic expression?

b Find the five missing expressions.

c Kate's dad is coming up for his 33rd birthday.

She reckons she can put the magic square on his card so that the numbers add up to 33.

What value of x will she use?

d List the possible birthdays that this magic square could be used for, assuming we only use whole numbers.

Exercise 1.3B

1 Simplify each expression by collecting like terms.

a $6d - 3d + 7e$ **b** $7g + 3g - 6h$ **c** $2v - w + 4v$

d $4l + 5m - 3l + 3m$ **e** $5r + 9s - 2r - 6s$

2 Expand the brackets then collect like terms to simplify:

a $2(a + 3b) + 4b$ **b** $5(3x + 2y) + 5x$

c $7(2p + 3q) - 4q$ **d** $8(2r + 3s) - 6r$

e $3(2s - 3t) + 8t$ **f** $4(2k + l) - 5k$

g $10l + 5(l - 6m)$ **h** $7n - 3(m - 5n)$.

3 Remove the brackets from each expression, then collect like terms to simplify:

a $4(x + 3y) + 3(2x + 5y)$ **b** $3(3l - m) + 5(l + 2m)$

c $5(4j + 2k) + 4(j - 3k)$ **d** $10(3a - b) + 2(6a - b)$

e $3(7m + 10n) + 5(2m - 6n)$ **f** $5(p - 3q) - 4(3p - 5q)$

g $-7(r + 3s) + 2(r + 6s)$ **h** $-2(4a + 3b) - 5(2a - b)$.

4 Two cuboids are as shown.

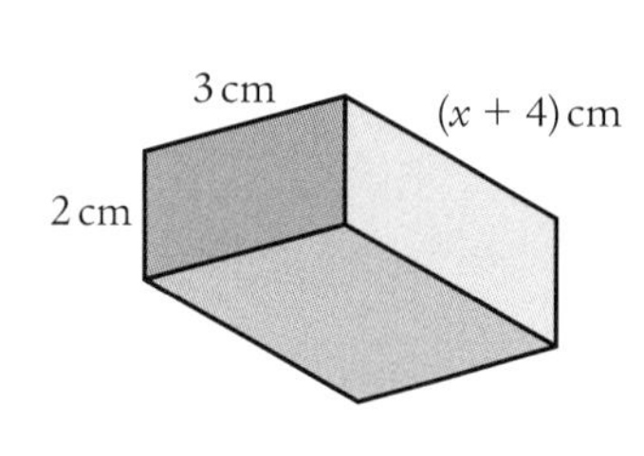

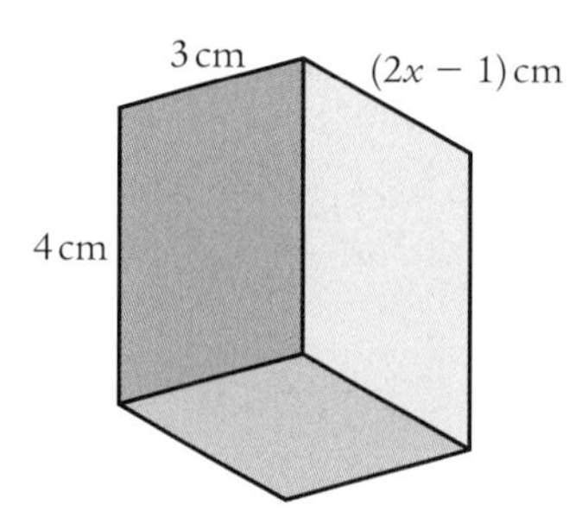

a Find an expression for the volume of each.

b For what value of x will the volumes be the same?

c Give the dimensions of each cuboid.

5 When money is placed in the bank it attracts interest.

This interest is a fraction of the money in the bank.

When £A is placed in the bank for a year, the rate of interest is 4% p.a.

After a year the bank puts £0·04A interest into your account.

a Write down an expression for the amount in the bank at the end of a year.

b Simplify this by taking out a common factor.

c By what factor could you multiply the amount in the bank by to find how much will be in the bank next year?

6 We learn in science that a metal rod will expand when we heat it.

An iron rod of length L units will expand by 0·00001L units for every degree it is heated up.

a Write down an expression for the new length of a rod of length L cm after it has been heated by 10 °C.

b Simplify this expression by taking out a common factor.

c A railway track 1 km long was laid when the temperature was 0 °C.

How long would the length of track be in a summer temperature of 25 °C?

7 A mathematician was studying the number patterns he got by following simple rules.

He drew a triangle of 'bricks', and put four numbers in the bottom row.

Thereafter, each brick held the sum of the two bricks below it.

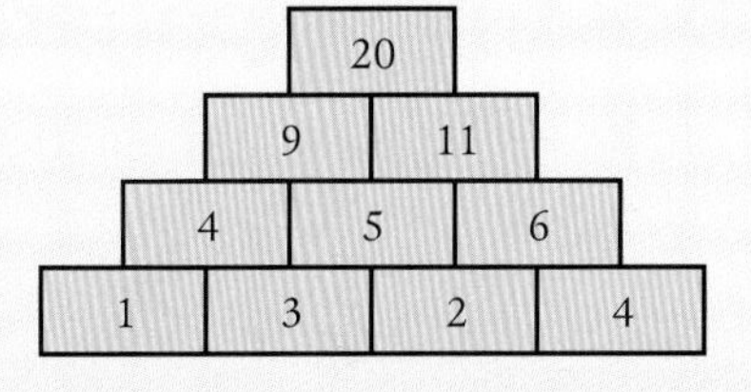

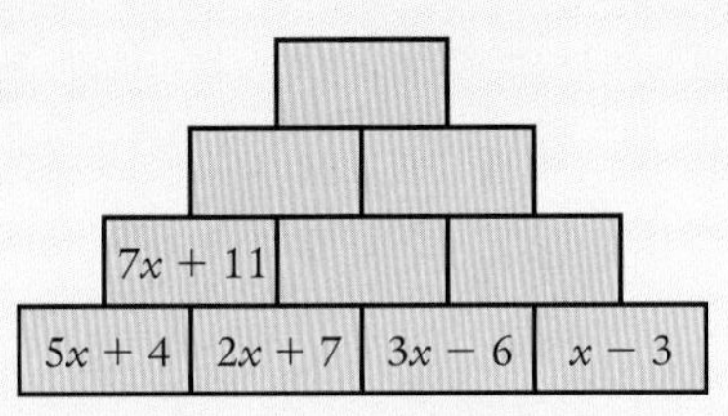

He then considered the effect it had on algebraic expressions instead.

a **i** Find the expression for the top brick when the bottom bricks hold the expressions shown.

ii What would x have to be worth so that the top number is 88?

iii What would be the four bottom numbers?

iv Is it possible to find a value of x that will let you start with four whole numbers and end up with 100?

b The mathematician wanted to see how the four numbers at the bottom combined to form the number at the top.

Let the four numbers at the bottom be represented by p, q, r and s.

What expression in p, q, r and s represents the top number?

c Without going through the whole wall of bricks,

i write an expression for the top brick when the bottom bricks hold $(x + 1)$, $(2x + 1)$, $(2x - 1)$ and $(5x + 2)$

ii expand this expression and simplify

iii explain why, whatever whole number you chose for x, the top number will be divisible by 3.

d Here are a couple of puzzles to try.

Find the expression at the top of each triangle of bricks.
(You don't have to do every brick.)

i

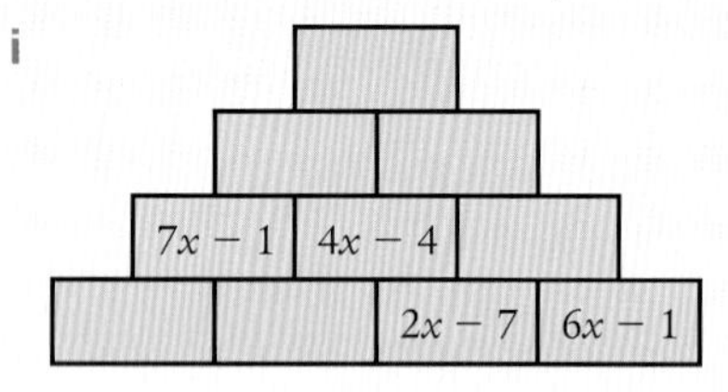

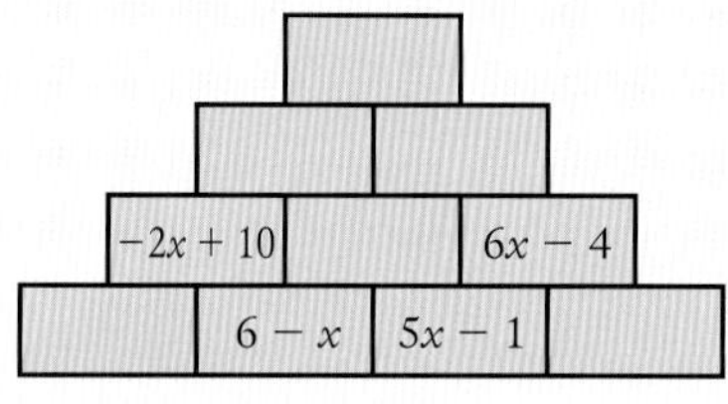

ii For what value of x will the top numbers be the same?

1.4 Patterns from expressions

Describing pattern

You have already studied number patterns where numbers come in a list following certain rules.

You've seen:

Odd numbers: 1, 3, 5, 7, ...

Multiples of 4: 4, 8, 12, 16, 20, ...

Squares: 1, 4, 9, 16, 25, ...

Cubes: 1, 8, 27, 64, 125, ...

Prime numbers: 2, 3, 5, 7, 11, 13, ...

Fibonacci numbers: 1, 1, 2, 3, 5, 8, 13, 21, ...

In order to discuss them further, we need a way of describing them.

One way is to provide a formula for each term based on where it is in the list.

We talk about the ***n*th term** from the notion of 1st, 2nd, 3rd, 4th, 5th, 6th, etc.

For most of the lists above we can find a formula for the nth term.

Odd numbers: 1, 3, 5, 7, ..., $2n - 1$

Multiples of 4: 4, 8, 12, 16, 20, ..., $4n$

Squares: 1, 4, 9, 16, 25, ..., n^2

Cubes: 1, 8, 27, 64, 125, ..., n^3

Prime numbers: 2, 3, 5, 7, 11, 13, ..., famously no one's found a formula.

Example 1

Write down the first five terms of the pattern of numbers whose nth term is $5n + 3$.

1st term: $5 . 1 + 3 = 8$ Remember that . is used instead of × here

2nd term: $5 . 2 + 3 = 13$

3rd term: $5 . 3 + 3 = 18$

4th term: $5 . 4 + 3 = 23$

5th term: $5 . 5 + 3 = 28$

So the pattern begins with 8, 13, 18, 23, 28.

Example 2

Find a formula for the nth term of each sequence where each term differs from the next by a fixed amount:

a 9, 13, 17, 21, ...

b 1, 7, 13, 19, ...

c 40, 37, 34, 31, ...

a Each term is 4 more than the one before it.

This leads us to compare our sequence with 4, 8, 12, 16, ..., $4n$.

We would need to add 5 to each of these terms to get ours.

The nth term is $4n + 5$.

b Each term is 6 more than the one before it.

So, compare our sequence with 6, 12, 18, 24, ..., $6n$.

We would need to subtract 5 from each of these terms to get ours.

The nth term is $6n - 5$.

c Each term is 3 less than the one before it.

This leads us to compare our sequence with $-3, -6, -9, -12, ..., -3n$.

We would need to add 43 to each of these terms to get ours.

The nth term is $-3n + 43$ or $43 - 3n$.

Exercise 1.4A

1 Each of the following expressions is the nth term of a sequence. List the first five terms in each case.

a $3n + 4$ **b** $7n - 1$ **c** $8n$

d $9n - 5$ **e** $10 - n$ **f** $20 - 3n$

2 The year 2012 is an Olympic year. We can calculate the nth Olympic year after this, using as an nth term, $4n + 2012$.

a What is the first Olympic year after 2012? Show your working.

b Use the formula to calculate the 10th Olympic year after 2012.

c Football World Cup years after 2010 can be calculated using as an nth term $4n + 2010$.

What is the 8th World Cup year after 2010?

3 In a science experiment a spring is gradually loaded with weights, one gram at a time.

The length in millimetres from the top of the spring to the pan is measured.

Interestingly, the list of lengths forms a sequence whose nth term is $100 + 6n$.

a What is the length of the spring when the

i 4th weight

ii 7th weight

is placed on it?

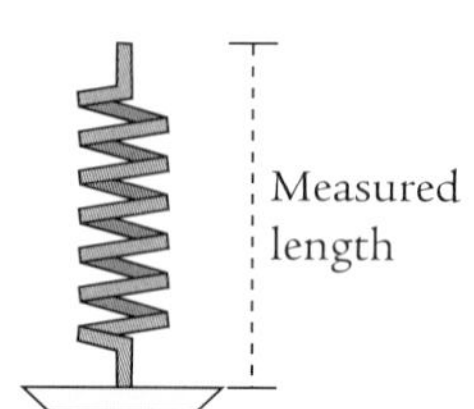

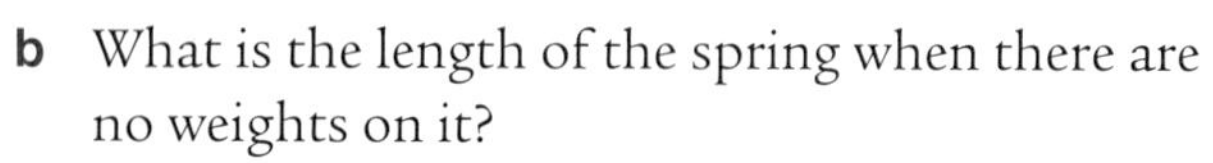

b What is the length of the spring when there are no weights on it?

c Write down the first four terms of the sequence.

4 The sequence whose nth term is $9n + 23$ is a pattern of Fahrenheit temperatures.

a Work out the first five Fahrenheit temperatures in the list.

b The sequence whose nth term is $5n - 5$ is a pattern of Celsius temperatures.

Work out the first five terms of this pattern of Celsius temperatures.

c These two sequences can be paired up to give the Celsius equivalent of a Fahrenheit temperature.

The 10th term of one is 113 °F and the 10th term of the other is 45 °C. These are the same temperature.

i What value of n gives a temperature of 140 °F?

ii Use this value of n to find the equivalent Celsius temperature.

5 **i** Find a formula for the nth term of each sequence. (The sequences are of the type where each term differs from the next by a fixed amount.)

ii Use the formula to find the 100th term in each case.

a 5, 8, 11, 14, ...

b 7, 16, 25, 34, 43, ...

c 50, 43, 36, 29, ...

d 4, 6, 8, 10, ...

e 1, 9, 17, 25, ...

f 100, 98, 96, 94, ...

6 In the History department the students were taught about the water clock.

They began an experiment to study the problem.

Having filled a calibrated 2-litre plastic bottle, they punctured the bottom with a needle and recorded the diminishing volume each minute in millilitres:

2000, 1989, 1978, 1967, ...

The readings formed a simple sequence.

a By what constant amount is the volume diminishing?

b Calculate the nth term of the sequence.

c How much water will be left after an hour ($n = 60$)?

7 The 1st January 2014 is a Wednesday. So the 1st Sunday will be on the 5th day of the year. Thereafter the Sundays form a sequence, 5, 12, 19, 26, 33, ...

a Work out the nth term of this sequence.

b On which day of the year will the 15th Sunday occur?

c An estimate of the dates of the full moons in 2014 also forms a sequence:

The 1st is on the 16th day of the year. The sequence continues, 16, 46, 76, 106, ...

i Work out the nth term of this sequence.

ii On which day is the 7th full moon of the year?

d The 6th full moon falls on a Sunday.

i On which day of the year is the 6th full moon?

ii Hence calculate on which Sunday the full moon falls.

Example 3

Rectangular numbers are formed from the areas of rectangles as shown:

1st term:	2nd term:	3rd term:	4th term:
$1 \times 2 = 2$	$2 \times 3 = 6$	$3 \times 4 = 12$	$4 \times 5 = 20$

a **i** The first factors form a sequence. State the nth term.

ii The second factors also form a sequence. State the nth term.

b **i** Find an expression for the nth rectangular number.

ii What is the 10th rectangular number?

a **i** The sequence is 1, 2, 3, 4, ..., so the nth term is n.

ii The sequence is 2, 3, 4, 5, ..., so the nth term is $n + 1$.

b **i** The nth rectangular number = first factor × second factor = $n(n + 1)$.

ii The 10th rectangular number = $10 \times 11 = 110$.

Exercise 1.4B

1 The first four triangular numbers are as shown:

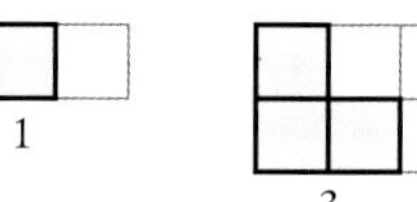
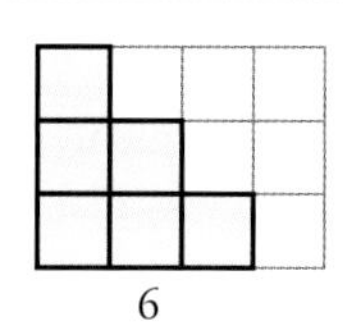
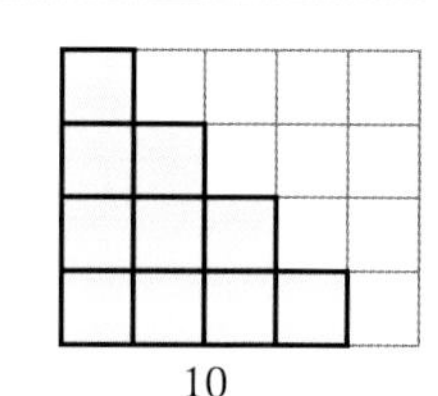

The white squares are not part of the sequence but they do show that the nth triangular number is half the nth rectangular number.

a Write down the formula for the nth triangular number.

b Calculate the 10th triangular number.

c Calculate the difference between the 20th and 21st terms of the sequence.

2 The nth square number can be written as $n \times n$.

a Write down the first six square numbers.

b Write down the first five terms of the sequence whose nth term is created by subtracting the nth square from the square after it.

c Write down the nth term of the new sequence.
Comment.

3 The sequence of numbers 5, 14, 27, 44, ... seems difficult unless you see that it is $1 \times 5, 2 \times 7, 3 \times 9, \ldots$.

The factors form two simple sequences.

a Find the nth term of both sequences.

b Hence write down the nth term of the original sequence.

c Explore the pattern created by examining the differences between terms.

4 Experiments by skydivers show that the distance a man will fall can be predicted:

In the 1st second after leaving the plane he will travel 5 metres, by the 2nd second he has fallen 20 metres, by the 3rd second he has gone 45 metres, ...

The pattern continues 5, 20, 45, 80, 125, ...

a What is the common factor that the terms share?

b Divide each term by this common factor to create a simpler sequence.

c Write down the nth term of:

i the simpler sequence **ii** the original sequence.

d How far will the skydiver have fallen after 10 seconds?

e How far will he fall in the 15th second? (Careful)

5 The nth term of a sequence is given by $(n - 1) \times (n + 1)$.

a List the first six terms of the sequence.

b Can you see another expression for the nth term of this sequence?

c Write down the sequence created by looking at the difference between terms.
Comment.

Preparation for assessment

1 A fence is made up of panels of wood.

Each panel is made of 5 planks each of width w cm.

Each space between planks is s cm wide.

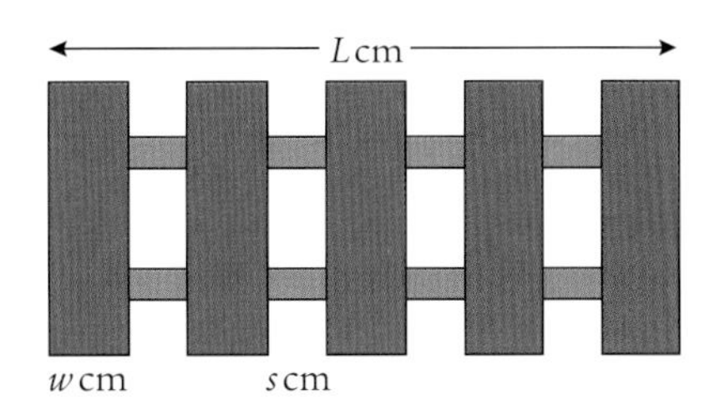

Seven such panels are put end to end.

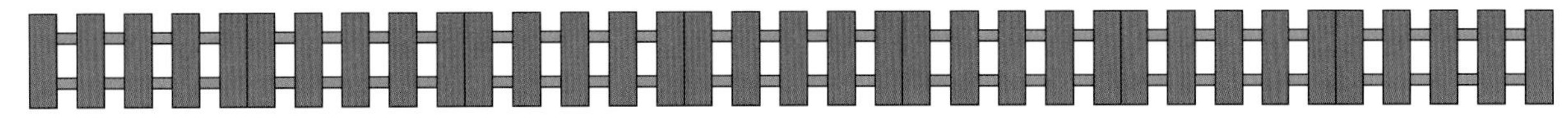

a Write an expression for the total length of the fence, making use of brackets.

b Use the distributive law to find another expression for the length of the fence.

2 Mrs Green is buying fish and chips for her choir group.

A portion of chips weigh c grams and a piece of fish weighs f grams.

The box itself weighs 25 grams.

A box of fish and chips contains one portion of chips and two pieces of fish.

a i Write down an expression for the weight of one box of fish and chips.

ii Mrs Green buys 30 boxes of fish and chips.

Write an expression for the total weight that Mrs Green buys.

b By multiplying out, find another expression for the weight she buys.

3 Distance is calculated using the formula $D = S \times T$, where S is the speed in km/h and T is the time in hours.

A ship goes up river at a rate of x km/h against a current of c km/h, so its overall speed is $(x - c)$ km/h.

a Write an expression for the distance it travels if it goes against the current for 3 hours.

b Going back downstream, it is sailing with the current, so its speed is $(x + c)$ km/h.

The journey downstream takes 2 hours.

Write an expression for the distance travelled downstream.

c i Write an expression for the total distance travelled.

ii Expand this expression and simplify.

4 Consul Cars employs salespeople to sell their cars.

The company uses a spreadsheet to work out the wages of the salespeople.

They get a basic pay of £200 a week plus a £40 bonus for every car they sell.

The formula to work out their wages is '=200+C2*40'.

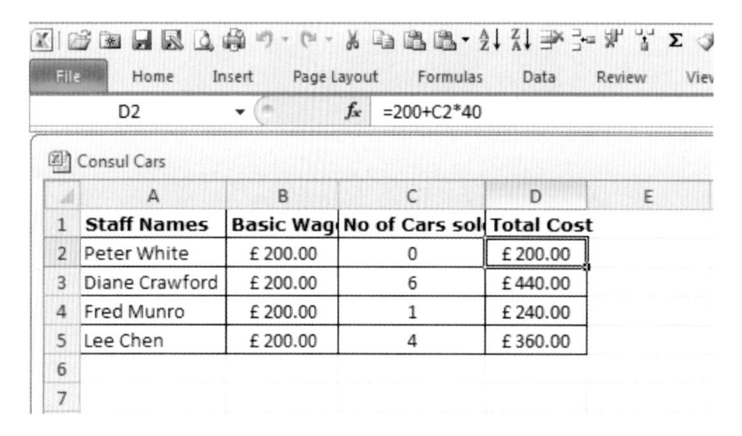

	A	B	C	D	E
1	Staff Names	Basic Wag	No of Cars sol	Total Cost	
2	Peter White	£ 200.00	0	£ 200.00	
3	Diane Crawford	£ 200.00	6	£ 440.00	
4	Fred Munro	£ 200.00	1	£ 240.00	
5	Lee Chen	£ 200.00	4	£ 360.00	
6					
7					

Write this expression in a different way using brackets.

5 An inspector monitors the use of a lift in a hotel.

He assumes the weight of each person is t kg.

He assumes that each suitcase weighs 22 kg.

a He records 7 trips each with 4 people and 6 suitcases.

Write an expression in t for the total weight carried in the 7 trips.

b He then logs another 3 trips where the lift carried 2 people and 4 suitcases.

Write an expression in t for the total weight carried in the 3 trips.

c i Write an expression for the total weight carried during the inspection.

ii Expand and simplify the expression.

6 A Rennie Mackintosh style box is made up of glass.

When flattened out, the net of the box is made up of 6 rectangles as shown.

Find the surface area of this shape in its simplest form.

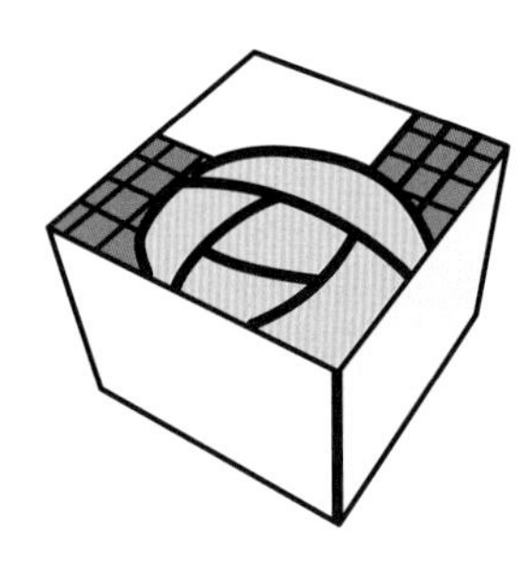

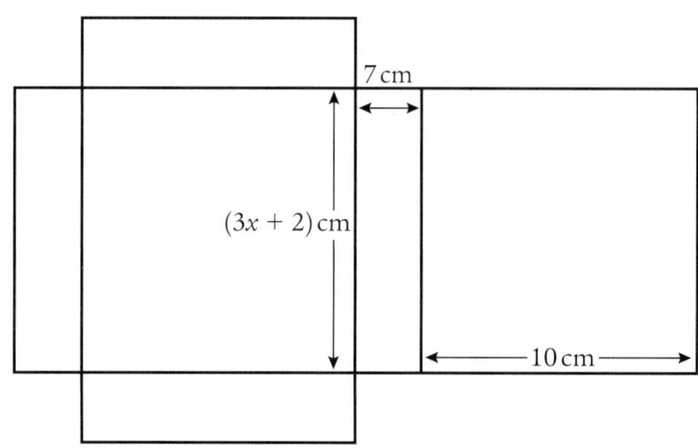

7 The nth term of a number pattern can be found from the expression $3n - 2$.

a List the first 5 terms of the sequence and describe the pattern in words.

b What is the sum of the 8th, 9th and 10th terms?

c Samuel used matchsticks to illustrate the pattern:

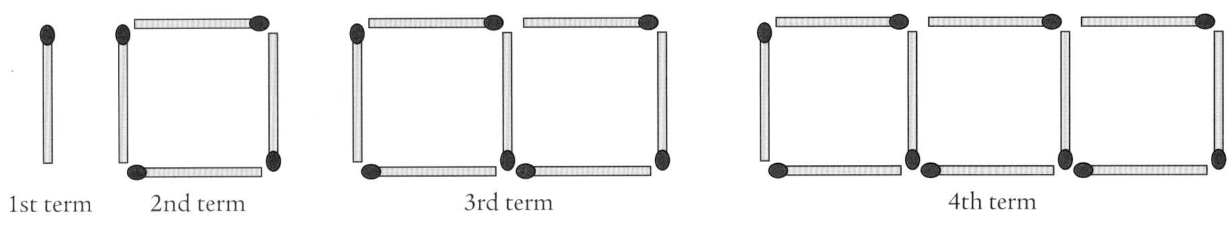

Mary felt that it would look a better pattern if the 1st term was not the solitary match, but the first square.
What would be the nth term of that sequence?

8 Remember the Charles Rennie Mackintosh fireplace?

Now can you quickly work out the area without a calculator?

$$\text{Area} = \tfrac{1}{2} \times 5{\cdot}39 \times 2{\cdot}14 + \tfrac{1}{2} \times 4{\cdot}61 \times 2{\cdot}14$$

... and now can you find a number with exactly 14 factors?

Can you find the smallest such number?

2 Perimeter, area and volume

Before we start...

At Grangemouth Oil Refinery there are 47 storage tanks.

Some of the tanks are shown above.

They are cylindrical and not all the same size.

The diameter of one of the tanks is 43·89 metres and it has a height of 9·91 metres.

Is this enough information to enable us to calculate the maximum volume (capacity) of the tank?

What you need to know

1 The kitchen floor is tiled with rectangles and squares.

Calculate the area of the rectangle and square indicated below:

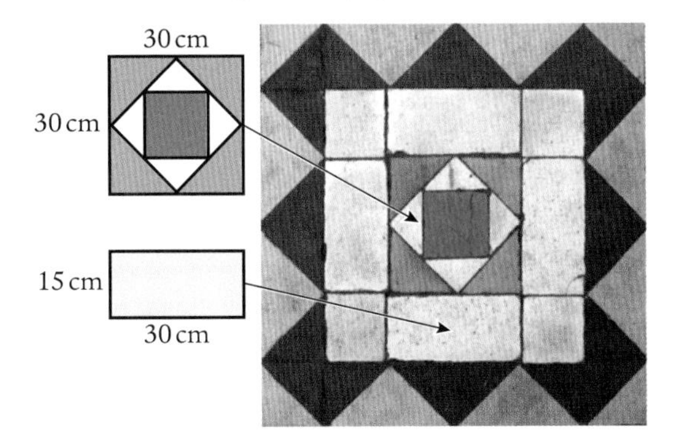

2 The sails of a dinghy are in the shape of triangular sheets. Given the base and altitude, find the area of each.

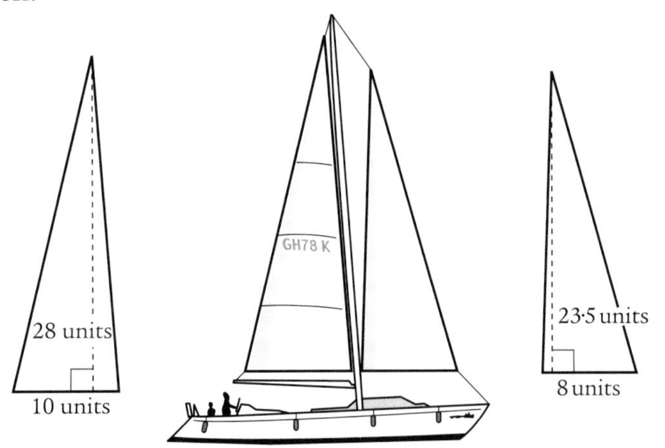

3 PQ and RS are parallel lines.

Find the area of ΔABC, ΔDBC and ΔRBC.

Comment on your answers.

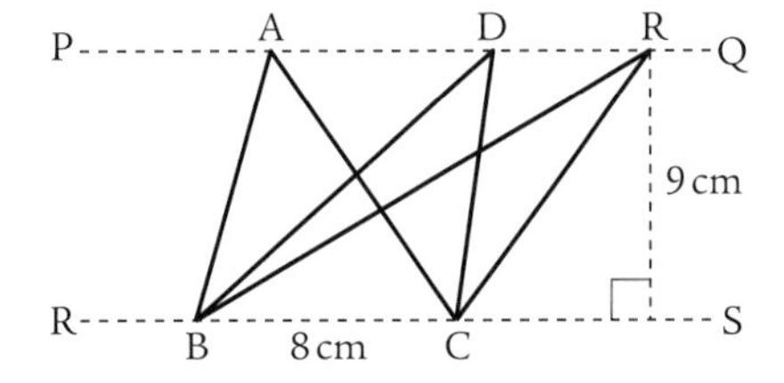

4 a Which of these wire shapes has the greater perimeter?

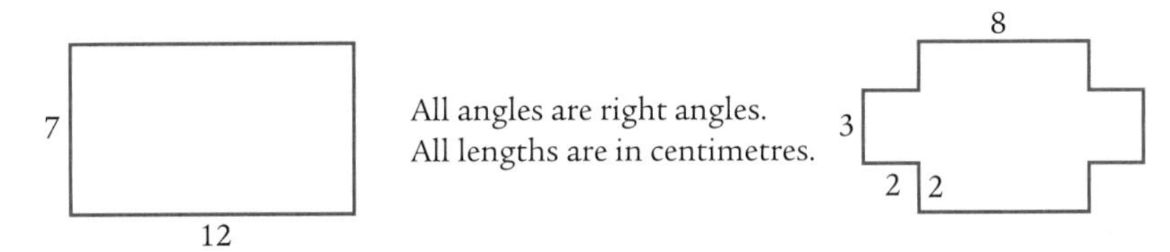

b Find the dimensions of a square whose perimeter is numerically the same as its area.

2.1 Taking the next steps

We can use the formula for the area of the triangle to find the areas of other shapes.

The trapezium

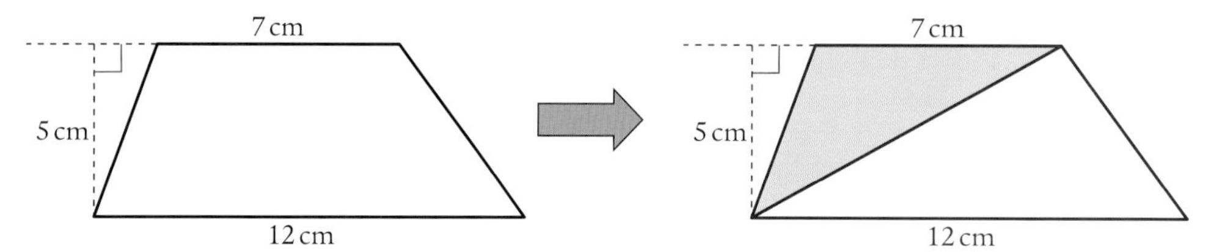

Drawing a diagonal divides the trapezium into two triangles.

The green one has a base of 7 cm and an altitude of 5 cm.

The yellow one has a base of 12 cm and an altitude of 5 cm.

The parallelogram

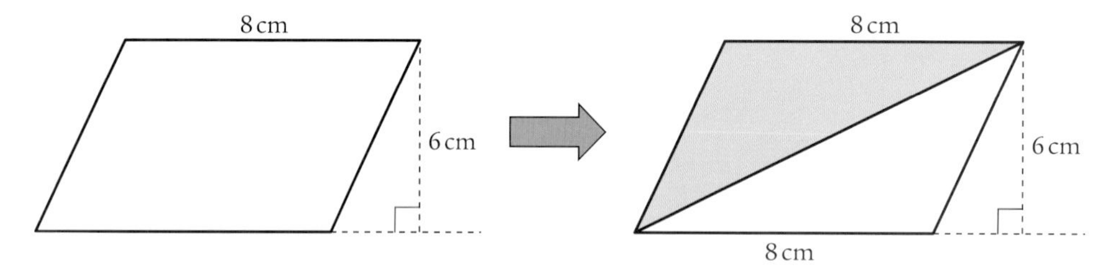

Drawing the diagonal divides the parallelogram into two congruent triangles.

Each has a base of 8 cm and an altitude of 6 cm.

The kite

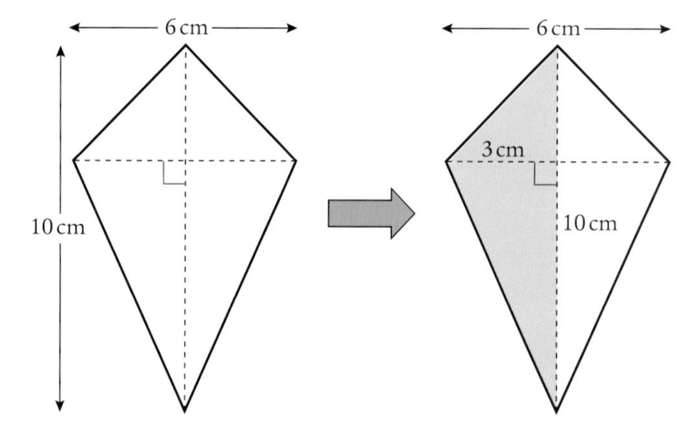

Drawing the diagonal which lies on the axis of symmetry divides the kite into two congruent triangles.

Each has a base of 10 cm and an altitude of 3 cm.

Using the result

Example 1

Popcorn is sold in the cinema in cardboard cartons.

The front face of the carton is a trapezium.

Find the area of the front face of the popcorn carton.

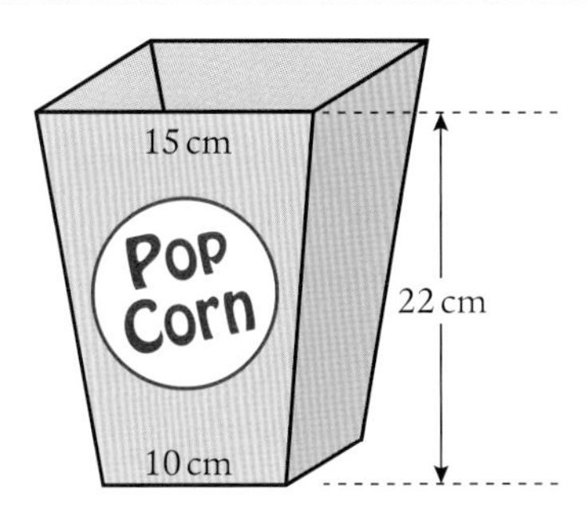

Sketch the trapezium and divide it into two triangles:

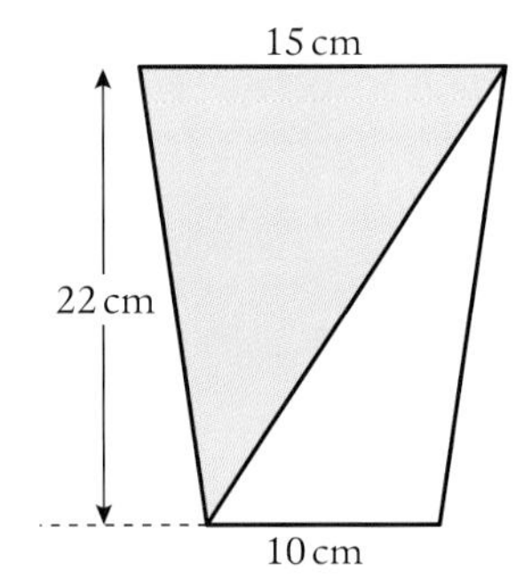

Area of trapezium $= \left(\frac{1}{2} \times 10 \times 22\right) + \left(\frac{1}{2} \times 15 \times 22\right)$

$= 110 + 165$

$= 275\ \text{cm}^2$

Example 2

A futuristic office block has a glazed section in the shape of a parallelogram with a base of 90 m and an altitude of 24 m.

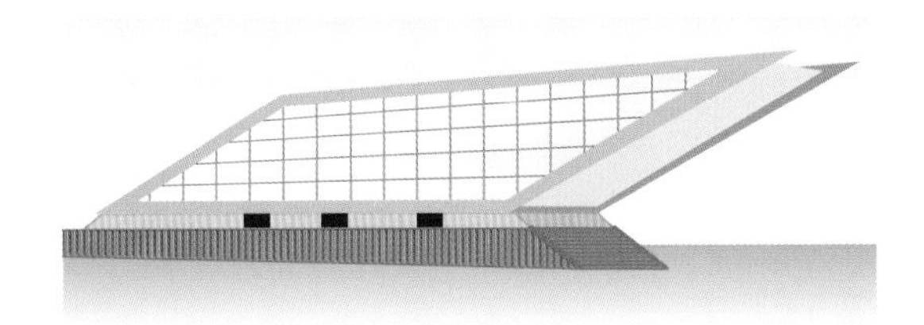

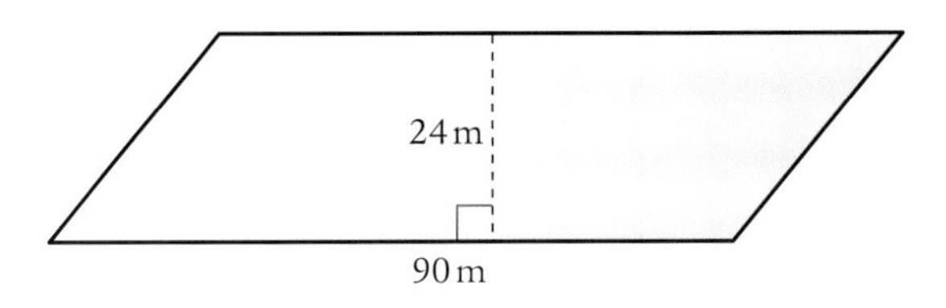

Calculate the area of glass in the front of the building.

Sketch the parallelogram and divide it into two triangles:

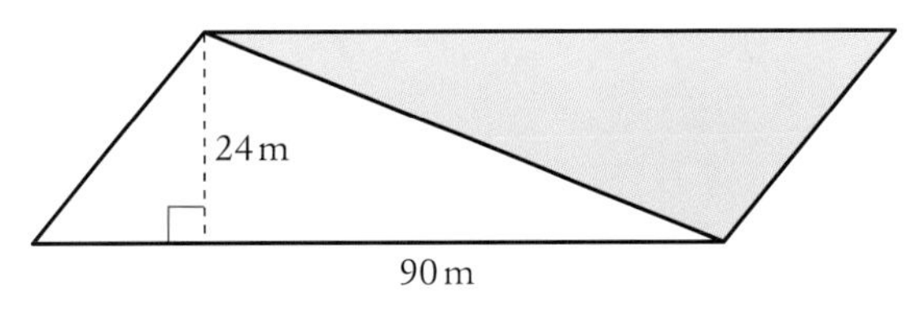

Area of parallelogram $= \left(\frac{1}{2} \times 90 \times 24\right) + \left(\frac{1}{2} \times 90 \times 24\right)$

$= 1080 + 1080$

$= 2160\ \text{m}^2$

Example 3

A child's kite is to be made from rods of lengths 120 cm and 80 cm.

What area of material is needed?

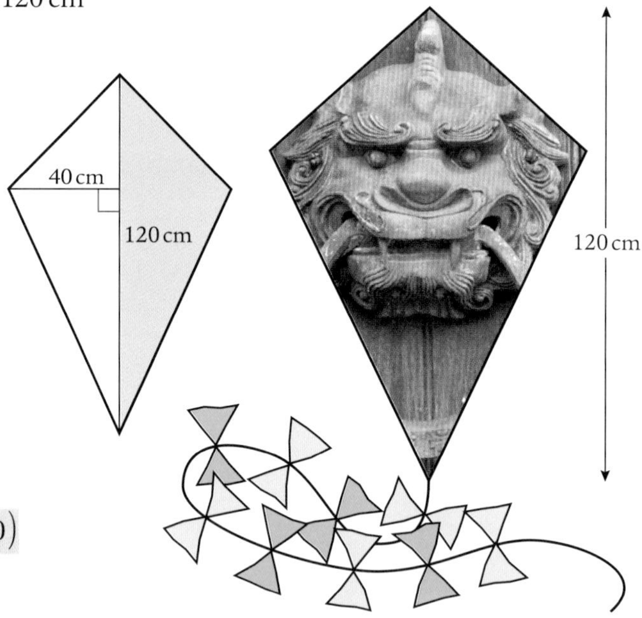

Sketch the kite and divide it into two triangles:

$$\text{Area of kite} = \left(\tfrac{1}{2} \times 40 \times 120\right) + \left(\tfrac{1}{2} \times 40 \times 120\right)$$
$$= 2400 + 2400$$
$$= 4800 \text{ cm}^2$$

Exercise 2.1A

1 Calculate the area of each quadrilateral. The units are all centimetres. A sketch will help.

a

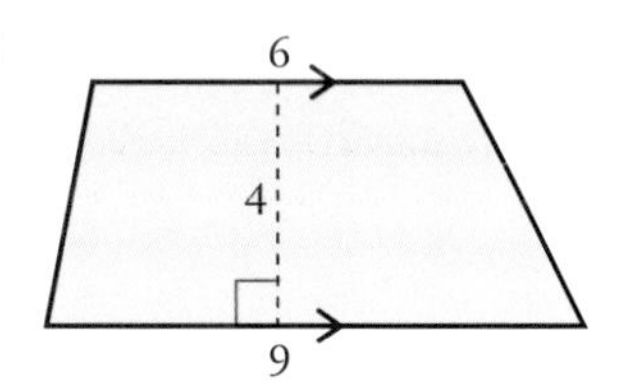

b

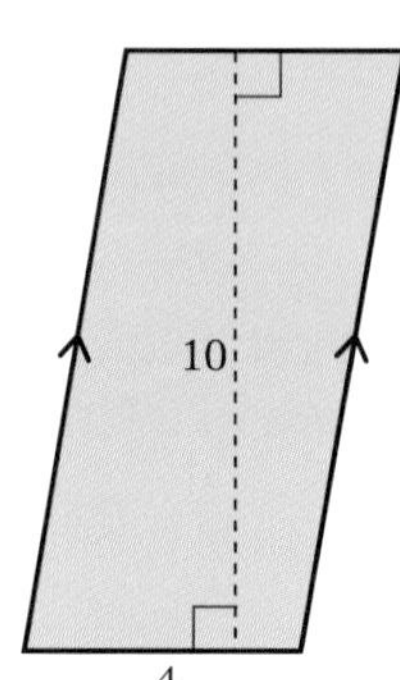

c

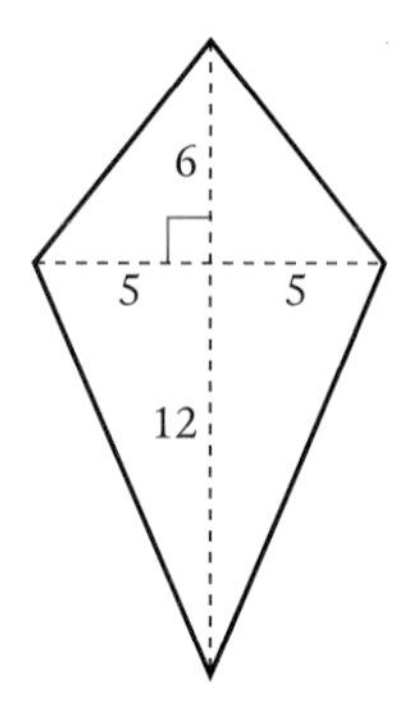

2 Calculate the area of each parallelogram. Note that in part **c** you are given more information than you need.

a

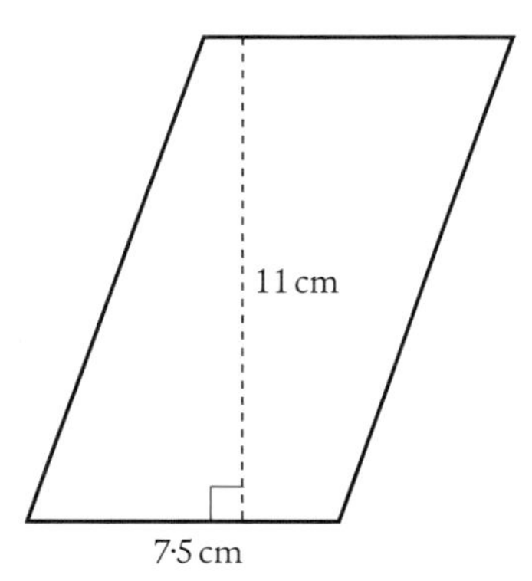

b

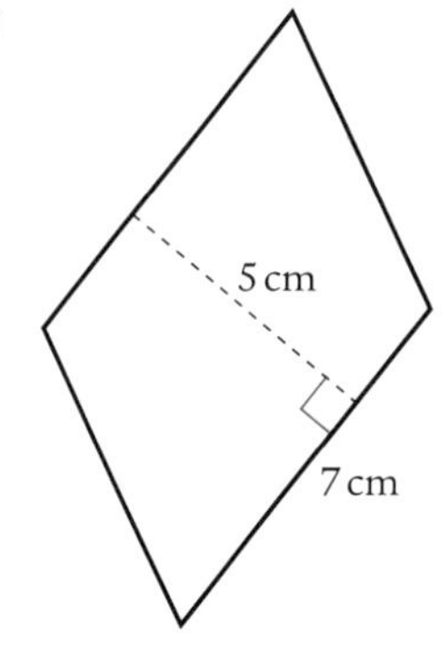

c

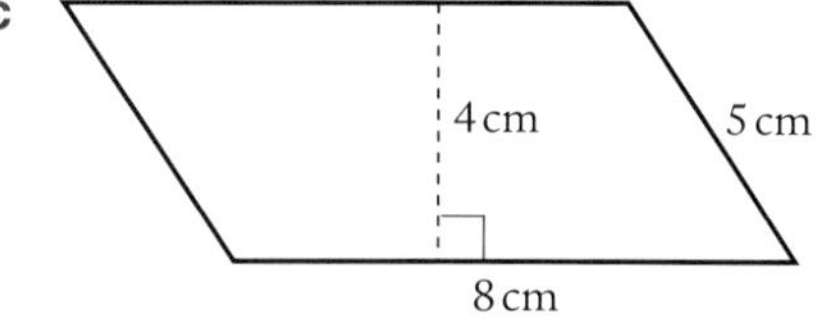

3 **a** Margaret uses two rods, one of 10 cm length and one of 15 cm to define both a trapezium and a kite in the sand.

Which shape has the bigger area?
How much bigger is it?

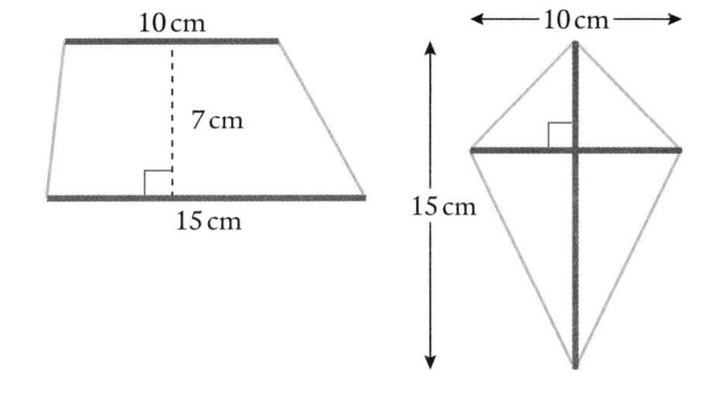

b She then repeats the exercise using rods of 6 cm and 12 cm.

She sets the rods 4 cm apart when making the trapezium.

Which shape now has the bigger area?

c She now has rods of 14 cm and 35 cm to define the kite and the trapezium.

How far apart should she set her rods when making the trapezium if she wants the two shapes to have the same area?

4 A railway station on the planned Waverley Line has an area that is to be turned into a car park.

The area is a parallelogram (see the surveyor's sketch shown).

a Calculate the area to be surfaced.

b Tar costs £12 per square metre.

What is the total cost of tarring the car park?

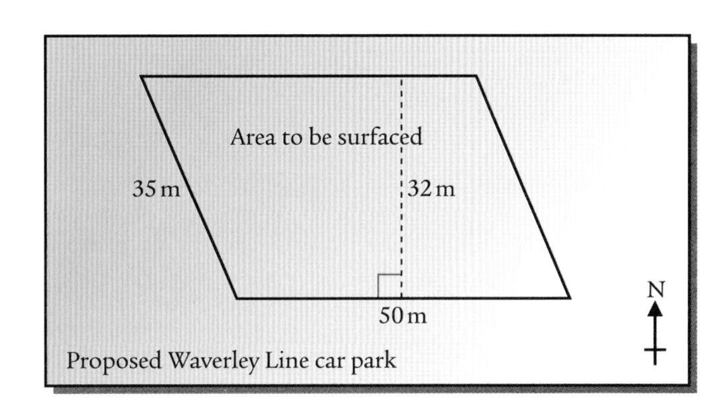

Making use of formulae

The trapezium

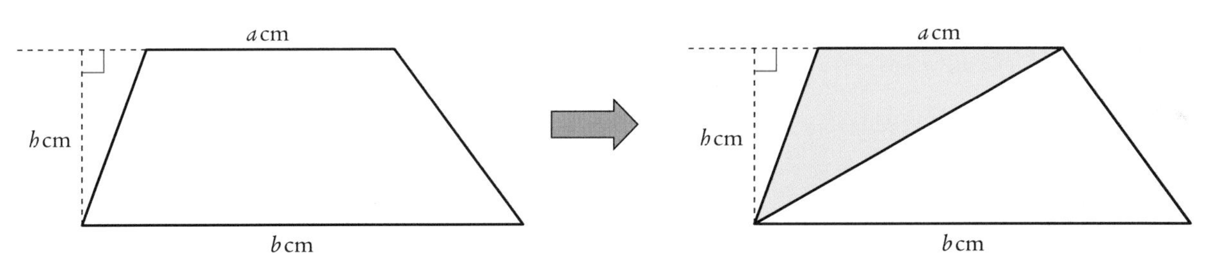

Area of trapezium $= \left(\frac{1}{2} \times h \times a\right) + \left(\frac{1}{2} \times h \times b\right)$

$= \frac{1}{2}h(a + b)$

Area of a trapezium = $\frac{1}{2}$ sum of parallel sides × distance between them

The parallelogram

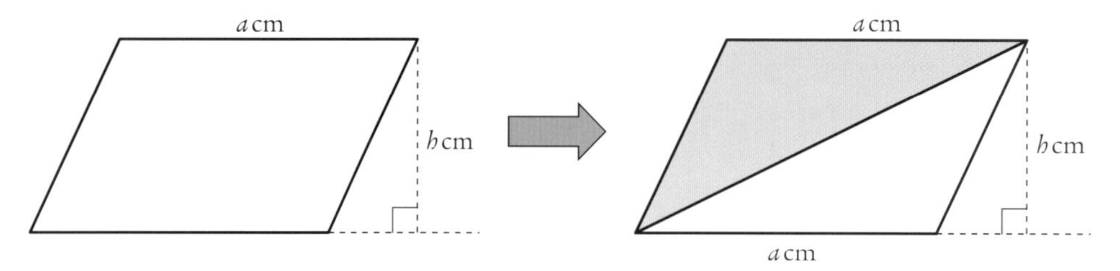

$$\text{Area of parallelogram} = \left(\tfrac{1}{2} \times a \times h\right) + \left(\tfrac{1}{2} \times a \times h\right)$$
$$= 2 \times \tfrac{1}{2} \times a \times h$$
$$= ah$$

Area of a parallelogram = length of its base × its altitude

The kite

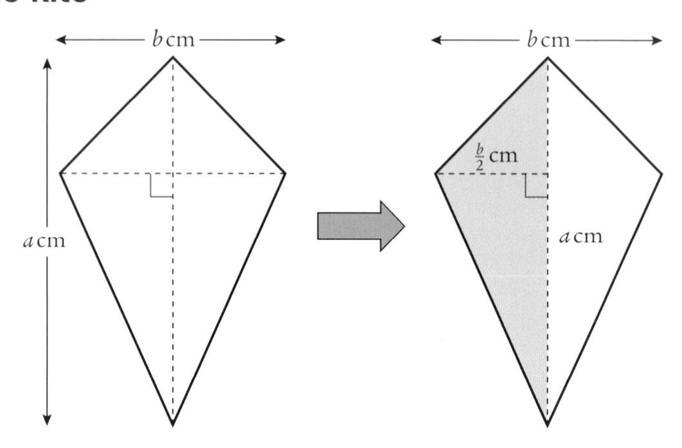

$$\text{Area of kite} = \left(\tfrac{1}{2} \times \tfrac{b}{2} \times a\right) + \left(\tfrac{1}{2} \times \tfrac{b}{2} \times a\right)$$
$$= 2 \times \tfrac{1}{2} \times \tfrac{b}{2} \times a$$
$$= \tfrac{1}{2}ab$$

Area of a kite = half of the product of its diagonals

Exercise 2.1B

1 The local ironmonger sells two types of snow shovel, A and B.

Both are trapezium shaped.

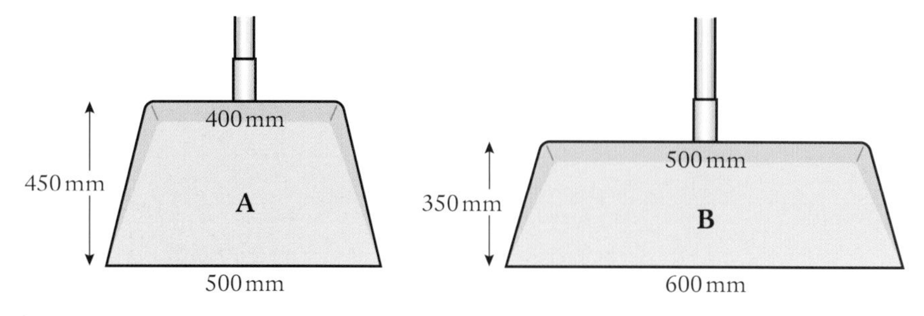

a **i** Which shovel has the greater area?

ii What is the difference between the two areas?

b To an extent, the number of calories burned up shovelling snow is proportional to the area of the shovel.

Mike burns up 300 calories clearing his drive using shovel A.

How many calories would you expect him to burn up using shovel B?

2 Marie, a jeweller, has designed a gold brooch.

It is in the shape of two congruent parallelograms.

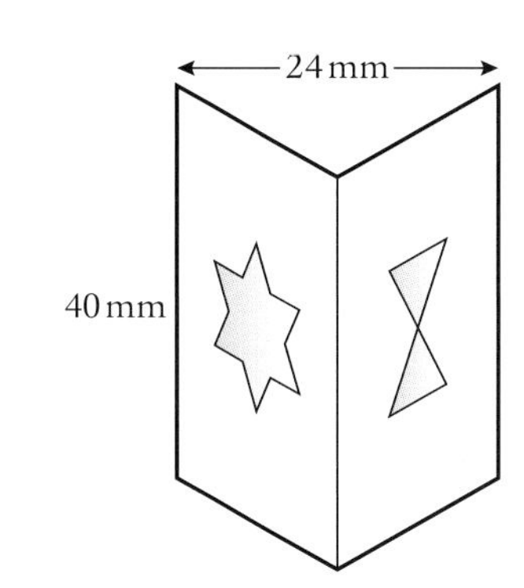

a Calculate the area of the brooch.

b Nine-carat sheet gold costs 3p per square millimetre.

Eighteen-carat gold costs 9p per square millimetre.

How much more will it cost Marie to make the 18-carat brooch than the 9-carat one?

c She wants to make 50% profit.

At what should she set the selling price of the 18-carat brooch?

3 Anwar, a kite manufacturer, has two types for sale.

The diagonals of one are 90 cm and 70 cm long.

The diagonals of the other are half these lengths.

a Calculate the area of each type of kite.

b **i** How many times greater than the smaller one is the area of the larger kite?

ii Would it be sensible to charge twice as much for the larger kite? Give a reason for your answer.

c If the diagonals had been three times greater, by what factor would the area increase?

4 A dressmaker cuts the front of a garment from a rectangle of cloth.

a What is the area of the material cut out?

b Calculate the area of wastage on both sides.

c If the material costs £20 per square metre, what is the cost of the wastage?

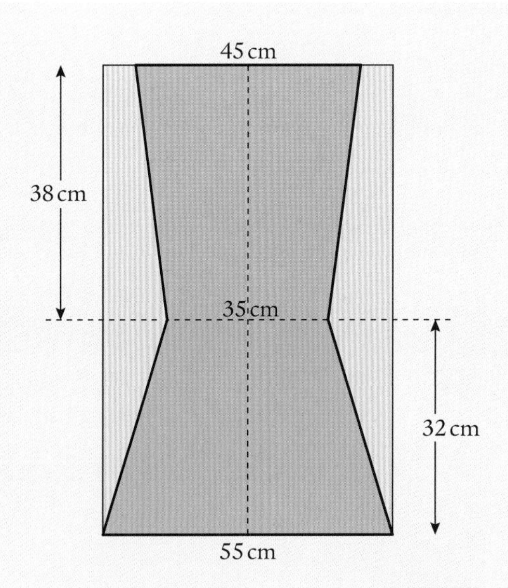

5 A stone mason charges £20 per square metre of wall that he repoints.

He inspects a job that is sited on a hill.

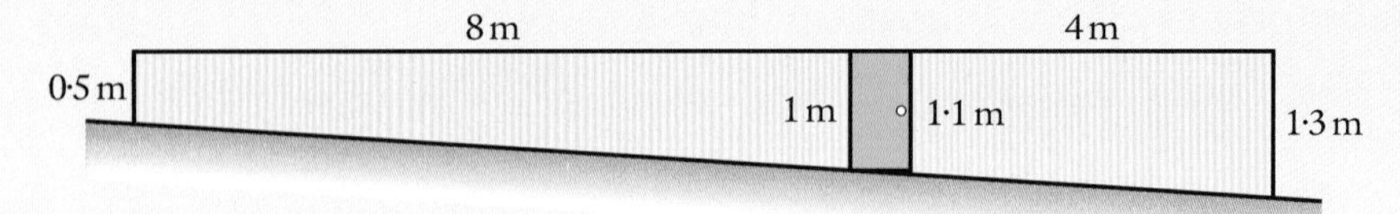

There are two sections of wall that need doing.

Both are in the shape of a trapezium.

a Calculate the area of each section of wall.

b Calculate the total estimate for the job.

6 A furniture manufacturer is designing a new line of tables.

The table top is a regular hexagon, veneered to enhance its appearance.

Each side of the table is 90 cm long.

The diagonal AE is 156 cm, to the nearest centimetre.

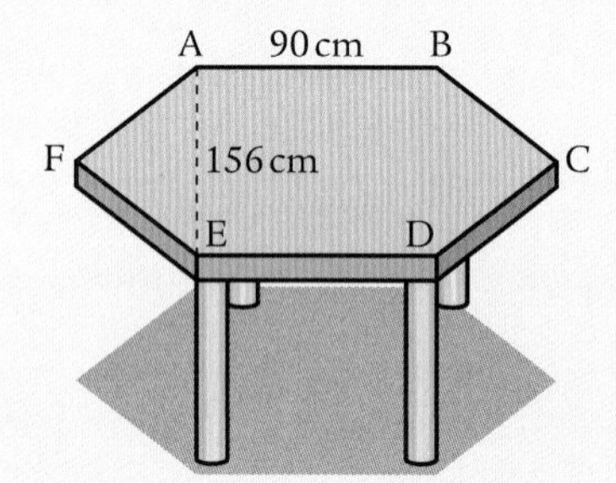

a Calculate the area of the table top, correct to three significant figures.

b The table shows the price per cm^2 of the different veneers.

Four possible designs are considered.

Veneer	Cost (pence per cm^2)
Ash	1·1
Mahogany	0·3
Teak	0·4
Walnut	2·2
Pine	0·1

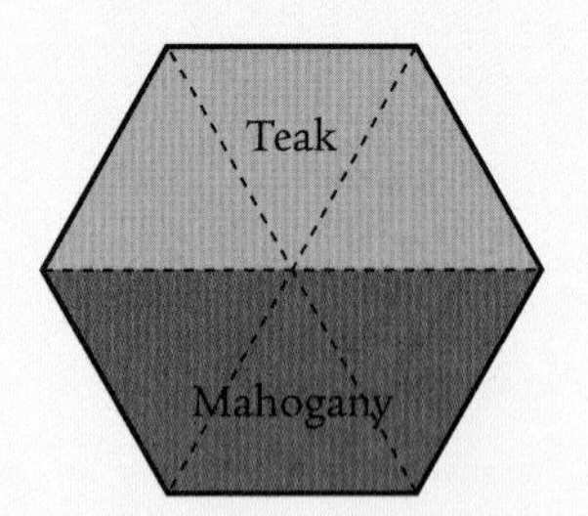

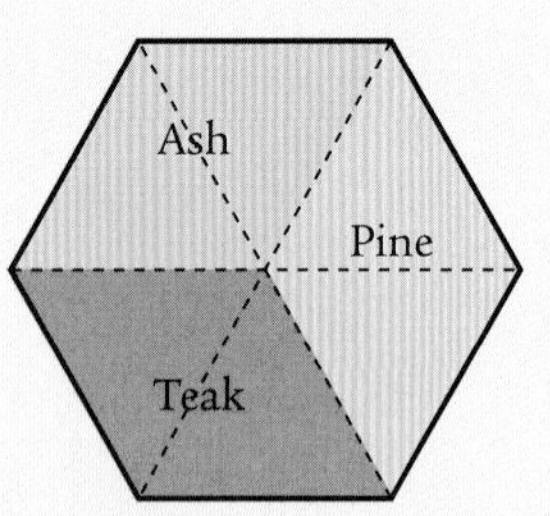

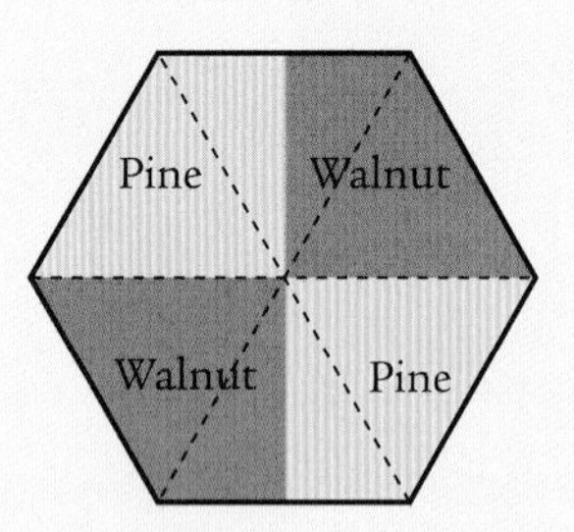

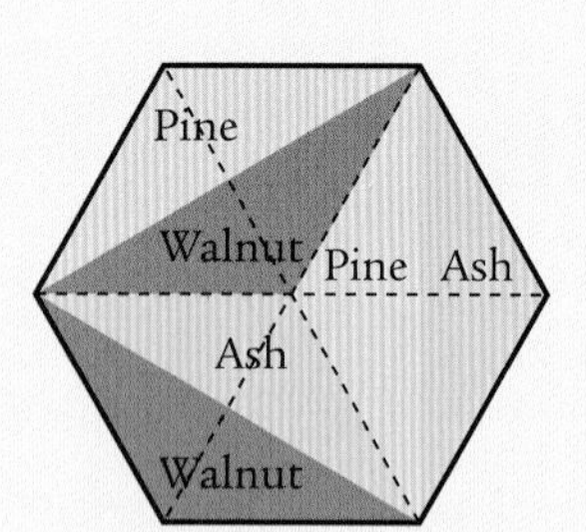

Calculate the cost of each design.

7 **a** A roof slope shaped like a trapezium has to be covered with solar cells.

Calculate the area to be covered.

b The owner is quoted £70 per square metre. How much will it cost the owner for the job?

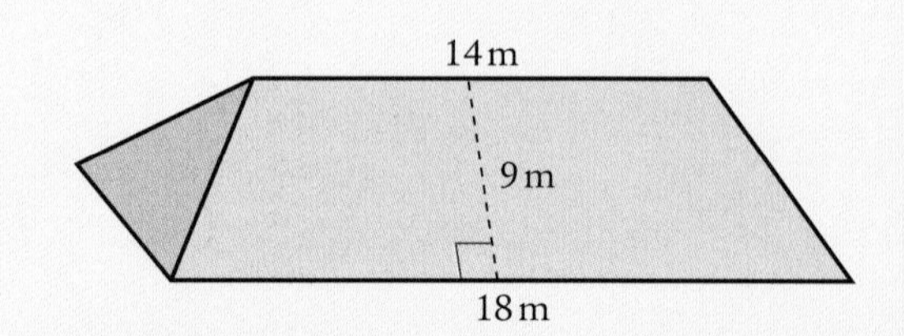

c A neighbour has solar cells fitted into an area shaped like a parallelogram on his roof.
It has the same overall area as the trapezium.
Its base is 20 metres long. Calculate its height.

d The customers have been promised that on average they should expect to get 105 watts of power per square metre. How much should each neighbour expect from his solar array?

8 **a** The area of parallelogram ABCD has the same numerical value as its perimeter.
Calculate the length of the sides AD and BC.

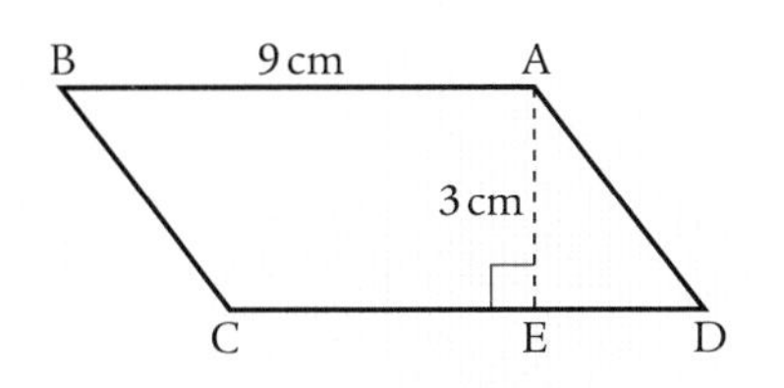

b A parallelogram has sides 3 cm by 5 cm.
One diagonal 4 cm long is at right angles to the side of length 3 cm.
What is the perpendicular distance between the sides of length 5 cm?

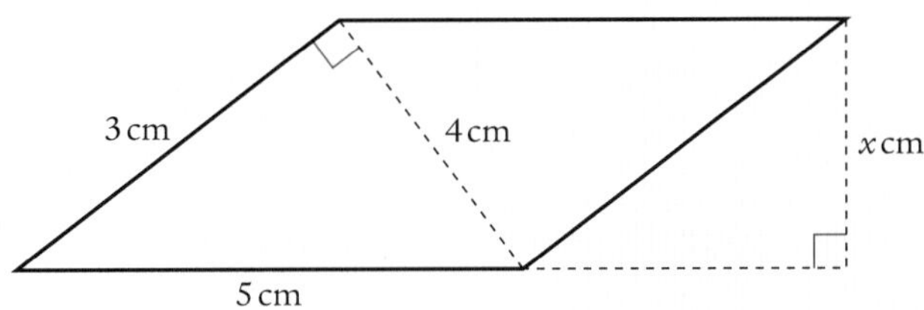

2.2 The circumference of a circle: a class discussion

Object	Diameter (cm)	Circumference (cm)
Pepper	4·6	14
Tin of spaghetti	7·5	24
Cheese box	11·2	35
Puzzle	5·5	17
£2 coin	2·9	9
10p coin	2·4	8
Salt cellar	6·7	21
Biscuit barrel	26	82
Cup	5·2	16
Saucer	15	47
Plate	22	69
Patterned plate	18·8	59

Make a collection of circular objects.

Make a table of diameters and circumferences.

Make a graph of diameter against circumference.

You'll find that whatever objects you use, you'll get a straight-line graph that goes through the origin.

This suggests that diameter (D) and circumference (C) are related by a simple formula of the form $C = kD$, where k is a constant.

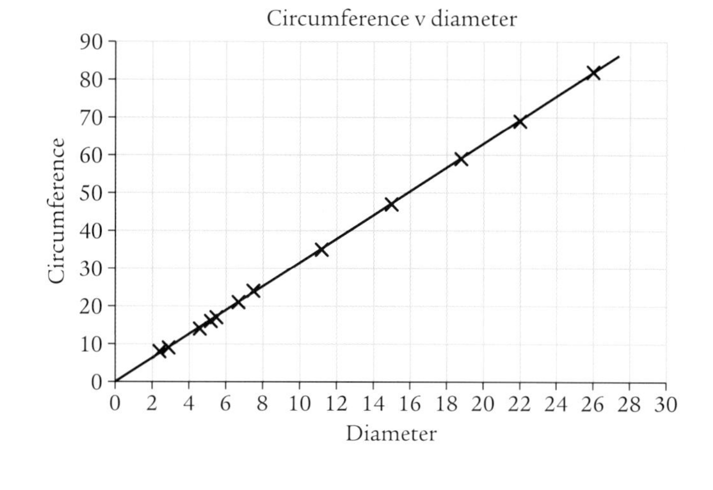

Go back to the table and make a column of $C \div D$.

The circumference (C) is approximately three times its diameter (D):

$$C \approx 3D$$

The results of each division should lie between 3·1 and 3·2.
The average value is 3·14.

$$C \approx 3{\cdot}14 \times D$$

The formula is most often written as $C = \pi D$, where π represents the number we multiply the diameter by to get the circumference.

Calculating π to a million decimal places is now used as a 'speed' test when seeing how good a computer is.

To seven decimal places, the value of π is 3·1415927.

Using the result

Example 1

Calculate the circumference of a circle with diameter 25 metres.

$C = \pi D$

$= \pi \times 25$ m

$= 78{\cdot}5398\ldots$ **Using the π button on the calculator** π

$= 78{\cdot}5$ m (to 3 s.f.)

π: a Greek letter pronounced 'pie' and spelt 'pi'. On a spreadsheet '= PI()' will give you its value.

Example 2

The Queen's Diamond Jubilee gold sovereign has a diameter of 22·05 mm.

Calculate its circumference to four significant figures.

$C = \pi D$

$= \pi \times 22{\cdot}05$ mm

$= 69{\cdot}272 \ldots$ mm π

$= 69{\cdot}27$ mm (to 4 s.f.)

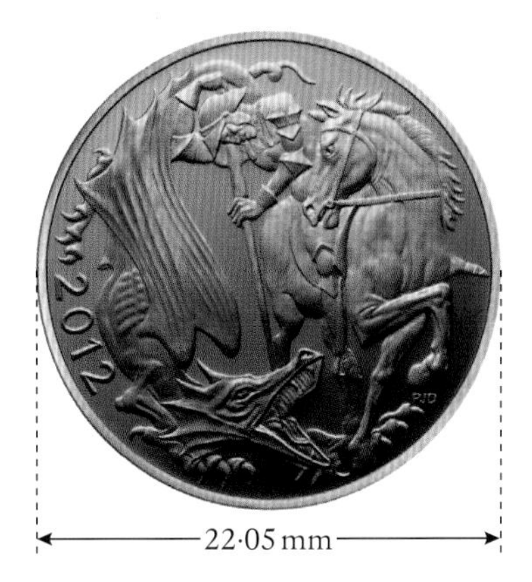

Exercise 2.2A

Give all answers correct to one decimal place.

1 Calculate the circumference of a circle with diameter:

a 10 metres **b** 35 cm **c** 7·5 mm.

2 Calculate the circumference of a circle with radius:

a 12·5 cm **b** 250 mm **c** 8·25 m.

3 $C = \pi D$. Rewrite this formula to express the circumference in terms of the radius of the circle.

4 Calculate the circumference of each of these objects:

a

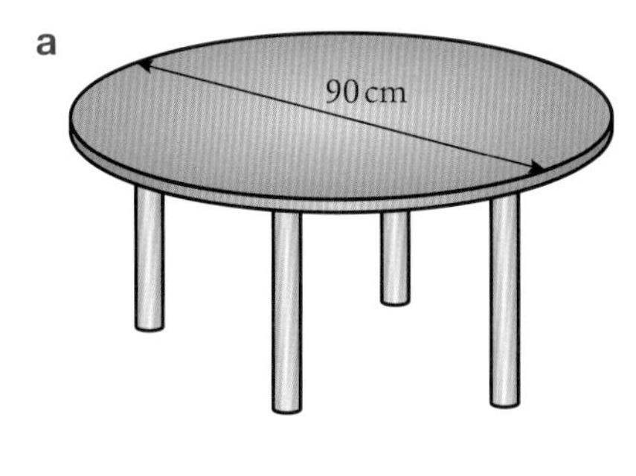

b

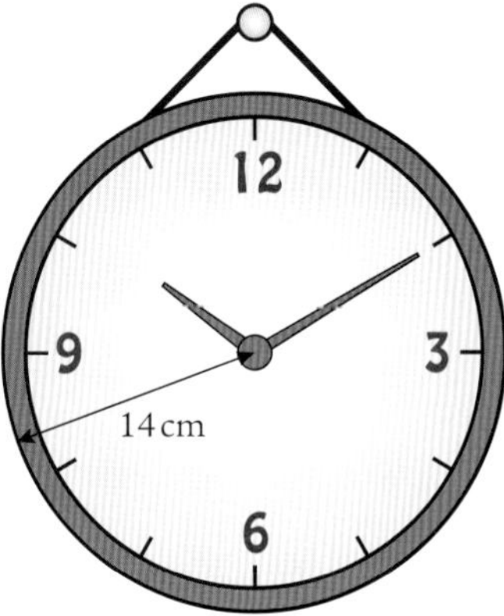

c

5 The wind turbine has blade lengths of 48 m.

Calculate the circumference of the circle swept out by the blades.

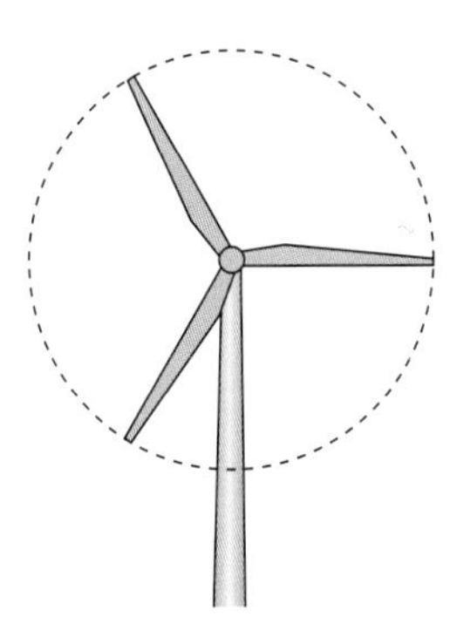

6 The London Eye Ferris wheel has a diameter of 120 metres.

a Calculate its circumference.

b It has 32 evenly spaced capsules to hold passengers.

How far apart are two neighbouring capsules as measured round the outside of the wheel?

c The wheel rotates at 26 cm per second.

How long does it take to make one complete turn?

7 The wheels of a bicycle have a diameter of 55 cm.

a Calculate their circumference, in metres.

b How far will the bicycle travel in 500 complete turns of the wheels?

c A Penny Farthing has a front wheel of diameter 135 cm and a rear wheel of 45 cm.

i How far will the bicycle travel when the front wheel has made one turn?

ii How many turns will the rear wheel make over this distance?

8 The radius of the ring at the top of a basketball net is 22·5 cm.

a Calculate its circumference.

b Will a ball whose circumference is 140 cm pass through the ring of a basketball net?

Exercise 2.2B

Give your answers correct to **three** significant figures.

1 The Drumness steam train runs on a track which is shaped like a figure of eight.

The two circles each have a radius of 250 metres.

Calculate the length of the track.

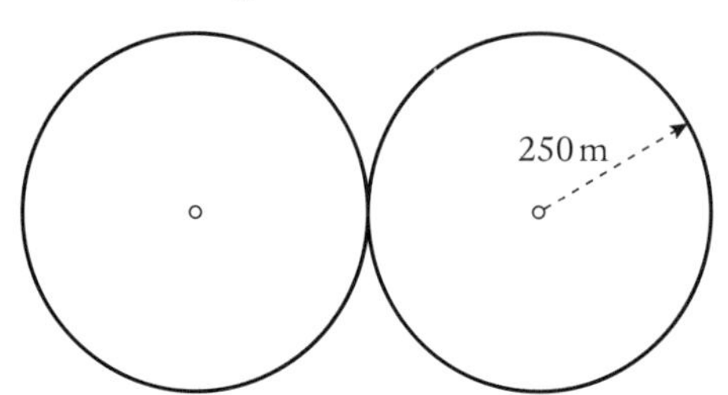

2 The inside lane of a running track is rectangular with semicircular ends.

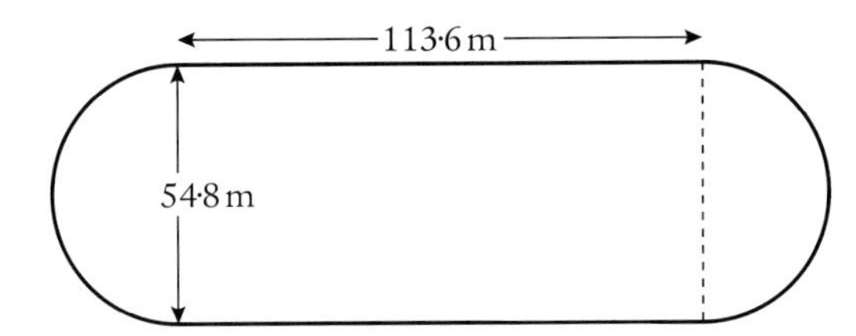

a What is the distance covered in one lap of the track?

b The next lane is an extra metre out. The straights are still 113·6 m but the diameter of each end is 56·8 m.
How much further does the runner in this track have to go?

c Assuming each lane is similarly formed, investigate how the start must be staggered to make the race fair.

3 Draw a circle with a circumference of:

a 26 cm **b** 8·5 cm.

4 A trundle wheel is used to find the distance between two objects.
We can see that the one illustrated measures a metre with every turn.

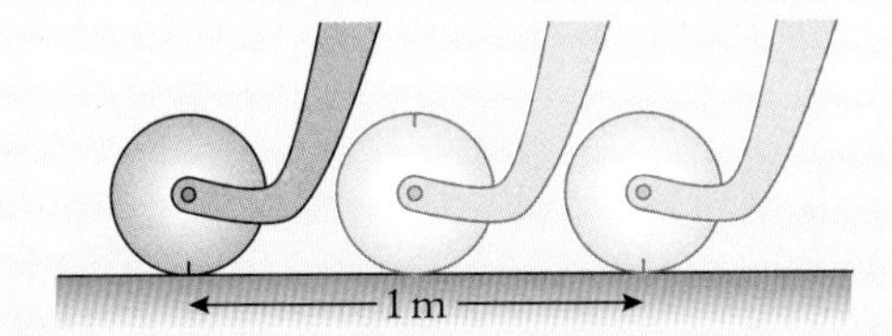

a If you made such a trundle wheel, what would the radius of your wheel be?

b Based on the same idea, you can buy a 'pen' to help you measure distances on a map.
As you 'draw' along a route, the pen measures out **centimetres** ... 1 centimetre per revolution.
Calculate the radius of the wheel in the pen, correct to 3 decimal places.

5 The Earth is not exactly a sphere.
Its diameter across the equator is 12 760 km, but from the North Pole to the South Pole its diameter is 12 710 km.
Calculate the Earth's circumference, correct to 4 significant figures:

a round the equator

b round the Poles.

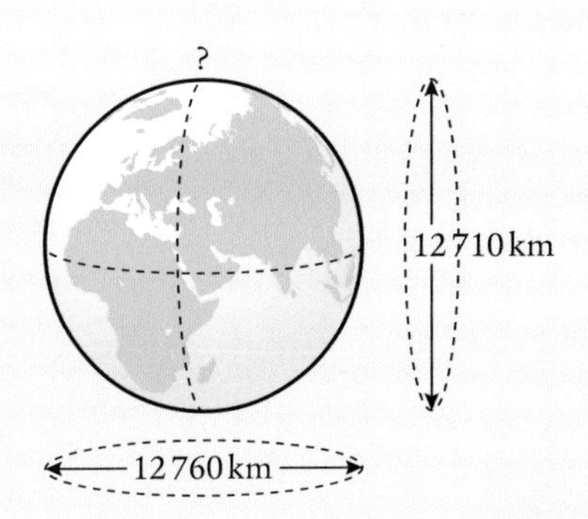

6 Imagine a band sitting tight round the equator.

Now imagine it is split and has a piece added so that all round the world the band is a centimetre off the ground.

How big is the piece that was added?

Now imagine a band round an orange of diameter 10 cm is broken and has a piece added so that all round the orange the band is a centimetre off.

How big is the piece that was added?

Comment.

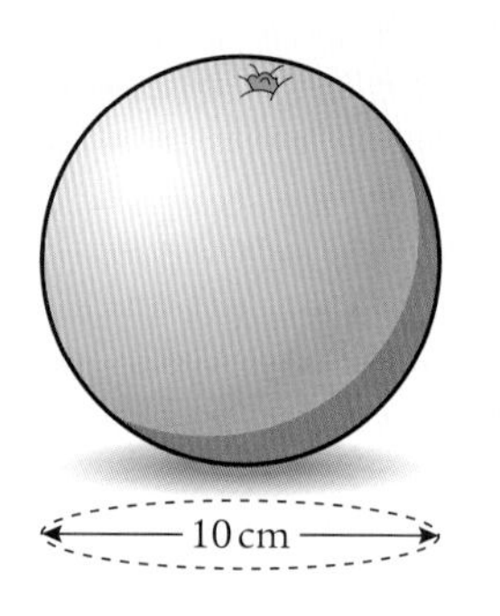

7 A satellite orbits the Earth, 760 km above its surface.

Taking the radius of the Earth to be 6350 km, calculate:

a the radius of the orbit

b how far the satellite travels in one orbit, correct to five significant figures

c how much further it would travel if the orbit was one kilometre higher.

2.3 The area of a circle

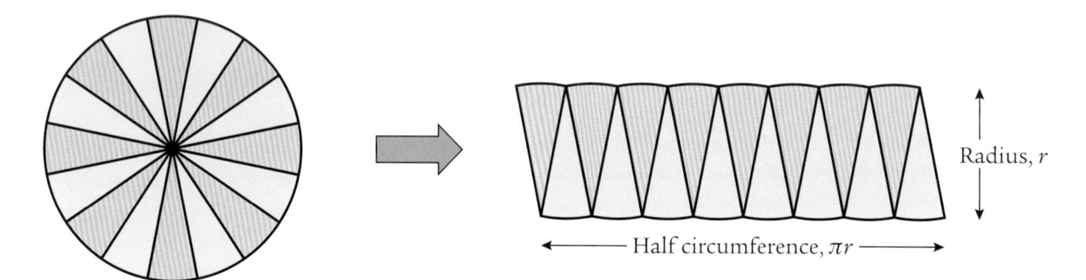

The circle with radius r is cut into 16 congruent sectors.

These sectors can be cut out and re-formed as a shape that's very nearly a parallelogram.

If we cut the circle into thinner sectors we'd get more, and the closer the shape would be to a parallelogram.

The base of the parallelogram is half the circumference of the circle $= \pi r$.

The altitude of the parallelogram is the radius of the circle $= r$.

The area of the parallelogram = base × altitude $= \pi r \times r = \pi r^2$.

So we have a formula for the area, A, of the circle:

$A = \pi r^2$ where r is the radius of the circle.

Using the result

Example 1

Calculate the area of a circle with radius 7·8 metres.

$$\text{Area} = \pi r^2$$
$$= \pi \times 7{\cdot}8 \times 7{\cdot}8\ \text{m}^2$$
$$= 191{\cdot}1344\ldots\ \text{m}^2$$
$$= 191\ \text{m}^2\ (\text{to 3 s.f.})$$

Example 2

Pizzas come in three sizes: small, medium and large.

The medium pizza has a diameter of 25 cm.

Calculate the area of a medium pizza.

Diameter = 25 cm ⇒ Radius = 12·5 cm

$$A = \pi r^2$$
$$\Rightarrow A = \pi \times 12{\cdot}5 \times 12{\cdot}5$$
$$= 490{\cdot}87\ldots\ \text{cm}^2$$

$$= 491\ \text{cm}^2\ (\text{to 3 s.f.})$$

Exercise 2.3A

Unless told otherwise, give your answers correct to three significant figures.

1 Calculate the area of a circle of radius:

a 10 cm **b** 7·5 m **c** 54 mm.

2 Calculate the area of a circle of diameter:

a 36 m **b** 9·8 cm **c** 1 km.

3 Calculate the perimeter and area of each shape:

Radius 1·6 cm

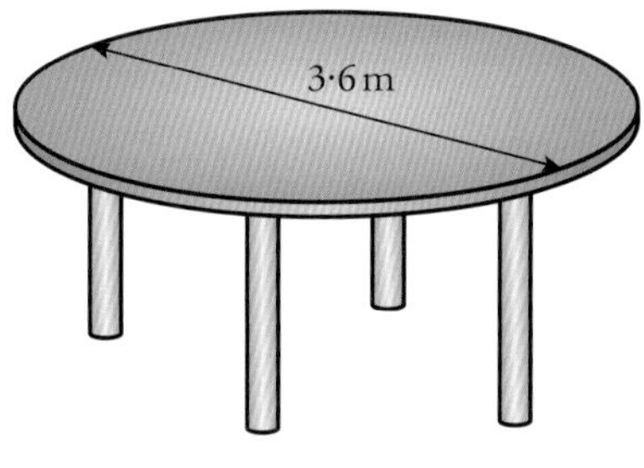

Diameter 3·6 m

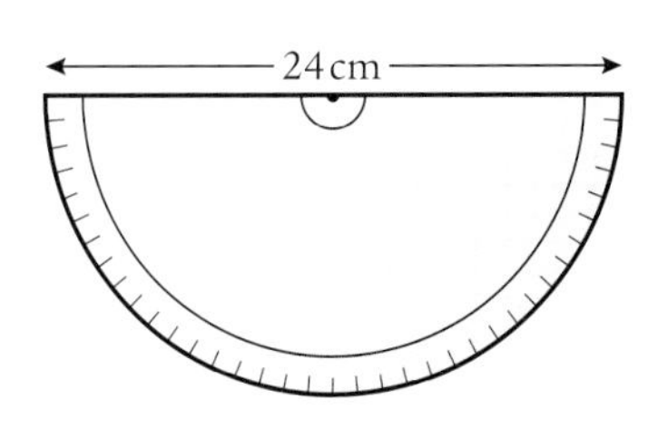

Blackboard protractor

4 An archery target has a radius of 61 cm.

The gold centre of the target has a radius of 12·2 cm.

a Calculate:

i the area of the target

ii the area of the gold centre.

b Each ring is 6·1 cm wide.

What is the area of the inner red ring?

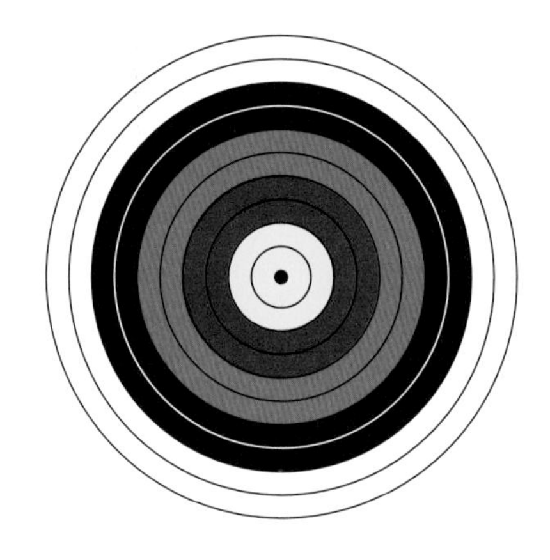

5 **a** The clock face has a radius of 15 cm.

Calculate its area.

b The minute hand is 12 cm long.

Calculate the area it sweeps through every hour.

c The hour hand is 6 cm long.

Calculate the area it sweeps through every 12 hours.

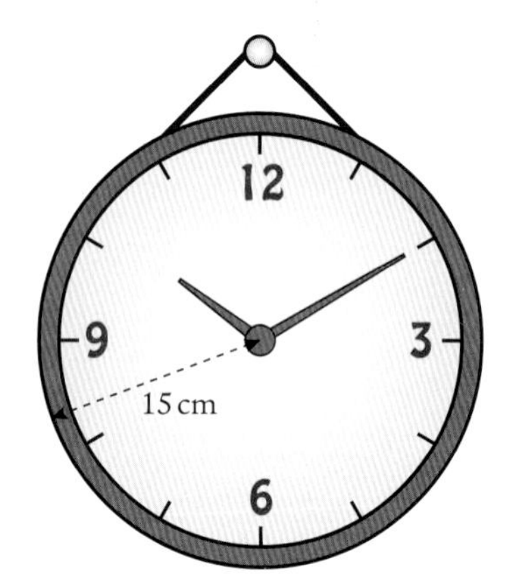

6 The energy you get from a wind turbine is proportional to the area swept out by the blades.

If you double the area, you get double the energy.

a Calculate the area swept out when the rotor blades are 20 m long.

b What area is swept out if the length of the rotor blades is doubled?

Comment.

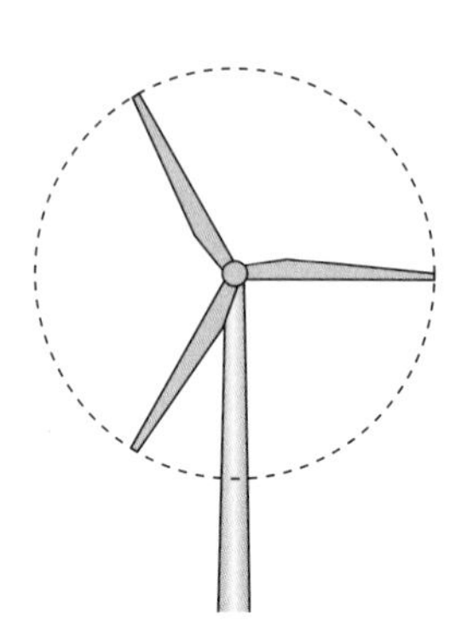

7 A rectangular swimming pool of length 25 m and breadth 8 m has three circular fountains in it as shown in the diagram.

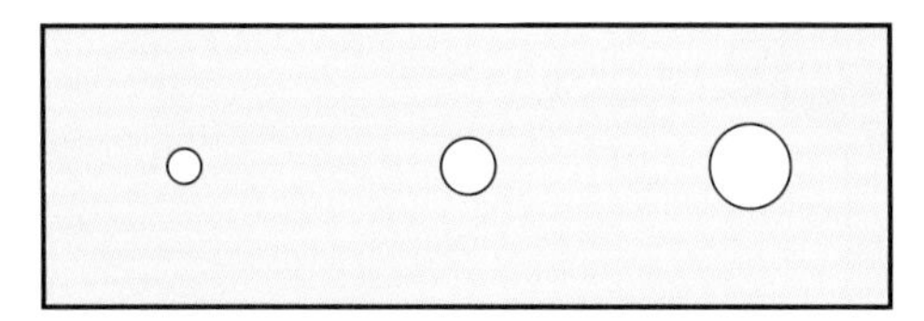

The radii of the fountains are 0·5 m, 0·8 m and 1·2 m.

Calculate the surface area of the water in the swimming pool.

8 On a hockey pitch there is a semicircle round the goals called the shooting circle.

This has a radius of 14·63 m with the goal at the centre.

a What is the area of this semicircle?

b Five metres beyond this there is another semicircle, drawn on the playing surface as a broken line.

What is the area enclosed between the broken semicircle and the shooting circle?

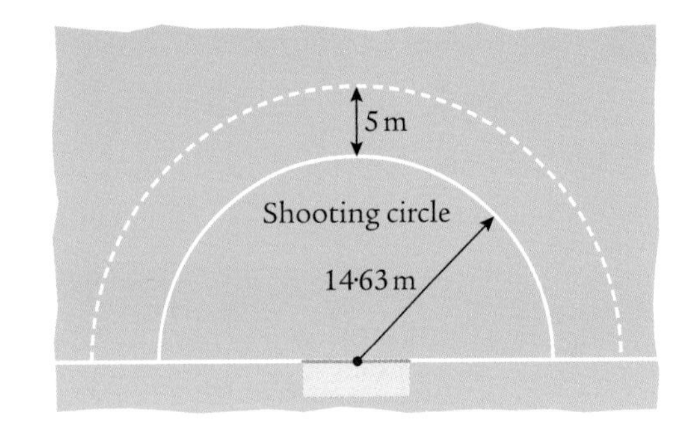

Exercise 2.3B

Unless told otherwise, give your answers correct to three significant figures.

1 In athletics the discus circle has a diameter of 2·5 m.

The diameter of the shot put circle is 2·135 m.

By how much is the area of the discus circle greater than the area of the shot put circle?

2 The sports arena is rectangular with two congruent semicircles, one at each end.

Calculate the area of the sports arena.

Give your answer to four significant figures.

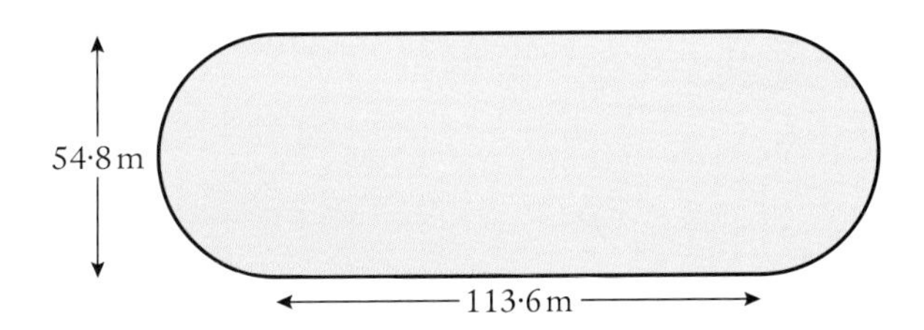

3 The top of a corner unit is a quarter circle.

It has a radius of 38 cm.

Calculate the area of the top of the unit.

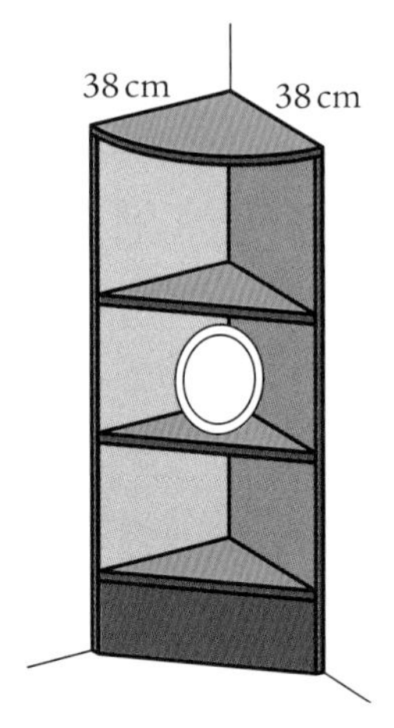

4 A circle has an area of 264 cm^2.

Calculate the radius of the circle.

5 A £1 coin has a diameter of 22·5 mm.

a Calculate its: i circumference ii area.

b For charity, the class is asked to collect a square metre of pound coins.

This could mean one of two things.

A Coins arranged to form a square at least 1 metre by 1 metre.

B Enough coins so that their total area is at least 1 square metre.

i How much is collected by method A?

ii How much more is collected by method B?

6 A circular tablecloth has a diameter of 150 cm.

It is edged with lace all the way round.

a What is the length of lace round the tablecloth?

b The lace trim adds 2 cm to the diameter of the tablecloth.

What is the area of the tablecloth with the trim?

7 Gary the goat is tied to a post in the middle of a field.

The length of the rope connecting Gary to the post is 16 m.

a Calculate the area of the grass on which Gary can graze.

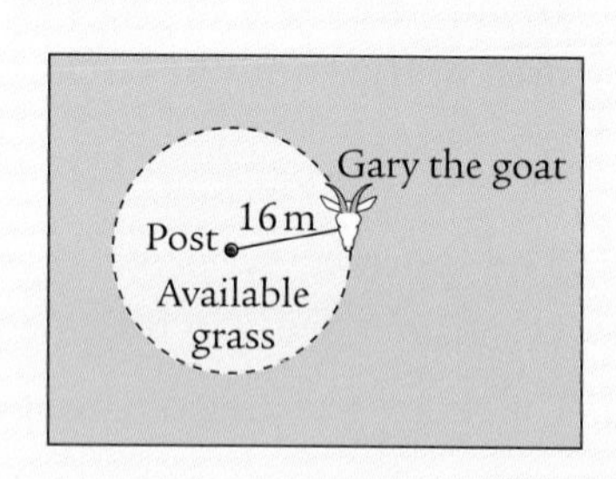

b On a second occasion Gary is tethered to the corner of a barn by an 8-metre rope. The barn is 10 m by 6 m.

Calculate the area of grass available to Gary.

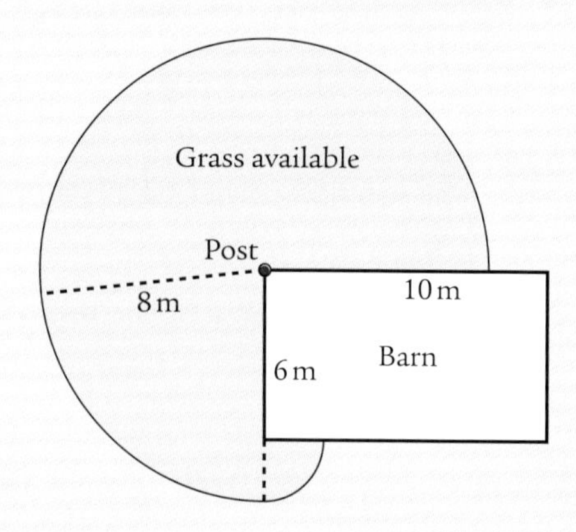

8 Farmer Spence has 144 m of flexible wire fencing.

He wants to enclose the largest possible area with his fencing.

a Which would give him the larger enclosed area: a square or a circle?

b Calculate the difference between the two areas.

2.4 Prisms

Reminders:

A prism is a solid with a uniform cross-section.

If it is cut parallel to its end (its base), the cross-section is congruent to the base.

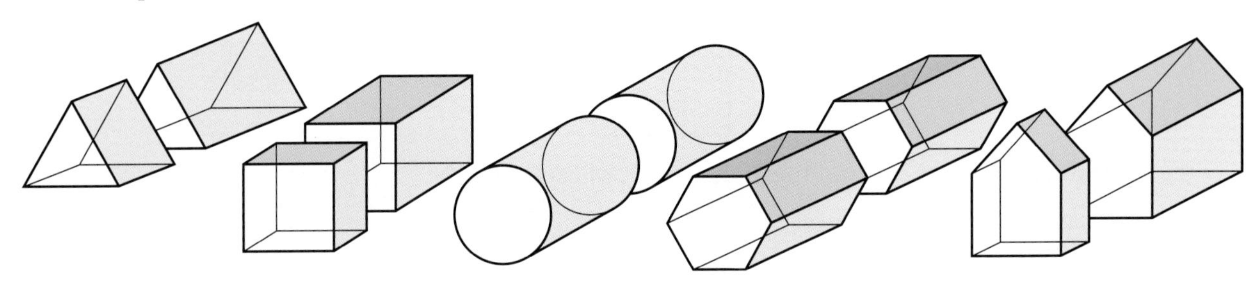

A prism will take its name from the shape of its base.

So we have, from left to right:

a triangular prism, a rectangular prism (a cuboid), a circular prism (a cylinder), a hexagonal prism and a pentagonal prism.

Each solid we study is bounded by surfaces or faces.

Where two faces meet, edges are formed.

Where edges meet, vertices are formed. (The singular of 'vertices' is 'vertex'.)

Exercise 2.4A

1 The diagram shows a net for a packet of salt: it's a cuboid.

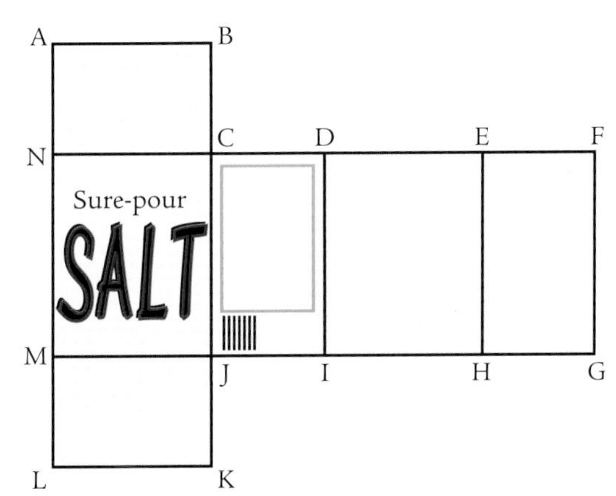

a Write down the lengths of:

i AB **ii** BC **iii** CD **iv** FG **v** IH **vi** JK.

b Hence calculate the total area of the six faces of the packet of salt.
(This total area is called the surface area of the packet.)

c Draw a different net for the packet of salt, marking in the lengths of the edges.

d Calculate the surface area of the packet from this net.

e What do you notice about your answers to **b** and **d**?

2 Here is one possible net of this triangular prism.

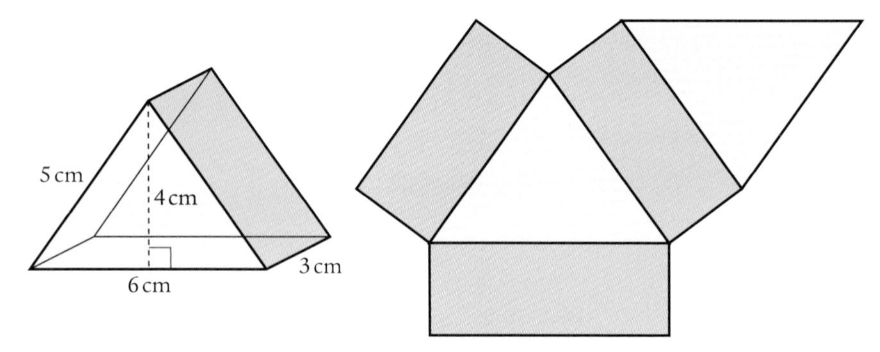

a Make a sketch of the net and mark the length of each edge.

b Calculate the area of each face.

c Hence find the surface area of the triangular prism.

3 a How many faces does a triangular prism have?

b What types of shapes are the faces?

c Sketch a net of the triangular prism below, marking the length of each edge.
All lengths marked are in centimetres.

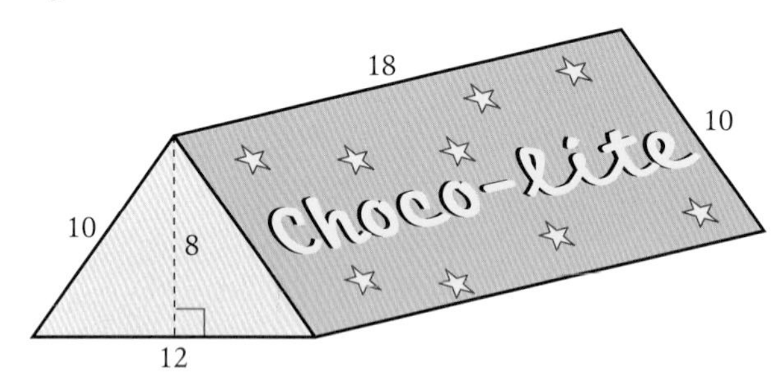

d Calculate the area of cardboard required to make the chocolate box (i.e. calculate the surface area of the box).

4 A tin of beans has been opened top and bottom and the label taken off.

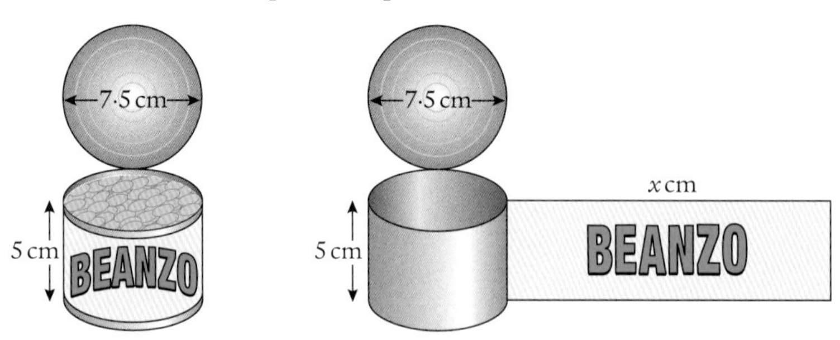

a Explain why 23·6 cm is a good approximation for the length of the label.

b The label represents the curved surface area of the cylinder. Calculate its area.

c Hence calculate the total surface area of the cylinder.

5 A garden roller is a heavy cylinder with a handle.

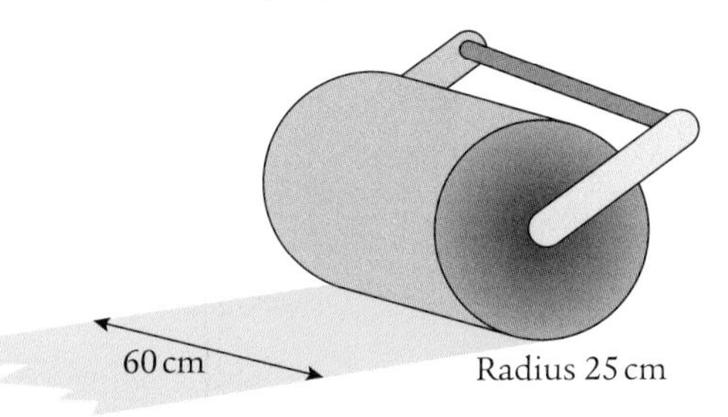

a Calculate the curved surface area of the cylinder.

b Calculate its total surface area.

6 The inner tube of a kitchen roll is a cardboard cylinder with no ends.

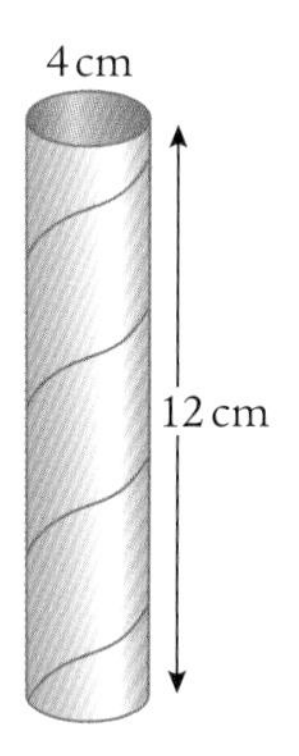

a Describe the net of a cylinder that has no lid and no base.

b The tube is 12 cm long with a diameter of 4 cm.
Calculate its surface area.

7 The cost of heating a house is related to its surface area.

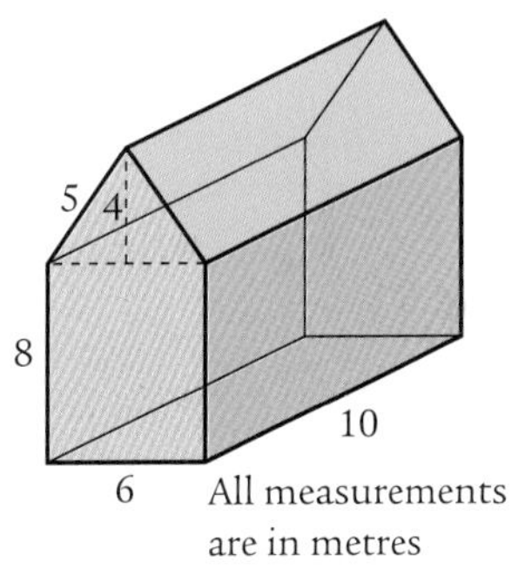

a Sketch the net of the house, ignoring the floor.

b Calculate the area of:

i a green face

ii a pink face

iii a brown face.

c Calculate the surface area of the house, again ignoring the floor.

Exercise 2.4B

1 A £2 coin has a diameter of 28·4 mm and is 2·50 mm thick.
Calculate its surface area.

2 An aluminium trough is a prism with ends that are symmetrical trapeziums.

Its dimensions are as shown.

a Sketch its net.

b Calculate its surface area.

c Aluminium costs 0·2 pence per square centimetre.

What will the materials to make the trough cost?

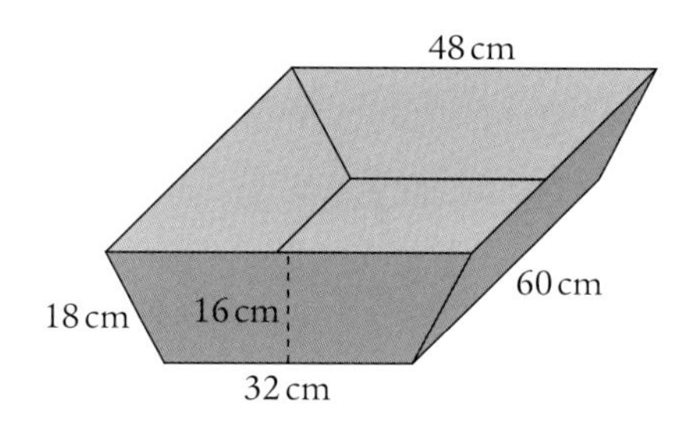

3 A bale of silage is cylinder-shaped.

Its diameter is 1·4 m and its length is 1·4 m.

It is to be wrapped in black polythene to preserve it.

Calculate the area of wrap required to cover the whole bale.

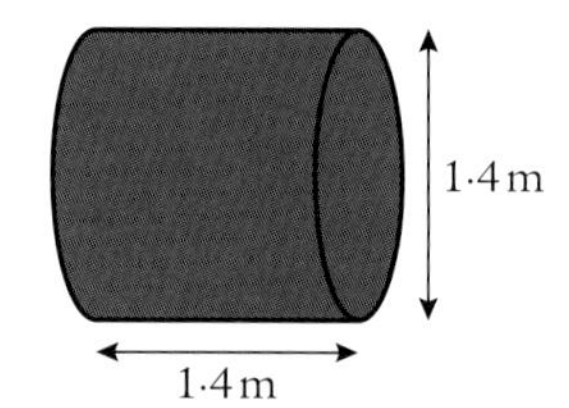

4 The greenhouse is in the shape of a triangular prism.

Calculate the area of glass in the greenhouse.

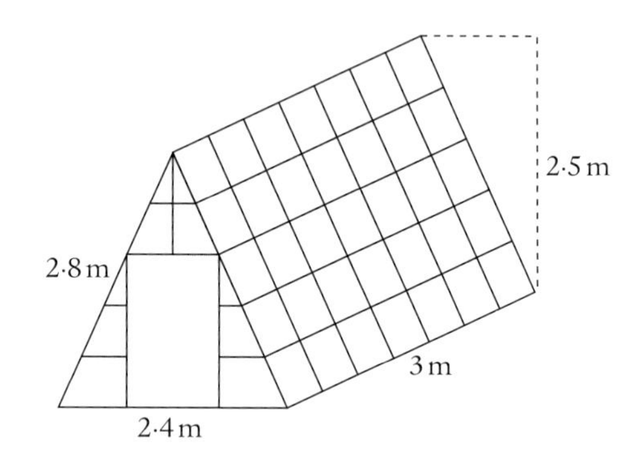

5 The speed that a chemical dissolves depends on the surface area of its crystals.

One particular crystal is a prism with ends that are parallelograms. All of its other faces are rectangles.

a Sketch the net of the crystal.

b Calculate its surface area.

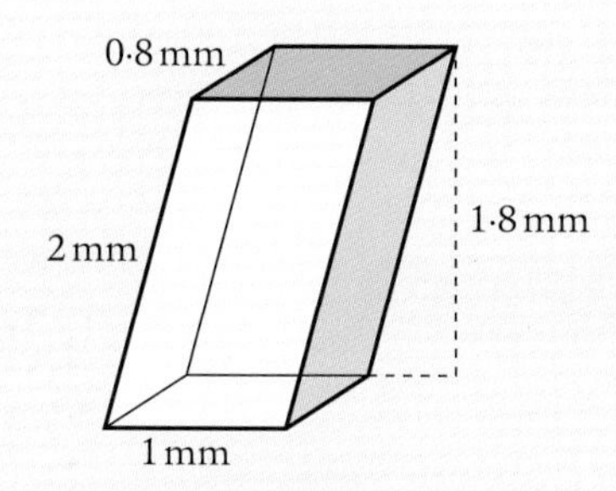

6 A company manufactures metal drums for use in storing chemicals.

The outside and the inside of the drum (top, base and side) need to be coated with a special paint to prevent rusting.

a Calculate the area to be coated.

b If a tin of this paint will cover 1 m^2, how many tins are needed to treat 100 drums?

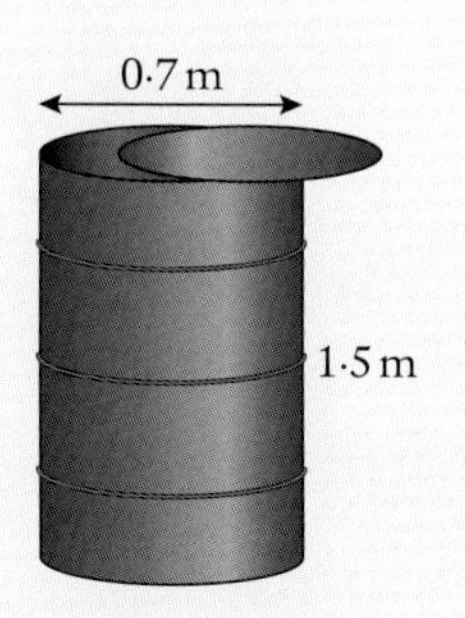

7 A tent is in the shape of a triangular prism.
The triangles are equilateral, with sides of 2 m.

The height of the tent is 1·73 m, correct to 2 decimal places.

The ends of the tent are 5 metres apart.

- **a** Calculate the area of nylon needed to make the tent. (The tent does not have a base.)
- **b** Waterproofing for the nylon costs £15 per litre and a litre will treat 18 m^2 of material.

  What will be the cost of treating 100 such tents?

8 A lady's hat is made from a cylinder (A) and an outer circular brim (B).

Both parts are made out of velvet.

- **a** Calculate the area of velvet required for the brim.
- **b** Calculate the area of velvet required for the centre of the hat.

  (Note there is no base to the cylinder.)

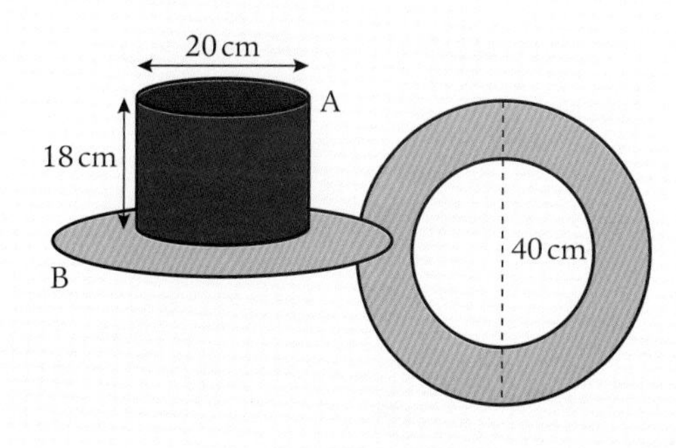

9 Tins of pears are to be packed in cuboid cartons holding 36 tins.

The tins have to be arranged so that a minimum area of cardboard is used for a carton.

Consider the possible arrangements and calculate which one uses the least cardboard.

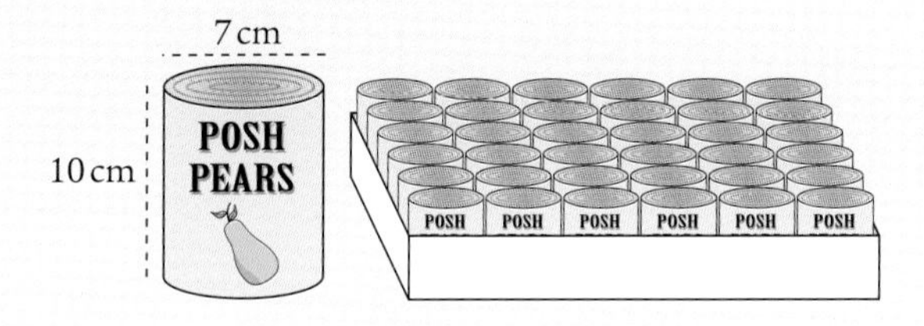

2.5 Volume of a prism

Making a cuboid from 1 cm cubes lets us see that the volume of a cuboid (a rectangular prism) is:

$$V = \text{length} \times \text{breadth} \times \text{height}$$
$$\Rightarrow \quad V = l \times b \times h$$
$$\Rightarrow \quad V = (l \times b) \times h$$
$$\Rightarrow \quad V = \text{area of base} \times h$$
$$\Rightarrow \quad V = Ah$$

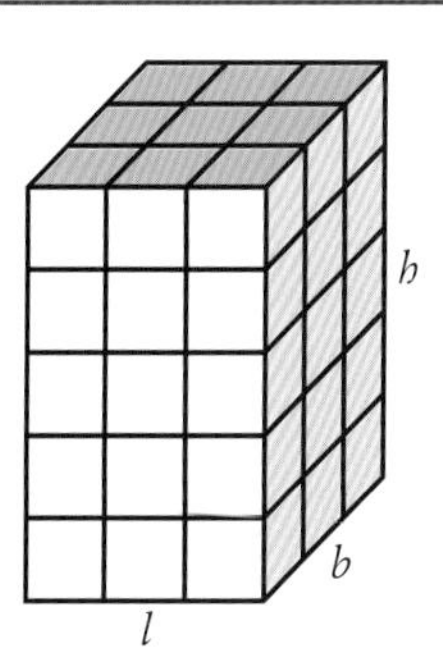

i.e. **volume of a rectangular prism = area of base × height.**

Cut the cuboid through the diagonal plane.

Two triangular prisms are obtained, each half of the original.

Each has a volume of $V = \frac{1}{2}lbh = \left(\frac{1}{2}lb\right)h$ = area of base × height.

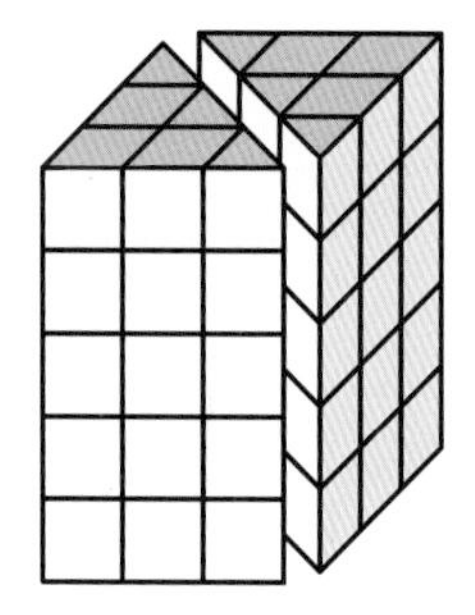

Fit them together to form a triangular prism whose base has the same area as the base of the original cuboid, and the same height.

So again we get $V = Ah$.

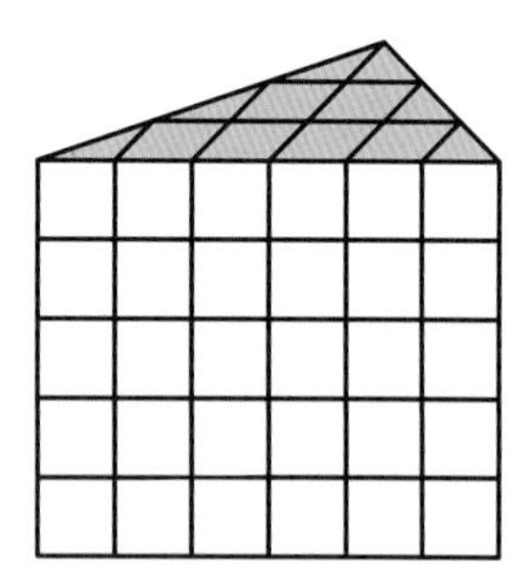

Six congruent triangular prisms can be placed together to form a hexagonal prism ...
... and again we get $V = Ah$.

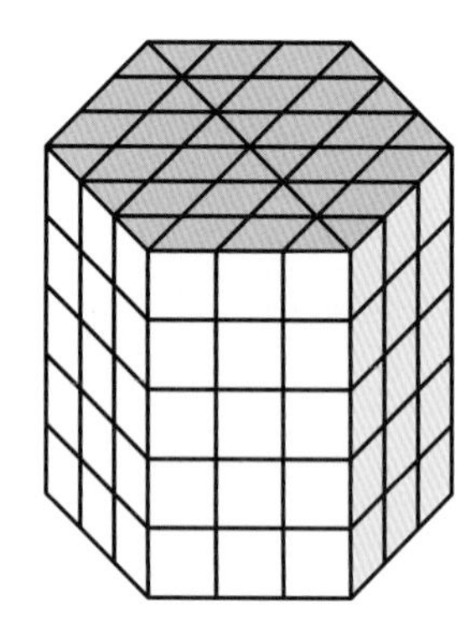

Imagine a cylinder as made up of lots of very thin triangular prisms. We can see that the volume of a cylinder (a circular prism) can also be found by the formula $V = Ah$.

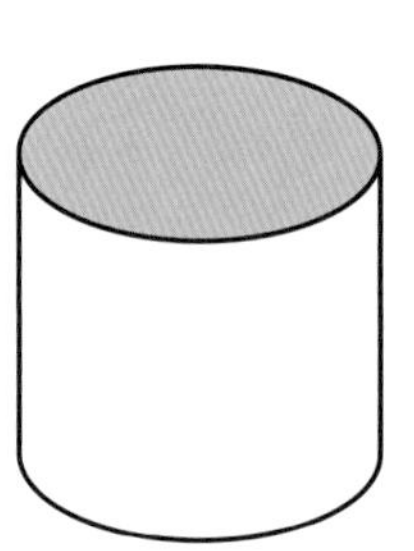

For a cylinder, the area of the base is πr^2.

So, for a cylinder, $V = \pi r^2 h$.

In fact, we can show that, for all prisms:

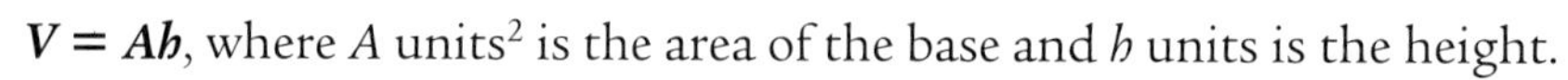

$\boldsymbol{V = Ah}$, where A units^2 is the area of the base and h units is the height.

We can also say:

volume = area of cross-section × length.

Using the result

Example 1

Calculate the volume of a triangular prism, 20 cm in length, whose base is a right-angled triangle, with shorter sides 9 cm and 8 cm.

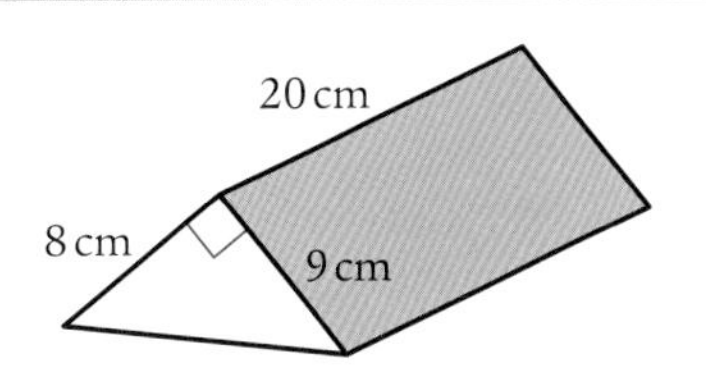

Use the formula $V = Ah$

A = area of base $= \frac{1}{2} \times 9 \times 8 = 36\ \text{cm}^2$

$h = 20$ cm

$\Rightarrow \quad V = 36 \times 20$

$\Rightarrow \quad V = 720\ \text{cm}^3$

Example 2

An oil drum is three-quarters full of oil.

Its diameter is 0·8 m and its height is 1·2 m.

Calculate the volume of oil in the drum.

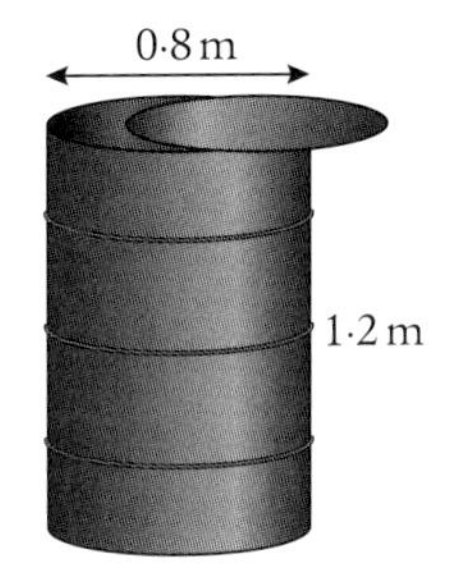

A = area of base $= \pi r^2 = \pi \times 0{\cdot}4 \times 0{\cdot}4$

$= 0{\cdot}503\ \text{m}^2$ (to 3 s.f.)

Volume when full: $V = Ah$

$\Rightarrow \quad V = 0{\cdot}5026 \ldots \times 1{\cdot}2 = 0{\cdot}60318 \ldots\ \text{m}^3$

$\Rightarrow \quad \frac{3}{4}V = \frac{3}{4} \times 0{\cdot}60318 \ldots\ \text{m}^3$

$\Rightarrow \quad$ Volume of oil $= 0{\cdot}45238 \ldots\ \text{m}^3$

$= 0{\cdot}452\ \text{m}^3$ (to 3 s.f.)

Exercise 2.5A

1 Calculate the volume of these prisms:

a area of base $5{\cdot}6\ \text{m}^2$, height 3 m

b area of base $7{\cdot}4\ \text{cm}^2$, height 9 cm.

2 Calculate the volume of each prism:

a

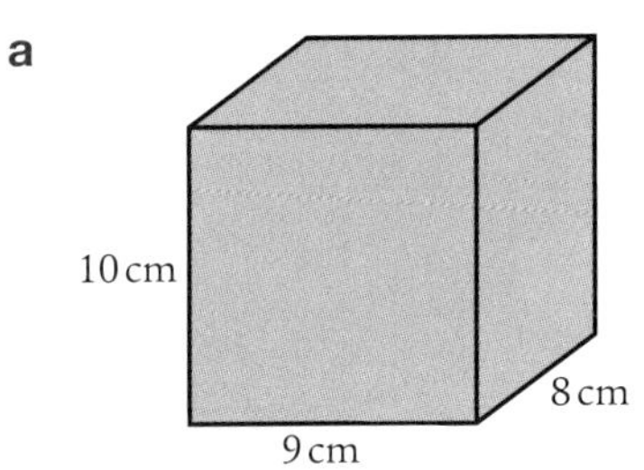

b

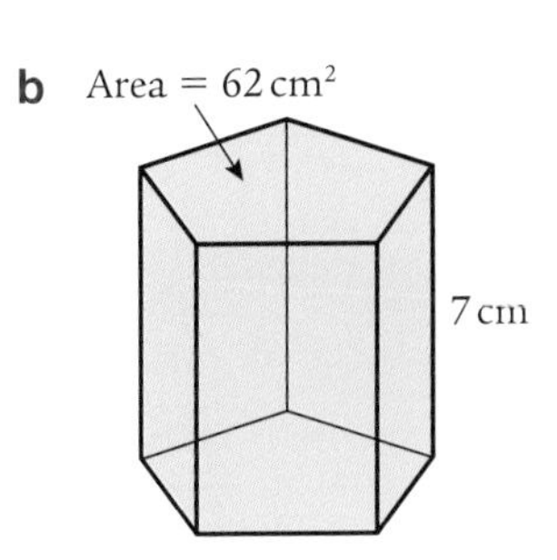

c

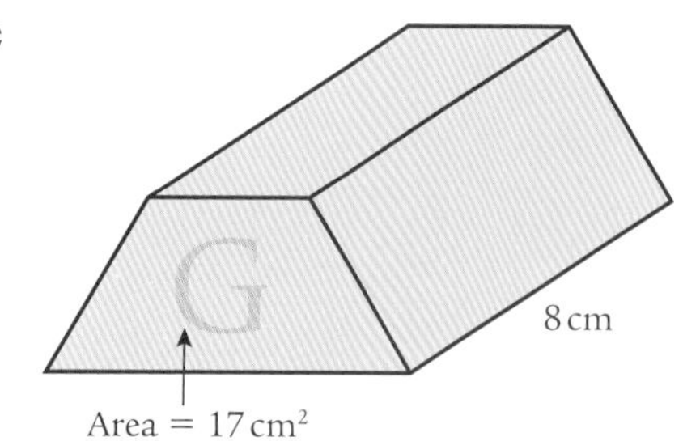

d

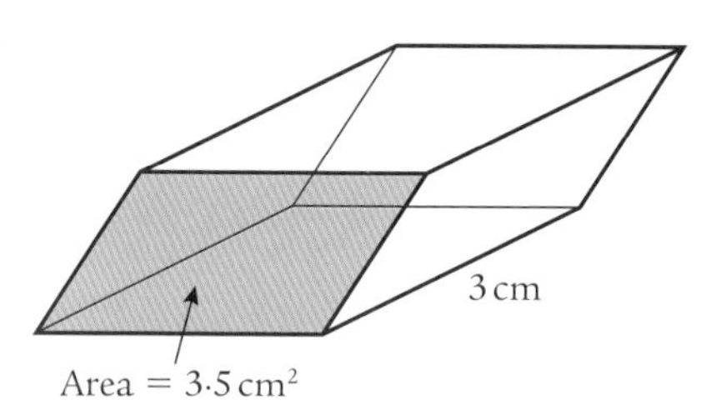

3 Calculate the volume of these prisms:

a

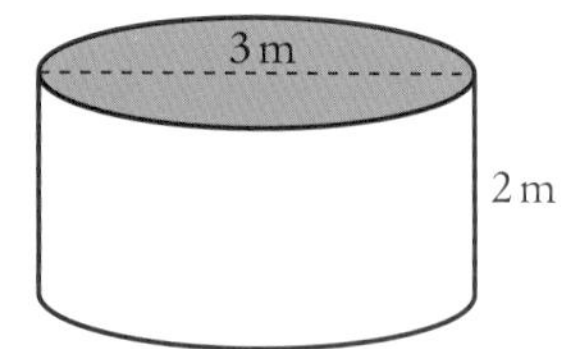

b

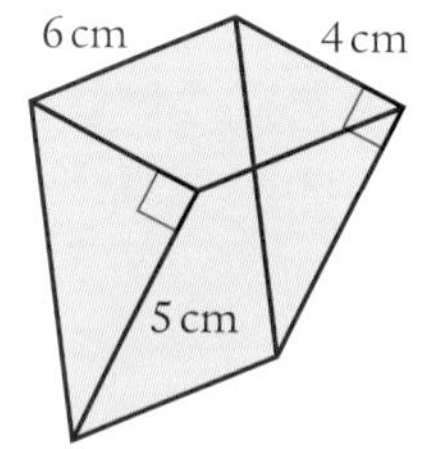

c

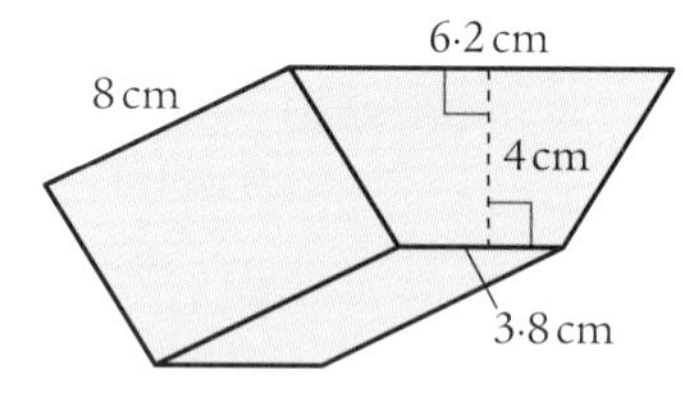

4 An airline sells sandwiches in right-angled triangular prism containers.

a Calculate the surface area of the container.

b Calculate its volume.

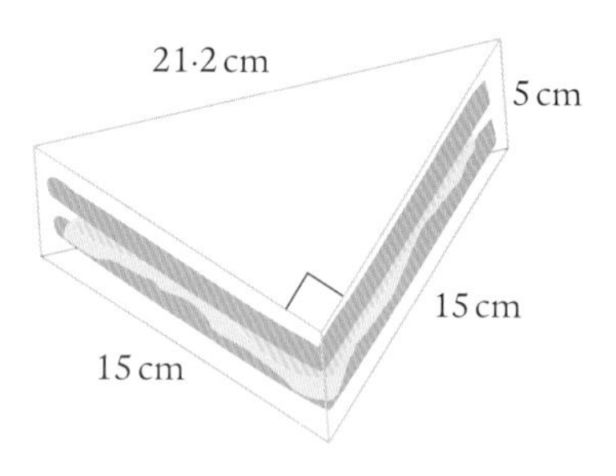

5 Beans are sent to the supermarket in boxes.

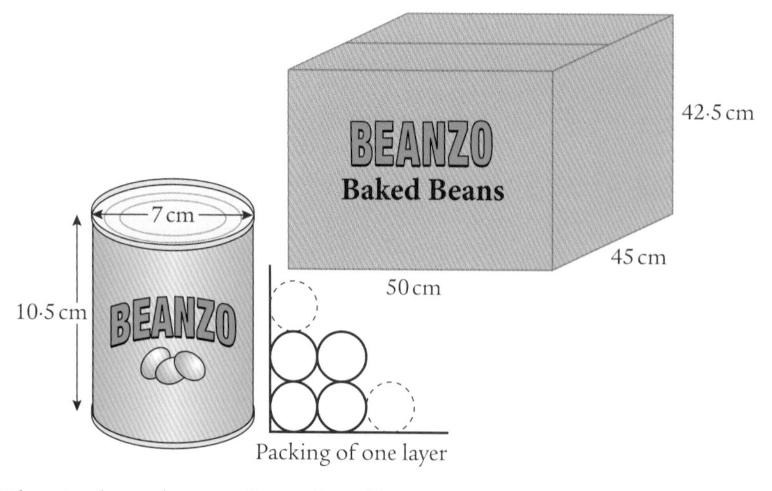

a What is the volume of one tin of beans?

b What is the volume of the box?

c How many tins can be packed in the box?

d What volume of the box is empty space?

6 A candlemaker is costing three models of candle.

Each is a prism and each has a symmetrical cross-section.

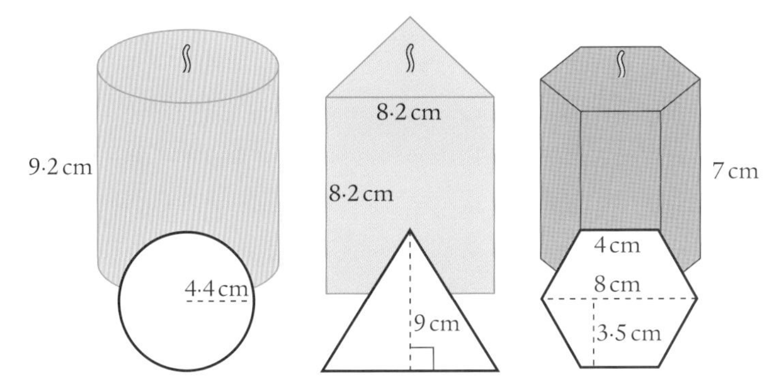

a Calculate the volume of wax needed to make each candle.

b A cubic centimetre of wax weighs 0·9 grams.

What weight of wax is needed to make each candle?

c **i** What weight of wax is needed to make ten cylindrical candles?

ii Wax costs £4·40 per kilogram. What is the cost of the wax for these candles?

iii What price should the candlemaker set for each type of candle to break even?

7 The diameter of a £2 coin is 28·4 mm.

The inner ('silver') circle has a diameter of 20 mm and is made from a mixture of copper and nickel.

The outer ('gold') ring is made from a mixture of nickel and brass.

The coin is 2·5 mm thick.

a Calculate the volume of metal in the inner 'silver' section.

b Calculate the volume of metal needed for the outer 'gold' ring.

Exercise 2.5B

1 Calculate the volume of this box of paper tissues.

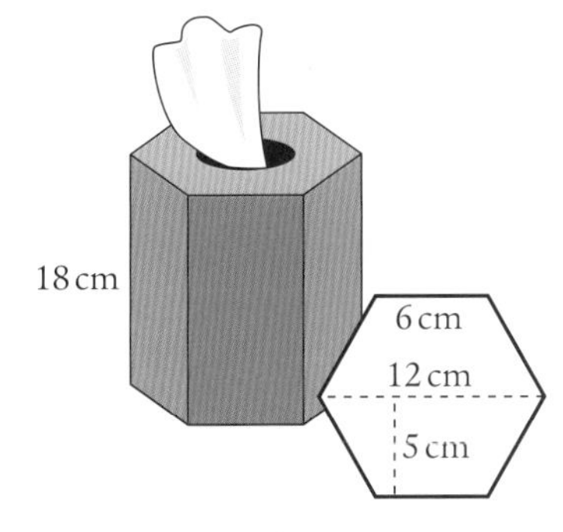

The dimensions of its symmetrical top are shown.

Its height is 18 cm.

2 **a** Which salt container has the greater volume, and by how much?

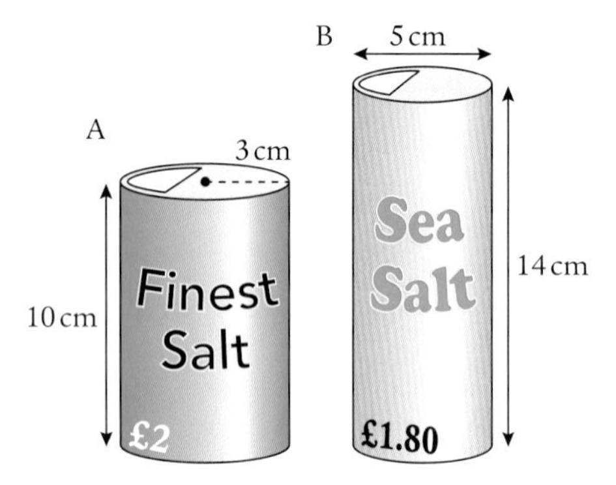

b Which is the better buy?

3 A child's paddling pool is cylindrical with a diameter of 2·4 m.

It can be filled to a depth of 0·5 m.

What is the weight of the water in the paddling pool when it is full?

(1 m^3 of water weighs 1 tonne.)

4 The Aroma Soap Company's research department found that if two soaps have the same volume, the one with the smaller surface area lasts longer.

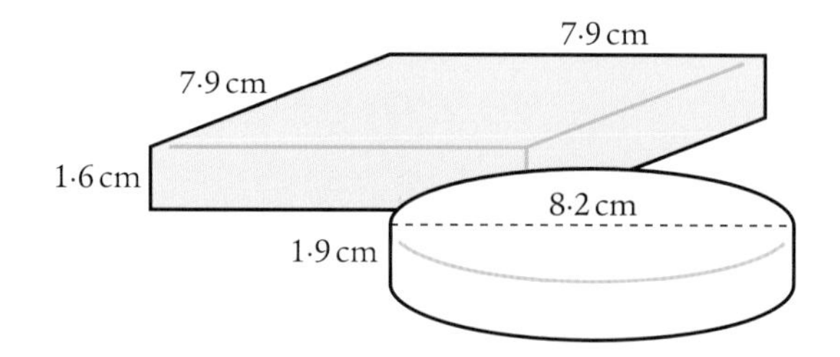

a Check that the volumes of these two soaps are the same when calculated to the nearest cubic centimetre.

b According to the research, which soap should last longer?

Explain your answer.

5 The confectioner who designs a box of chocolates, the Choccy Mixtures, aims to make each chocolate the same volume.

Three of the chocolates below have the same volume, when calculated to the nearest cubic centimetre.

Calculate the volumes of the four chocolates and find the one whose volume is different. (All measurements are in cm. The tops are in the shape of a parallelogram, a trapezium, a circle, and a kite.)

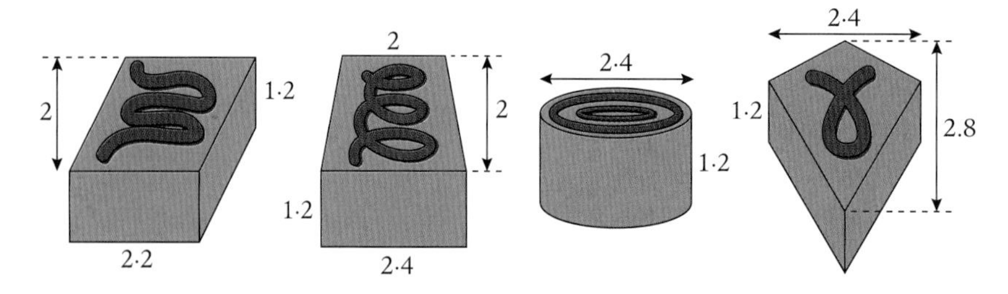

6 The Roads Department provides plastic bins filled with grit for treating ice on the roads and pavements.

The sides of a bin are congruent trapeziums and the bin is a prism.

Calculate the volume of grit the bin will hold.

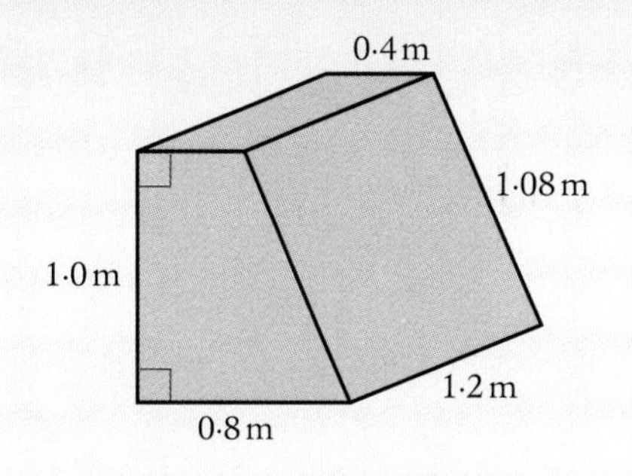

7 An oil tank is cylindrical.

It is 150 cm long and has a radius of 45 cm.

It leaks oil at a rate of 80 ml per minute.

a How long would it take a full tank to empty?

(1 ml = 1 cm^3.)

b The leak was spotted immediately and the contents put into a cuboid tank whose base was 0·9 m^2.

What is the depth of oil in the cuboid tank?

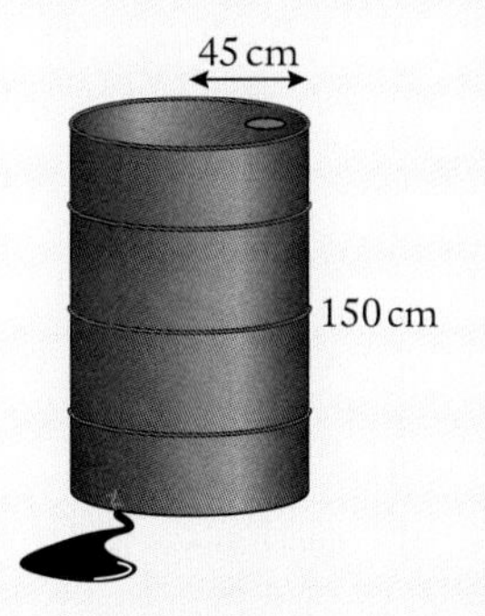

8 Guttering, 4 metres long, has semicircular ends of diameter 20 cm.

a Calculate its capacity in litres. (1000 cm^3 = 1 litre.)

b Rainwater is running off the roof and into the guttering at a rate of 8 litres per minute.

It is being taken down the drainpipe at a rate of 6 litres per minute.

How long will it be before the guttering overflows?

Preparation for assessment

1 Calculate the area of the parallelogram, the kite and the trapezium.

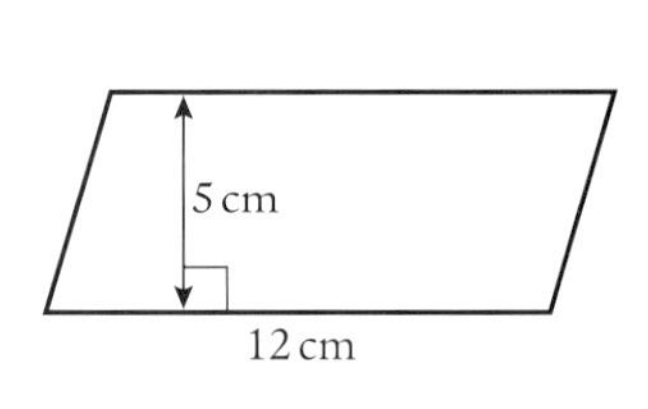

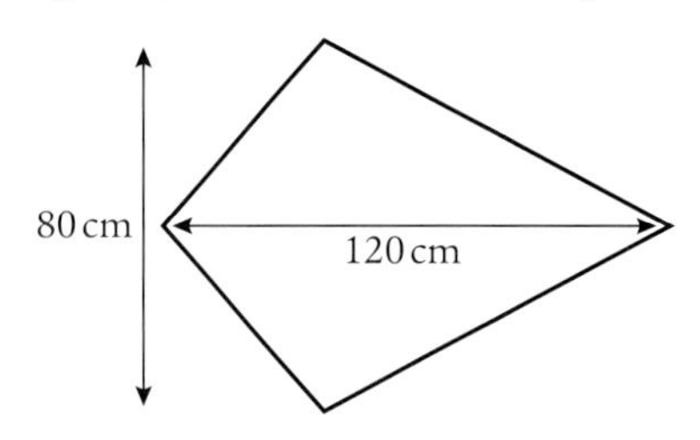

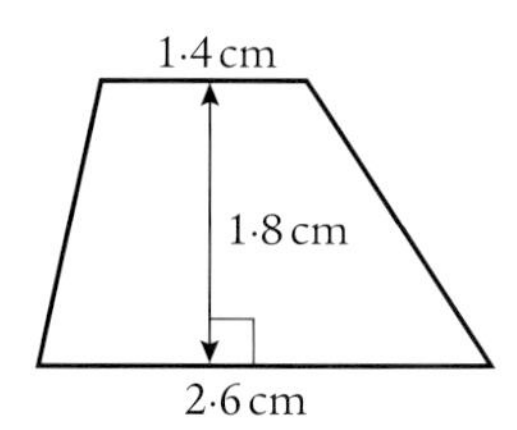

2 a Calculate the circumference of a circle whose radius is 25 cm.

b Calculate its area.

3 Calculate the area of a kite whose diagonals are both 1 metre long.

4 a Sketch the net of each of these shapes, marking the radius of any circle, the length and breadth of any rectangle and the base and altitude of any triangle.

All measurements are in centimetres.

b Name each prism.

c Calculate the surface area of each prism.

d Calculate the volume of each prism.

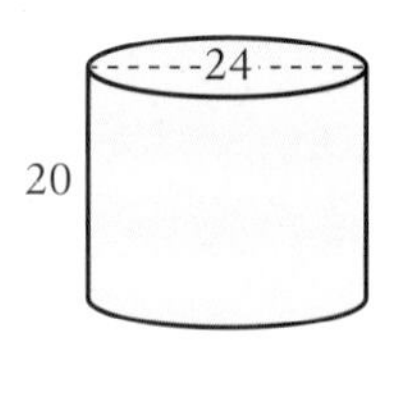

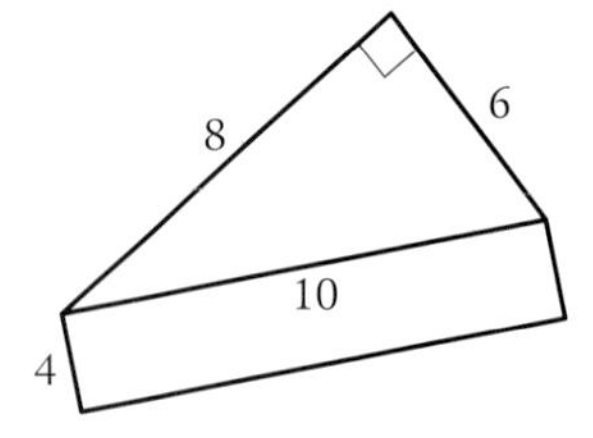

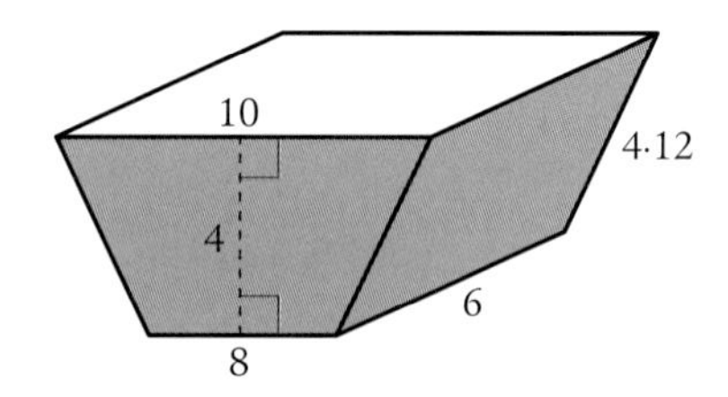

5 a A semicircular table top has a diameter of 110 cm.

Calculate:

i the perimeter of the table top

ii its area.

b If the top is 0·7 cm thick, what is its volume?

6 The inner circumference of a gold ring is 54 mm.

a Calculate its inner diameter, correct to one decimal place.

b The outer diameter is 17·5 mm. The ring is cylindrical and 5 mm thick.

Calculate the volume of gold.

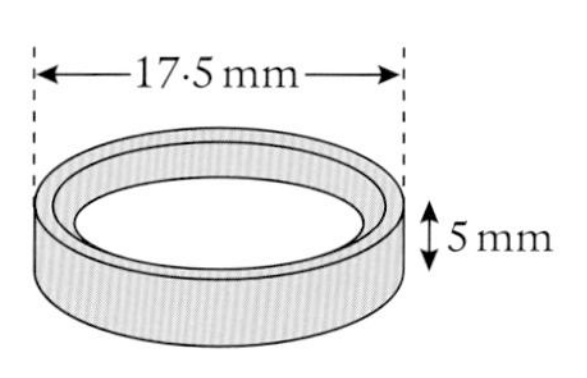

7 A cylindrical tea-urn has a height of 80 cm and a base with diameter 50 cm.

a Calculate its capacity in litres, correct to one decimal place.

b How many cups of 200 ml can it fill? (1 ml = 1 cm^3.)

8 a This is the net of a solid. Which solid?

b Calculate the volume of the solid formed from the net. (The measurements are in metres.)

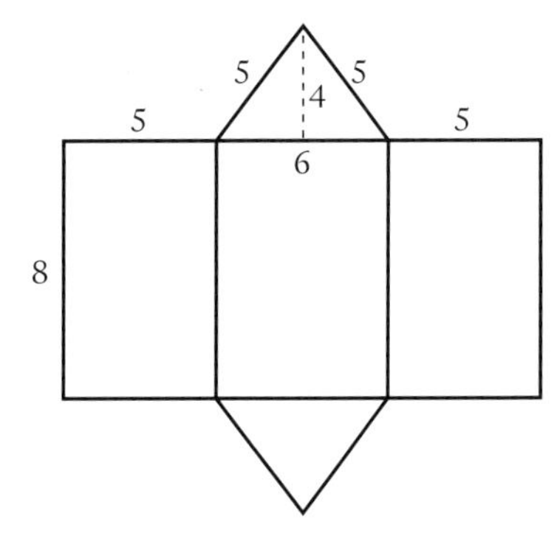

9 Remember the oil storage tanks at Grangemouth?

The diameter of the cylindrical tank is 43·89 m. Its height is 9·91 m.

a How much liquid can the tank hold?

b In reality, no tank is filled to the very top for safety reasons.

Calculate the capacity of the tank if it can only be filled to a height of 9 m.

3 Gradients and straight lines

Before we start...

Below is a picture of the Tradeston Bridge in Glasgow.

When building the bridge, engineers had a number of things to consider.

The bridge had to get high enough to allow river traffic under.

However, it was a pedestrian bridge so had to have a 'walker-friendly' gradient.

Its start and finish points were predetermined by the fact that the city is built up to both sides of the river at that point. The engineers couldn't take a 'run up'.

How did they make sure that the gradient was low enough for pedestrians but the middle was high enough for boats to get under?

What you need to know

1 Express each fraction in its simplest form:

a $\frac{42}{50}$ **b** $\frac{15}{21}$ **c** $\frac{124}{300}$.

2 Express each common fraction as a decimal fraction:

a $\frac{2}{5}$ **b** $\frac{1}{8}$ **c** $\frac{7}{4}$.

3 Write each common fraction **i** in decimal form **ii** as a percentage.

a $\frac{9}{10}$ **b** $\frac{4}{5}$ **c** $\frac{4}{9}$ (to 3 s.f.)

4 In a recent traffic survey, 12 out of 20 cars entering a roundabout turned left.

Write down the ratio of cars entering the roundabout to cars turning left in its simplest form.

5 **a** State the coordinates of A and B.

b How much to the right of A is B?

c How much above A is B?

d What do the arrowheads pointing right and up indicate?

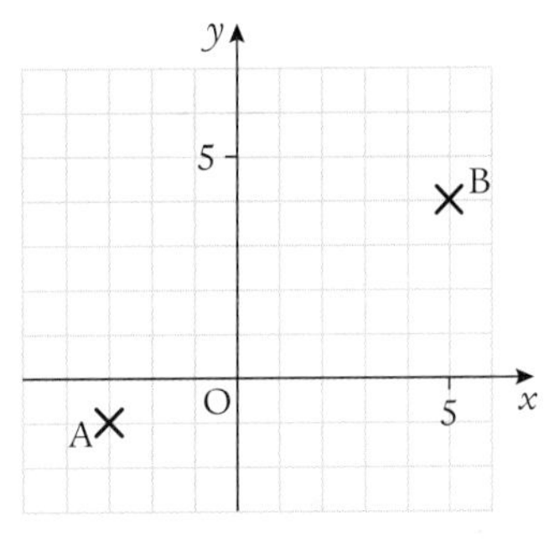

3.1 Gradient

Calculating the gradient

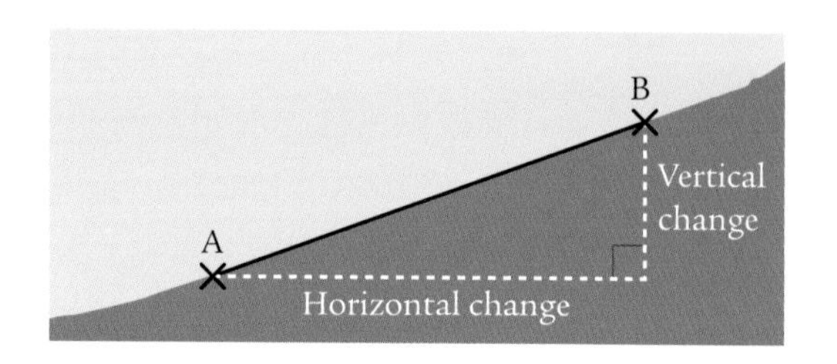

In mathematics the **gradient** is a measure of the 'steepness' of a slope. It is defined as the ratio of the vertical change to the horizontal change as you move between two points on the slope.

The gradient between the points A and B is defined as:

$$\text{gradient}_{AB} = \frac{\text{vertical change}}{\text{horizontal change}}$$

Example 1

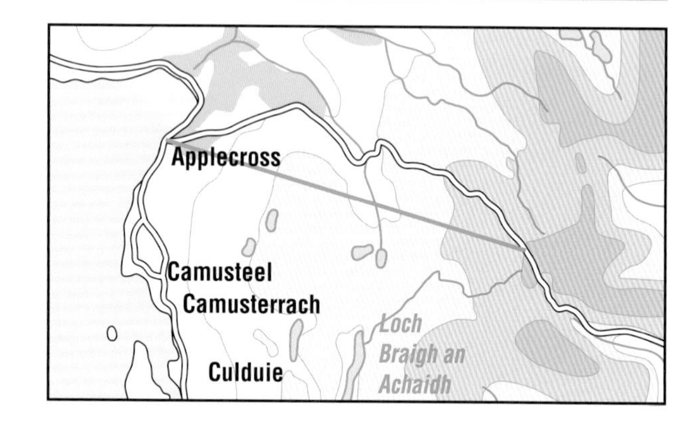

The Bealach na Ba near Applecross in the west of Scotland is one of the steepest roads in the country and is popular with many cyclists.

It rises from sea level to 620 metres over a distance of 6600 metres.

a What is the average gradient for this stretch of road?

b It is claimed that one section has a gradient of 1 in 5. Express this as a percentage.

a From the 'side' we can see that some parts are steeper than others.

But we are just concerned about the changes from start to finish ...

Viewpoint (V)
620 m
Applecross (A)
6600 m

$$\text{gradient}_{AV} = \frac{\text{vertical change}}{\text{horizontal change}}$$

$$= \frac{620}{6600}$$

$$= \frac{31}{330}$$

This is a gradient of $\frac{31}{330}$ or 31 in 330 or 9·4%.

b A gradient of 1 in 5 is $\left(\frac{1}{5} \times 100\right)\% = 20\%$.

Example 2

When fitting solar panels to a roof the fitter needs to be aware of the gradient of the roof.

Calculate the gradient of the roof below.

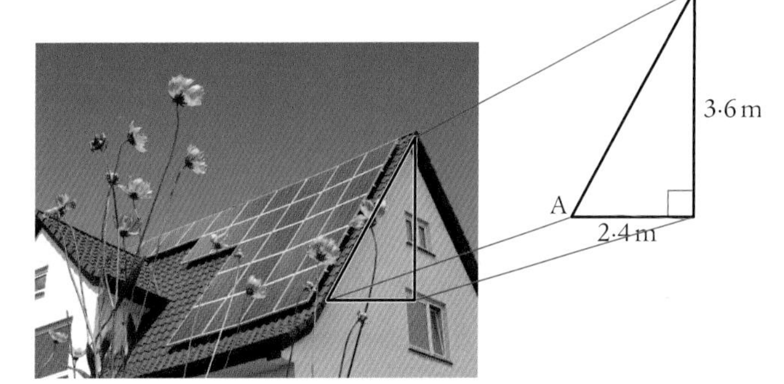

$$\text{gradient}_{AB} = \frac{\text{vertical change}}{\text{horizontal change}}$$

$$= \frac{3{\cdot}6}{2{\cdot}4} = \frac{36}{24} = \frac{3}{2}$$

$$= 1{\cdot}5$$

The roof has a gradient of 1·5 or 3 in 2 or 150%.

Exercise 3.1A

1 Calculate the gradient for each of the following, leaving your answer as a common fraction and simplifying where appropriate:

a

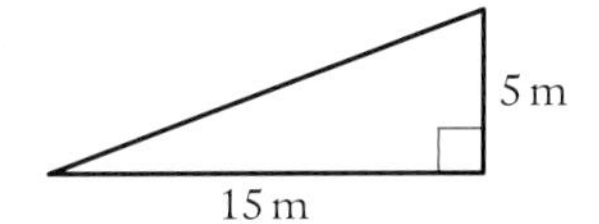

b

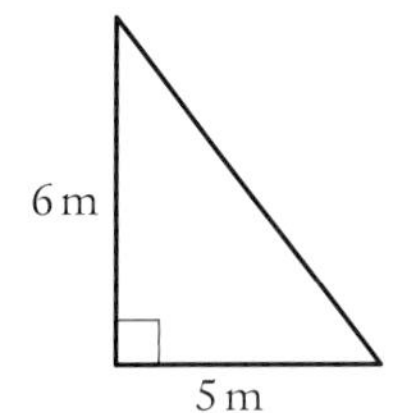

c

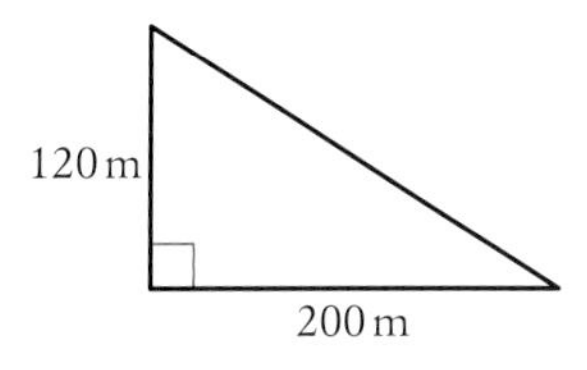

2 Calculate the gradient of the following as a decimal fraction correct to 2 decimal places where appropriate:

a

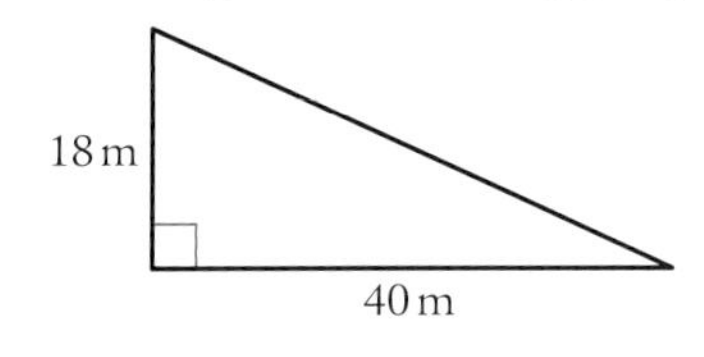

b

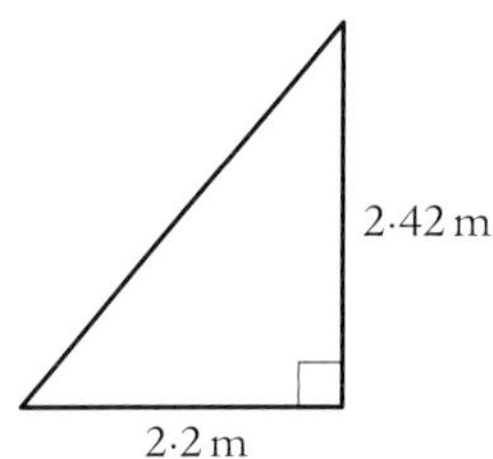

c

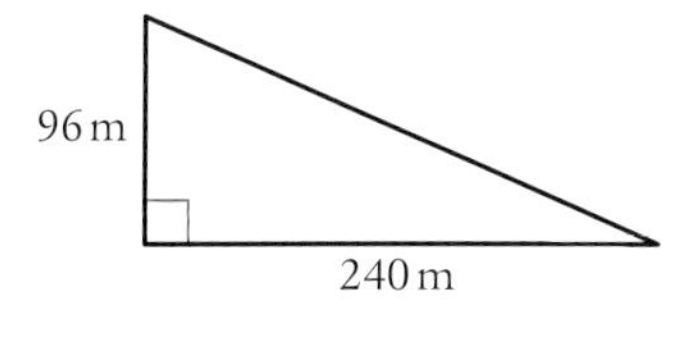

3 **a** An extension ladder leans against a wall.

The diagram shows it is in various positions.

W represents the foot of the wall.

AW = 1 m, BW = 1·5 m,

CW = 2 m, DW = 2·5 m.

Calculate the gradient of the ladder in each of these positions.

b Health and Safety Regulations state that 'for every 4-up, place the ladder base 1-out from the wall'.

Any steeper would be considered dangerous.

Which ladder positions in **a** are not safe?

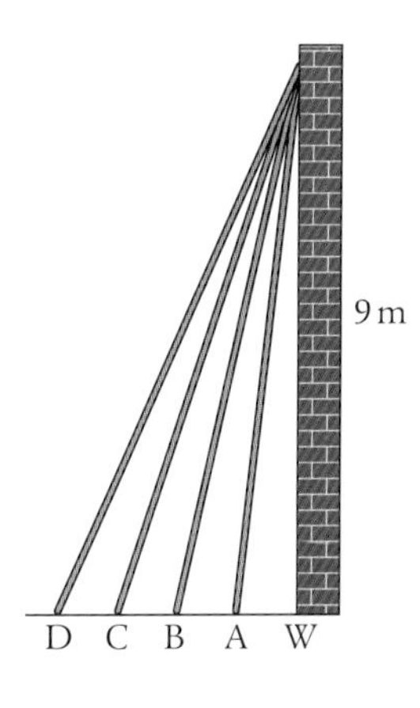

4 Calculate the gradient of the mountains in the mountain range below.

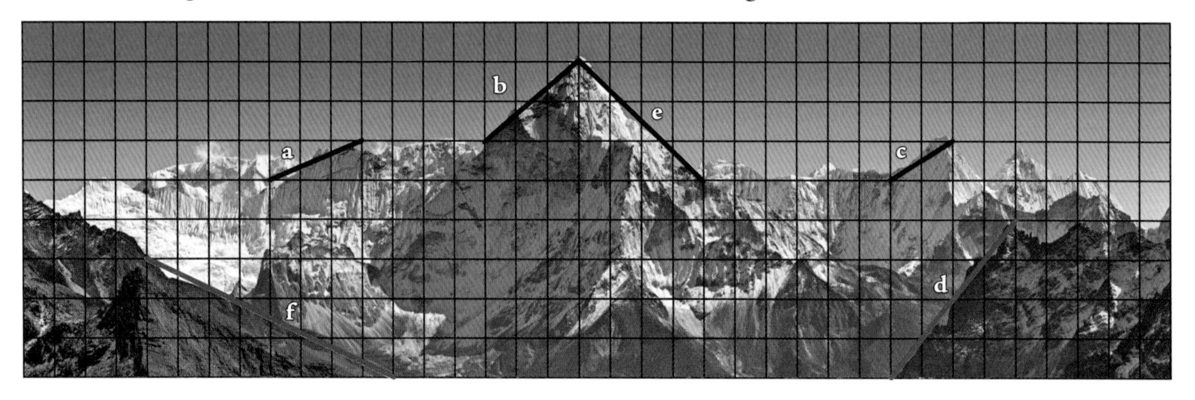

5 Splash Water Park boasts the steepest water slide in Rivertown.
On the other side of Rivertown, Ocean Play Park is building a new water slide.

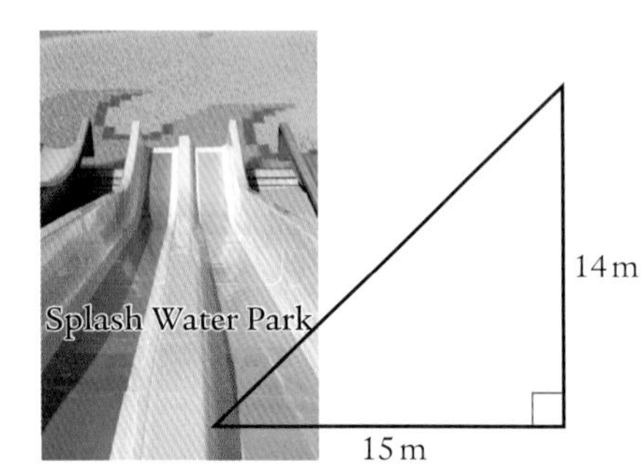

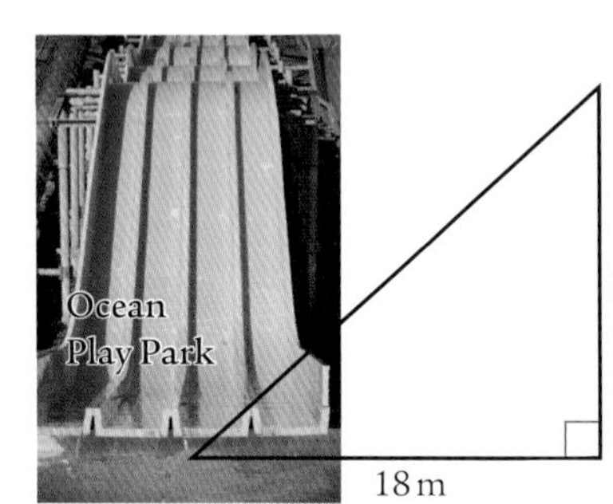

a Calculate the gradient of the slide at Splash Water Park.

b Calculate the gradient of the slide at Ocean Play Park.

c Can Splash keep the title 'Steepest water slide in Rivertown'?

Example 3

Road signs are used to warn motorists of steep hills.
They give the gradient as a percentage.
This sign warns of a 17% slope.

a How far has a car climbed on this hill when it has travelled 15 m horizontally?

b How far horizontally has it gone for a rise of 1 metre?

a Vertical change is 17% of horizontal change.
Vertical change $= 0{\cdot}17 \times 15 = 2{\cdot}55$ metres.

b Since we multiply by 0·17 to get the vertical change, we will divide by 0·17 to get the horizontal change:
horizontal change $= 1 \div 0{\cdot}17 = 5{\cdot}88$ metres (to 2 d.p.).

Example 4

A road sign at the start of a hill indicates its steepness using the ratio 1 : 5.

a Express this as: **i** a common fraction **ii** a percentage.

b Is a 1 : 4 hill steeper or shallower than a 1 : 5 hill?

a **i** $1:5 = \frac{1}{5}$ **ii** $\frac{1}{5} = \left(\frac{1}{5} \times 100\right)\% = 20\%$

b Express 1 : 4 as a percentage to make comparison easier:

$\frac{1}{4} = \left(\frac{1}{4} \times 100\right)\% = 25\%$.

Comparing the percentages: 25% > 20% ... so, yes, a 1 : 4 hill is steeper than a 1 : 5 hill.

Exercise 3.1B

1 **a** Which of the following roads is the steepest?

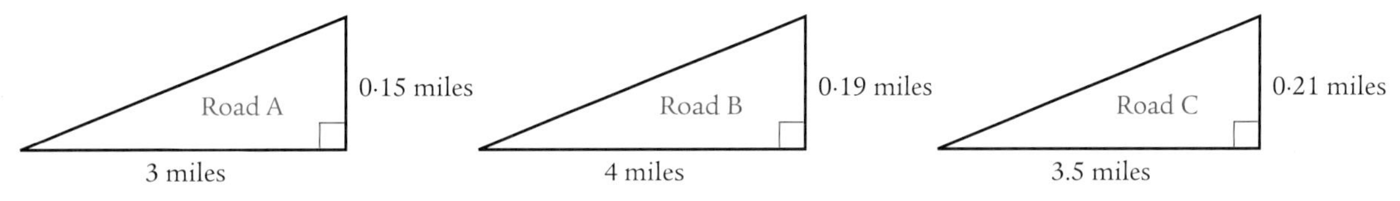

b Express these gradients as percentages.

c A hill will only merit a road sign if it has a slope of 5% or more. Which road will not have a road sign?

2 The Bougilie Hill has an incline of 1 : 7.

a Rewrite this gradient as: **i** a common fraction **ii** a percentage.

b A hill on the Arran coast road has a gradient of 17%. How does this hill compare with the one in the question?

c How much height would a car gain if its horizontal change was 200 m as it drove up the Bougilie?

3 The Health and Safety Guidelines for wheelchair ramps state that 'the maximum permissible gradient is 1 : 12'.

a From the measurements given, does this ramp fit with the guidelines?

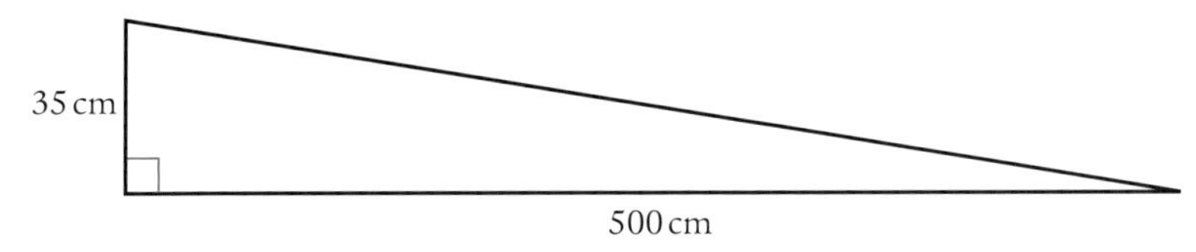

b For practical reasons the height, 35 cm, is fixed. However, because the ramp is 500 cm long, it interferes with the pavement. How short can you make this distance and still meet the regulations?

4 Ski slopes have different grades to identify the level of difficulty.

One of the measures for deciding the grade is the gradient of the slope.

In Europe the slopes are graded roughly as follows:

Green: Gentle slope. *Blue*: No steeper than 25%.

Red: No steeper than 40%. *Black*: Steeper than 40%.

a What is the gradient of the slope in the picture?

b What grade would this slope be rated at?

c When designing an indoor ski slope the architect has a horizontal distance of 200 m to use for one of the runs.

She wants to make it a Red run.

How high above the end of the 200 m run can she build the start?

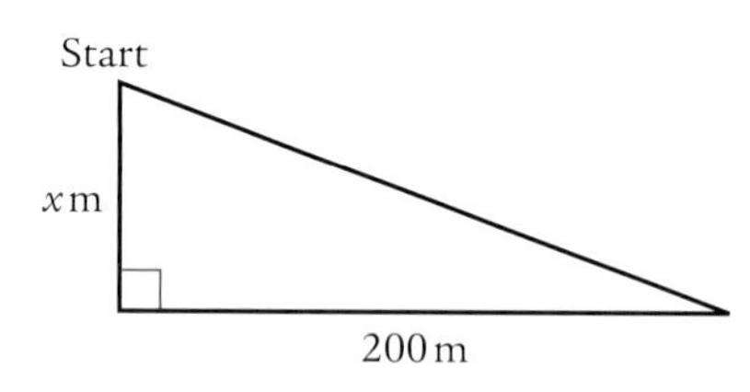

5 For an exciting 'Death Slide', the gradient from start to finish needs to be between 10% and 15%.

A new 'Death Slide' has been proposed for Ambles Adventures.

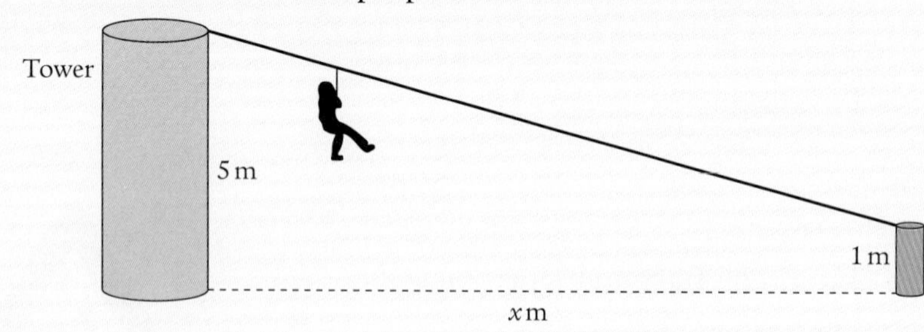

To make it an exciting 'Death Slide', what would be:

a the closest horizontal distance from the tower that you would need to put the end support

b the furthest horizontal distance from the tower that you would need to put the end support?

6 In science they can measure how slippy one surface is against another using gradient.

They sit a block of the material they wish to test on a plane. They slowly tilt the plane until the block just starts to slip. They measure the gradient of the plane.

This number is used as a measure of the slippiness of the material. It is called the **coefficient of static friction**.

a What do you think will be more slippy, a material with a high coefficient of static friction or one with a low coefficient?

b For which of the following will the gradient be at its highest before it starts to slip: ice, steel or wood?

7 Find out about local building plans in your area and investigate the rules about gradients of different things about the house, e.g. stairs, roofs, ramps, paths, drainpipes.

Do the same rules apply for public buildings?

Go on the internet and research the safe use of ladders.

Local planners have decision rules in place as to when a hill merits a road sign. It's not just as simple as 'over 5%'. Find out more.

3.2 Calculating gradient from a coordinate diagram

Positive and negative gradients

The coordinate system was invented by **René Descartes**.

The arrow on the end of the x-axis indicates the direction of increasing x.

Similarly, increasing y-values are shown by an arrow on the y-axis.

We no longer have a vertical and horizontal. Instead we talk of the y-direction and the x-direction.

The lines in this diagram do each have a 'slope' and we would like to talk about it.

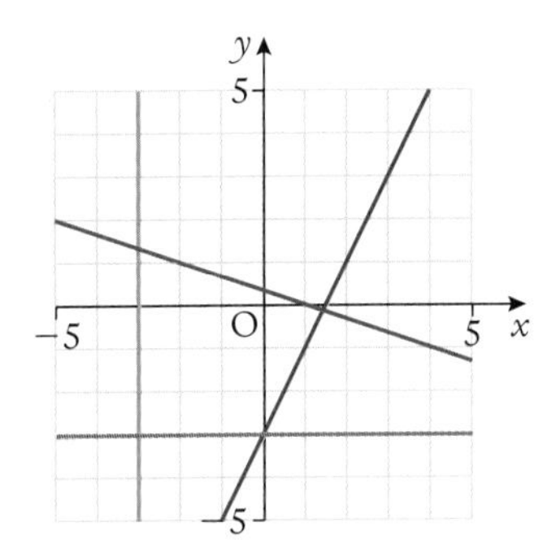

We continue the idea of 'gradient' ... but have a new definition that suits coordinates.

By tradition the letter m has been used to represent the gradient when we work with coordinates.

If we wish to measure the gradient between two points A and B,

$$m_{AB} = \frac{\text{change in the } y\text{-direction}}{\text{change in the } x\text{-direction}} \text{ as you go from A to B.}$$

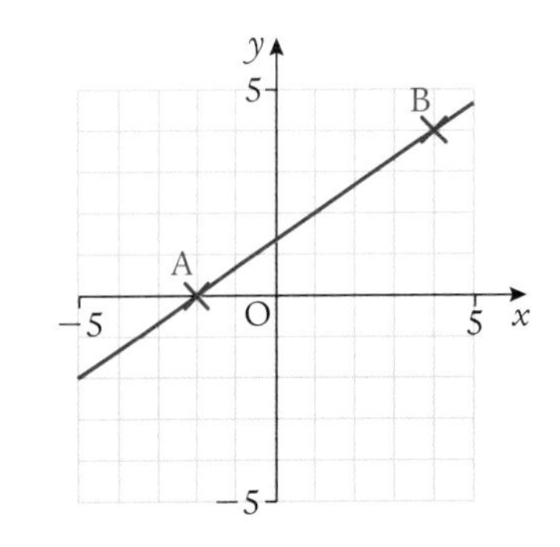

René Descartes: a Frenchman, who first came up with the coordinate system and it is believed that he used the letter *m* for the gradient, taking it from the French word *monter*, which means 'go up' or 'climb'.

Example 1

Calculate the gradient of the four coloured lines in the diagram.

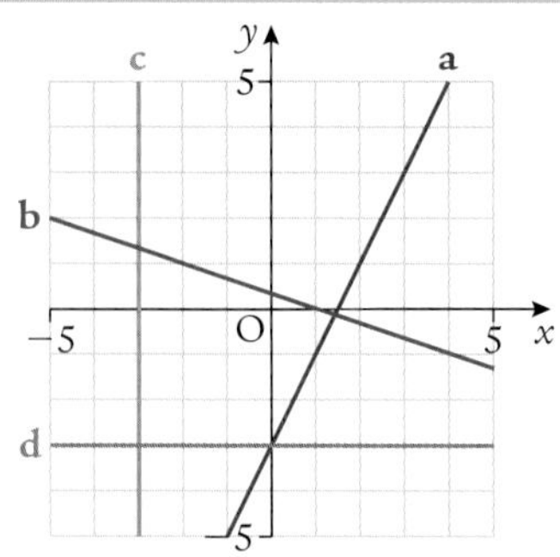

a The red line:

Pick any two convenient points on the line.

Note the shift in the x-direction and the y-direction.

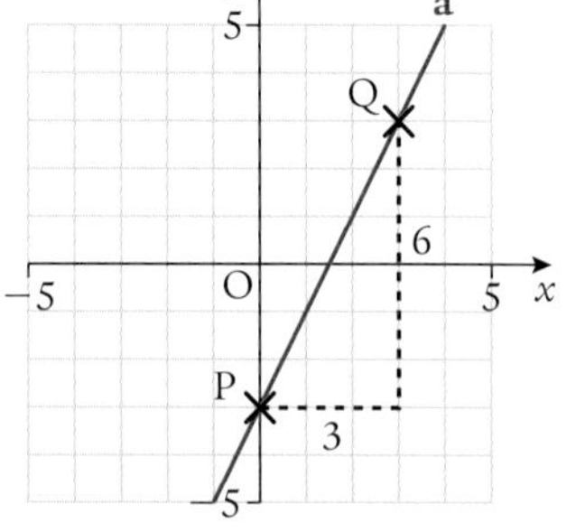

$$m_{PQ} = \frac{\text{change in the } y\text{-direction}}{\text{change in the } x\text{-direction}} = \frac{6}{3} = 2$$

The gradient of the red line is 2.

b The blue line:

Pick any two convenient points on the line.

To go from T to V we have to go 2 down and 6 to the right ... the change in the y-direction is negative.

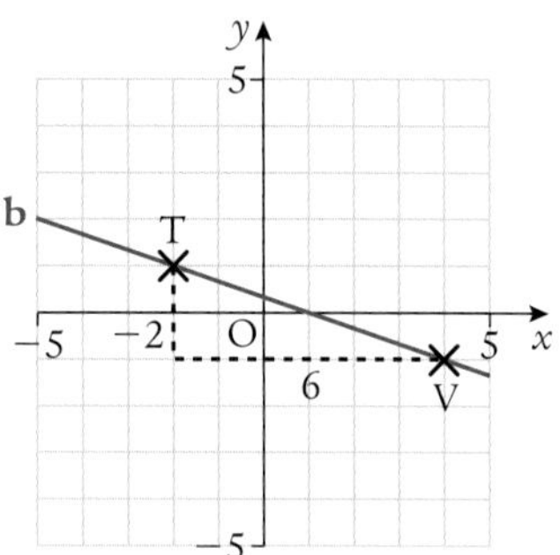

$$m_{TV} = \frac{\text{change in the } y\text{-direction}}{\text{change in the } x\text{-direction}} = \frac{-2}{6} = -\frac{1}{3}$$

The gradient of the blue line is $-\frac{1}{3}$.

c The brown line:

This line runs in the y-direction. There is no change in the x-direction (change = 0), and, since we can't divide by zero, we can't calculate a gradient.

We say the gradient of the brown line is *undefined*.

d The grey line:

This line runs in the x-direction.
The change in the y-direction = 0.

So the gradient of the grey line = 0.

Exercise 3.2A

1 Calculate the gradient of the lines in the following coordinate diagrams.

Leave your answers as common fractions in their simplest form where necessary.

a

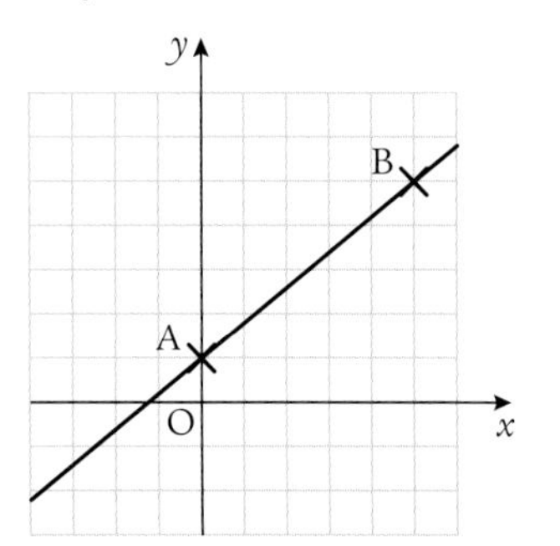

b

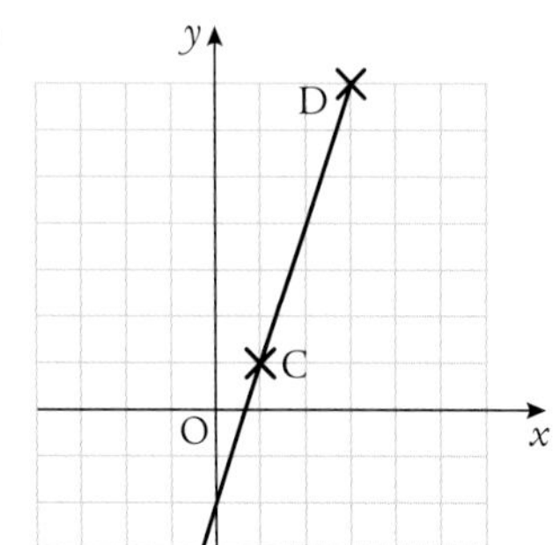

c

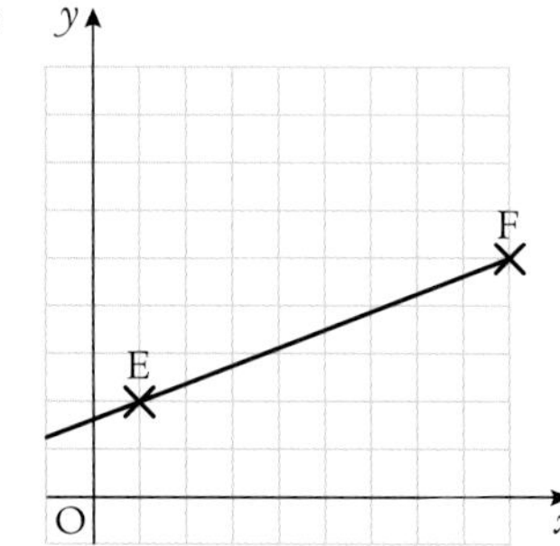

2 **i** Plot the set of points and draw the line that passes through each.

ii Calculate the gradient of each line.

a (0, 0), (3, 1), (6, 2) **b** (0, 1), (3, 3), (6, 5) **c** (0, 2), (5, 5)

d (5, 1), (5, 4) **e** (1, 3), (5, 3)

3 Calculate the gradient of each of the red lines.

a

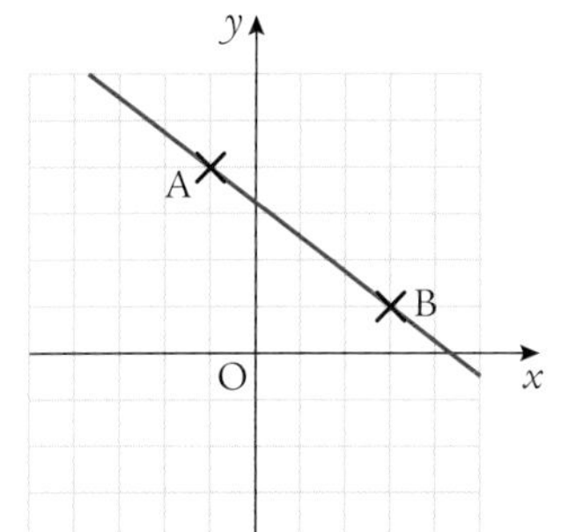

b

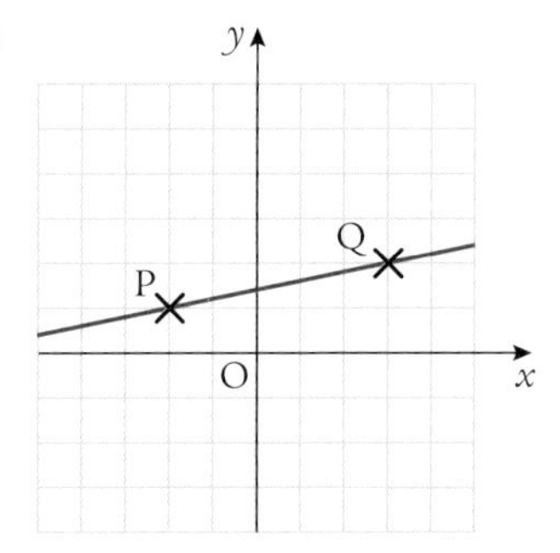

c

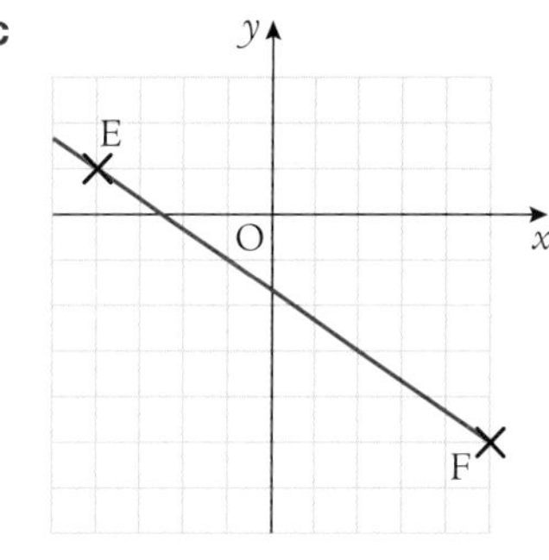

d

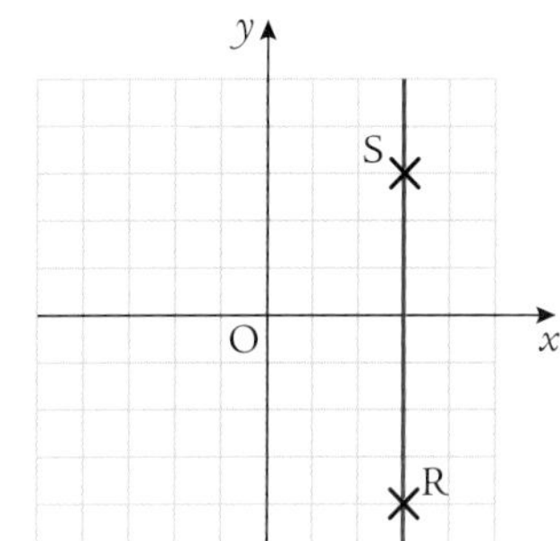

e

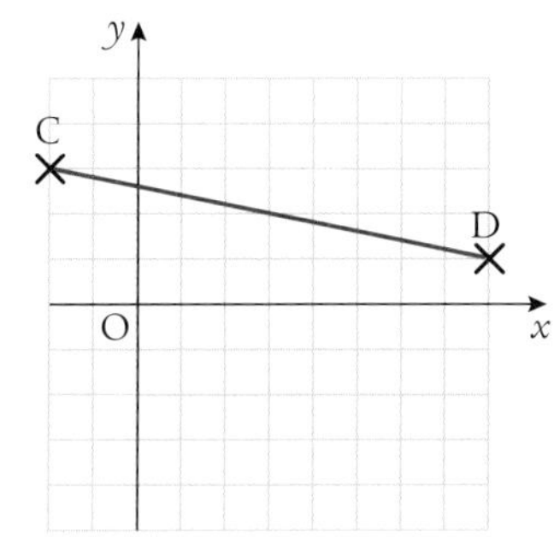

f

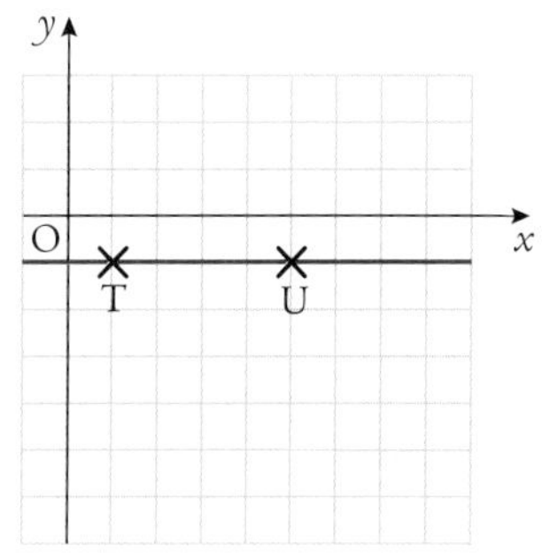

4 Four lines intersect, trapping a square as shown.

a State the coordinates of the four vertices.

b Calculate the gradient of each side of the square.

c Comment on the gradients of the opposite sides of the square.

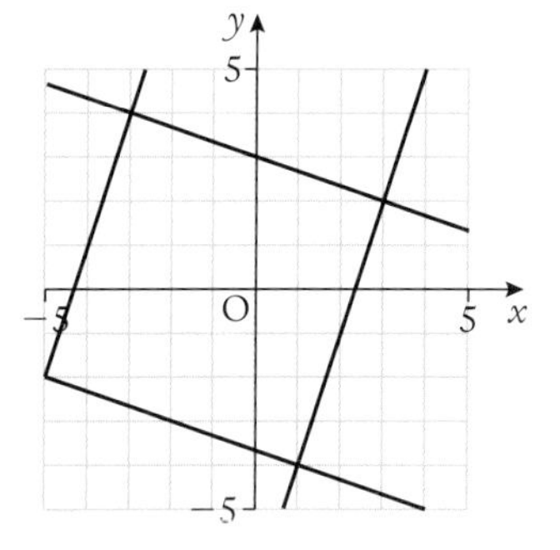

5 Computer graphics are used to generate images in a driving simulator.

It takes advantage of the fact that all parallel lines seem to vanish into one point on the horizon.

The x-axis represents the horizon. The origin represents the vanishing point.

Some important points are plotted. Use these to calculate the gradients of:

a The top of the embankment, two lines passing through: **i** (−6, 1) **ii** (6, 1)

b The grass verge passing through (−4, −1)

c The white line passing through: **i** $(-1, -1)$ **ii** $(2, -1)$

d The central reserve passing through $(8, -0{\cdot}8)$

e The horizon (represented by the x-axis)

f The y-axis.

6 A Duke of Edinburgh's Award group is calculating the time it will take them to complete a walk. Their walking speed depends on the gradient of the slope. The steeper the slope the longer it takes them.

Calculate the gradient of each part of the hill below.

They are walking from left to right in the picture ... it's important to know whether they are going uphill or downhill.

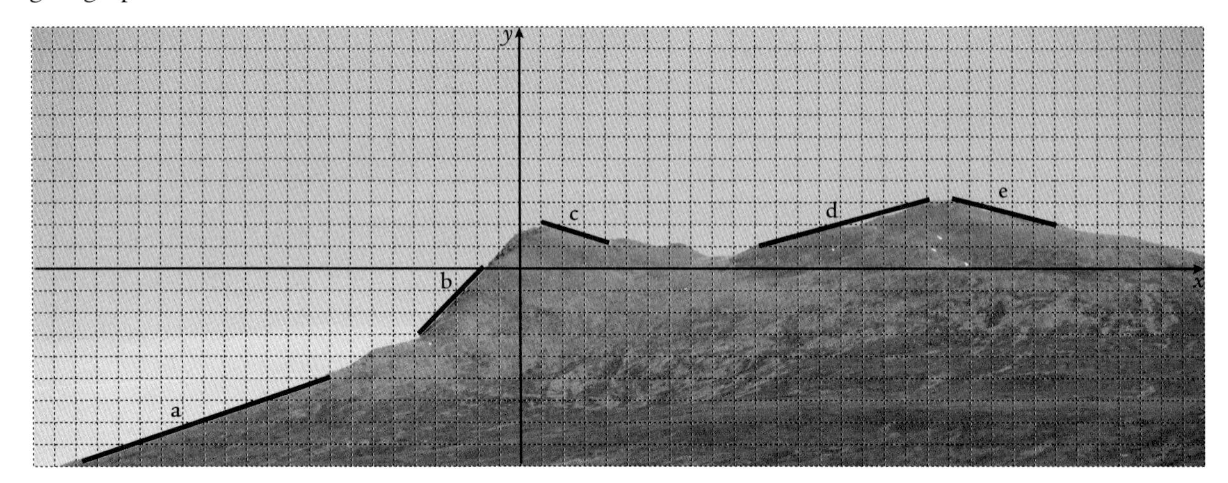

7 An architect photographs the profile of a house to consider the structure of the roof. He marks three sections, A, B and C.

a In each section two convenient points have been identified.

i State their coordinates. **ii** Calculate the gradient of each section.

b **i** Comment on your answers to sections B and C.

ii What do you think this means about these two sections?

Example 2

Without drawing a coordinate grid, find the gradient of the line joining:

a T(2, 5) and V(6, 13)

b A(−4, 6) and B(6, −9).

a Imagine you're moving from V to T.

$$m_{VT} = \frac{\text{change in the } y\text{-direction}}{\text{change in the } x\text{-direction}} = \frac{13 - 5}{6 - 2} = \frac{8}{4} = 2$$

b Imagine you're moving from A to B.

$$m_{AB} = \frac{\text{change in the } y\text{-direction}}{\text{change in the } x\text{-direction}} = \frac{6 - (-9)}{-4 - 6} = \frac{15}{-10} = -\frac{3}{2}$$

Example 3

Luxury Cabs use a graph as a ready-reckoner of their prices.

On such a graph, the rate they charge in £ per mile can be found by calculating the gradient.

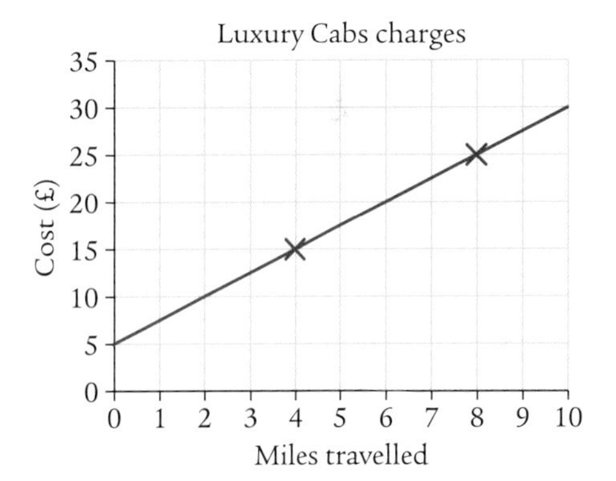

a Find the gradient of the line (be careful with the scales).

b How much does it cost to hire the cab initially (before you clock up any miles)?

a Pick two convenient points on graph, say (4, 15) and (8, 25).

$$m = \frac{\text{change in the } y\text{-direction(£)}}{\text{change in the } x\text{-direction (miles)}}$$

$$= \frac{25 - 15}{8 - 4} = \frac{10}{4} = 2{\cdot}5$$

They are charging £2·50 per mile.

b When the miles travelled = 0, the cost reads £5. The cost to hire the cab initially is £5.

Exercise 3.2B

1 **a** Calculate the gradient of each of the straight lines on the graph:

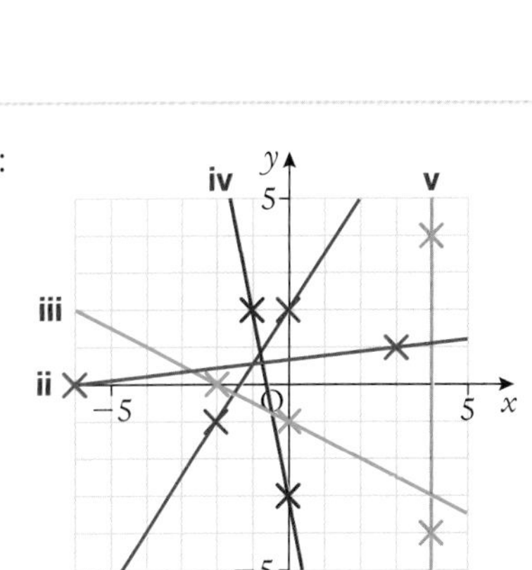

b **i** A point lies on the red line with an x-coordinate of 4. What is its y-coordinate?

ii A point lies on the green line with an x-coordinate of 6. State its y-coordinate.

iii A point lies on the violet line with a y-coordinate of −8. What is its x-coordinate?

2 a Without plotting the points, calculate the gradient of the line that passes through:

i A(2, 3), B(7, 9) ii C(−7, −3), D(1, −9) iii I(−10, 4), J(−3, 10)

iv E(0, 0), F(10, −2) v G (−10, −1), H(−3, 5).

b Two of the lines in part **a** are parallel. Identify them. Explain your answer.

c What is always true about parallel lines?

3 PQRS is a quadrilateral with vertices P(−2, 1), Q(2, 2), R(3, 0), S(−1, −1).

a Calculate the gradients of PQ, QR, RS and PS.

b What kind of quadrilateral is PQRS?

4 Calculate the gradient of each line. (Be careful: the scales on the axes are not the same!)

a 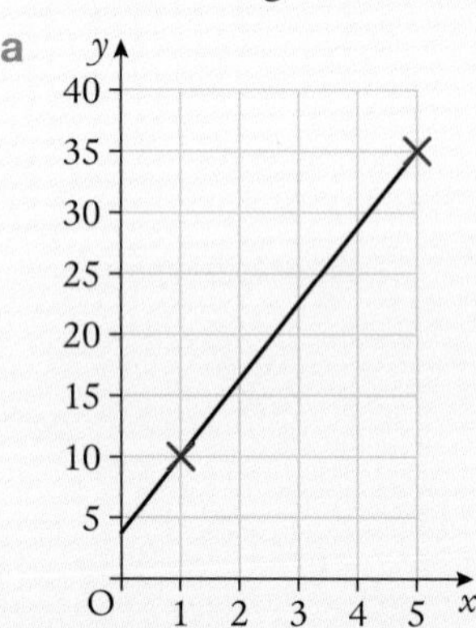

b 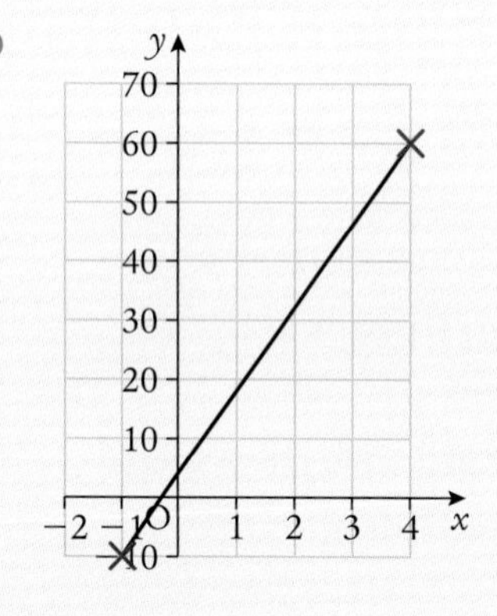

c 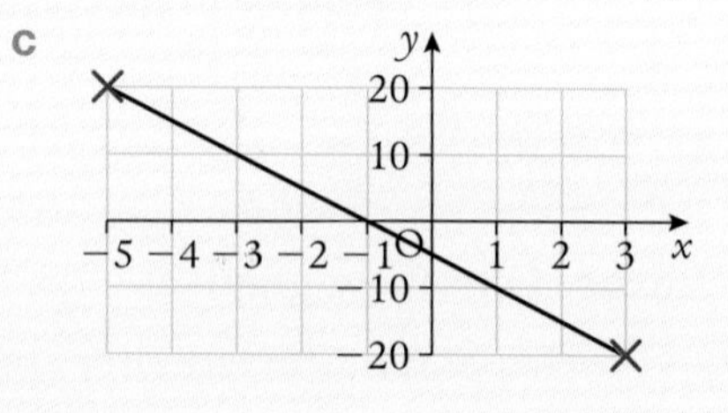

5 Isaac Newton studied motion. He knew that when he drew a graph of distance against time he got the speed by finding the gradient.

Robert took a video of the *Waverley* as it left harbour. From it he managed to get data to draw a graph.

a Calculate the speed of the *Waverley* by finding the gradient of the line.

b How far from the pier was the *Waverley* when the video started?

Waverley leaving

Distance (metres)

Time (seconds)

6 Helen bought a £12 000 dinghy. She paid a deposit, then regular monthly instalments for two years.

The graph shows how much she still owed at the end of each month.

a What was the original deposit?

b i Calculate the gradient of the line.

ii What is this measurement telling you?

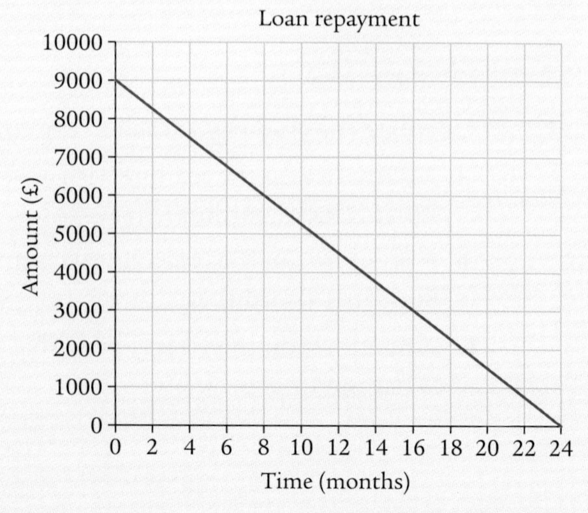

3.3 Drawing straight lines

x	$y = 3x + 2$	(x, y)
−1	−1	(−1, −1)
0	2	(0, 2)
1	5	(1, 5)
2	8	(2, 8)

If, when plotting points, we use a rule that lets us work out the y-coordinate from the x-coordinate, we find that a pattern shows up in the graph.

We might use the rule, say, $y = 3x + 2$ (the y-coordinate is three times the x-coordinate plus two).

To draw the graph we first pick convenient values for x (our choice) and use the rule to work out the corresponding values for y.

Now we plot the points and join them in order of increasing x.

In this case we get a straight line, which we call 'the line $y = 3x + 2$'.

In fact, any time we draw a graph where the rule is of the form, $y = mx + c$, where m and c are constants, we get a straight line.

We want to explore such rules ... and their graphs.

How does changing the value of m or c affect where the line goes?

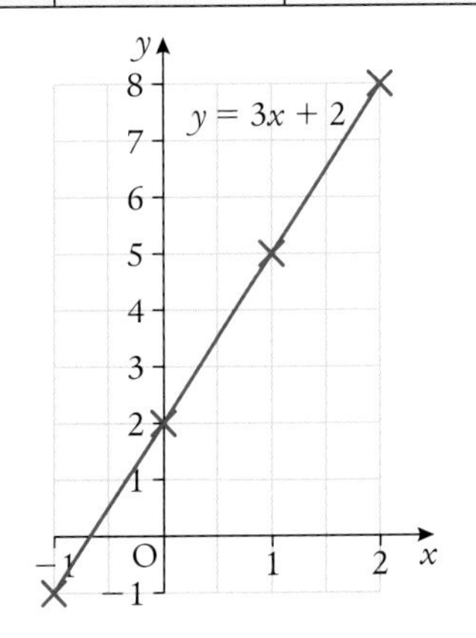

Example 1

Draw the line $y = 2x + 1$.

a Complete a table of values.

b Plot the points on a coordinate diagram and draw a straight line through them.

c **i** Calculate the gradient of the line. **ii** Read the **y-intercept** from the graph.

a

x	$y = 2x + 1$	(x, y)
0	1	(0, 1)
1	3	(1, 3)
2	5	(2, 5)
3	7	(3, 7)
4	9	(4, 9)

b

c **i** Consider any two points on the line, say A(1, 3) and B(3, 7).

$$m_{AB} = \frac{\text{change in the } y\text{-direction}}{\text{change in the } x\text{-direction}} = \frac{7-3}{3-1} = \frac{4}{2} = 2$$

ii Notice that the line cuts the y-axis at (0, 1), so 1 is the y-intercept.

(We could also have found this by setting x to 0 in the equation: $y = 2 . 0 + 1 = 1$.)

The **y-intercept** is the name given to the point where the line crosses the y-axis. We find it by setting $x = 0$ in the 'rule'.

Example 2

Scuba divers need to know the **ambient pressure** as they go deeper ... the amount of breathing gas they need is related to this pressure. Pressure is measured in 'bars'.

The relationship between ambient pressure (P bar) and the depth below sea level (D metres) is given by the formula:

$$P = 0{\cdot}1D + 1$$

a Create a table of values for this equation.

b Plot the points and draw a straight line to represent this relationship.

c Give **i** the gradient and **ii** the y-intercept of the line.

d From the graph what would be the ambient pressure at 25 metres below sea level?

a $P = 0{\cdot}1D + 1$

D	$P = 0{\cdot}1D + 1$	(D, P)
0	1	(0, 1)
5	1·5	(5, 1·5)
10	2	(10, 2)
15	2·5	(15, 2·5)
20	3	(20, 3)
25	3·5	(25, 3·5)

b

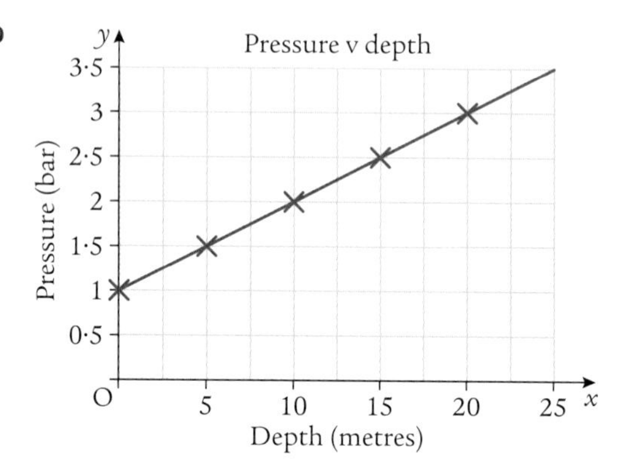

c **i** Consider any two points on the line, say A(0, 1) and B(15, 2·5).

$$m_{AB} = \frac{\text{change in the } y\text{-direction}}{\text{change in the } x\text{-direction}} = \frac{2{\cdot}5 - 1}{15 - 0} = \frac{1{\cdot}5}{15} = 0{\cdot}1$$

ii Notice that the line cuts the y-axis at (0, 1), so 1 is the y-intercept.

d From the graph you can see that at 25 m the ambient pressure would be 3·5 bar.

Exercise 3.3A

1 **a** Copy and complete the following table of values for the equation $y = 3x + 1$.

x	0	1	2	3	4
$y = 3x + 1$					
(x, y)					

b Plot the points on a coordinate diagram and draw a straight line through them.

c **i** Calculate the gradient of the line.

ii Identify the y-intercept.

2 **a** Copy and complete the following table of values for the equation $y = 2x$.

x	-2	-1	0	1	2
$y = 2x$					
(x, y)					

b Draw the graph of $y = 2x$ (it should be a straight line).

c **i** Calculate the gradient of the line. **ii** Identify the y-intercept.

3 For each of the following straight lines:

i $y = 2x + 5$ **ii** $y = x + 3$ **iii** $y = 4x + 2$ **iv** $y = -2x - 4$

a Complete a table of values, using convenient values of x between 2 and -2.

b Hence draw the line representing the equation.

c Calculate the gradient of the line and identify the y-intercept.

4 The equation of a straight line can be written as $y = mx + c$.

a By comparing the values of m and c with the gradients and y-intercepts of the lines already drawn:

i suggest the 'role' played by m in the equation of a line

ii suggest the 'role' played by c in the equation of a line.

b Hence, for the line $y = 6x - 5$, state the gradient and the y-intercept.

5 Write down **i** the gradient and **ii** the y-intercept of the straight lines with equations:

a $y = x$ **b** $y = 2x + 6$ **c** $y = -3x - 5$

d $y = 5x$ **e** $y = -0{\cdot}5x + 2{\cdot}4$ **f** $y = \frac{3}{4}x - 1$

6 A car hire company charges a basic charge of £20 and then a further £15 a day.

Let d days represent the number of days for which the car has been rented.

The cost, £C, can be expressed in terms of d days by the formula $C = 15d + 20$.

a Copy and complete this table of values:

d (days)	1	2	3	4	5
C (pounds)	35	50			

b Plot the points and hence draw a line to represent the relationship between C and d.

c Write down the gradient and y-intercept of the straight line.

d **i** What does the gradient represent in the context?

ii What does the intercept represent?

e Use the graph to calculate how much it would cost to hire the car for a week.

7 **a** In a time trial, Mark is cycling at a constant speed. Copy and complete the table:

Time, T (hours)	0	1	2	3	4	5
Distance, D (km)	0	35	70			

b Draw a graph of the table.

c **i** Calculate the gradient of the line (be careful of the scale).

ii Write down the value of the y-intercept of the line.

d Hence write down the equation of the line.

e Use the formula to calculate how far Mark would cycle, if he could maintain his speed, in:

i $2\frac{1}{2}$ hours **ii** $4\frac{3}{4}$ hours.

f At what constant speed is he travelling?

g What does the gradient of the line represent in the context?

Exercise 3.3B

1 A group of students is carrying out an experiment with hanging weights on an elastic band.

The students want to find the connection between the lengths of an elastic band and the number of weights hung on it.

Here is a table of their results

Number of weights, w	0	1	2	3
Length of elastic band, L (cm)	9	12	15	18

a Draw a graph of the data in the table.

b **i** Calculate the gradient of the line.

ii State the y-intercept.

c Write down the equation of the straight line in the form $L = mw + c$.

d What would the length of the elastic band be if you added six weights to it?

2 A gas engineer charges for coming out (a call-out charge). He also charges an hourly rate.

On a weekday during normal working hours he asks for £30 call-out and then £8 an hour.

a Make a table to show his total charge for jobs lasting 1 hour up to 5 hours.

b Draw a graph to represent the data.

c Outside normal working hours he charges more: £50 call-out and £10 an hour.

On the same diagram as part **b**, draw a line to represent these dearer charges.

d Write down the equation of each line in the form $C = mh + c$.

e Calculate the cost of a job on Thursday evening which took two hours outside of working hours.

3 Fertiliser is used to help plants grow bigger.

The table below shows the results of trials to test a fertiliser's effectiveness.

Amount of fertiliser, f (grams)	0	10	20
Height of plant, h, after 8 weeks (cm)	6	7·5	9

a Draw a straight-line graph to illustrate the data.

b Work out the equation of the line in the form $h = mf + c$.

c If you wanted your plant to be twice the size it would be without fertiliser, how much fertiliser should you use?

4 a The Yellow Mobile Phone Company charges a basic £10 a month plus 25p a minute for all calls.

The Blue Mobile Phone Company doesn't have a monthly charge but charges 30p a minute for all calls.

Copy and complete the table:

Calls per month, N (minutes)	0	20	40	60	80	100
Yellow costs, Y (£)	10	15				
Blue costs, B (£)	0	6				

b Draw graphs on the same coordinate diagram to illustrate and compare the data.

c From the graph, which company would be the cheaper if you used:

i 140 minutes a month ii 240 minutes a month?

d For how many minutes is the cost the same?

e Write down an equation to express both Y and B in terms of N.

5 The life expectancy of women has increased over the last hundred years.

In 1900, the average life expectancy for a woman was 44 years. By the year 2000 it had risen to 80 years.

This information can be used to model the life expectancy for women and predict what the life expectancy will be in the next few decades if we assume the relation between life expectancy and time since 1900 is **linear**.

We call a relation linear if, when we draw a graph of it, we get a straight line.

a i By how much has life expectancy improved in a century?

ii By how much will it improve in a quarter century?

b Use these answers to help you copy and complete the table:

Year	1900	1950	2000	2025	2050	2075	2100
Time since 1900, t (years)	0	50	100	125	150	175	200
Life expectancy, L (years)	44	62	80				

c Draw a straight-line graph representing the relation between t and L.

d Using your graph, in which decade would the life expectancy reach 100 years?
(Remember that 0 represents 1900.)

e By first finding the gradient and y-intercept, write down the equation of the straight line in the form $L = mt + c$.

f Using the formula, calculate the predicted life expectancy in 2045.

g Could this model be used beyond 2100? Explain your answer.

Preparation for assessment

1 a Calculate the gradient of each of the following lines:

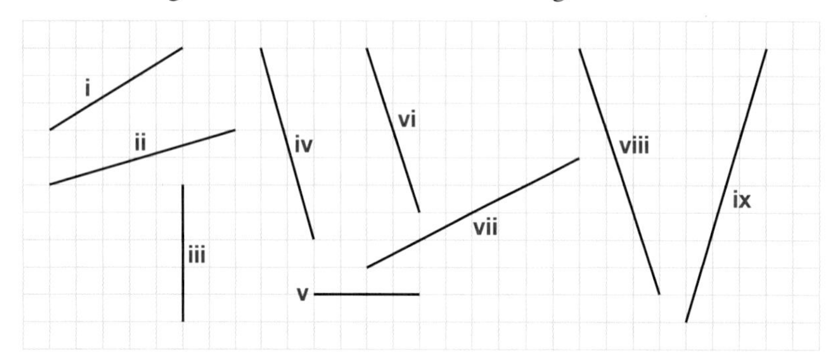

b Which two lines are parallel?

2 a A quadrilateral has vertices A(1, 5), B(2, 10), C(11, 1) and D(5, 1).
Calculate the gradient of each of its sides.

b What kind of quadrilateral is ABCD? Give a reason.

3 a There are two routes to the top of Ben Fasgaidh.
The route from A to X via B covers a distance of 1400 m. A to B is 800 m and B to X is 600 m.
All distances are measured from the map ... and therefore are horizontal distances.
Calculate the gradient for each section of this route.

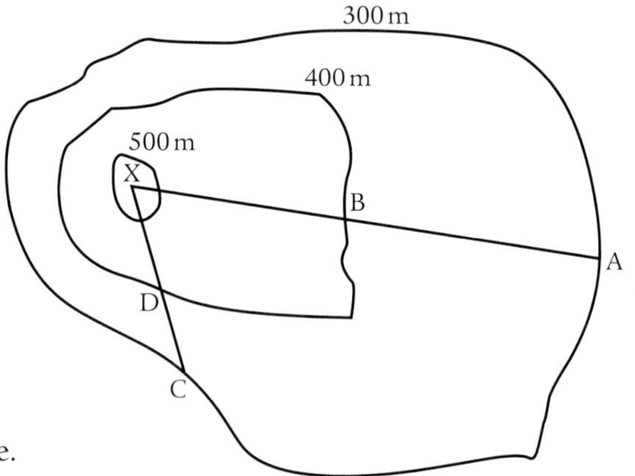

b The route from C to X via D is only 600 m long.
C to D is 250 m and D to X is 350 m.
Calculate the gradient for each section of the route.

c Describe one benefit and one drawback of each route.

4 a Copy and complete the table for the equation $y = \frac{1}{2}x - 2$.

x	−2	−1	0	1	4
y	−3				

b Plot the points (x, y) on a coordinate diagram.

c Write down the gradient and y-intercept of the line on which all of the points lie.

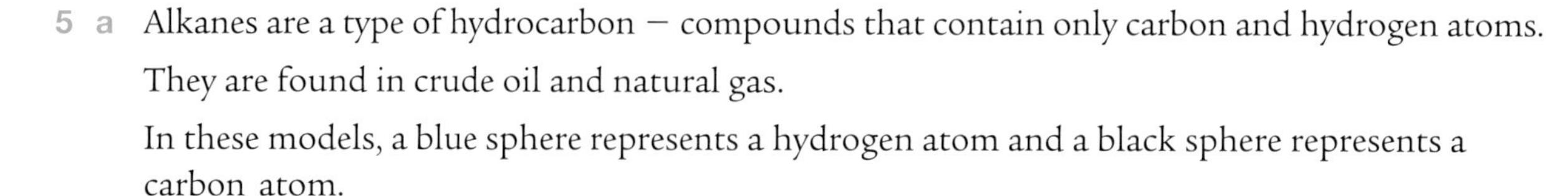

5 a Alkanes are a type of hydrocarbon – compounds that contain only carbon and hydrogen atoms.
They are found in crude oil and natural gas.
In these models, a blue sphere represents a hydrogen atom and a black sphere represents a carbon atom.

Use these models to help you copy and complete the table:

Number of carbon atoms, C	1	2	3	4	5
Number of hydrogen atoms, H	4				

b Draw a straight-line graph to illustrate the relationship between the number of hydrogen atoms and the number of carbon atoms.

c Calculate: i the gradient of the straight line ii the y-intercept.

d Write down an equation for H, the number of hydrogen atoms in terms of C the number of carbon atoms. The form of the equation will be $H = mC + c$.

e How many hydrogen atoms would the alkane with 10 carbons have?

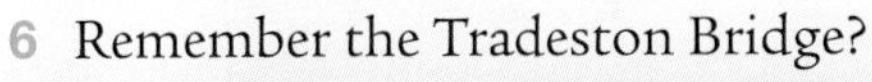

6 Remember the Tradeston Bridge?

The bridge had to get high enough to allow river traffic under.

However, it was a pedestrian bridge so had to have a 'walker-friendly' gradient.

Its start and finish points were predetermined by the fact that the city is built up to both sides of the river at that point. The engineers couldn't take a 'run up'.

How did they make sure that the gradient was low enough for pedestrians but the middle was high enough for boats to get under? What would you do?

4 Transformations

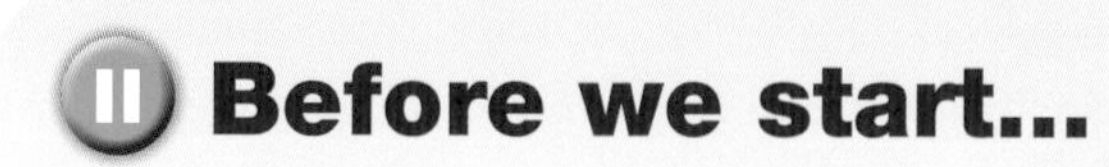

Before we start...

Give a computer the coordinates and it will draw the picture.

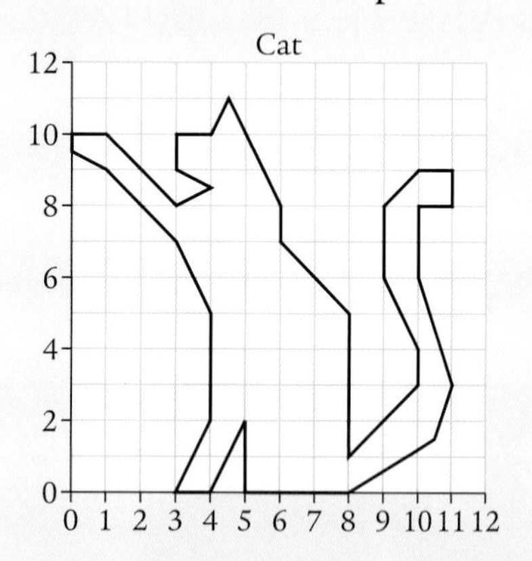

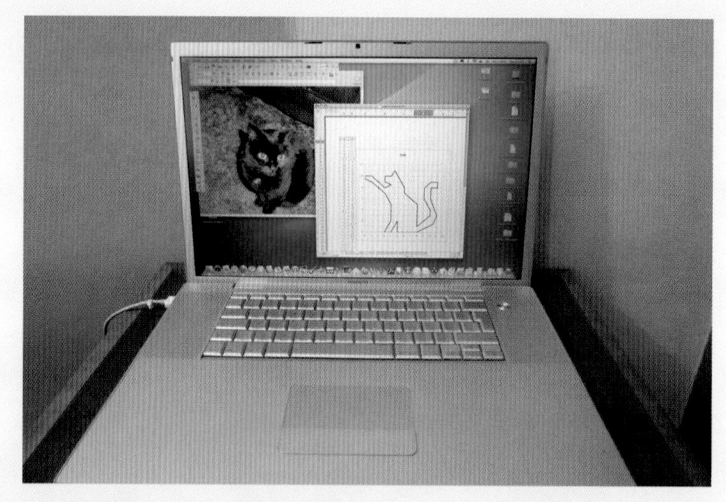

The cat above has been drawn using the set:

(3, 0), (4, 0), (5, 2), (5, 0), (8, 0), (10·5, 1·5), (11, 3), (10, 6), (10, 8), (11, 8), (11, 9), (10, 9), (9, 8), (9, 6), (10, 4), (10, 3), (8, 1), (8, 5), (6, 7), (6, 8), (5, 10), (4·5, 11), (4, 10), (3, 10), (3, 9), (4, 8·5), (3, 8), (1, 10), (0, 10), (0, 9·5), (1, 9), (3, 7), (4, 5), (4, 2), (3, 0).

Using formulae we can transform the cat's coordinates to make the computer ...

- reflect it in the y-axis
- reflect it in the x-axis
- turn it upside down
- rotate it anticlockwise through 90°.

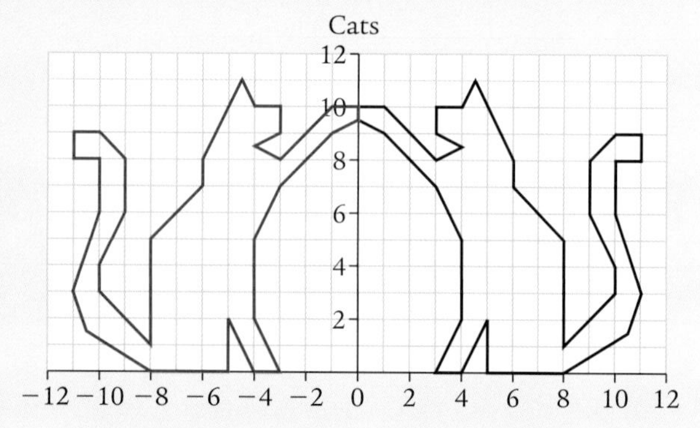

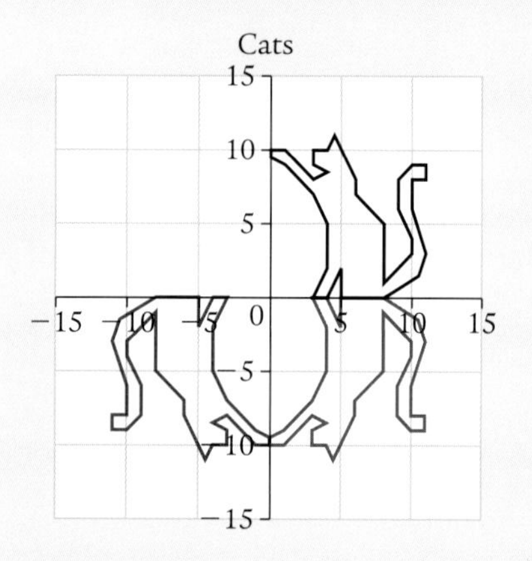

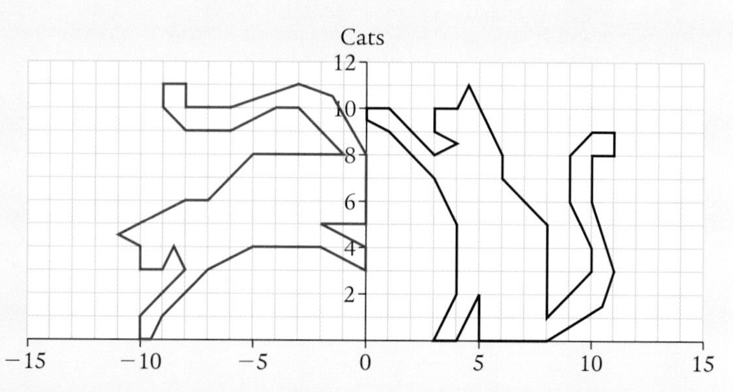

Can you find the formulae that do these jobs?

What you need to know

1 These designs are based on car badges. How many lines of symmetry does each have?

a 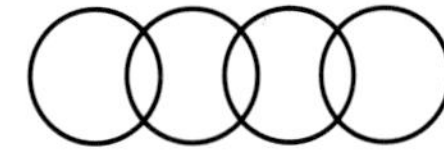

b 

c 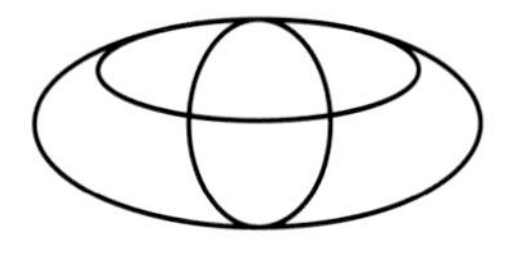

2 a Plot the points A(0, 3), B(4, 5) and C(2, 0) and join them up to form triangle ABC.

b Reflect the triangle in: **i** the x-axis **ii** the y-axis.

3 A particular car company gives computer instructions for drawing their logo.

Carry out their instructions.

a Join A(−2, 5) to B(0, 0) to C(2, 5).

b Move each point 2 left and 5 down.

i State the coordinates of the new points A′, B′ and C′.

ii Draw the shape again.

c Reflect the new image in the y-axis.

i State the coordinates of the new points A″, B″ and C″.

ii Draw the shape again.

d Can you name the type of change made at step **b**?

4 a Draw the line $y = 2x - 1$ on a coordinate diagram.

b B lies on the line. Its x-coordinate is 2. What is its y-coordinate?

c C lies on the line. Its y-coordinate is 5. What is its x-coordinate?

d When B is reflected in the x-axis its image is B′.
What are the coordinates of B′?

e What are the coordinates of C′, the image of C under reflection in the x-axis?

f Draw the image of the line $y = 2x - 1$ after reflection in the x-axis.

4.1 Transformation and symmetry

Exercise 4.1A

Class discussion

1 Examine each picture and say how the red image has been transformed into the black.

2 Some of the above transformations were done using this menu on a computer.

a Which transformation goes with which option?

b Two involved a change in size, **e** and **f**. What's the difference between them?

c Two involved a change in one direction only.

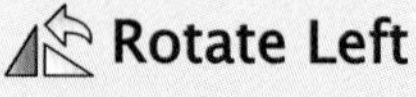

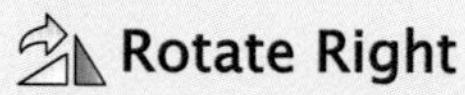

This type of transformation can be used to make writing readable from a very shallow angle.

- **i** What does the writing on the right say?
- **ii** The next time you watch a rugby match ask yourself how they manage to get adverts painted on the grass to look rectangular on the screen.
- **iii** Why are 'GIVE WAY' signs on the road elongated?

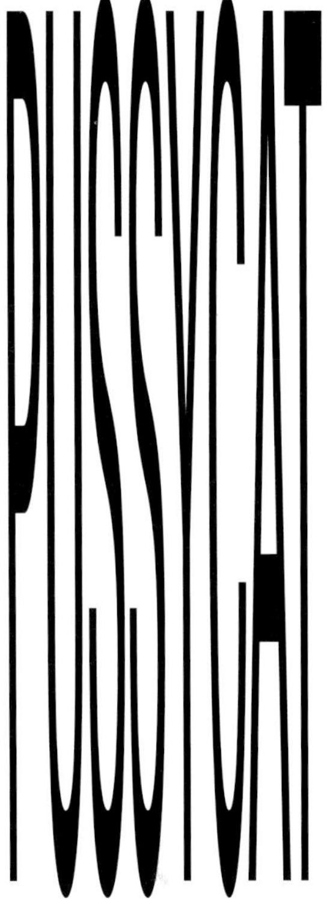

d One involves two simple transformations performed one after the other.
Which one ... and what was the effect?

3 When giving instructions to a computer we must be precise.

When asking for a reflection, it is important to be exact about where the axis is ... it matters.

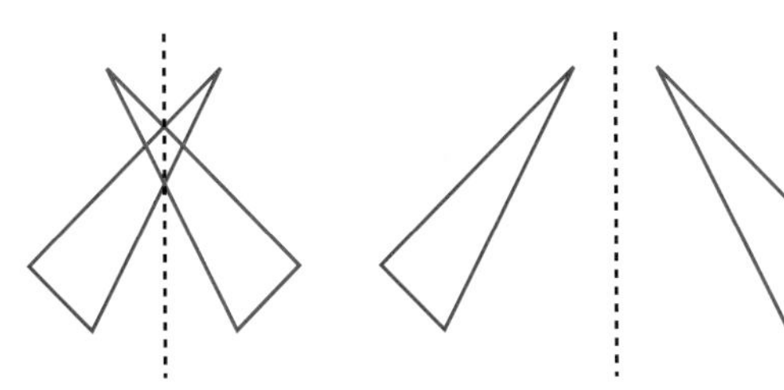

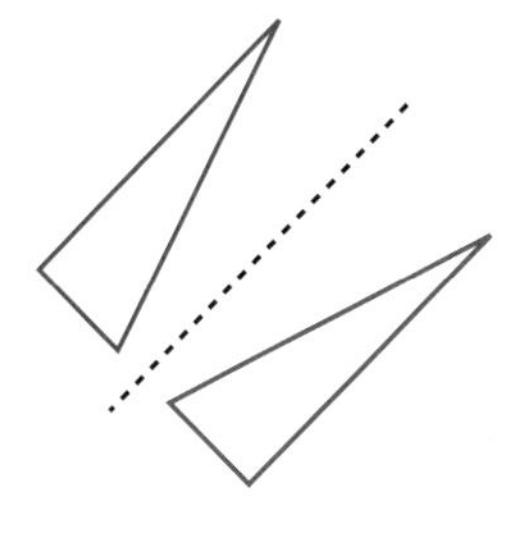

What information would you think is essential for describing:

a a translation **b** a rotation?

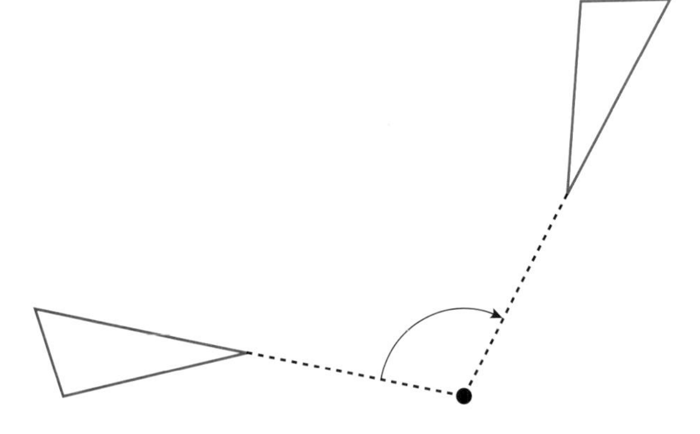

4 Sometimes when a shape is transformed it doesn't look different.

When this happens, the shape is said to possess **symmetry**.

The type of symmetry is usually named after the type of transformation used.

a Discuss the symmetry of:

- **i** an isosceles triangle ... what transformation would leave it looking the same?
- **ii** a kite
- **iii** a rhombus
- **iv** a parallelogram
- **v** a straight line that goes on forever.

b Can you think of anything that has translational symmetry?

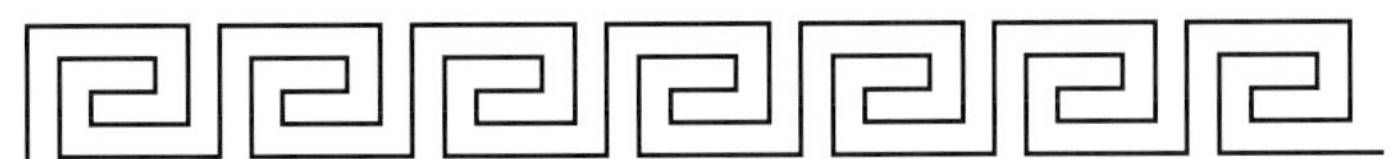

5 Some transformations distort the original shape, turning straight lines into curves.

Here is a person photographing himself in the hall of mirrors.

There should be a rectangular wall hanging to his right and a vertical straight pipe to his left.

Why is this picture no use for figuring out the height of the photographer?

We will only study transformations that keep straight lines straight.

Some useful notation

When the point A(2, 5) is reflected in the y-axis its image A′ is the point (−2, 5).

This can be quite tidily written as A(2, 5) → A′(−2, 5).

In the same way B(−3, 2) → B′(3, 2).

Since straight lines have images that are straight lines then the image of the line AB is the line A′ B′ ... AB → A′ B′.

Example 1

The triangle PQR has vertices P(2, 5), Q(−3, 2) and R(3, −2).

a Plot the triangle on a coordinate grid.

b Draw its image under reflection in the y-axis.

c State how each vertex maps onto its image.

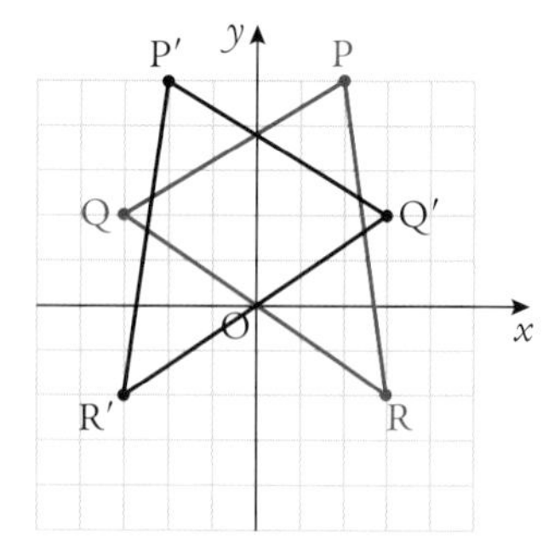

a, b Points plotted in red ... images in black.

c P(2, 5) → P′(-2, 5)

Q(−3, 2) → Q′(3, 2)

R(3, −2) → R′(−3, −2)

Example 2

a Show how the point P(x, y) is transformed by reflection in the y-axis.

b Express the coordinates of its image, P′, in terms of x and y.

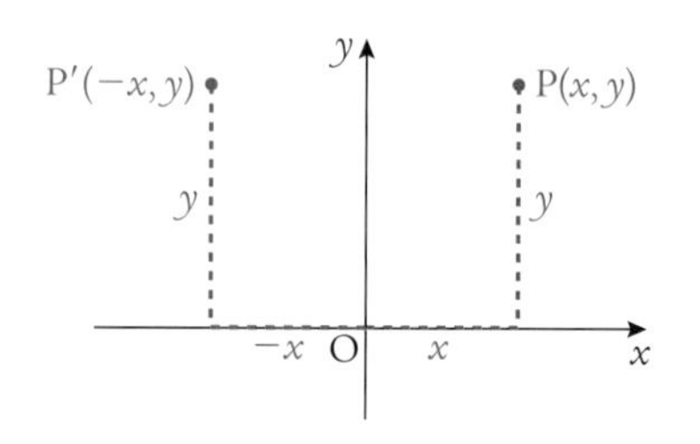

a Original point in red ... image in blue.

b Given that P has coordinates (x, y), the formulae for working out the coordinates of P′ are:

x-coordinate of P′ $= -1 \times x$

y-coordinate of P′ $= y$.

You may have done the above in an earlier course.

If so, you can skip this exercise ... but the practice might be useful!

Exercise 4.1B

1 Karen is cutting out fabric to make a skirt.

She folds the fabric in half lengthwise so that the pattern on the fabric will have reflective symmetry when the front of the skirt is cut out.

The finished skirt should be of the dimensions shown.

a Draw a diagram to show the shape Karen should cut out on the folded material, so that when she opens out the material the piece of fabric is the correct shape.

b What are the dimensions of the smallest rectangle you could fold to get the required shape?

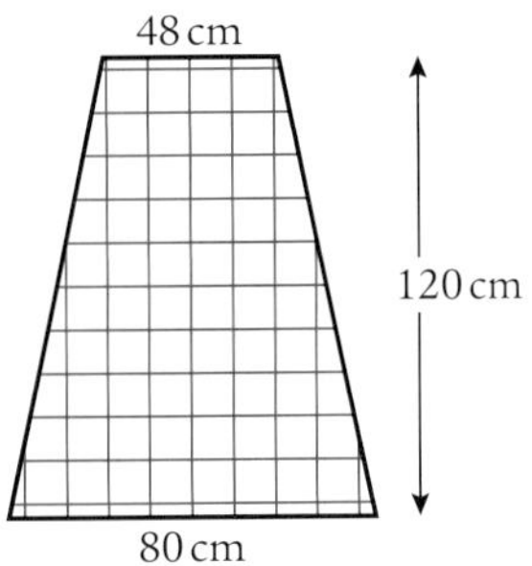

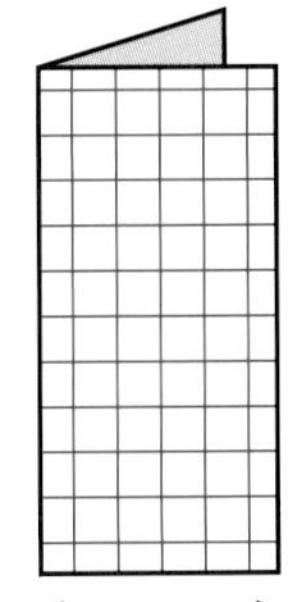

2 For an animation, a swan has been represented by 12 control points on a grid.

A computer draws the swan by joining the points.

It can work out the swan's image in the water by reflecting these points in the x-axis.

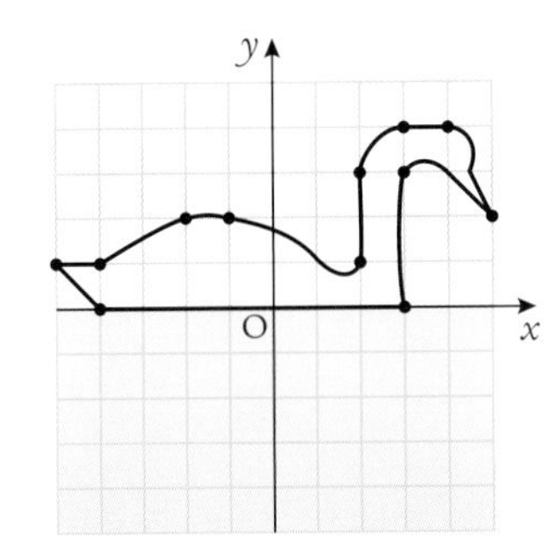

a List the coordinates of the 12 control points.

b Give the coordinates of their image after reflection in the x-axis.

c Write down the formulae for working out the coordinates of P′, the reflection in the x-axis of a point P(x, y).

3 The markings on the moth's wings are being studied in the Biology department.

To save time, having noted the reflection symmetry in the diagonal line $y = x$, only points on one wing are recorded. Those on the other wing are calculated.

Nine control points have been noted.

a Give the coordinates of the control points and their image under reflection in $y = x$.

b Write down the formulae for working out the coordinates of P′, the reflection in $y = x$ of a point P(x, y).

c A dark patch occurs between (−4, 1) and (−2, −1).
Where will its image in $y = x$ occur?

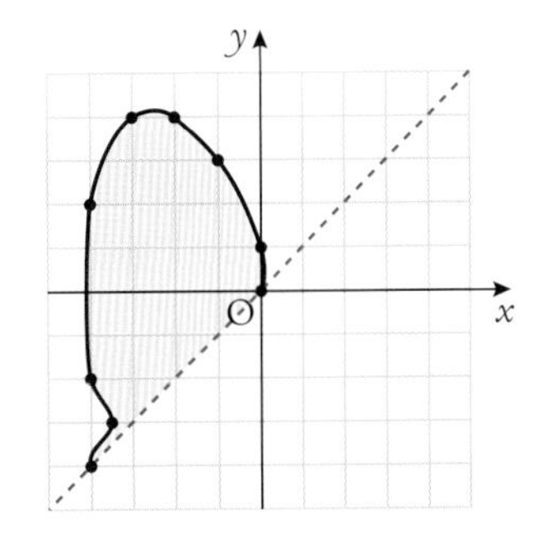

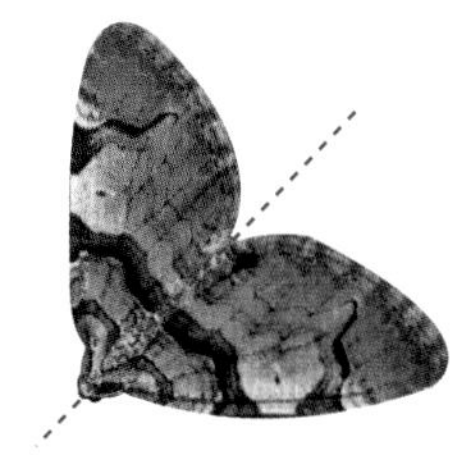

4 The isosceles triangle gets its name from two Greek words: **isos** (same) and **skelos** (leg).

It is defined as a triangle with an axis of symmetry.

Triangle ABC has an axis of symmetry AD, where D lies on the base BC.

Under reflection in this axis, B → C.

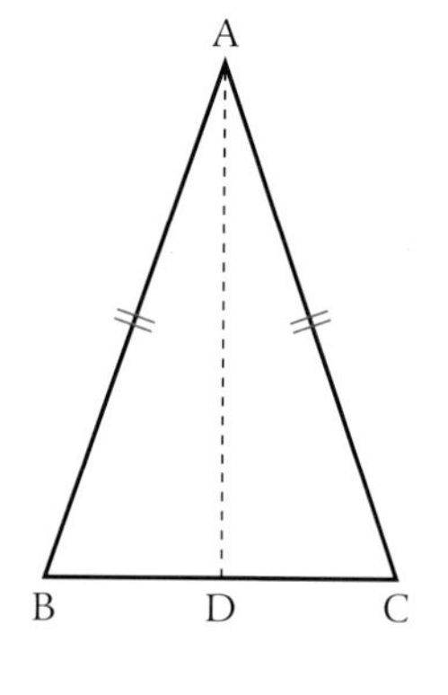

- **a** Indicate the image of **i** A **ii** C **iii** D under this reflection.
- **b** What is the image of **i** $\angle ABC$ **ii** $\angle BAD$ **iii** $\angle ADB$?
- **c** Name an angle equal to **i** $\angle ABC$ **ii** $\angle BAD$ **iii** $\angle ADB$.
- **d** Using symmetry in your argument, explain why $\angle ADB = 90°$.
- **e** What does this tell you about the line AD in relation to the triangle?

5 A kite is a quadrilateral with an axis of symmetry passing through two of its vertices.

The quadrilateral ABCD has an axis of symmetry AC and a diagonal BD. BD cuts AC at E.

- **a** Give the images of A, B and C under reflection in AC.
- **b** $\angle ABD = 40°$ and $\angle DBC = 70°$.
  - **i** State the image of both $\angle ABD$ and $\angle DBC$.
  - **ii** Hence give the size of $\angle ADC$.
- **c** Using symmetry in your argument explain why:
  - **i** the angles round E are all 90°
  - **ii** the diagonal AC bisects the diagonal BD.
- **d** Calculate the size of $\angle BAC$ and $\angle BCA$.
- **e** Name and give the size of the four angles of the quadrilateral.

6 A tanker carrying milk has a cross-section as shown.

It is a circle of radius 10 units and a chord PQ of length 12 units.

XY is an axis of symmetry of the diagram, cutting PQ at R.

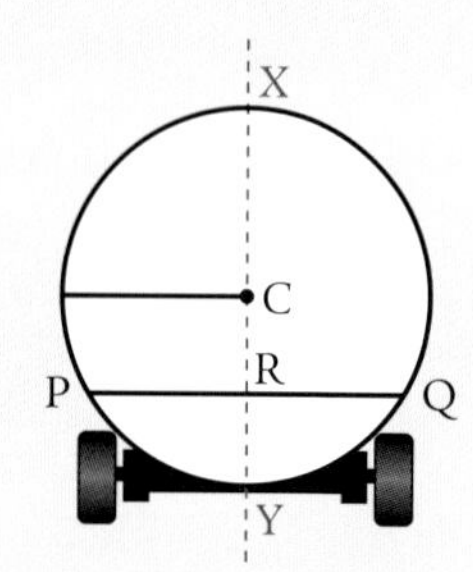

- **a** State how you know that the centre of the circle, C, lies on XY.
- **b** How do we know that XY is perpendicular to PQ?
- **c** Deduce, using symmetry, the length of PR.
- **d** Calculate the distance of R from C.

7 A translation is described by the formulae:

x-coordinate of the image of $P(x, y) = x + 4$

y-coordinate of the image of $P(x, y) = y - 3$.

- **a** State the image of the point (2, 1) under this translation.
- **b**
  - **i** Sketch the point and its image.
  - **ii** Use Pythagoras' theorem to calculate the distance actually moved in the translation.
- **c** The triangle ABC has vertices A(0, 0), B(−2, 5) and C(1, 3).
  - **i** State the coordinates of A′, B′ and C′, the images of A, B and C under the translation described above.
  - **ii** State the coordinates of A″, B″ and C″, the images of A′, B′ and C′ under the same translation.

8 A translation maps triangle 1 onto triangle 2 so that $(-3, 1) \rightarrow (2, 4)$.

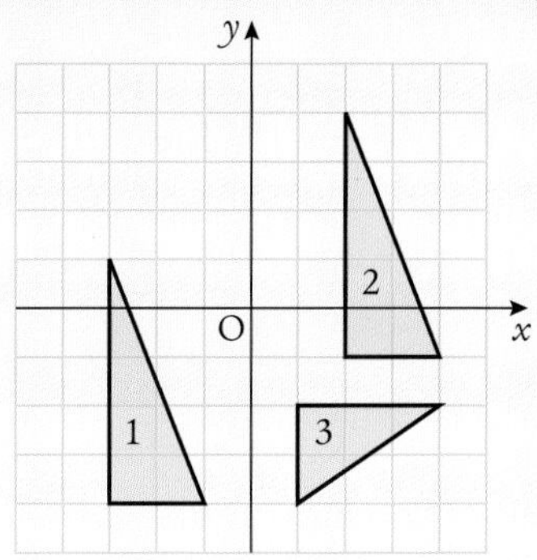

a Describe the shift of the other two vertices of triangle 1.

b Use formulae to describe the translation of $P(x, y)$ to its image P' under this translation.

c State the coordinates of the vertices of the image of triangle 3 under the same translation.

9 The diagram shows the line $y = 2x + 1$.

a Check that the points $A(-1, -1)$, $B(0, 1)$, $C(1, 3)$ and $D(2, 5)$ all lie on the line.

b A translation is described by the formulae:

x-coordinate of the image of $P(x, y) = x + 1$

y-coordinate of the image of $P(x, y) = y - 2$.

i Find the images of A, B, C and D under this translation.

ii Draw the line which is the image of $y = 2x + 1$ under this translation.

iii By considering its gradient and y-intercept, find the equation of the image of the line under the translation.

c Another translation is described by the formulae:

x-coordinate of the image of $P(x, y) = x + 1$

y-coordinate of the image of $P(x, y) = y + 2$.

i Find the images of A, B, C and D under this translation.

ii By considering its gradient and y-intercept find the equation of the image of the line under this translation.

iii Comment, considering the idea of translational symmetry.

10 A jewellery designer is designing a brooch from strands of silver wire.

She uses her computer to assist in the design.

She uses the lines with equations $y = 6 - x$, $y = 6 - 2x$ and $y = 6 - 3x$ to represent three of the strands.

She realises that the necklace would not hang properly as it is not symmetrical.

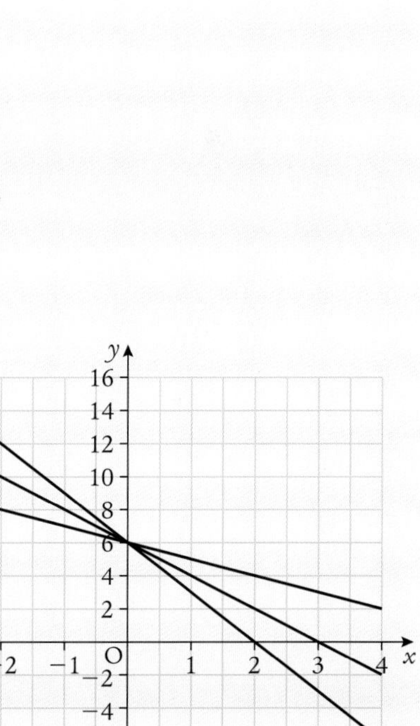

a Using reflection in the y-axis, draw another three lines which make the diagram symmetrical.

b State the equation of the three new lines and comment on the equations.

11 The line $y = x$ and the points A(3, 3), B(2, 0), C(−4, −4), D(0, 2) are drawn.

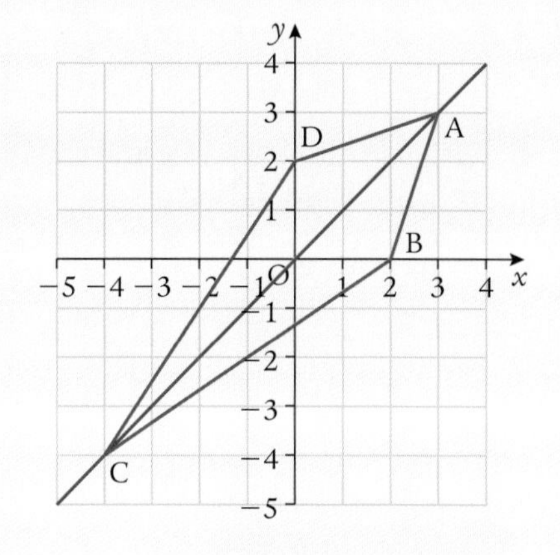

a Reflect the quadrilateral ABCD in the line $y = x$, stating the coordinates of the image of each point.

b Comment on the results.

12 In chess, the moves that pieces make can be recorded by treating them as translations.

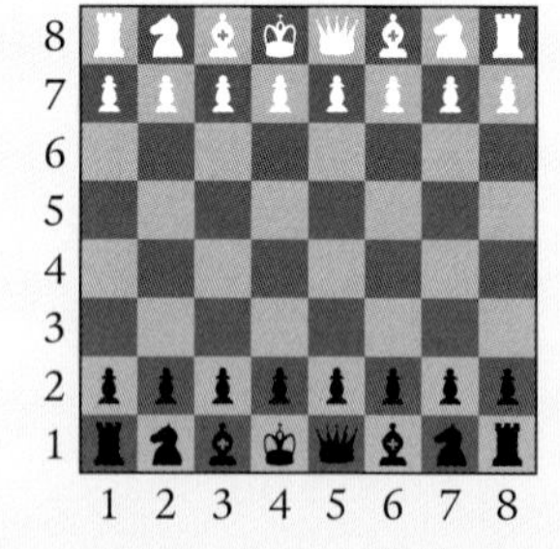

Each square can be addressed using a coordinate system.

By this means the game can be programmed into a computer.

The most awkward mover is the knight (horse).

He may change his x-coordinate by ±2 and his y-coordinate by ±1 OR his x-coordinate by ±1 and his y-coordinate by ±2.

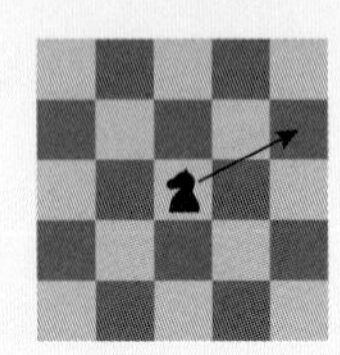

The diagram shows one possible translation the knight can make: $[x + 2, y + 1]$.

a Copy the grid and indicate all of the squares to which the knight can move, indicating the actual translations as above.

b During the game the knight is on square (7, 5).

Which translations are possible, ignoring any other pieces but keeping in mind that he must stay on the board?

4.2 Rotation

Describing rotation

When an object is turned around a fixed point we call such a transformation **rotation**.

The fixed point is known as the **centre of rotation**.

A particular rotation can be described by giving its centre and the angle turned through.

Example 1

The diagram shows the rotation of a shape around a point P.

Through what angle has the shape been rotated?

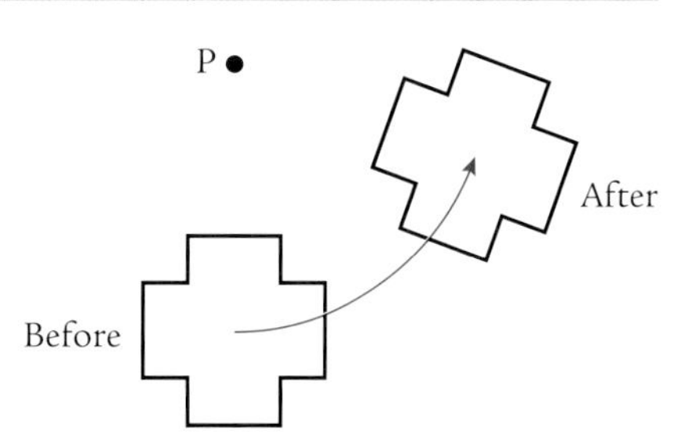

P is the centre of rotation.

Join any point A, on the original shape to P.

Join its image A′, to P.

Measure the angle ∠APA′
... it measures 70°.

The rotation is one of 70° anticlockwise about P.

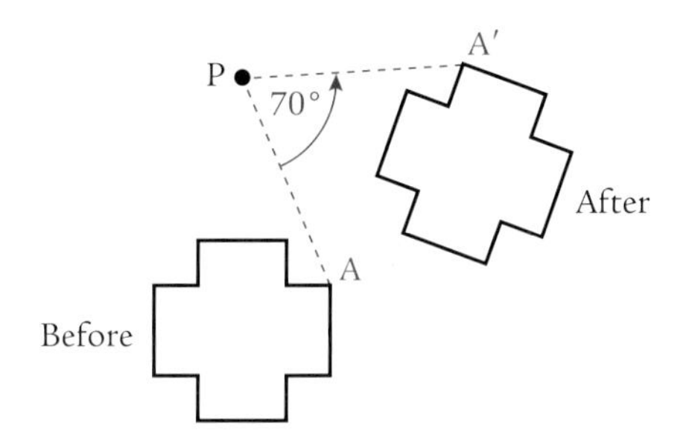

Example 2

A rectangle has been rotated from the red position to the blue.

a Find the centre of rotation.

b Describe the rotation.

a 1. Trace the diagram.

2. Join a point A to its image A′
   ... draw the perpendicular bisector of this line.
   The centre must lie on this bisector.
3. Repeat step 2 for another point B.
4. Where the two bisectors intersect is the centre of rotation.

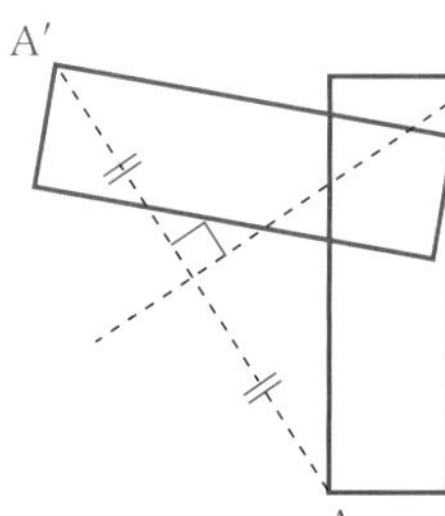

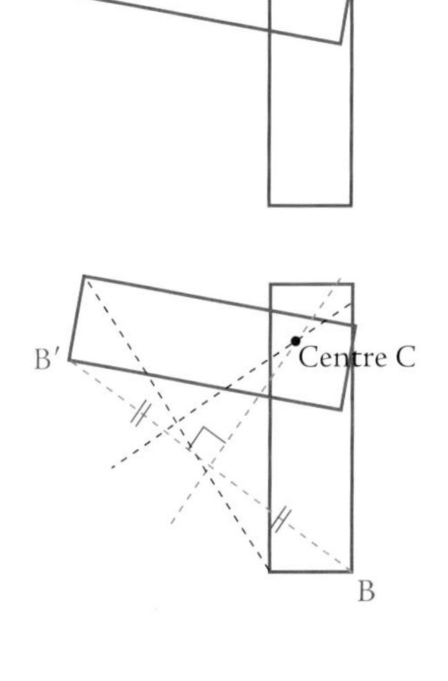

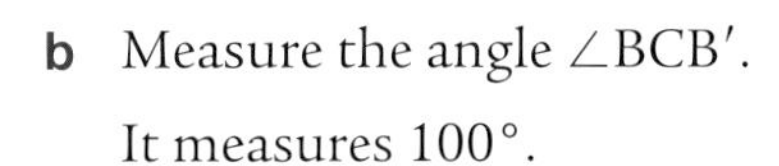

b Measure the angle ∠BCB′.

It measures 100°.

The rectangle was rotated 100° clockwise about C.

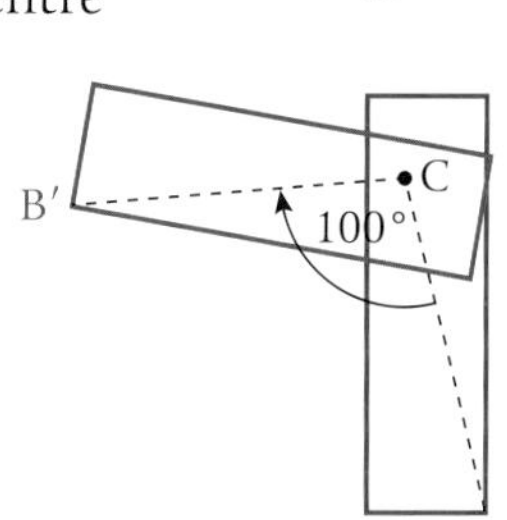

Rotational symmetry

If a shape looks the same after a rotation then the shape possesses **rotational symmetry**.

The number of times it looks the same in a complete revolution is called the **order** of the symmetry.

The regular polygons exhibit rotational symmetry about their centres:

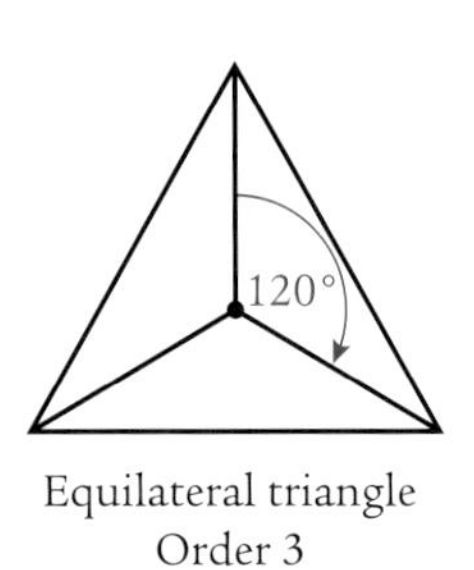

Equilateral triangle
Order 3

Square
Order 4

Pentagon
Order 5

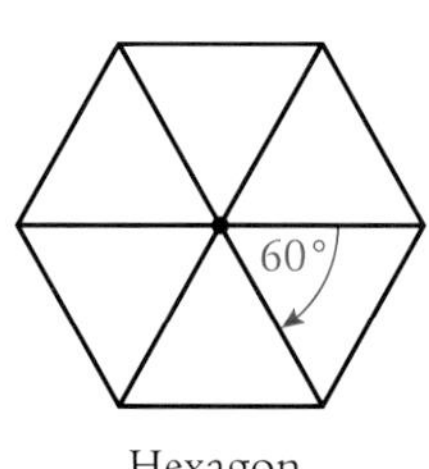

Hexagon
Order 6

Example 3

Parallelogram ABCD has rotational symmetry about a centre E.

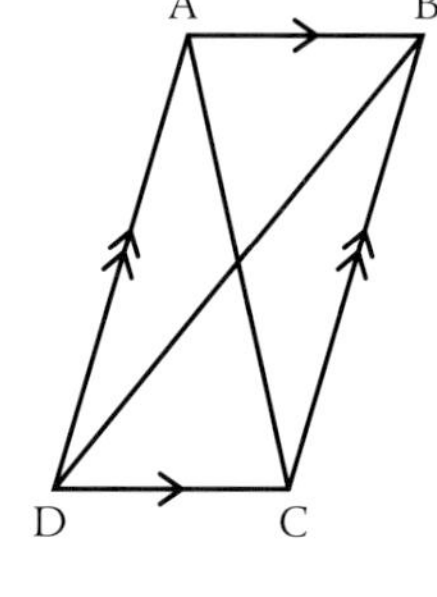

a Describe the location of E.

b **i** What is the order of the symmetry?

ii How much does it rotate before it first looks the same?

c What is the image of:

i A **ii** AB **iii** $\angle ABC$ **iv** $\angle BEC$

under the rotation?

d Name an angle equal to $\angle BDC$.

a E is at the intersection of the diagonals.

b **i** The symmetry is of order 2.

ii It will rotate through 180° before it looks the same.

c **i** $A \rightarrow C$ **ii** $AB \rightarrow CD$ **iii** $\angle ABC \rightarrow \angle CDA$ **iv** $\angle BEC \rightarrow \angle DEA$.

d By rotational symmetry around E, $\angle BDC \rightarrow \angle DBA \Rightarrow \angle BDC = \angle DBA$.

Note: rotational symmetry of order 2 is often called **half-turn symmetry**.

Example 4

The diagram shows a shape in the 1st quadrant.

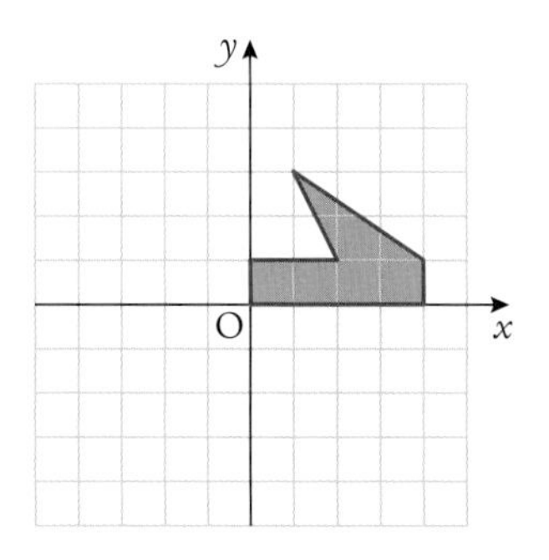

a Copy the diagram and draw its image after a quarter turn anticlockwise about the origin.

b Draw the image of the image under the same rotation.

c Repeat the process to create a shape which has rotational symmetry of order 4 with the origin as the centre of symmetry.

a
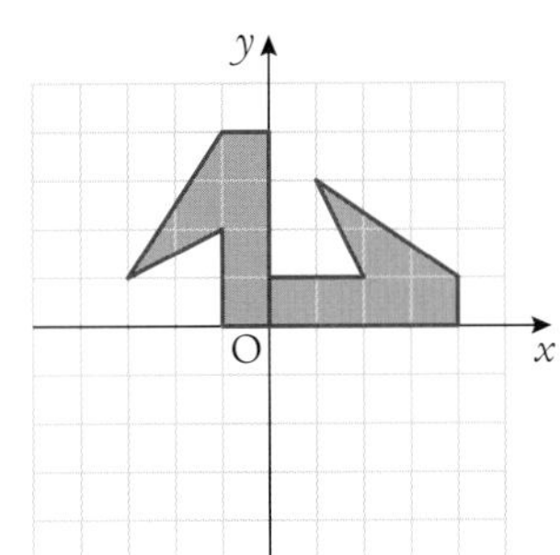

b
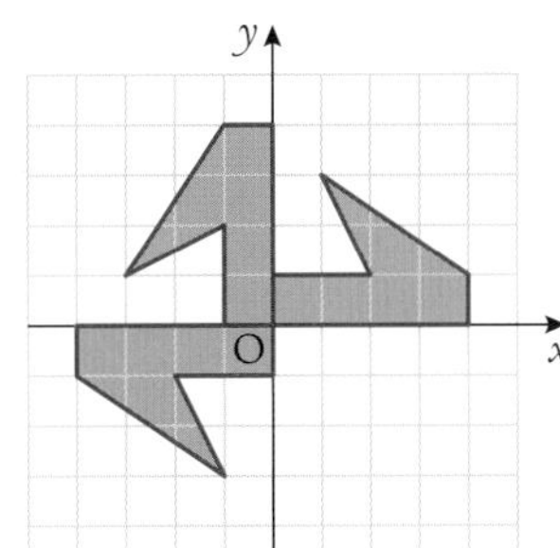

c
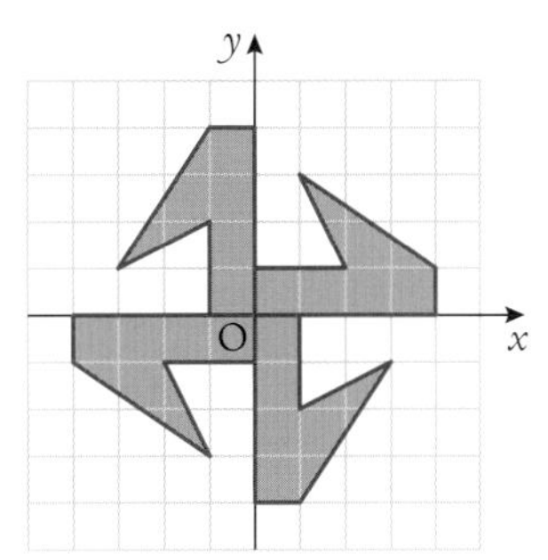

Note: rotational symmetry of order 4 is often called **quarter-turn symmetry**.

Exercise 4.2A

1 **i** Trace each of the diagrams.

ii Identify the centre of rotation.

iii Describe the rotation that maps the red shape onto the blue.

a
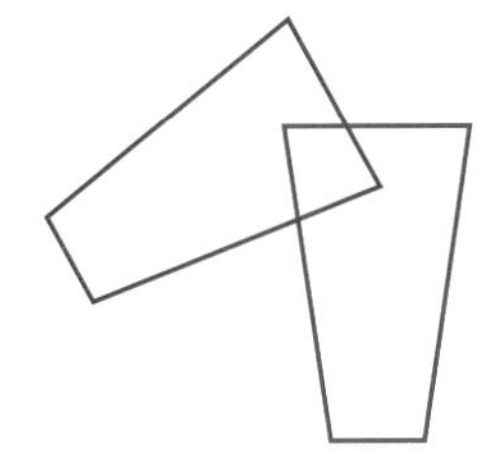

b
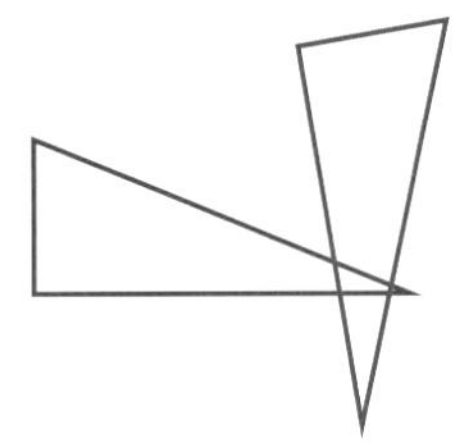

c
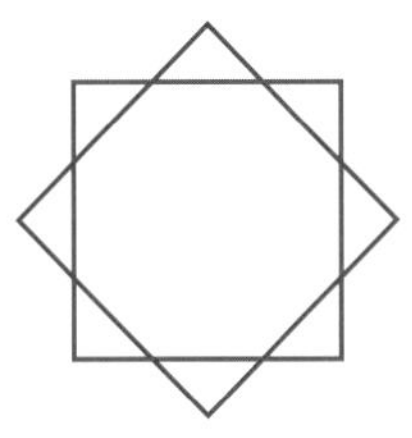

2 Over a 24-hour period the stars in the heavens seem to revolve once round Polaris, the Pole Star.

In 6 hours they make a quarter turn anticlockwise.

The seven stars of the constellation called 'The Plough' are shown.

Polaris is used as the origin.

Taking each square of the grid as 1 unit,

a list the coordinates of the 7 stars

b give their coordinates after 6 hours have elapsed.

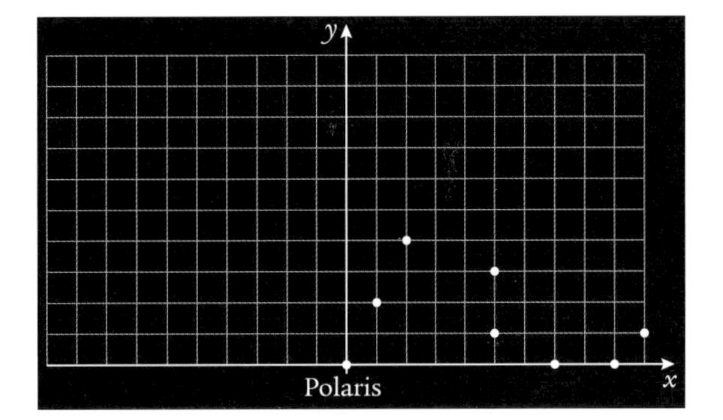

3 The SQA logo is based on the tick (✓) ... the symbol universally used for 'correct'.

a Copy the diagram and draw its image after a quarter turn anticlockwise about the origin.

b Draw the image of the image under the same rotation.

c Repeat the process to complete the logo that has rotational symmetry of order 4.

d Can you see a cross (✗) in the logo?

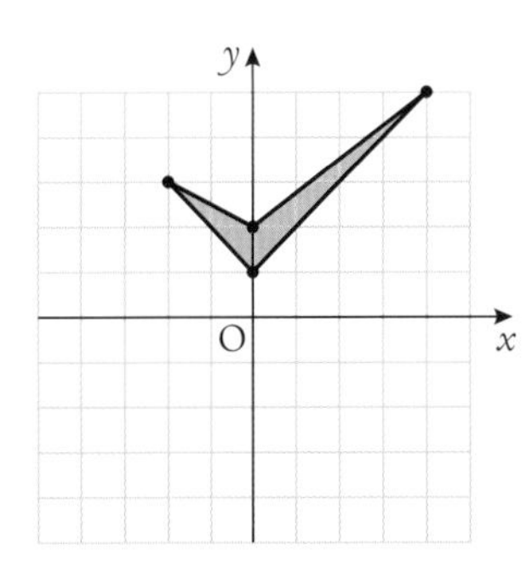

4 For each diagram state:

i the coordinates of the centre of rotational symmetry

ii the order of the rotational symmetry

iii the size of the smallest rotation required to show the symmetry.

a
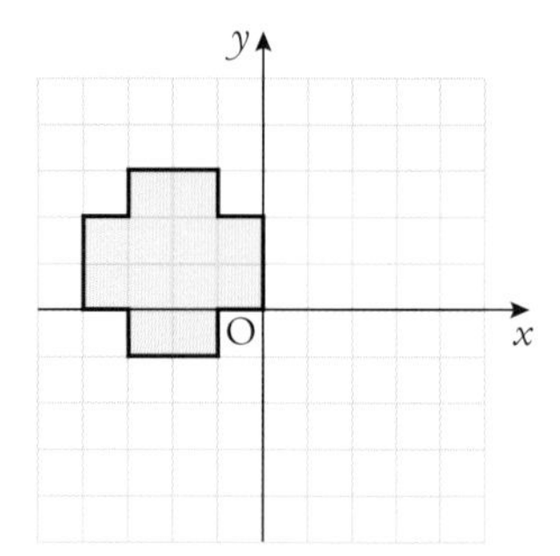

b
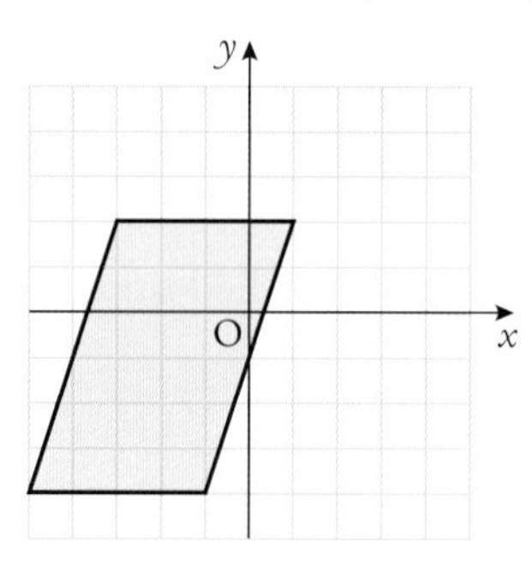

c
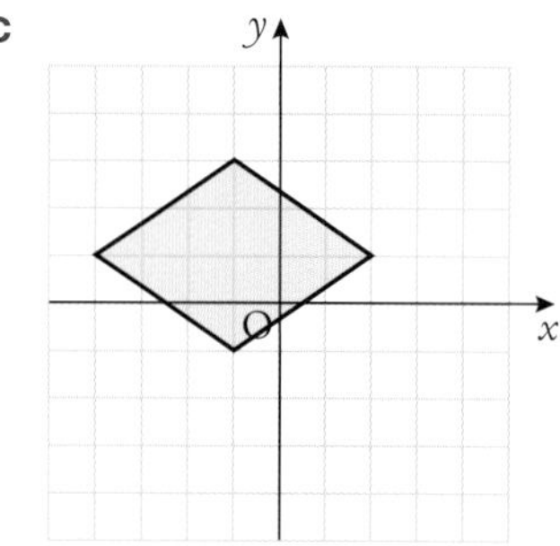

d
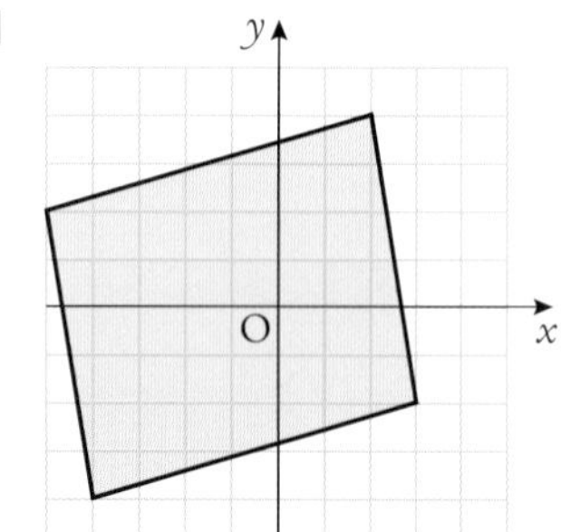

e
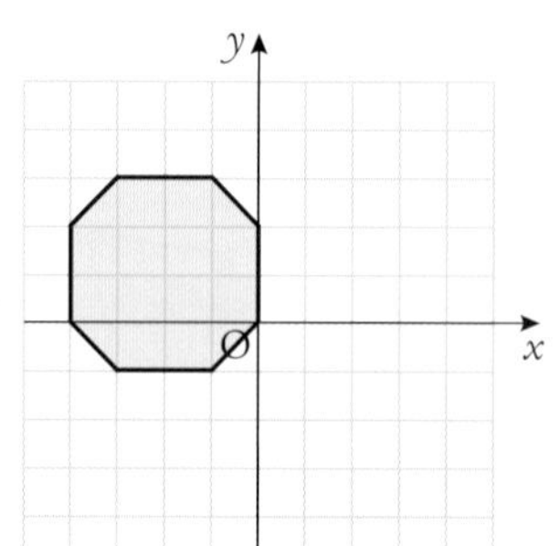

f
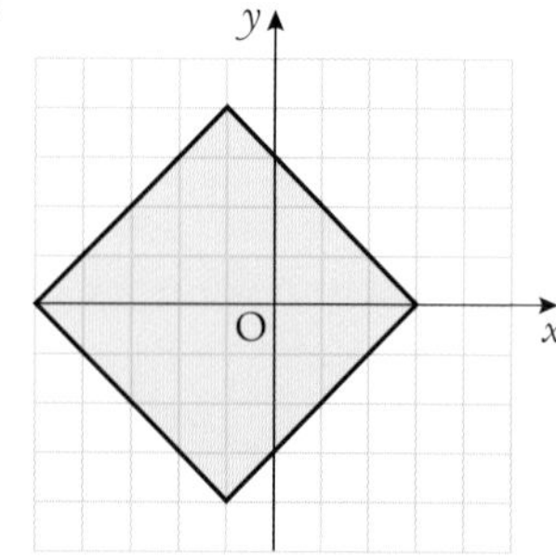

5 A company wishes to design backs of cards to have rotational symmetry of order 2.

Copy and complete each design so it has half-turn symmetry.

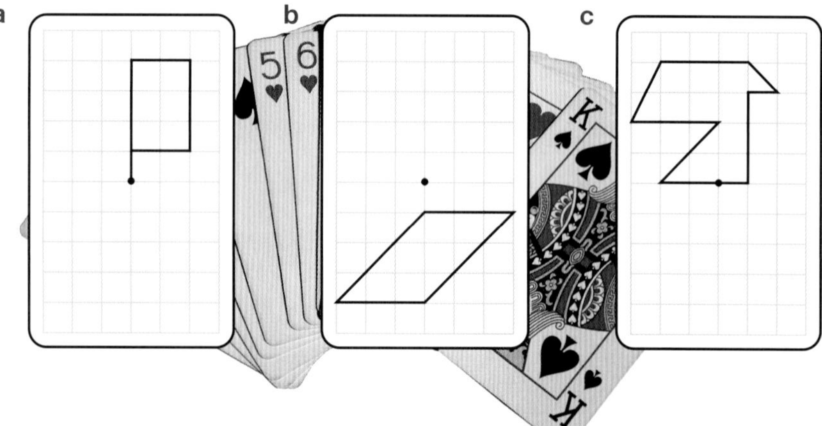

6 During a dig, an archaeologist comes across partially covered floors that used white and green tiles. His theory is that originally they possessed quarter-turn symmetry.

a Complete each of the four floor patterns so that they have quarter-turn symmetry about the point indicated.

i
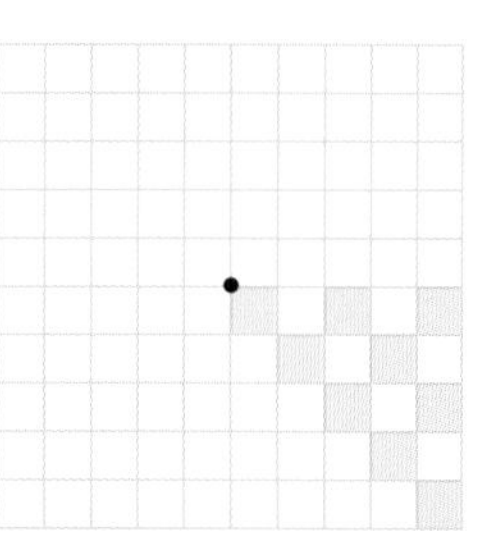

ii
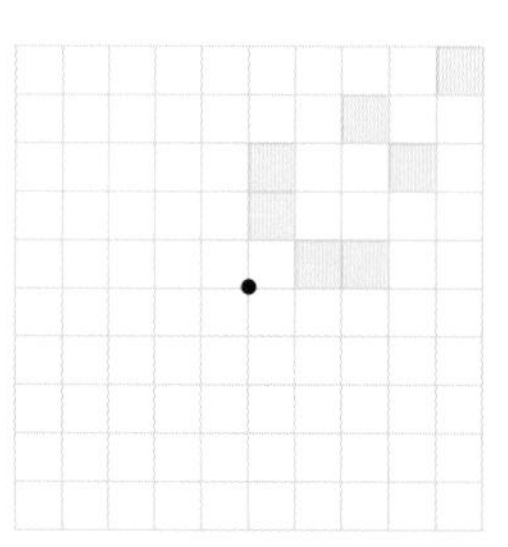

iii
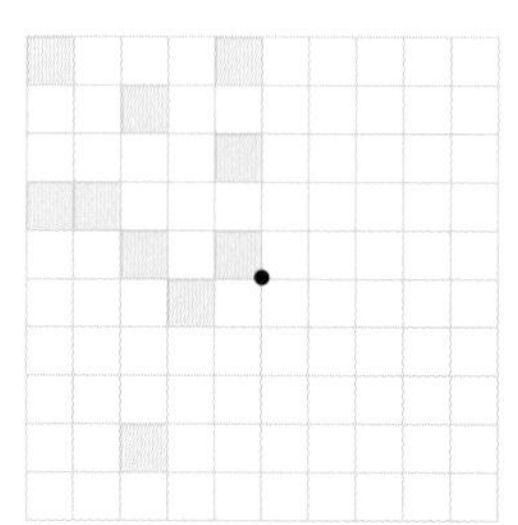

iv
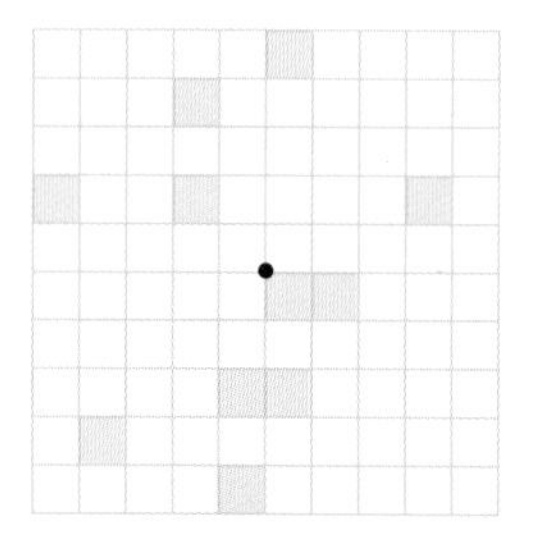

b On one of these four sites all of the 100 tiles were recovered.
The ratio of green to white was 2 : 3.
Which of the four floors was this?

7 The regular pentagon has rotational symmetry of order 5.

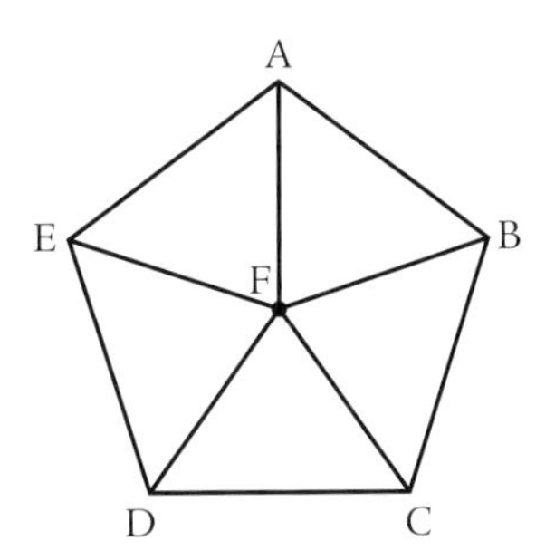

Figure 1

A B C D E F

Figure 2

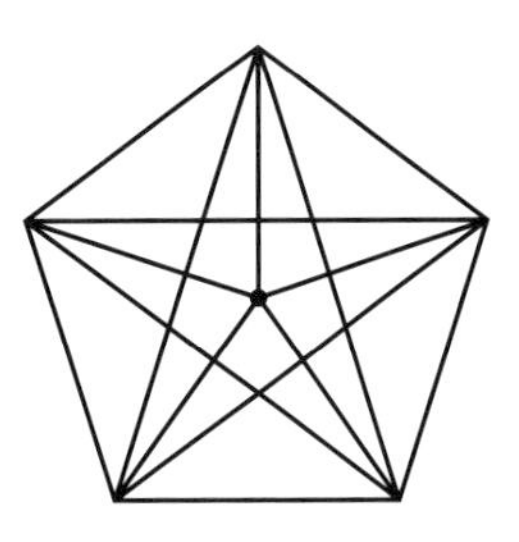

Figure 3

Figure 4

a Use the symmetry to calculate all the angles in Figure 1.

b A diagonal is added as in Figure 2. Use reflective symmetry about AF to:

i identify a kite with both diagonals drawn

ii work out the size of the angles in ΔABE and ΔEBF.

c All five diagonals define a star (Figures 3 and 4). Use symmetry to find all the angles in the star.

8 A line of unit length is drawn.

Using its right-hand end as a centre of rotation, a second line of unit length is rotated through 40°.

Using the end of the second line as the centre the process is repeated again and again until we end up where we started.

With the rotation set at 40° we end up with a nonagon.

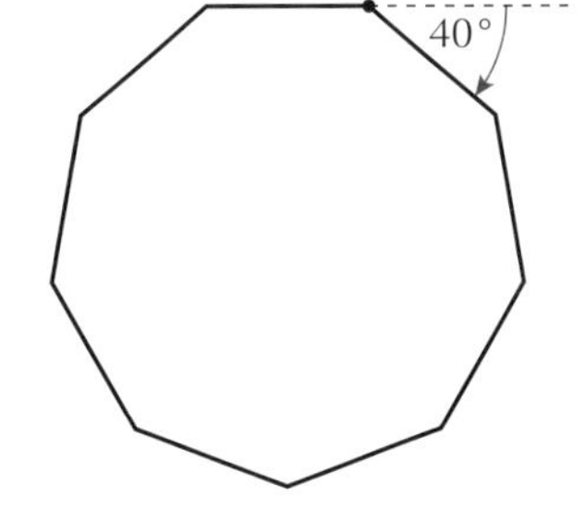

a What will we end up with if we rotate through **i** 90° **ii** 120°?

b Investigate what happens when we set the rotation at 10°.

i How many times must we rotate before we end where we began?

ii What order of rotation will the finished article have?

Exercise 4.2B

1 **i** Plot each set of coordinates, joining them up to form a shape.

ii Rotate the shape through an angle of 180° about the origin.

iii State the coordinates of the image points.

a A(0, 6), B(3, 2), C(0, 4), D(−3, 2)

b O(0, 0), P(4, 2), Q(4, 5), R(0, 3)

c S(−1, −2), T(−3, −1), U(−5, −2), V(−3, −4)

2 By spotting patterns between the coordinates of a point and its image, devise formulae for transforming the coordinates of P(x, y) into the coordinates of its image under a half turn about the origin:

x-coordinate of image of P(x, y) =

y-coordinate of image of P(x, y) =

3 **i** Plot each point, joining them up to form a shape.

ii Rotate the shape about the origin through an angle of 90° anticlockwise.

iii State the coordinates of the image points.

iv Complete the diagram so that the finished drawing has quarter-turn symmetry.

a (0, 2), (3, 4), (2, 0)

b (−3, 0), (−3, 2), (−2, 3), (0, 3)

c (−1, 0), (−2, 1), (−1, 3), (0, 1)

4 After a quarter turn anticlockwise, the red L-shaped polygon maps onto the blue.

a Make a table like the one shown and:

i enter the coordinates of A, B, C, D, E and F

ii enter the coordinates of their images under the rotation.

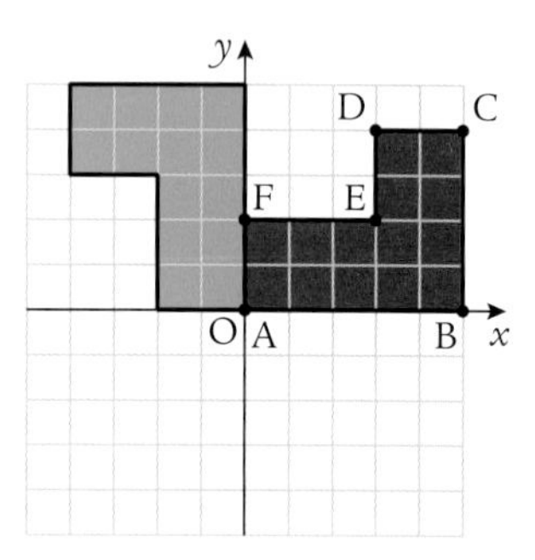

Red	x	y	Blue	x	y
A			A′		
B			B′		
C			C′		
D			D′		
E			E′		
F			F′		

b By spotting patterns in the table, devise formulae for transforming the coordinates of P(x, y) into the coordinates of its image under a quarter turn anticlockwise about the origin:

x-coordinate of image of P(x, y) =

y-coordinate of image of P(x, y) =

5 When planning the road layout in a new housing estate the engineers design a major crossroads. This is modelled on a coordinate grid.

The origin is placed at the intersection of the roads.

One road runs from the junction at O(0, 0) to a dead end at A(400, 800).

The other road runs at right angles to this.

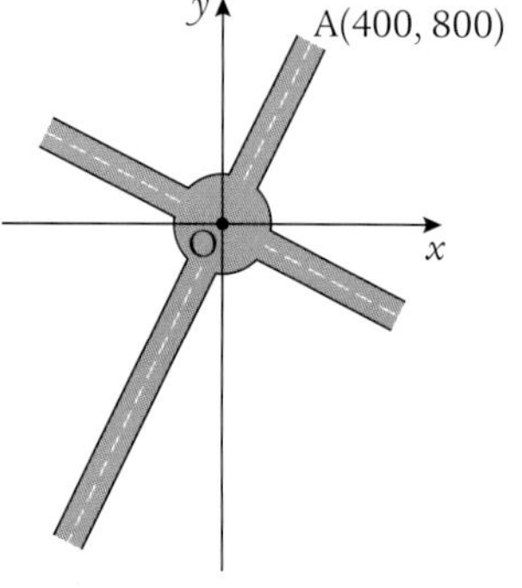

a State the equation of the road that stops at (400, 800).

b By considering a quarter turn anticlockwise around the origin,

i find the image of A

ii find the equation of the second road.

6 a On a coordinate diagram plot the triangle with vertices (1, 1), (3, 1) and (5, 4).

b Reflect the triangle in the x-axis and then reflect that triangle in the y-axis.

c Rotate the original triangle by 180° about the origin.

d Comment on your results.

7 Roger Penrose, a famous mathematician, devised two shapes known as the Penrose kite and dart. These fit together to make a rhombus.

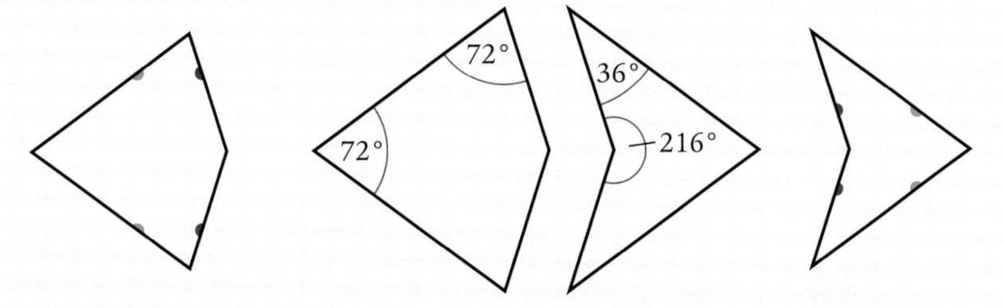

Copies of these can be used as tiles to form shapes that exhibit various symmetries.
These tiles must be fitted together according to a simple rule.
The coloured half-dots must match up exactly.

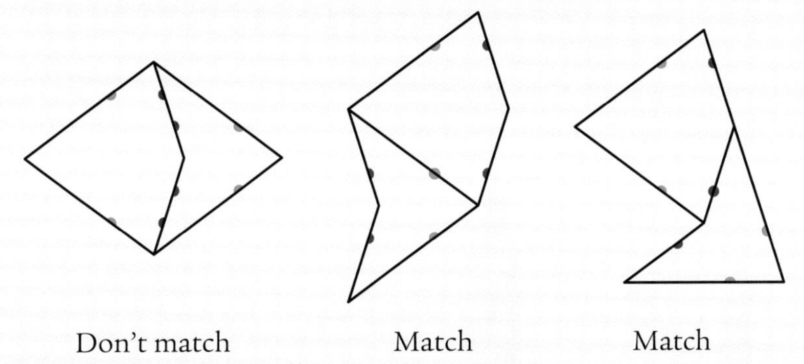

Don't match Match Match

They can be fitted together to make a tiling with rotational symmetry ... or reflective symmetry

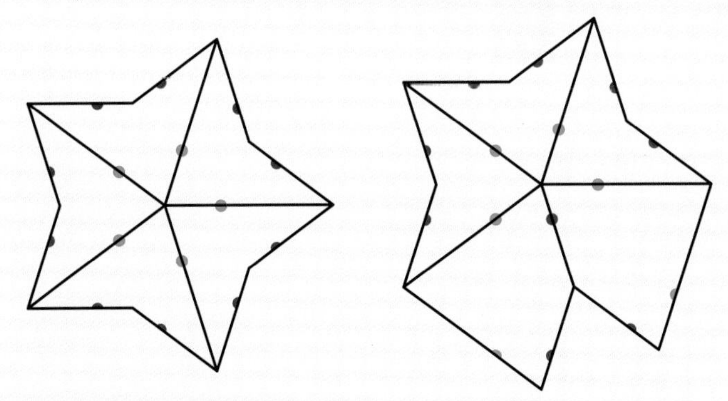

... but they cannot be fitted to make translational symmetry.

a Find the missing angles in the kite and V-kite.

b Investigate the way they fit together.

4.3 Enlargement and reduction

A biologist will look at an enlarged diagram of an insect to study detail.

A geographer will look at a map, a reduced image, to see the whole picture.

These scale drawings are something with which you are familiar.

When we study these changes of scale, we treat them as the same type of transformation ... just that the scale factor is different.

Just as rotation needed a centre, so does **enlargement**.

During the Renaissance, the artists used this transformation to create the illusion of depth and distance through perspective.

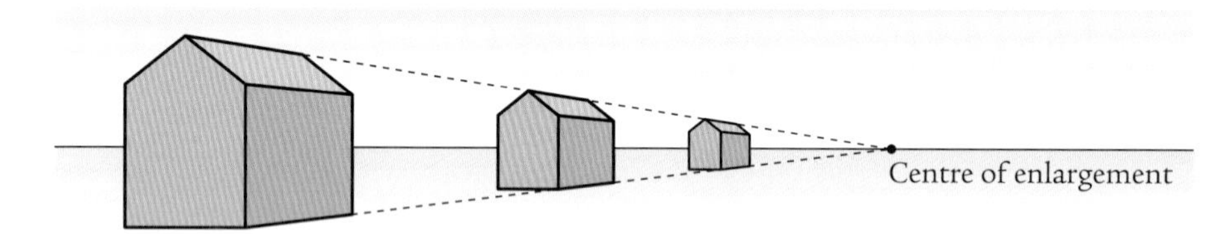

Each point on a shape is a certain distance from the centre.

If you double the distance of each point from the centre, each length in the shape doubles ... you get an enlargement scale factor 2.

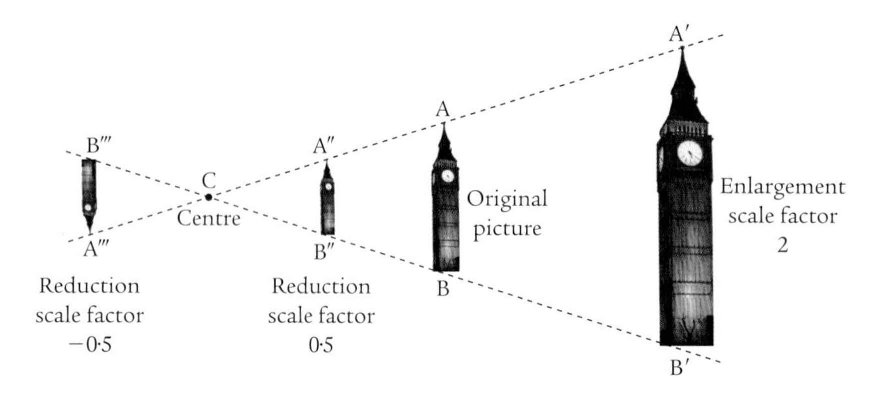

A photo of the Elizabeth Tower in London is 2 cm tall (AB).

The distances CA and CB are measured.

- If we extend these lines for twice the distance so that CA′ = 2CA and CB′ = 2CB, we get a tower twice as tall, viz. 4 cm.
- If we mark points A″ and B″ on the line so that CA″ = 0·5CA and CB″ = 0·5CB, we get a tower half as tall, viz. 1 cm.
- Taking going to the left as negative, we can even see the effect of a negative factor:

  CA‴ = −0·5CA and CB‴ = −0·5CB

  and we get a tower half as tall, viz. 1 cm ... which is upside down.

Example 1

Enlarge this shape by a scale factor of $\frac{3}{2}$.

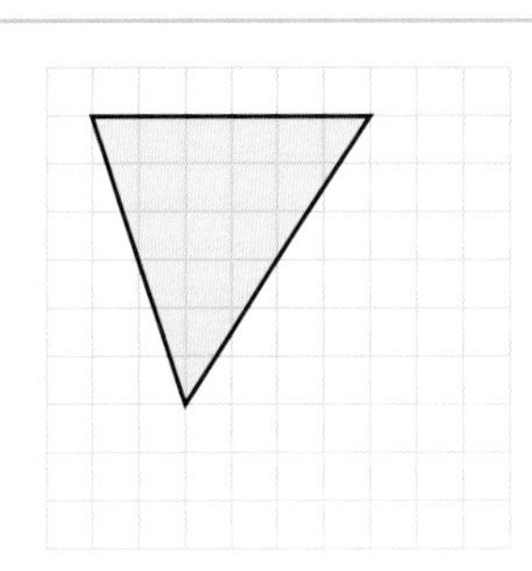

Select some point within the shape to act as a centre of enlargement.

Use it as the origin.

The coordinates of the point can then be multiplied by the scale factor $\frac{3}{2}$ to find its image under the enlargement.

For example:

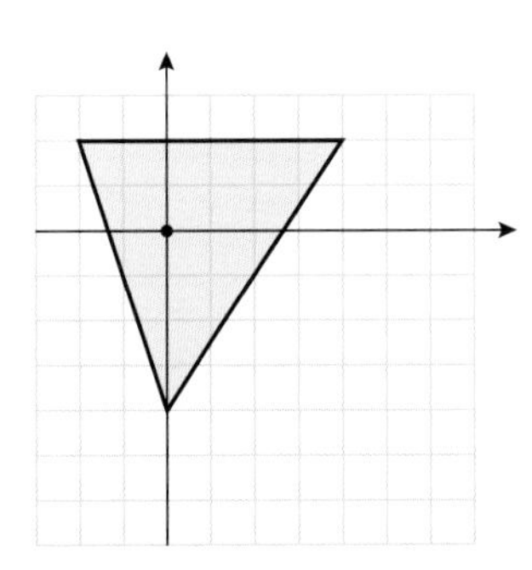

$(4, 2) \rightarrow (6, 3)$

$(0, -4) \rightarrow (0, -6)$

$(-2, 2) \rightarrow (-3, 3)$

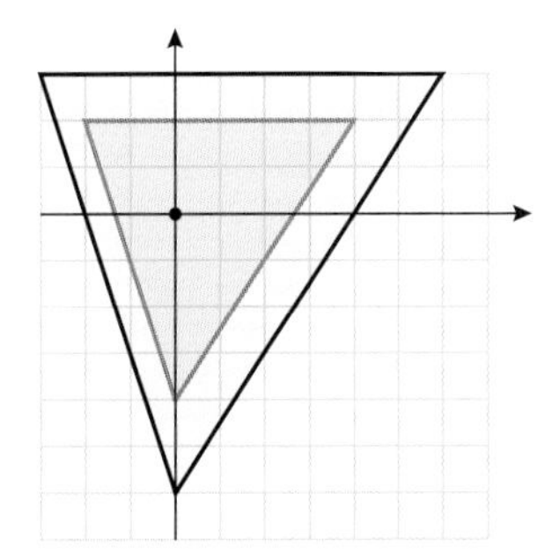

Example 2

A class task in the technical department is to design and build a bird box.

The class start by drawing a model with scale factor $= \frac{1}{2}$.

Distances shown are in cm.

What size will the front panel of the scale model be?

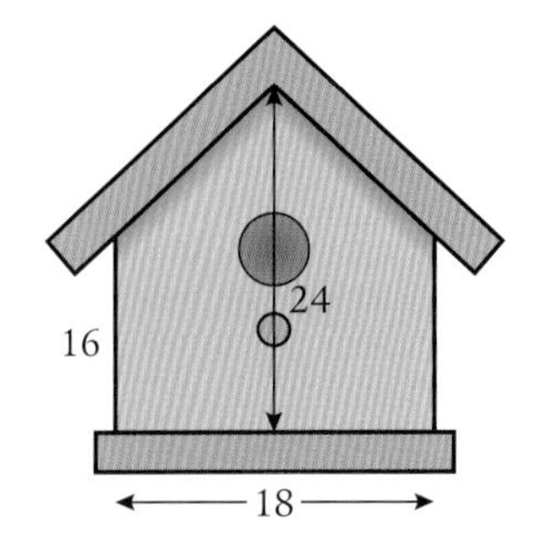

Each dimension of the actual bird box has to be multiplied by the scale factor $\frac{1}{2}$ to get the new length.

Width $= 18 \times \frac{1}{2} = 9$ cm

Height $= 24 \times \frac{1}{2} = 12$ cm

Height of side $= 16 \times \frac{1}{2} = 8$ cm

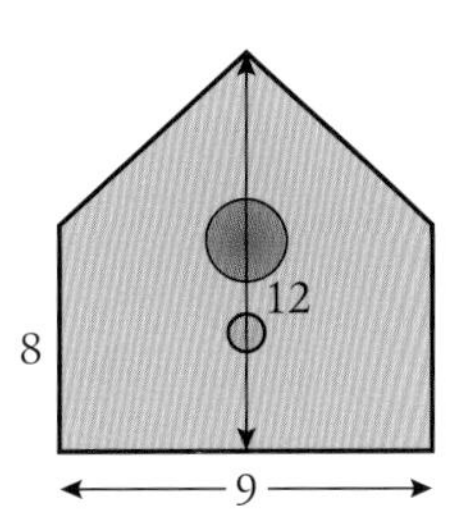

Example 3

Enlarge this shape by a factor of 1·5 using the centre of enlargement C.

Imagine C is the origin and list the coordinates of the control points.

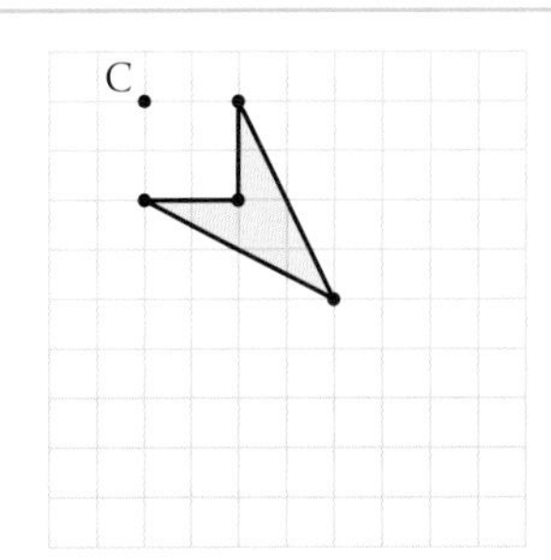

Multiply each coordinate by the scale factor to get the required images.

Join these images to get the enlargement.

$(2, 0) \rightarrow (3, 0) \qquad (1{\cdot}5 \times 2, 1{\cdot}5 \times 0)$

$(2, -2) \rightarrow (3, -3)$

$(0, -2) \rightarrow (0, -3)$

$(4, -4) \rightarrow (6, -6)$

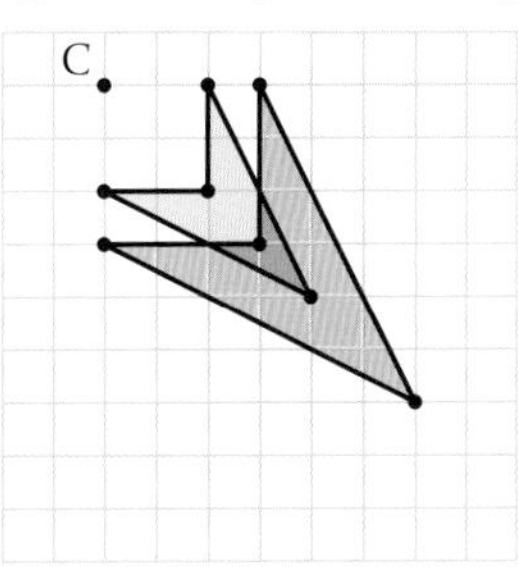

Exercise 4.3A

1 Enlarge these shapes by the scale factor shown, choosing your own centre.

a

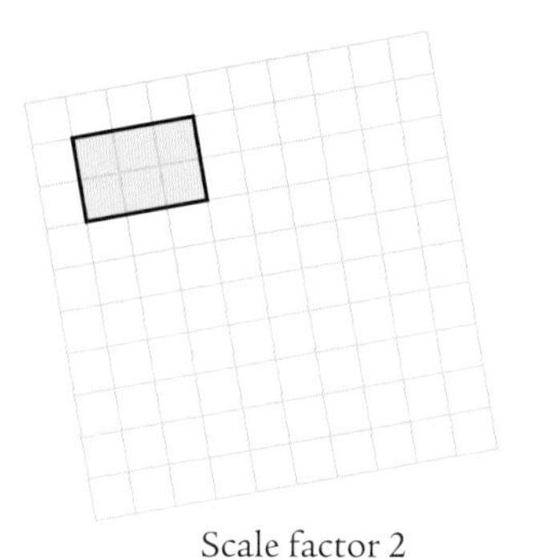

Scale factor 2

b

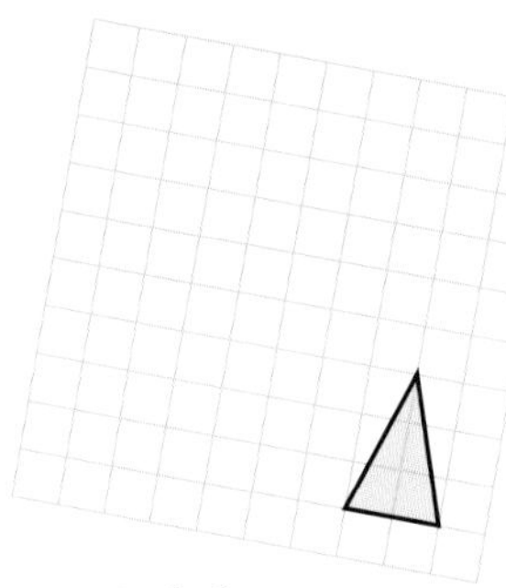

Scale factor 3

c

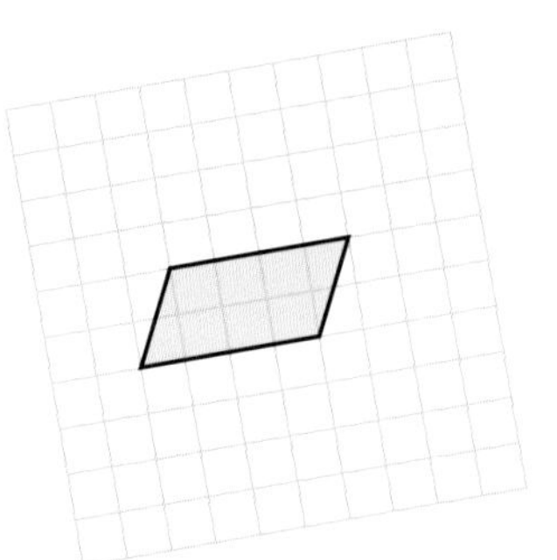

Scale factor 1·5

2 Enlarge these shapes using the given centre and scale factor.

a

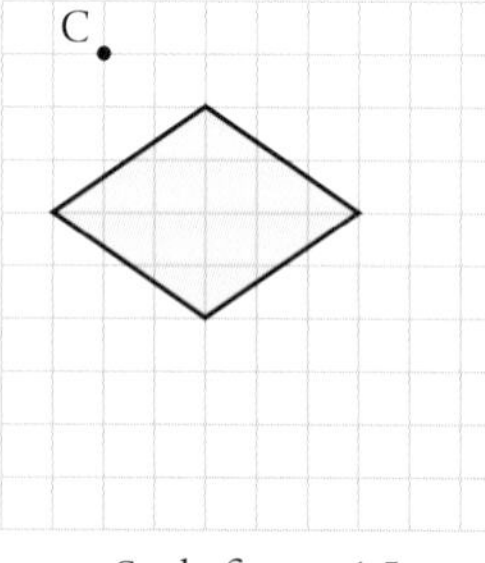

Scale factor 1·5

b

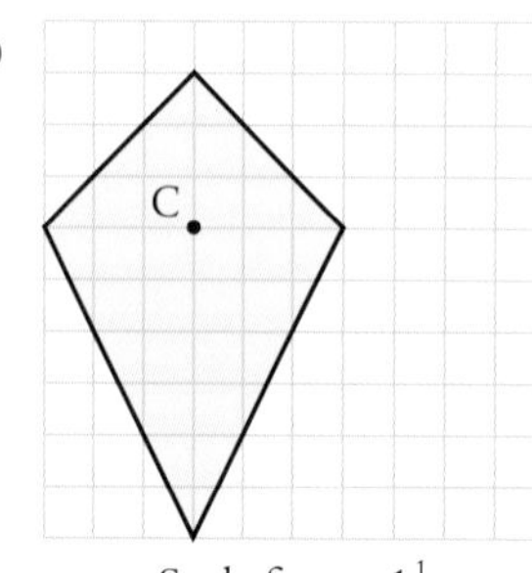

Scale factor $1\frac{1}{3}$

c

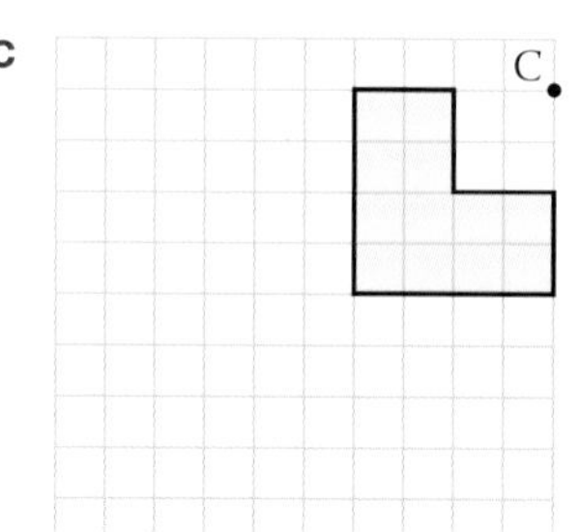

Scale factor 2·5

3 Reduce these shapes by the scale factor shown.

a

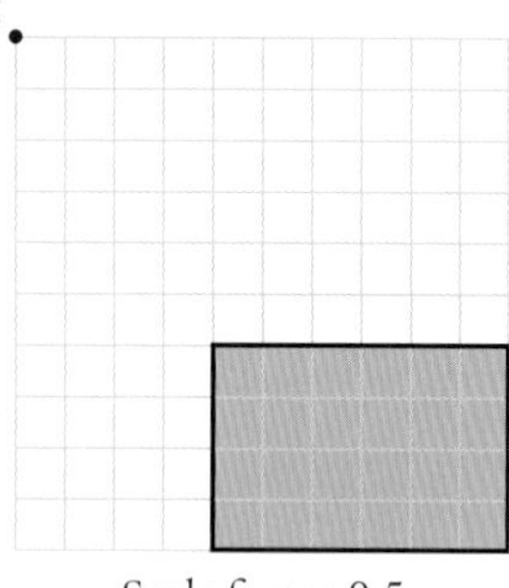

Scale factor 0·5

b

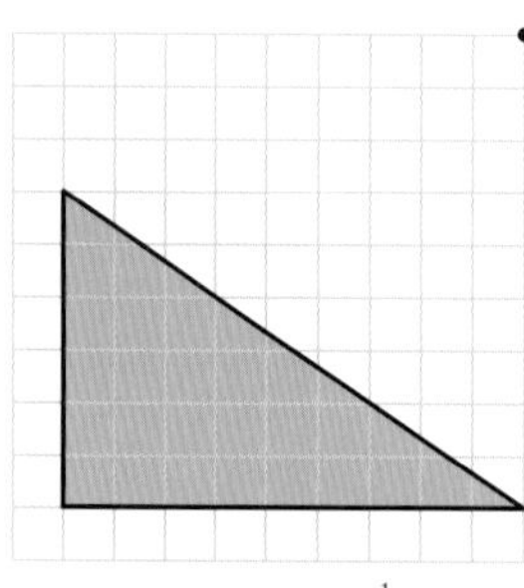

Scale factor $\frac{1}{3}$

c

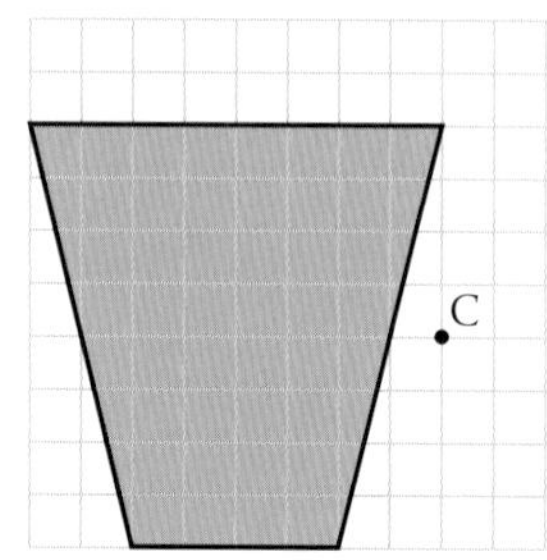

Scale factor 0·25

4 Here the scale factors are negative.

Draw the shape and its image using the centre and scale factor given.

a

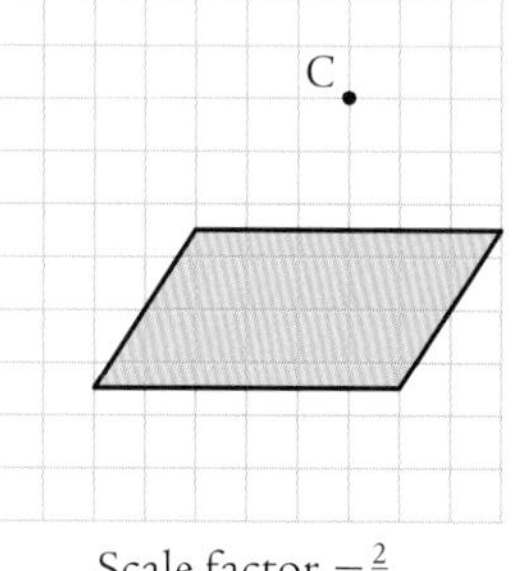

Scale factor $-\frac{2}{3}$

b

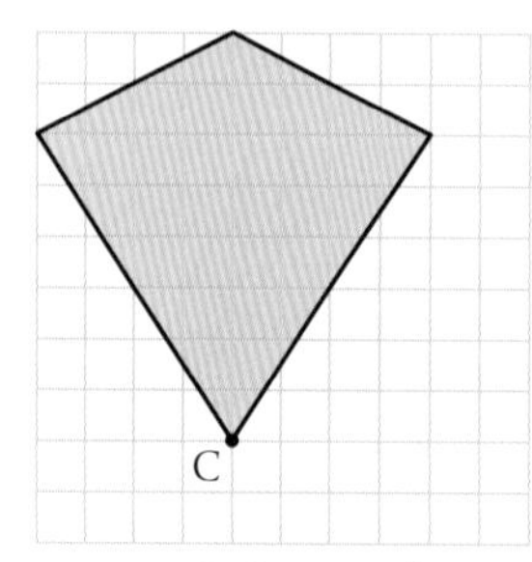

Scale factor $-\frac{3}{4}$

c

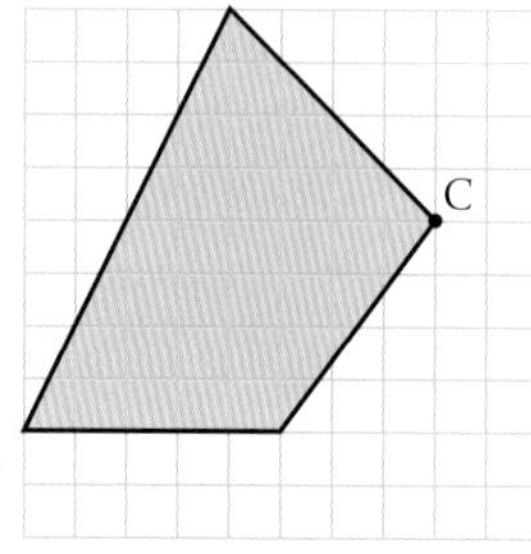

Scale factor −0·25

5 In the Peruvian Desert there are gigantic drawings of birds, monkeys and other animals.

Archaeologists still argue about how such large drawings could have been made 200 metres across in AD 400.

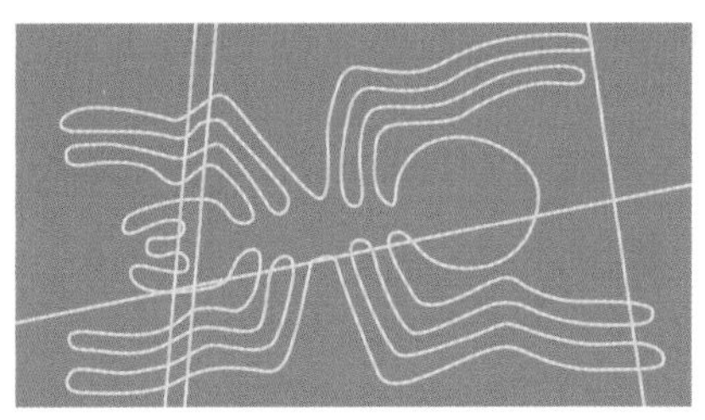

Perhaps people used the method above ... drawing a small animal and enlarging it using a centre and control points.

Examine how easy this is by making a cat twice as big using the origin as the centre of enlargement.

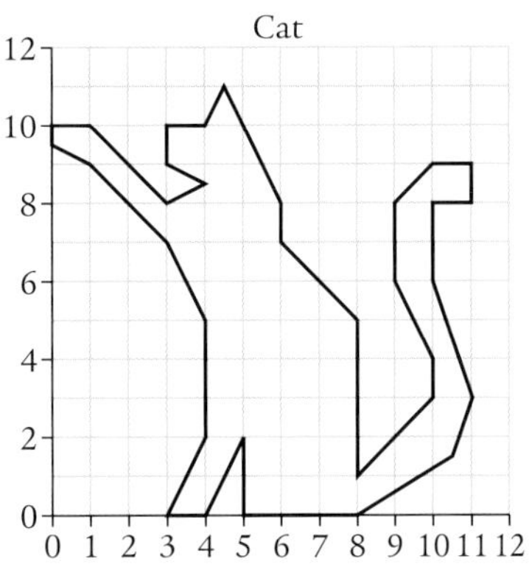

6 Mrs Lawson is designing a kitchen for a doll's house to include a washing machine, fridge, dishwasher and kitchen cabinets.

She uses a reduction scale factor of $\frac{1}{10}$ in the plan.

a The kitchen is 4·2 metres by 3·35 metres.

What will be the size of the kitchen in the doll's house? (Give your answer in centimetres.)

b The fridge she has picked is 1700 mm tall. How tall will the fridge be in the model?

c The washing machine is 600 mm wide and 850 mm deep. What size will the scale drawing be?

7 In Physics they study the pin-hole camera.

Rays of light pass through the pin-hole from an object outside the box and a smaller image is produced inside the box.

A man who is 1·8 m tall has an inverted image of 9 cm.

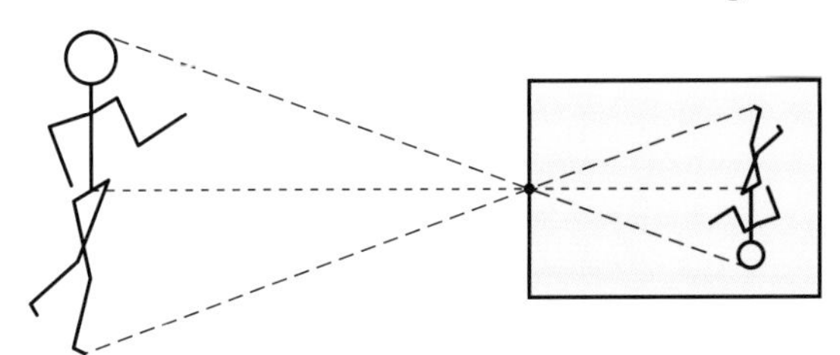

a What is the scale factor produced by the camera?

b The man is 3 m from the pin-hole. How far from the pin-hole is the image?

c Beside him there is a child of height 0·96 metres.
Describe its image.

d The window behind the man has an inverted image of height 12 cm.
What is the height of the actual window?

Inverted means 'upside down'.

Exercise 4.3B

1 A jeweller designs a necklace and matching earrings.

The motif on the earrings is a reduction of the one on the necklace.

A reduction factor of $\frac{1}{3}$ is used.

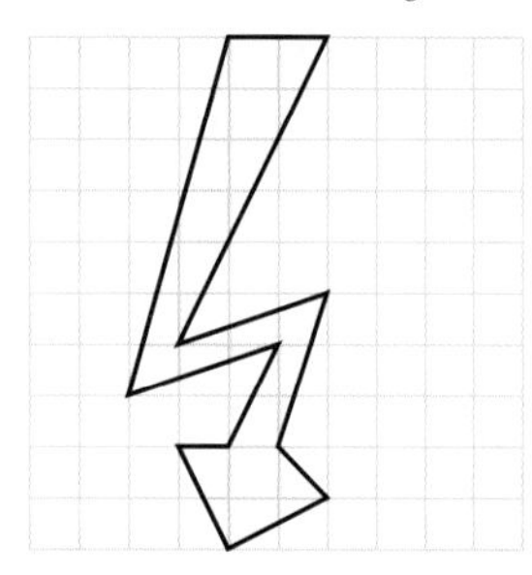

a What is the enlargement factor that scales the earring to the necklace?

b The diagram shows the size of the earring motif. Copy this onto squared paper.

Draw the necklace motif on the same paper.

2 Mr Evans the Geography teacher shows a map of Scotland on his whiteboard.

He explains that the projector has enlarged the original map by a scale factor of 5.

a Chris comes out to the board to measure the distance between Glasgow and Edinburgh and finds it is 40 cm.
What would the distance be on the original map?

b The map has a scale of 1 : 1 000 000.
What is the actual distance between Glasgow and Edinburgh?

3 Consider the simple L-shape made by joining the points (2, 6), (2, 2) and (4, 2).

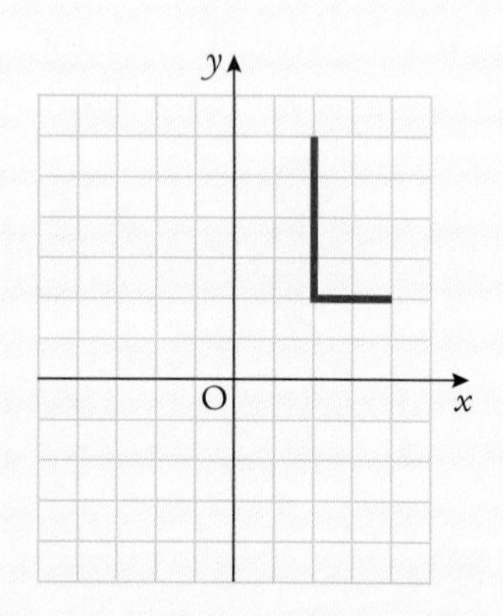

Take the origin as the centre of enlargement.

a List the coordinates of the images of the three points when the scale factor is:

i 2 ii 1·5 iii 0·5 iv 3 v −1 vi −0·5.

b By spotting patterns, devise formulae for transforming the coordinates of $P(x, y)$ into the coordinates of its image under an enlargement scale factor n where the origin is the centre of enlargement:

x-coordinate of image of $P(x, y) =$

y-coordinate of image of $P(x, y) =$

4 A5 paper is a scale reduction of A4 paper.

Cut a piece of A4 in half as shown to get two pieces of A5.

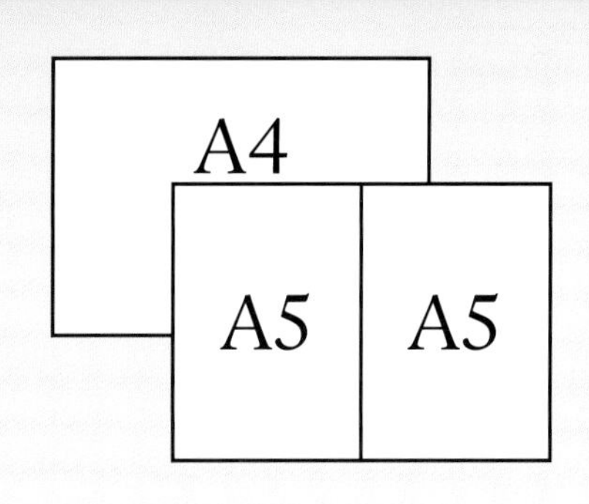

a Measure the length and breadth of:

i a piece of A4 paper

ii a piece of A5 paper.

b Calculate the enlargement factor that enlarges A5 to A4.

c Investigate the sizes of various standard paper sizes with a special regard to enlargement.

5 Using A as the centre of enlargement, △ABC has been enlarged to △ADE.

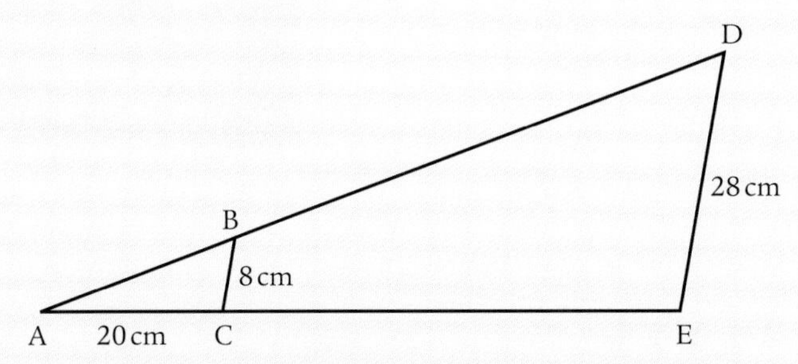

a Calculate the enlargement factor.

b Given that AC = 20 cm, calculate the length of AE.

c Given that AD = 77 cm, calculate the length of BD.

6 Enlargement is a transformation.

Does it have a symmetry associated with it?

Can you enlarge something so that it looks the same as it did before the enlargement?

The computer equivalent to enlargement and reduction is 'zoom in' and 'zoom out'.

If you zoom in by just the right amount into this picture, it will look unchanged.

Investigate 'self-similarity' on the internet.

Preparation for assessment

1 Cross stitch embroidery is used to make pictures, cards and other craft objects.

Cross stitch patterns use symbols in the square where a stitch should be made.
Different symbols are used for different colours.

Corrine decides to embroider Christmas trees to make into cards or gift tags.

Copy and complete the pattern so that the diagram has reflection symmetry about the vertical line.

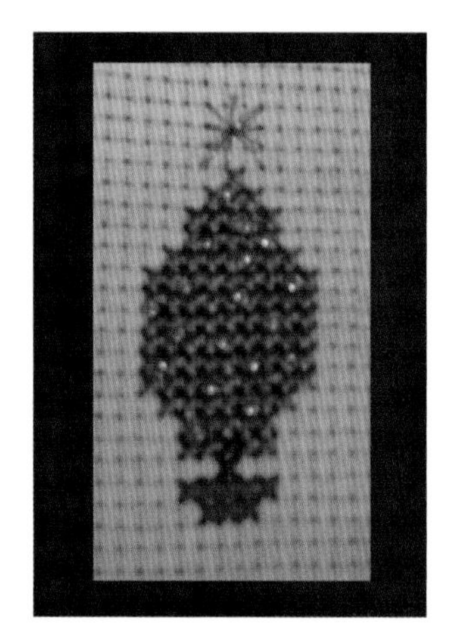

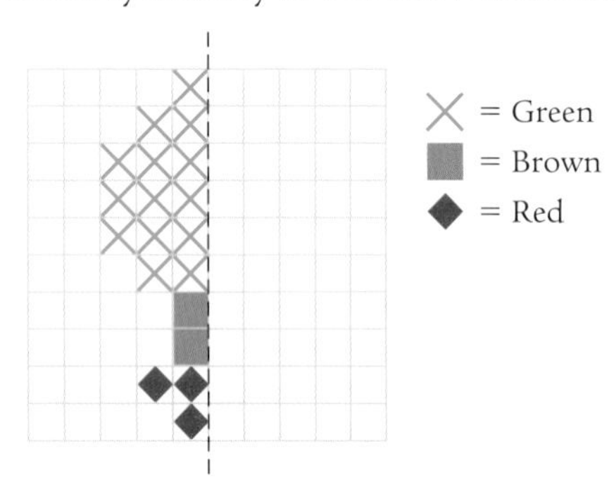

2 Describe the symmetries to be found on each face of a normal dice.

3 Copy the diagram.
Reflect the second quadrant contents into the first.

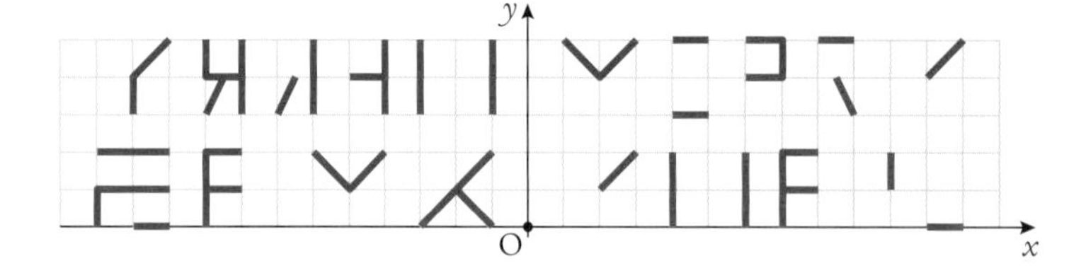

4 A staircase has a gradient of 0·6.

a Write this gradient as a common fraction.

b The riser in these stairs is 15 cm.
What is the size of the tread?

c There are 14 steps from the bottom to the top.
Describe the translation that maps the bottom step to the top one.

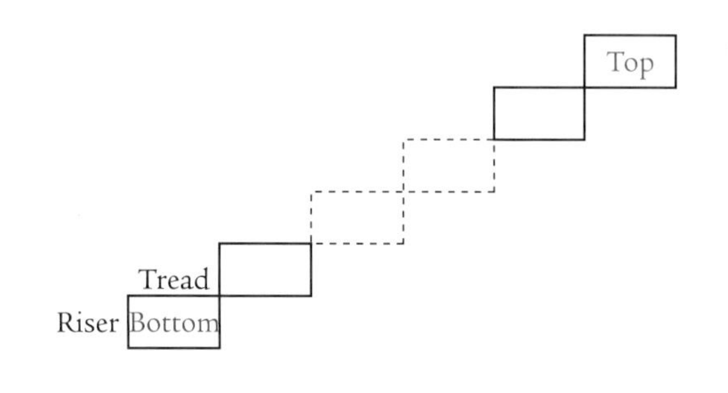

5 The image of the red triangle after a rotation is the blue triangle.

a Trace the design onto paper.

b Find the centre of rotation.

c What size and in what direction is the rotation?

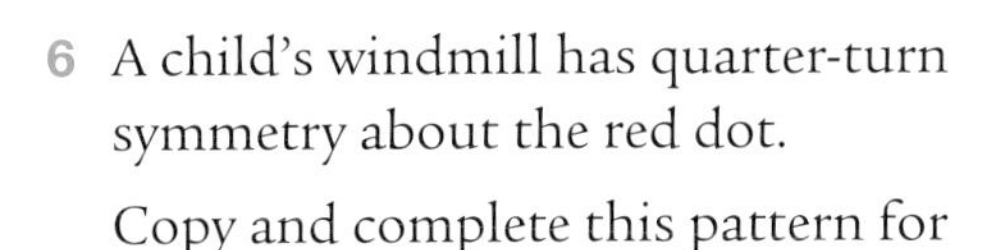

6 A child's windmill has quarter-turn symmetry about the red dot.

Copy and complete this pattern for the windmill.

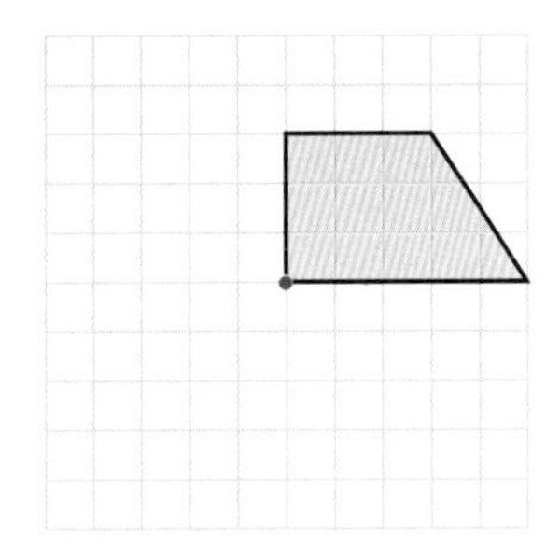

7 A triangle has vertices A(1, 3), B(3, −2) and C(−4, −3).

State the coordinates of the images of these vertices under:

a a half turn about the origin

b a quarter turn clockwise about the origin

c a translation of 3 to the right and 1 down

d a reflection in $y = x$

e an enlargement scale factor 2 about the origin.

8 a Show the images of the V-kite under enlargement about the origin with:

i scale factor 2

ii scale factor −2.

b How else might the enlargement about the origin with scale factor −1 be described?

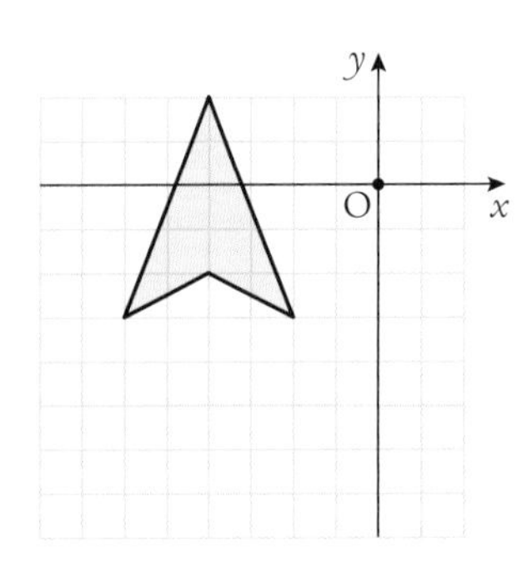

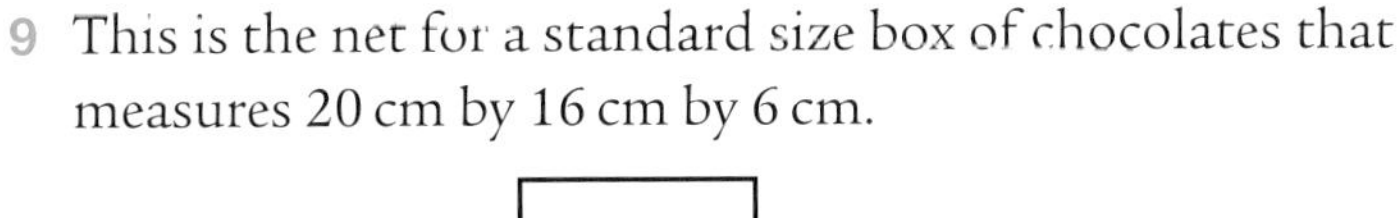

9 This is the net for a standard size box of chocolates that measures 20 cm by 16 cm by 6 cm.

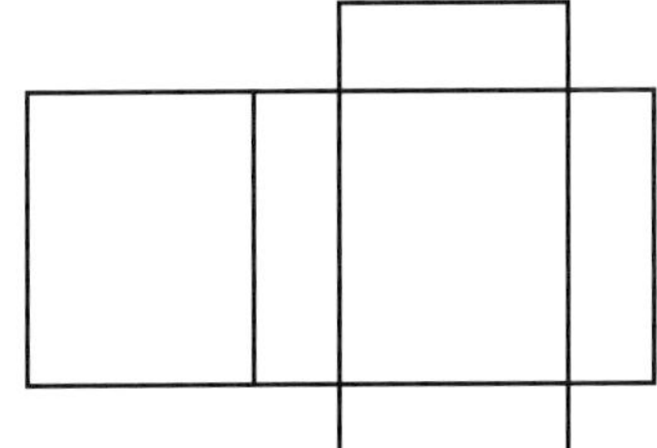

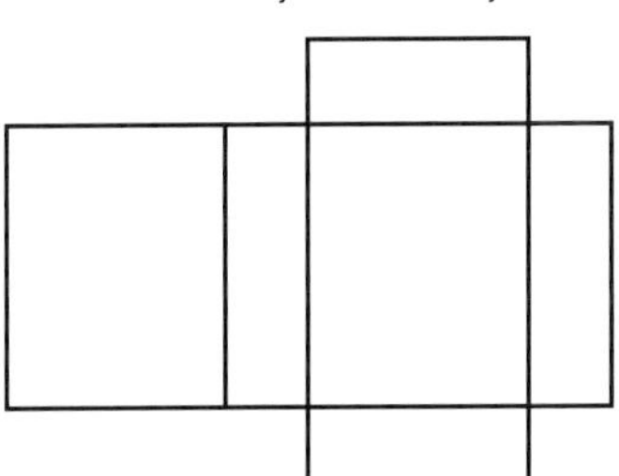

a Sketch the net and mark in the sizes of its edges.

b The manufacturer wants a bigger box and enlarges the design of the small one by a factor of 1·5.

What will be the dimensions of the new box?

c By what factor has the area of card needed increased?

d By what factor has the volume increased?

10 Remember the cat?

Give a computer the coordinates and it will draw the picture.

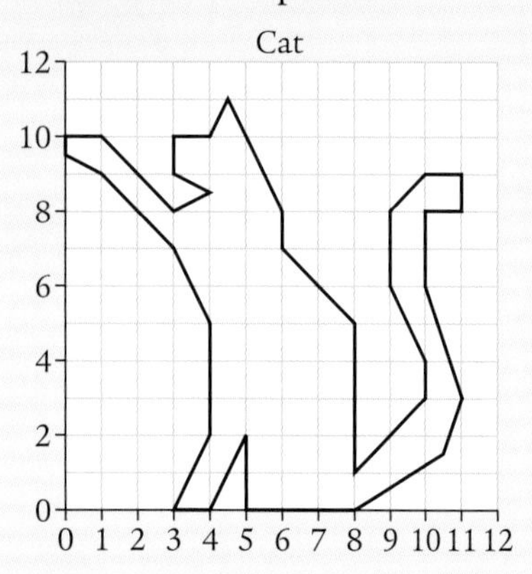

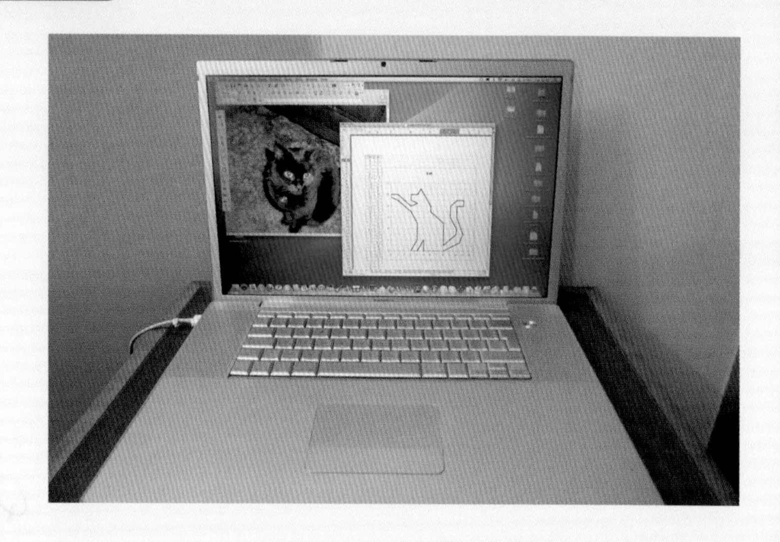

The cat above has been drawn using the set:

(3, 0), (4, 0), (5, 2), (5, 0), (8, 0), (10·5, 1·5), (11, 3), (10, 6), (10, 8), (11, 8), (11, 9), (10, 9), (9, 8), (9, 6), (10, 4), (10, 3), (8, 1), (8, 5), (6, 7), (6, 8), (5, 10), (4·5, 11), (4, 10), (3, 10), (3, 9), (4, 8·5), (3, 8), (1, 10), (0, 10), (0, 9·5), (1, 9), (3, 7), (4, 5), (4, 2), (3, 0).

Using formulae we can transform the cat's coordinates to make the computer ...

- reflect it in the y-axis
- reflect it in the x-axis
- turn it upside down
- rotate it anticlockwise through 90°.

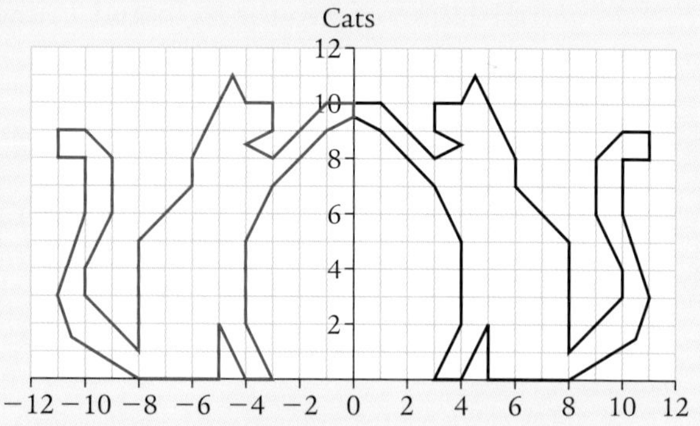

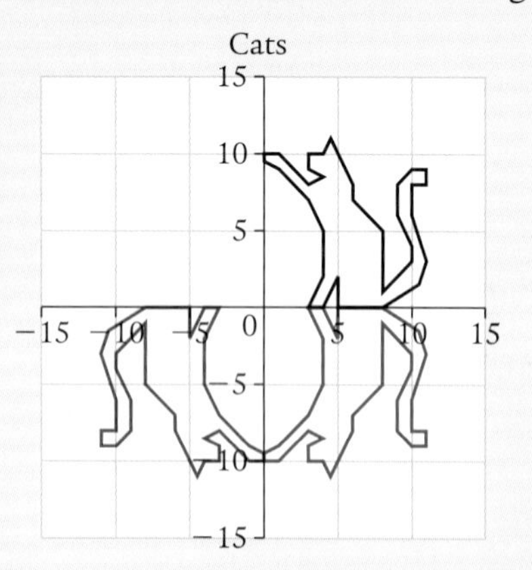

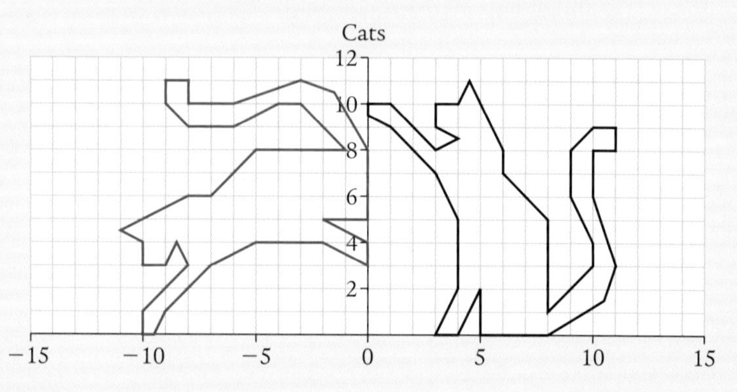

Can you find the formulae that do these jobs?

5 Summary statistics

Before we start...

The aim of the Formula 1 team is to improve their consistency for 'pit stop times'.
Although they are keen to have a quick pit stop, they are more concerned with being **consistent** with an average time of 3 seconds.

Last month's pit stop practice times (seconds):

3·4	2·8	5·9	3·3	2·8	3·0	2·6	2·8	2·3	2·9	4·0	3·4
3·0	2·5	2·9	3·0	3·1	2·7	3·0	4·2	3·7	2·5	3·1	3·1

This month's pit stop practice times (seconds):

3·0	3·7	4·2	2·9	2·8	3·0	2·7	2·8	2·9	2·7	3·2	3·7
3·2	2·5	2·6	2·8	3·0	2·8	3·0	3·0	3·4	3·2	3·4	3·1

Can the team claim a consistent average time of 3 seconds?

What you need to know

1 The speeds (mph) of 20 vehicles were recorded at a particular stretch of road:

31, 32, 28, 30, 31, 38, 35, 29, 32, 30, 28, 32, 29, 31, 30, 35, 38, 31, 29, 37.

Draw a frequency table for the data.

2 Put the following numbers in numerical order from lowest to highest:

23·1 0·768 5·01 5·006 4·2 143·1

3 Calculate the following, without the aid of a calculator:

a 20·3 + 23·6 + 15·9 + 21·7 + 24·8 + 22·5 + 19·8

b $\frac{365{\cdot}98}{9}$, rounding your answer to 2 decimal places.

5.1 Central tendency

I just scored 99! ... Good.

Not so good. Some others scored 200. ... Bad.

Not so bad. The average score was 79. ... Good.

Not so good... I was playing golf!

If you don't provide a context for data, then that data has no meaning.

It's useful to give a 'typical' score, the range of marks possible and a storyline before deciding whether a particular piece of data is 'good' or 'bad'.

Information such as highest score, lowest score, typical score, range of scores, etc. are known as **summary statistics**.

We will consider two types of summary:

- measures of **central tendency** – these provide a guide as to what is 'typical' ... an **average**
- measures of **spread** – these give an indication of the **range** of the data.

Measures of central tendency

Average: mean, median and mode

We consider three different measures of average: the **mean,** the **median** and the **mode.**

Which one we use depends on which one gives the best measure of what is a typical score.

Mean: This is what most people refer to as the 'average'.

$$\text{Mean} = \frac{\text{total of pieces of data}}{\text{number of pieces of data}}.$$

Median: This is the 'middle' number when the data or measurements are put in numerical order.

$$\textbf{Position} \text{ of the median} = \frac{\text{number of pieces of data} + 1}{2}.$$

If this turns out to be, say, 11·5, then the median lies between the 11th and 12th scores.
We then use the mean of these two scores as the median.

Mode: This is the piece of data that occurs most often, i.e. has the highest frequency.

This is the only one of the three averages you can use if the data is not numerical.

For our purposes, if more than one score holds this honour we say the data set doesn't have a mode.

Example 1

A group of 11 students is asked how long it takes them, in minutes, to get to school.

They give the following answers:

12, 12, 10, 2, 30, 15, 22, 25, 8, 6, 12.

For this set of data:

a calculate the mean time to get to school

b find the median

c identify the mode.

a $\text{Mean} = \dfrac{\text{total of pieces of data}}{\text{number of pieces of data}}$

$\text{Mean} = \dfrac{(12 + 12 + 10 + 2 + 30 + 15 + 22 + 25 + 8 + 6 + 12)}{11}$

Mean = 14 minutes

Mean time to walk to school for this group of students is 14 minutes.

b First arrange the data in numerical order.

There are 11 pieces of data so the 'middle' is the 6th piece of data [(11 + 1) ÷ 2 = 6].

2, 6, 8, 10, 12, 12, 12, 15, 22, 25, 30

median = 12 minutes

c The mode is 12 minutes as this is the answer that occurs most often.

Example 2

Here are the qualifying times, in seconds, from the Women's Individual Pursuit at Beijing in 2008.

221 215 208 224 212 220 218 226 209 216 214 227

For this set of data:

a calculate the mean

b find the median

c identify the mode.

a $\text{Mean} = \dfrac{(221 + 215 + 208 + 224 + 212 + 220 + 218 + 226 + 209 + 216 + 214 + 227)}{12}$

Mean = 217·5

b First put the data in order:

208 209 212 214 215 216 218 220 221 224 226 227

There are 12 pieces of data ... (12 + 1) ÷ 2 = 6·5,

so the middle value lies between the 6th and 7th number.

$\text{Median} = \dfrac{(216 + 218)}{2} = 217$

c There is no mode for this set of data.

Example 3

The mean foot-length of a class of 10 students is 250 mm.
Another student joins the class and the mean goes up to 255 mm.

a What is the new student's foot-length?

b A news report said that the average size of a woman's foot is 24·5 cm and of a man's foot is 27 cm. From your answer to part **a**, make a statement about the new student.

a The total foot-length for 10 students is:

mean $\times$ number of students $= 250 \times 10 = 2500$ mm.

The total for 11 students is:

mean $\times$ number of students $= 255 \times 11 = 2805$ mm.

The foot-length of the new student is the difference between the two totals:

$2805 - 2500 = 305$ mm.

The new student has a foot-length of 305 mm.

b The student is more than likely a male as their foot-length is above the average for a male.

Exercise 5.1A

1 Find: **i** the mean **ii** the median and **iii** the mode for each list of numbers.

a 90, 92, 93, 88, 95, 88, 97, 87, 98

b 2·3, 3·4, 2·4, 5·6, 3·8, 4·4

c 134, 220, 169, 268, 321, 220, 220, 170

2 The times of the first six finishers in the men's category of The Loch Ness Marathon were:

2·3 hours, 2·35 hours, 2·42 hours, 2·72 hours, 2·82 hours, 2·83 hours.

a Calculate: **i** the mean **ii** the median for the first six runners.

b For all the competitors in the men's category, the mean = 4·17 hours and the median = 4·07 hours.

i Compare the mean and median for the first six with those for all runners.

ii Why are they so different?

3 While on a caravan holiday round Scotland the Simpson family noted the fuel price per litre in each area they visited.

Inverness – 128·9p, Aberdeen – 124·9p,
Edinburgh – 126·9p, Perth – 127·7p, Peebles – 132·9p,
Stornoway – 146·9p, Glasgow – 127·7p

a **i** Where was the most expensive place to get fuel?

ii Why do you think it is so expensive there?

b Calculate the mean, median and mode of the fuel prices.

c **i** Excluding the most expensive place, calculate the mean, median and mode.

ii Which of the three measures of central tendency was affected by not considering the most expensive place?

4 A B&B recorded the number of guests who stayed each night during the month of May.

	May						
	M	T	W	T	F	S	S
		1	2	3	4	5	6
Number of guests		**2**	**2**	**4**	**3**	**5**	**2**
	7	8	9	10	11	12	13
Number of guests	**5**	**1**	**1**	**3**	**4**	**4**	**2**
	14	15	16	17	18	19	20
Number of guests	**0**	**1**	**3**	**3**	**3**	**3**	**0**
	21	22	23	24	25	26	27
Number of guests	**2**	**2**	**3**	**3**	**6**	**6**	**6**
	28	29	30	31			
Number of guests	**2**	**3**	**5**	**5**			

a What was the maximum number of guests who stayed in the B&B in any one night?

b Calculate the mean, median and mode of the number of guests per night for May.

5 In Olympic diving events, seven judges each award a mark out of 10.

The highest and lowest marks are ignored and the mean of the other five is calculated to provide a score.

Competitor	Judge 1	Judge 2	Judge 3	Judge 4	Judge 5	Judge 6	Judge 7
Cameron	8·2	7·9	6·9	8·4	7·8	8·5	8·0
James	8·0	6·7	9·2	7·0	8·1	7·5	8·0

Who had the highest score?

6 A student obtains marks of 87, 95, 76 and 88 for his assessments in the first year of his course.

In order to pass this year's course he needs to obtain a mean score of 85 or better.

What is the minimum mark he must get in the final assessment in order to pass?

7 Ten people work in a factory. Their rates per hour are:

£3, £4, £5, £6, £7, £8, £9, £20, £20, £100.

a Calculate: **i** the mean **ii** the median **iii** the modal rate.

b Which of these three do you think reflects the typical wage rate at the factory?

8 In a study of the common cold, two questions were asked of 20 children:

A For how many days did your symptoms last?

B How many colds do you get per year?

Question A				
10	10	7	9	10
7	9	8	10	8
8	9	19	9	18
8	9	9	7	7

Question B				
7	6	9	6	6
9	11	12	10	6
7	7	7	11	12
10	6	12	6	8

a What is the typical duration of a cold in the group?

b How many colds does a child in the group typically catch in a year?

9 Different liquids have different thicknesses or viscosity.

In a science experiment the viscosity of a liquid can be compared to water by timing how long it takes a marble to sink to the bottom of a column of both liquids.

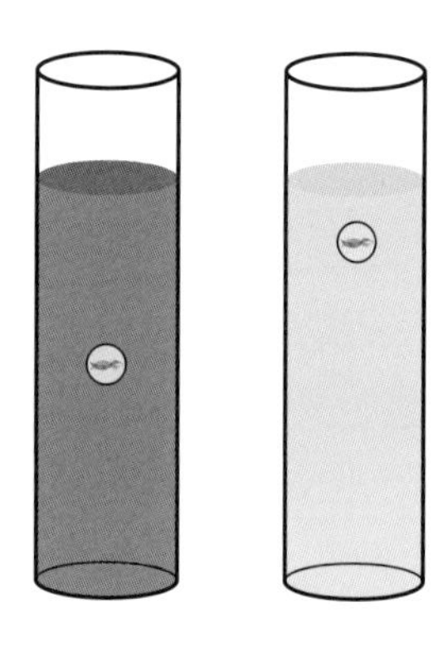

Water (tenths of a second)				
7	8	6	7	6
8	6	8	8	6
9	6	9	9	7
9	6	6	7	7

Oil (tenths of a second)				
14	15	12	12	11
13	14	11	13	12
12	15	14	12	12
15	12	13	14	13

a Find the mean times for the marble to drop down in both liquids.

b Comment on which liquid is thicker.

c Would the result of the experiment be different if the modes had been compared instead?

Calculating the mean, median and mode from a frequency table

Data are often organised in a frequency table.

The definitions for the mean, median, and mode are exactly the same, but we need to think carefully about what the table is telling us.

Example 1

A dice is thrown 20 times and the results are recorded in this table:

Value	Frequency
1	3
2	5
3	2
4	4
5	3
6	3

Calculate: **a** the mean **b** the median **c** the mode of the scores.

The table has been generated from this list of data:

1, 1, 1, 2, 2, 2, 2, 2, 3, 3, 4, 4, 4, 4, 5, 5, 5, 6, 6, 6.

a To calculate the **mean** you would add the list up and divide by 20.

The total would be equal to:

three 1s + five 2s + two 3s + four 4s + three 5s + three 6s ...

The quickest way to add this up is to add an extra column to the table:

Value, x	Frequency, f	$x \times f$
1	3	$1 \times 3 = 3$
2	5	$2 \times 5 = 10$
3	2	$3 \times 2 = 6$
4	4	$4 \times 4 = 16$
5	3	$5 \times 3 = 15$
6	3	$6 \times 3 = 18$
	20	**68**

The sum of the frequency column gives us **the number of throws**.

The sum of the 'value × frequency' column ('$x \times f$') gives us **the total of the scores**.

The mean can then be calculated by:

$$\text{mean} = \frac{\text{total of scores}}{\text{number of scores}} = \frac{\text{total of } (x \times f) \text{ column}}{\text{total of } f \text{ column}}$$

$$\text{mean} = \frac{68}{20}$$

$$\text{mean} = 3{\cdot}4$$

The mean of the scores is 3·4.

b The **mode** is the value which occurs most often.

The score of 2 has the highest frequency, namely 5.

The modal score is 2.

c The **median** for 20 sets of data would lie in the $(20 + 1) \div 2 = 10{\cdot}5$th position, i.e. between the 10th and 11th value.

The 10th value was 3 and the 11th value was 4, so the median is $(3 + 4) \div 2 = 3{\cdot}5$.

Example 2

The modern pentathlon consists of fencing, swimming, riding and finally a combined shooting and running event. In the fencing event, each athlete fights every other athlete once. Each competitor has 1 minute to score a single hit.

The table opposite shows the number of wins each athlete scored in a recent fencing event.

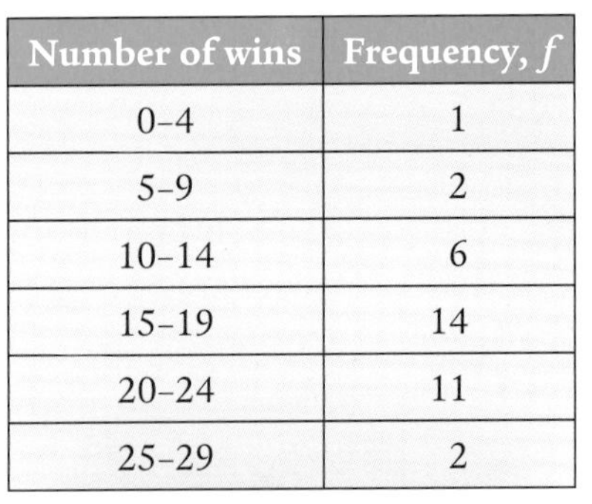

Number of wins	Frequency, f
0–4	1
5–9	2
10–14	6
15–19	14
20–24	11
25–29	2

a How many athletes took part in total?

b Write down the modal class interval.

c Which class interval contains the median?

d Estimate the mean number of wins scored.

Add two columns to the table:

- the mid-value of the class intervals (to represent the class)
- an $(f \times x)$ column to help you **estimate** the total of the scores.

Number of wins	Frequency, f	Mid-value, x	$f \times x$
0–4	1	2	$1 \times 2 = 2$
5–9	2	7	$2 \times 7 = 14$
10–14	6	12	$6 \times 12 = 72$
15–19	14	17	$14 \times 17 = 238$
20–24	11	22	$11 \times 22 = 242$
25–29	2	27	$2 \times 27 = 54$
Totals	**36**		**622**

a The number of athletes in total = sum of frequency column = 36 athletes.

b The modal class is the class with the highest frequency ... modal class is 15–19 wins.

c There are 36 'scores' ... $(36 + 1) \div 2 = 18{\cdot}5$ ... the median lies between the 18th and 19th.

Start a running total of the frequencies

1 ... 1st score lies in the 1st class,

$1 + 2 = 3$ ... 3rd score lies in the 2nd class,

$1 + 2 + 6 = 9$ ... 9th score lies in the 3rd class,

$1 + 2 + 6 + 14 = 23$ ... 23rd score lies in the 4th class.

So both the 18th and 19th score lies in the 4th class.

So the median lies in the 4th class.

Use the mid-values to represent the class ... this will let you estimate the total number of wins... and hence to estimate the mean.

$$\text{mean} = \frac{\text{sum of } (f \times x) \text{ column}}{\text{sum of frequency column}}$$

$$\text{mean} = \frac{622}{36}$$

mean $= 17{\cdot}3$ wins per athlete (to 1 d.p.)

Exercise 5.1B

1 The RSPB in Scotland monitor the population of black grouse in the Cairngorm area over a year.

The number of eggs that were found in each nest was recorded.

Number of eggs, x	Frequency, f	$f \times x$
6	5	
7	8	
8	7	
9	10	
10	5	
11	5	

a How many nests were monitored?

b Find:

i the mean

ii the median

iii the mode

for the number of eggs in a nest.

c Last year the mean was 8·5 eggs per nest.

What can you say about the population of black grouse in the Cairngorms from this information?

2 A class collected data on their shoe sizes.

Shoe size, x	Frequency, f	$f \times x$
4	4	
5	3	
6	3	
7	5	
8	7	
9	8	

a How many students were in the class?

b What was the most common shoe size?

c Calculate the median shoe size.

d Calculate the mean shoe size.

e What fraction of the class had a shoe size less than the mean?

3 VisitScotland survey how many 'stars' each bed and breakfast establishment has in a particular neighbourhood.

Stars	Frequency
1	4
2	10
3	43
4	15
5	8

a Calculate the mean, median and mode of the results.

b What effect would ignoring the 1-star establishments have on these measures?

4 As part of an investigation into mobile phone usage, students in a class were asked to take a note of how many texts they sent in a day.

No. of texts, x	Frequency, f
0–19	6
20–39	8
40–59	9
60–79	4
80–99	2

a What is the modal class interval?

b In which class interval does the median lie?

c Copy and complete the table and use it to calculate the mean number of texts sent in a week.

No. of texts	Frequency, f	Mid-value, x	$f \times x$
0–19	6	9·5	$6 \times 9{\cdot}5 = 57$
20–39	8		
40–59	9		
60–79	4		
80–99	2		
Totals			

5 At a fitness club, the weights of the members were taken.
A histogram records the findings.

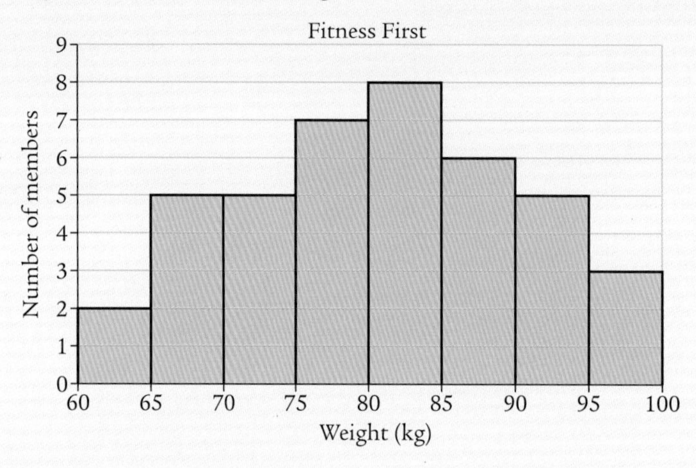

a What was the modal class?

b In which class interval did the median occur?

c Calculate the mean weight of the club members.

6 A science experiment is set up to investigate the reaction rates of sodium thiosulphate and potassium iodide. A cross is marked at the bottom of a beaker and, as the two chemicals react, the liquid in the beaker clouds and the cross can no longer be seen. The students time how long it takes for the cross to disappear.

These are the times in seconds:

45·3	46·5	47·3	48·2	48·4	48·7	48·6	48·5	48·9	49·1
48·1	49·2	48·0	49·0	47·7	48·8	46·8	50·5	50·1	49·8

a Arrange the figures in a frequency table using a suitable class interval.

b Use the mid-values of the class intervals to calculate the mean.

c What is the modal class interval?

5.2 Necessary notes

1 Data types

We deal with two forms of data, **qualitative** and **quantitative**.

Qualitative data

Qualitative data includes categories that are:

- not ordered, such as crisp flavours, named objects, attributes, etc. This type of data is called **nominal** data.
- ordered, such as a verbal scale, e.g. strongly disagree, disagree, have no opinion, agree, strongly agree. This type of data is called **ordinal** data.

Quantitative data

Quantitative data is numerical and includes:

- **discrete** data where there are gaps between values of the variable, e.g. shoe size, number of children in a family, number of moons round a planet, spanner sizes. With discrete data, if we declare one value as the 1st value then we can identify the 2nd, 3rd, 4th, etc.
- **continuous** data where between any two values of the variable you can always find another, e.g. length of foot, time of day, distance, weight. With continuous data, even if we declare one particular value as the 1st value, we find it impossible to identify the next bigger value.

2 Sample and population

We usually examine a **sample** because of the difficulty or expense of examining the whole lot – the **population**. We would test a sample of light bulbs to destruction to decide how long a light bulb lasts. It would be pointless testing all the bulbs that way, for then we'd have none to sell.

We make an assumption that what we discover about the sample, we can apply to the population.

We would like to treat the 'average' of the sample as an estimate of the 'average' of the population.

The bigger the sample size, the closer this is to being true.

3 Typical scores

How heavy is an apple?

A small apple weighs about 100 g, a large one is about 200 g and an average apple is around 150 g.

According to the *Guinness Book of Records*, the biggest apple ever weighed is 1·85 kg.

This, of course, is not what to expect of apples in general, and would be considered an **outlier** – not really worth considering if trying to determine what is typical.

The *Guinness Book of Records* is aimed at producing the 'wow! factor'.

Sampling apples

Using a computer simulation, we randomly 'picked 100 apples' in the range 100 g to 200 g.

We computed the mean, median and mode of our 100 apples.

We did this 10 times.

The table shows the mean, median and mode for each of the 10 samples.

The bottom row also gives the mean of each column.

Note how each of these agree around the 150 g mark.

Looking at how the three averages vary, we see that the mean is the least variable and the mode is the most variable as we select new samples.

	Mean	Median	Mode
	148·7	146	141
	152	147·5	147
	154·2	158	176
	150·6	151·5	131
	153·7	153	135
	148·5	147	140
	149·6	146·5	120
	149·7	150	194
	151	151·6	180
	154·6	156·5	156
Mean	151·26	150·76	152

Outliers are data values at the extremes of a group. They are non-typical of the group.

4 Which average?

The mean, median and mode are three measures of central tendency, but why have three different averages?

- The mean is the best when the data is continuous and 'symmetrical'.
- The mode is the best (only one appropriate) when the data is nominal.
- The median is the best when the data is 'skewed'
  ... or there are a few outliers in the list
  ... it could even be used when the data is ordinal.

Exercise 5.2A

1 Discrete and continuous data.

The bar chart is useful for displaying one type of data and the histogram for another.

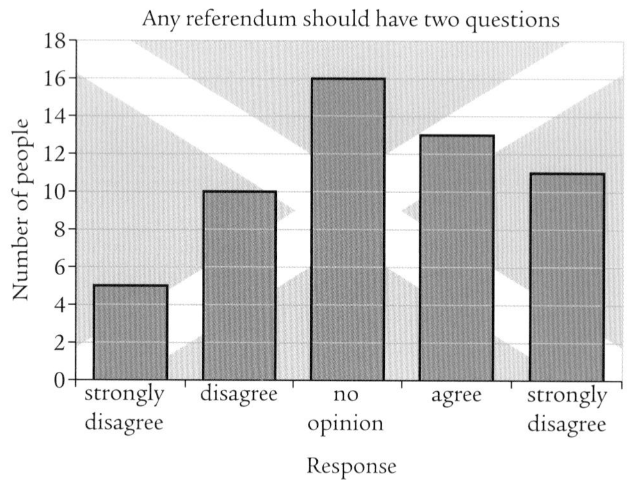

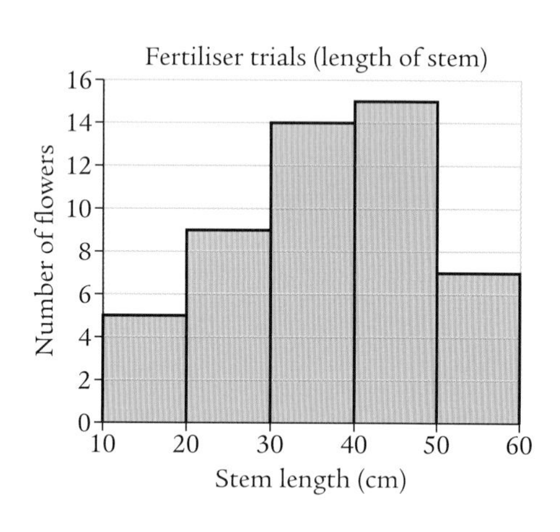

a Which diagram is which and why? Mention the horizontal axis on both types.

b For which of the types of data is the mean and median inappropriate?

2 One hundred students were asked, 'What is your favourite Olympic sport?'

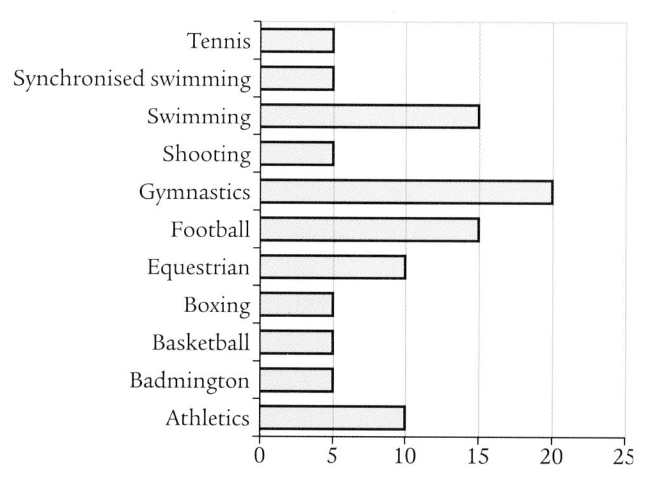

Which of the measures of central tendency can you obtain from the chart?

3 If two people picked two different samples, it is unlikely they would both come up with the same estimate for the average of the population.

Would they differ by much?

a The balls in the National Lottery run from 1 to 49. Let this be the population under examination.

i What is the mean of the numbers 1 to 49?

ii State the median.

b A 'Lucky Dip' is a ticket for the lottery containing six numbers randomly picked, **without repetition**. Let this be your sample.

i Do you expect the mean of these six numbers to be the same as the mean of the population?

ii Investigate for at least 10 sets of six, considering not just the mean of each individual sample, but the mean of the means.

You can produce random numbers on your calculator or, if you have access to a spreadsheet, follow these instructions:

1 In column A type the numbers 1 to 49.

2 In B1 type: =RAND() and Fill down to row 49.

3 Select columns A and B down to row 49.

4 Choose Data > Sort ... and sort by Column B.

5 Use the first six numbers in column A as your 'lucky dip'.

6 Each time you repeat 4, the list 'shuffles' up another six numbers.

7 In D1 type: =AVERAGE(A1:A6) to get the average of your sample.

4 One week last season the Rovers finished the game with more players than the average team.

The referee noticed it and said nothing.

Comment.

5 A company has nine employees each on an annual salary of £25 000 and a project manager on an annual salary of £95 000.

If you were asked what the typical salary for this company was, what would your answer be if:

a you were trying to give an objective impression of what you'd be likely to earn if you joined the company

b you were trying to give the best possible impression of what you'd be likely to earn if you joined the company?

6 Investigate the size of an apple (see above) using a simulation.

a You can do this simulation yourself:

Open a spreadsheet.

In A1 type: =INT(100*RAND()+100+0.5)

Fill right to J1.

Fill this row down to row 10.

You now have '100 apples'.

In L1 type: =AVERAGE(A1:J10)

In L2 type: =MEDIAN(A1:J10)

In L3 type: =MODE(A1:J10)

Each time you double-click A1 and RETURN, you sample another 100 apples.

b Suppose we included the 1850 gram apple in one of our samples.
Which of the three averages would record the biggest change?

Exercise 5.2B

1 For each of the following, state whether you would prefer to be:
i average **ii** below average **iii** above average.
Justify your answer in each case.

a A mark in a maths test

b Golf score

c Error count in exam

d Income

e Cholesterol count

f Number of ears

g Temperature

h Debt

i Blood type

2 Blood types in the UK have been summarised in the table below.

	United Kingdom blood groups							
Type	O+	A+	B+	AB+	O−	A−	B−	AB−
Percentage of population	37	35	8	3	7	7	2	1

a Assuming the population of the UK is 60 million,
calculate the number of people who are:

i A+

ii AB−.

b **i** Which measure of central tendency is appropriate for this type of data?

ii What is the 'average' blood group?

3 The table gives some statistics for the Scottish rugby squad who toured Australia in 2011.

Age (years)	Caps	Club
28	58	Edinburgh
31	37	Glasgow
30	32	Gloucester
29	13	Edinburgh
26	0	Glasgow
27	15	Glasgow
31	44	Newcastle
22	21	Glasgow
31	41	Glasgow
27	0	Gloucester
22	0	Glasgow
21	0	Edinburgh
26	16	Edinburgh
29	25	Gloucester
25	37	Glasgow

Age (years)	Caps	Club
25	18	Sale
31	80	Edinburgh
30	59	Glasgow
26	7	Edinburgh
21	1	Glasgow
28	33	Edinburgh
22	0	Glasgow
24	3	Worcester
21	1	Edinburgh
26	9	London
28	28	Castres
31	65	Scarlets
25	0	Edinburgh
22	0	Edinburgh
20	4	Glasgow

a What might be the most suitable measure of central tendency for:

i their ages

ii the number of times they have been capped

iii the club they belong to?

b Calculate the most suitable measure of central tendency for each.

c This diagram has plotted the age of each player against the number of caps they have.

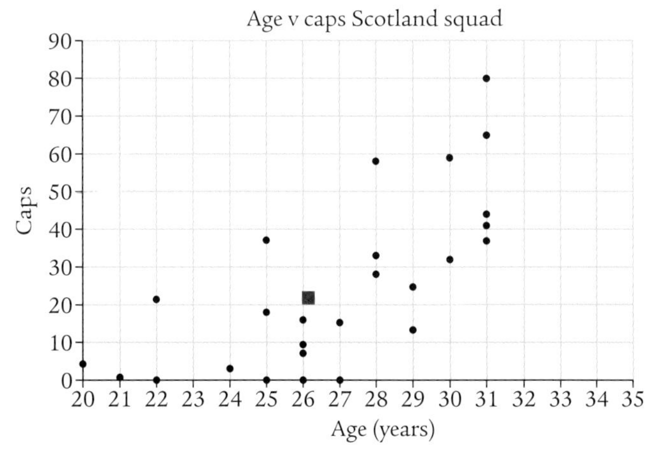

i One point has been plotted in red. Can you say what it might be?

ii Is there any relationship between age and number of caps?

4 Here are the heights of a class of 14- and 15-year-olds.

169	155	161
146	156	162
170	158	163
150	158	167
153	158	165
190	147	110
166	164	167

a Calculate the mean, median and mode of the data.

b Which of mean, median and mode would be most appropriate to describe the 'typical' height of the class?

c If you removed the shortest and tallest student from the data, what average would be most affected?

5 The attendance figures at a local football ground given over a period of six months were recorded below.

3812	3188	5888	3348	3249	3060	6416	3044	3274
6623	4457	3848	3036	3326	3540	3035	3142	6435

a Calculate the mean for this data.

b Find the median.

c Draw a frequency table using class intervals of 3001–4000, 4001–5000, 5001–6000, etc.

d Write down the modal class for the attendance figures.

e Which of mean, median and modal class would be most appropriate to use for the 'average' attendance figures?

6 The average household income in the United Kingdom in 2009–10 was £29 000. It is said that more than half the households in the UK earned well below the average household income. Explain why this statement is true.

5.3 Measures of spread

The range and the quartiles

The **range** is a measure of how widely spread a set of numbers is.

The range is the difference between the highest value and the lowest value in the data set.

When the data set has been arranged in numerical order:

- the **median** is the value that splits the list into two equal parts, an upper part and a lower part
- the **lower quartile**, Q_1, splits the lower part of the list into two equal parts
- the **upper quartile**, Q_3, splits the upper part of the list into two equal parts.

For consistency, the median is often referred to as Q_2.

Five-figure summaries

When you have a **large amount of data** it can usefully be summarised by quoting five figures:

- the lowest value L
- lower quartile Q_1
- median Q_2
- upper quartile Q_3
- the highest value H.

There is no real point in summarising a list with 10 pieces of data by a five-figure summary. However, for the sake of examples and explanations, we'll make an exception.

Example 1

The following are the qualifying times in seconds for the Women's Individual Pursuit at Beijing in 2008.

215 208 224 212 220 218 226 209 216 214 227 221

a Calculate the range.

b Calculate the five-figure summary for this data.

a First put the data in numerical order:

208 209 212 214 215 216 218 220 221 224 226 227

Lowest value = 208, highest value = 227.

⇒ Range = 227 − 208 = 19.

b **i** There are 12 pieces of data $\frac{12+1}{2} = 6{\cdot}5$

... the median is between the 6th and 7th pieces of data ,

namely halfway between 216 and 218 ... $Q_2 = \frac{216+218}{2} = 217$.

The list splits in two:

208 209 212 214 215 216 217 218 220 221 224 226 227

ii There are six pieces of data in the lower part ... $\frac{6+1}{2} = 3{\cdot}5 = 3{\cdot}5$

... the lower quartile is between the 3rd and 4th pieces of data,

namely halfway between 212 and 214 ... $Q_1 = \frac{212+214}{2} = 213$.

iii The upper quartile lies between the 9th and 10th pieces of data (by adding 6 to the positions for the lower quartile),

namely halfway between 221 and 224 ... $Q_3 = \frac{221+224}{2} = 222{\cdot}5$.

So the five-figure summary is:

$L = 208$, $Q_1 = 213$, $Q_2 = 217$, $Q_3 = 222{\cdot}5$, $H = 227$.

Exercise 5.3A

1 Find: **i** the range and **ii** the quartiles for each data set below.

a 90, 92, 93, 88, 95, 88, 97, 87, 98

b 2·3, 3·4, 2·4, 5·6, 3·8, 4·4

c 134, 220, 169, 268, 321, 220, 220, 170

2 In Media Studies, a survey is done on the lengths of commercial breaks between the hours of 5 p.m. and 9 p.m.

Times are recorded in minutes to 1 decimal place.

The times have been arranged in order.

2·5	2·6	2·6	2·6	2·6	2·6
2·7	2·9	3·0	3·0	3·0	3·3
3·3	3·4	3·6	3·6	3·8	3·8
4·0	4·2	4·4	4·5	4·6	4·7
5·0	5·2	5·5	5·6	5·8	6·0

a Make a five-figure summary of the data.

b **i** The mean time is 3·8 minutes.
What fraction of the times is greater than the mean?

ii What is the mode?
What fraction of the times is greater than the mode?

3 In Echo Bay, the main local industry is the sale of cockles.

A survey is done because there is a worry that the shells are getting smaller.

Fifty cockles collected at random have their shells measured in millimetres.

20	20	21	21	22	23	24	24	25	28
28	29	29	30	31	31	33	34	34	35
35	36	37	37	38	38	39	40	40	40
41	41	41	42	43	43	43	43	43	45
45	46	48	50	50	50	51	51	54	57

a Make a five-figure summary of the data collected.

b Five years earlier the mean size of a shell was 40 mm.
Can anything be said about this sample?

4 At the Fiat Lux light bulb company, the quality control manager tests 50 bulbs to destruction.

The number of hours each lasted is recorded to the nearest 100.

1000	1100	1200	1200	1200	1200	1300	1300	1400	1600
1800	1900	2100	2400	2500	2600	2600	2600	2600	2800
2900	3000	3000	3000	3500	3600	3600	3700	3700	3800
3900	4000	4000	4100	4200	4300	4400	4400	4400	4500
4500	4600	4600	4700	4700	4800	4900	5000	5000	5000

a Make a five-figure summary of the data collection.

b If this sample is typical of the behaviour of the light bulbs,

i what could they claim is the average lifetime

ii what is the likelihood of a bulb lasting 4000 hours or more?

5 Twenty-five children were selected at random and their health studied over a year. The number of nose colds they each caught was recorded.

7	10	6	6	3
10	6	12	9	7
1	6	5	3	4
5	1	5	7	7
5	8	4	4	7

a Make a five-figure summary.

b If this sample is typical of the children in a primary school of 500 pupils, how many would you expect to catch seven or more colds next year?

6 A local charity has a sunflower growing competition to help raise funds. People who take part have to plant their seeds at the same time. Ten weeks later a judge comes to measure the plants to the nearest centimetre.

Here are the results:

80	65	58	62	72	90	89	88	85
65	105	108	98	78	74	82	59	

a Make a five-figure summary for this data.

b Martin grows a sunflower that is 94 cm high. In which quarter of the data list does his height lie?

Displaying a five-figure summary – box plots

It is often easier to make comments or comparisons when five-figure summaries are displayed as a box plot.

A box plot includes:

- a number line that contains the range of data
- five vertical lines placed to correspond with the five-figure summary
- horizontal lines joining the tops and bottoms of the lines representing the quartiles – these will form a box enclosing the line representing the median
- 'whiskers' drawn out joining the box to the lines representing the high and low scores
- a title to describe the context
- units on the number line to say what is being measured.

If felt necessary, the values of the five-figure summary can also be added.

Example 1

The performances in the Women's Individual Pursuit at Beijing in 2008 can be summarised as:

$L = 208\text{ s}, \quad Q_1 = 213\text{ s}, \quad Q_2 = 217\text{ s}, \quad Q_3 = 222\text{ s}, \quad H = 227\text{ s}.$

Illustrate this in a box plot.

The data goes from 208 to 227.

A number line that goes from 206 to 228 would contain this.

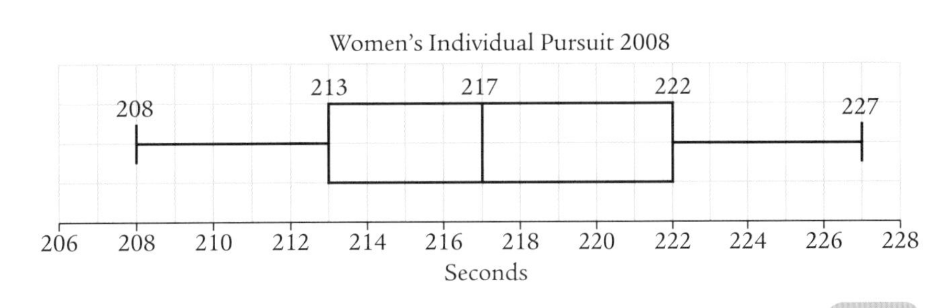

Example 2

The average rainfall in millimetres for Nairn for each month in 2011 is given below.

27·5, 44·4, 76·6, 29·6, 60·9, 66·8, 54·5, 112·2, 72·5, 47·3, 40·8, 94·4

a Make a five-figure summary of the data, working to one decimal place.

b Display this as a box plot.

a First place the numbers in numerical order:

27·5 29·6 40·8 44·4 47·3 54·5 60·9 66·8 72·5 76·6 94·4 112·2

⇧ Lower quartile (Q_1) ⇧ Median (Q_2) ⇧ Upper quartile (Q_3)

$L = 27{\cdot}5\text{ mm},\ Q_1 = \dfrac{40{\cdot}8 + 44{\cdot}4}{2} = 42{\cdot}6\text{ mm},\ Q_2 = \dfrac{54{\cdot}5 + 60{\cdot}9}{2} = 57{\cdot}7\text{ mm},$

$Q_3 = \dfrac{72{\cdot}5 + 76{\cdot}6}{2} = 74{\cdot}6\text{ mm},\quad H = 112{\cdot}2\text{ mm}$

b An appropriate number line might run from 25 to 115.

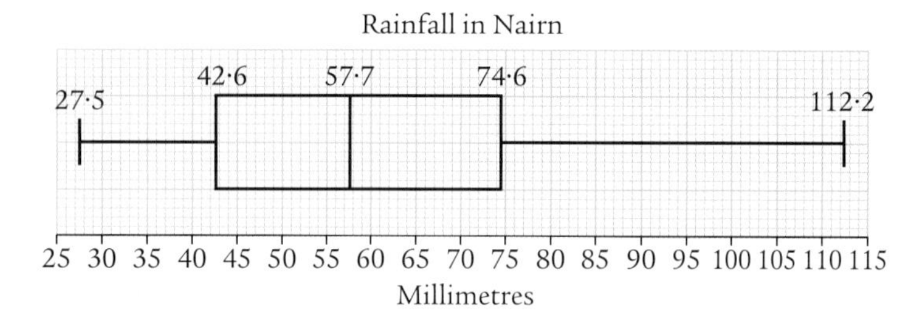

Example 3

The heights in centimetres of a class of 26 students were recorded.

169	155	161	146	156	163	170	158	163
150	158	167	153	158	166	198	147	110
166	164	167	165	163	164	162	166	

a The heights of two students stand out from the rest of the class.
What are their heights?

b The heights in **a** are referred to as outliers. These heights are too far away from the rest of the class and distort the 'picture' of what the class heights look like.

Make a five-figure summary of the data, but not including the outliers.

c Display the data as a box plot, showing where the outliers are in relation to the rest of the data.

First sort the data into numerical order:

110	146	147	150	153	155	156	158	158	158	161	162	163
163	163	164	164	165	166	166	166	167	167	169	170	**198**

a The heights 198 cm and 110 cm stand out from the rest of the class as they are quite a distance away from the rest of the data.

b $L = 146$ cm, $Q_1 = 157$ cm, $Q_2 = 163$ cm, $Q_3 = 166$ cm, $H = 170$ cm (to 3 s.f.)

c Representing the outliers by isolated dots, we get:

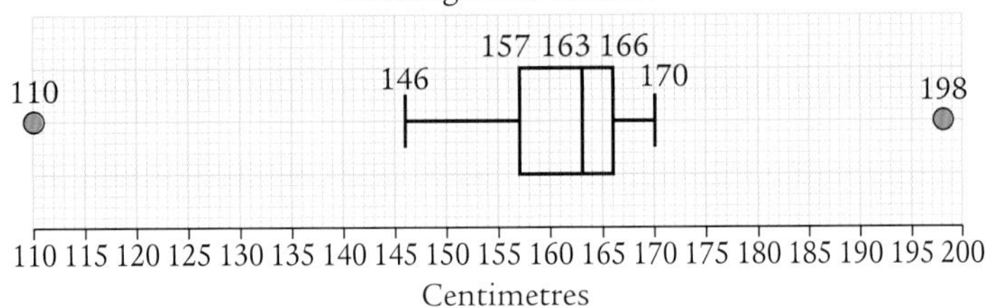

Exercise 5.3B

1 Draw a box plot for each of the questions answered in Exercise 5.3A.

2 Studying the checkout at a supermarket, a closed-circuit camera captures images of the queue at a particular till.

The number of people standing in the queue is recorded.

1	0	4	1	0	4	1	2
0	0	4	2	0	2	3	0
3	0	3	4	1	0	0	4
1	8	6	7	7	7	7	6
6	6	5	6	5	7	7	7
7	8	10	10	14	9	11	18

a Organise the data in a frequency table.

b Use your frequency table to help you devise a five-figure summary, treating any queue longer than 10 people as an outlier.

c Illustrate the distribution of the data in a box plot.

3 In country roads, during road works, temporary traffic lights are set up.

In one such site the traffic engineers want to make sure the phase of the lights doesn't cause too big a back-up.

The cars at the lights were counted as red turned to green.

0	5	5	6	11	12	7
7	9	9	2	2	3	12
6	10	8	5	12	12	6
12	2	5	4	11	3	9
9	3	5	8	3	20	3
8	11	6	11	3	2	5
2	7	3	6	7	6	8

a Treating the 0 and the 20 as untypical of the data set, make a five-figure summary of the data.

b Draw a box plot.

c The engineer altered the phasing of the lights and counted the cars at the next 10 changes of the lights:

4	6	7	6	4
7	7	7	3	6

By drawing a box plot of these 10 results say if you think changing the lights has had any effect on the size of the queues forming.

4 A roadside rescue organisation aims to arrive at the scene of a breakdown within an hour of receiving a call. A log of 30 calls and their response times in minutes was made.

15	40	40	35	15
35	20	15	45	35
50	40	35	45	15
40	40	30	30	85
20	30	50	25	50
25	40	50	40	40

a One of the times could be considered an outlier. Which one?

b Draw a box plot to illustrate the data.

c Comment on whether you think they are hitting their target.

5 A class wanted to investigate the time it took them to get to school.
They were all asked to record the time it took them in minutes. Here are their results:

5 1 10 14 15 18 18 20 20 62 22 30 30 25 25 6 7 20 23 35 60

a Calculate the five-figure summary for this data – treating 60 minutes and 62 minutes as outliers.

b Display the data as a box plot.

5.4 Comparing data sets

When comparing data sets it is usually appropriate to comment on the differences in:

- the average of the two sets, e.g. can an increase in average be interpreted as an improvement to the figures, or a deterioration of the situation?
- the range, e.g. can a narrowing of the range imply the score is being achieved more often, is one player less variable than another, can the word 'consistent' be applied?

Example 1

The prices for petrol in various filling stations in Glasgow and Inverness for a particular day are given below. The filling stations were selected at random.

Inverness	128·9p	128·9p	128·9p	128·9p	129·9p	130·9p	131·9p	131·9p	132·9p	135·9p
Glasgow	127·7p	127·7p	127·9p	127·9p	127·9p	127·9p	127·9p	127·9p	127·9p	127·9p

Calculate the mean, median, mode and range for each and comment on your findings.

Inverness

Mode = 128·9p, median = 130·4p, mean = 130·9p, range = 7p.

Glasgow

Mode = 127·9p, median = 127·9p, mean = 127.86p, range = 0·2p.

Comments

Petrol is, on all averages, more expensive in Inverness than it is in Glasgow.

The range of prices in Glasgow is very small compared to Inverness.

The fact that prices are more consistent (less variable) in Glasgow could be for many reasons, e.g. more competition in Glasgow than in Inverness.

Example 2

When Peter first started working in the office he went through typing trials.

Over several trials his speed, in words per minute, was measured.

A five-figure summary was produced.

$L = 13$, $Q_1 = 20$, $Q_2 = 36$, $Q_3 = 54$, $H = 69$.

After he had been on training, he went through another 25 trials.

Here are his results in words per minute.

63	73	68	72	64
64	73	75	63	72
62	72	63	69	60
69	72	70	60	71
62	73	60	62	61

a Make box plots of before and after the training.

b Comment on whether you think the training was effective.

a Inspecting the table we find a five-figure summary for after the training:

$L = 60$, $Q_1 = 62$, $Q_2 = 68$, $Q_3 = 72$, $H = 75$.

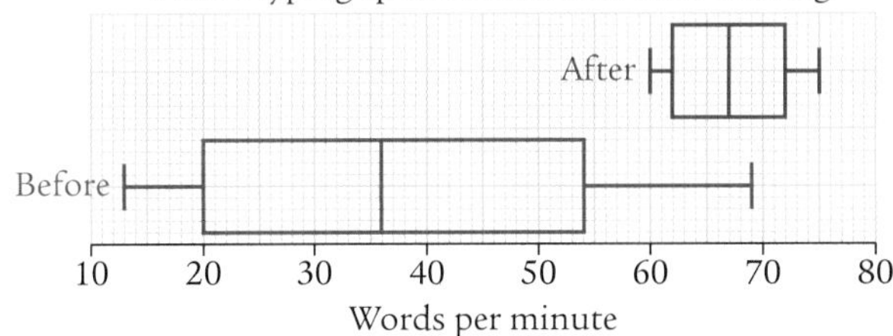

b Two things show up.

Peter's speed has improved, the median going up by 32 words per minute.

His consistency has improved too. Instead of the range being 56 words per minute it is now only 15 words per minute.

Exercise 5.4A

1 Two mobile phone companies' ratings from 1 (poor) to 6 (excellent) for the last 20 customers were recorded.

Company X	1	1	1	2	2	2	2	2	3	3	4	4	4	4	5	5	6	6	6	6
Company Y	2	2	2	3	3	3	3	3	3	3	4	4	4	4	4	4	5	5	5	5

a Calculate the median, mode and range of the ratings for each company.

b Which company would you choose to go with and why?

2 Two classes sat the same test and scored the following marks out of 100:

Class A: 94, 93, 90, 84, 81, 78, 75, 74, 70, 64, 61, 54, 53, 48, 40, 32

Class B: 100, 100, 97, 95, 92, 88, 85, 83, 83, 81, 80, 77, 75, 71, 70, 69, 66, 65, 63, 60, 58, 54, 51, 50

a Draw up a five-figure summary for each class.

b Display the results together using box plots.

c Compare the performances of the two classes.

3 A company has two machines that fill bottles of soft drinks.

Samples from each machine show the following number of millilitres per can.

Machine 1: 320, 319, 321, 318, 317, 316, 315, 320, 320, 318, 321, 317, 319, 315, 316, 320

Machine 2: 318, 321, 315, 314, 318, 317, 320, 313, 318, 321, 314, 315, 313, 317, 320, 318

a Draw a box plot for each machine on the same diagram.

b Comment on the performance of each machine.

4 A bus company runs a service at 10 a.m. and another at 11 a.m.

They have to make cut backs and one of the services will be cut.

They count the passengers on the bus each day for a month as it leaves the station.

10 a.m. service				
17	18	10	8	17
18	12	19	20	7
19	13	16	16	16
17	8	10	13	6
10	10	15	16	17
14	6	10	8	19

11 a.m. service				
15	14	12	14	15
12	15	14	15	13
13	15	12	13	15
13	15	13	14	12
14	15	15	12	15
12	15	12	12	12

a Compare the two services using the mean and range.

b Which of the two would you recommend to cut? Give your reasons.

c Draw a box plot to illustrate the comparison.

5 A soap company wishes to boost its sales and runs a TV ad campaign.

To test its effectiveness they counted the sales of their product the week before the ad campaign at 25 randomly chosen outlets. The week after the campaign, the sales from the same outlets were recorded again.

Sales before campaign				
36	9	22	28	30
32	7	48	28	30
12	18	35	42	33
15	11	11	10	14
47	7	13	28	6

Sales after campaign				
44	42	39	43	47
39	39	35	47	39
46	50	38	48	38
49	35	43	39	38
36	42	41	48	39

a Draw up a box plot to compare the sets of figures.

b Is there any evidence that the campaign was working?

Simple hypothesis testing

The same techniques can be used to help confirm whether a claim or hypothesis is correct or not.

Example 3

Hypothesis: Carbohydrate ingestion improves performance during a prolonged bout of exercise.

Researchers tested a group of seven trained runners over a 90-minute time trial twice, once with carbohydrate supplement and once with a placebo (placebo does not have any carbohydrate supplement). The runners did not know whether they were given the placebo or the carbohydrate at the time of running.

The distance travelled in km over the 90 minutes was recorded:

Carbohydrate supplement	19	23	17	18	18·5	16	20
Placebo	18	22	17	16	17	15·5	18

By calculating the mean for each set of data we can look at the hypothesis.

Carbohydrate: mean = 18·79 km, range 7 km.

This suggests that a runner on carbohydrates could be expected to run 18·8 ± 3·5 km.

Placebo: mean = 17·64 km, range 6·5 km.

This suggests that a runner not on carbohydrates could be expected to run 17·6 ± 3·3 km.

If a runner said he had run 18 km you couldn't tell whether he had been on carbohydrates or not.

At first analysis, because of the larger mean, it looks like the hypothesis is correct.

However, we generally work on the principal that the hypothesis is false unless the evidence is quite convincing.

A lot more trials or a bigger difference may make it more convincing.

Exercise 5.4B

1 In a science investigation students had to comment on the effect a temperature increase has on the reaction rates in a chemical experiment.

A claim is made that heating the solution leads to a faster reaction time.

At room temperature, 20 °C, the reaction times, in minutes, were recorded as follows:

45·3	46·5	47·3	48·2	48·4	48·7	48·6	48·5	48·9	49·1
48·1	49·2	48·0	49·0	47·7	48·8	46·8	50·5	50·1	49·8

At 30 °C the results were recorded as follows:

25·5	23·5	26·2	26·4	25·8	25·3	23·8	27·5	28·3	26·7
26·4	25·1	25·7	25·4	24·4	26·3	24·6	27·5	28·1	29·1

a Calculate the mean and the range for each set of data.

b Write a statement about the effect of temperature on reaction rate. Does the experiment support the claim?

2 A garden centre claims that their special compost leads to better foliage in plants.

An experiment was set up to test this.

Twenty-five plants were fed the compost and another 25 were not.

Leaf count – without compost				
5	6	4	5	5
4	7	3	6	7
6	8	8	4	6
6	8	3	8	5
7	3	6	8	7

Leaf count – with compost				
11	8	11	11	6
6	11	6	11	3
8	5	6	5	5
8	11	7	10	6
8	10	5	7	7

a Calculate the mean, median and range for each set of data.

b Comment on the claim made by the garden centre, using box plots to illustrate your argument.

3 A golfer playing on the practice green chipped a shot, which hit the flag and went down the hole. 'That was lucky,' said a spectator. 'Yes,' was the reply, '... and the funny thing is, the more I practise, the luckier I get.'

The spectator was on the practice green next. He recorded 30 chips onto the green, measuring how far it landed from the hole. After a week of solid practice, he did the same again.

Distance to hole before practice (cm)				
63	162	147	192	205
231	113	119	155	193
173	86	147	219	96
170	148	108	230	64
66	59	79	222	195
223	139	189	186	248

Distance to hole after practice (cm)				
52	37	77	24	67
55	48	69	99	33
71	49	66	16	98
59	62	28	63	73
87	68	33	49	50
47	55	96	17	31

Test the claim: 'The more I practise, the luckier I get' using these figures.
(Compare medians and ranges and use box plots.)

4 A printing company wants to convince a local restaurant that posting fliers though the doors of the neighbourhood will significantly boost their business. They show the restaurateur 'before' and 'after' figures for a restaurant in another town.

Before (customers)					
0	6	5	0	6	5
6	0	5	5	3	1
5	1	2	5	2	4
2	3	7	5	5	4
7	6	2	5	1	2

After (customers)					
10	6	8	7	4	10
11	6	11	11	5	7
10	12	10	8	4	7
7	12	9	12	9	10
6	10	12	11	5	5

a Calculate the median and the range for both the **before** and **after** figures.

b Would you be convinced by the figures that a flier campaign boosts custom?

Preparation for assessment

1 Kenny Miller is a Scottish footballer. Here is a record of the number of goals he scored each season from 1998 until 2012.

Season	1998–99	1999–00	2000–01	2001–02	2002–03	2003–04	2004–05	2005–06	2006–07	2007–08	2008–09	2009–10	2010–11
Number of goals	12	13	11	2	24	6	21	12	8	9	15	22	27

a In which season did he score the most number of goals?

b Why might you think something happened to him in the 2001–02 season?

c Calculate the mean number of goals Kenny scored from 1998 until 2012.

2 The LIBOR (London interbank offered rate) is the rate of interest that banks charge to lend money to each other. The British Bankers' Association calculates this daily for a variety of different currencies. The rate is calculated by finding the mean of the middle 50% of the interest rates given by the contributing banks.

Note that the middle 50% are the figures between the quartiles.

On a particular day, 12 banks contributed the following interest rates:

0·80%, 0·81%, 0·75%, 0·79%, 0·76%, 0·93%, 0·83%, 1·12%, 1·02%, 0·94%, 1·25%, 0·78%.

Calculate the LIBOR rate for this particular day.

3 The mean weight of a class of 18 students is 52 kg. A new student arrives in the class and the mean is now 51·5 kg.

What is the weight of the new student?

4 a Peter kept a note of his golf scores over one season. Here are his results:

Golf score	Frequency
69	4
70	3
71	2
72	6
73	3
74	1
75	1

What was his modal score?

Calculate the mean and median of his scores.

b Peter took up the opportunity for some golf coaching over the winter.

He then recorded his results for the following season. Here are his results:

Golf score	Frequency
65	5
69	8
70	2
72	3
74	2

Do you think the golf coaching improved Peter's scores? Justify your answer.

5 Here are the average monthly maximum temperatures in degrees Celsius for two cities:

Sydney, Australia	25·8	25·7	24·7	22·3	19·3	16·8	16·2	17·7	19·8	22·0	23·5	25·1
Boston, USA	2·1	3·1	7·7	13·3	19·2	24·6	27·7	26·6	22·7	17·1	11·2	4·7

a Draw a box plot for each city.

b Use the median temperatures and the range of temperatures to compare the two cities.

c Comment on the results.

6 Remember a Formula 1 team's aim to improve their consistency for 'pit stop times'? Although they are keen to have a quick pit stop, they are more concerned with being **consistent** with an average time of 3 seconds.

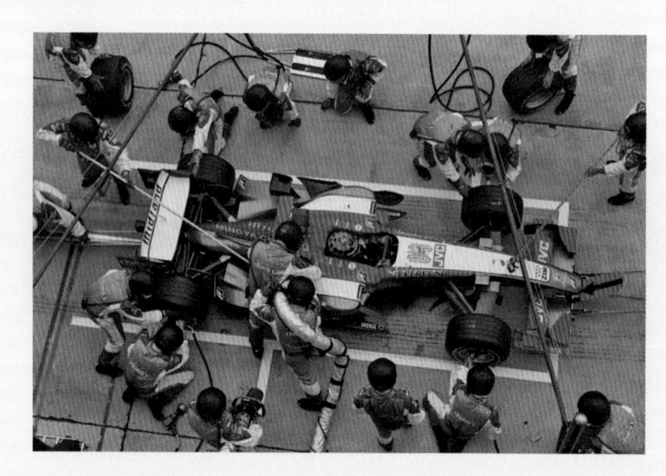

Here are their times in seconds for last month's pit stop practice:

3·4	2·8	5·9	3·3	2·8	3·0	2·6	2·8	2·3	2·9	4·0	3·4
3·0	2·5	2·9	3·0	3·1	2·7	3·0	4·2	3·7	2·5	3·1	3·1

Here are their times in seconds for this month's pit stop practice:

3·0	3·7	4·2	2·9	2·8	3·0	2·7	2·8	2·9	2·7	3·2	3·7
3·2	2·5	2·6	2·8	3·0	2·8	3·0	3·0	3·4	3·2	3·4	3·1

Have they achieved their aim?

6 Frequency tables, charts and graphs

Before we start...

Suppose you are the managing director of a company and are attending a meeting to decide on the company's sales policy for the next year. Two sales executives Ralph and Masoud make a presentation.

Ralph goes first. He tells the meeting that the company should target the sales of its No. 1 competitor. He says that if the company adopts his sales policy then it can expect to be market leader in its field. To support his argument he shows the meeting a bar chart.

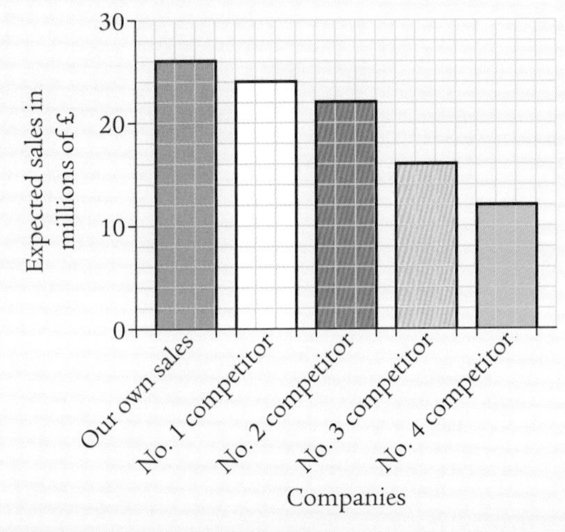

Next up is Masoud. He admits that it is unlikely that his sales policy will make the company into a market leader. He argues that its No. 1 competitor is very strong, and in order for the company to grow it should target the market share of its other three competitors. Masoud has produced a pie chart, but the fact that it shows the No. 1 competitor ahead of the company doesn't look good.

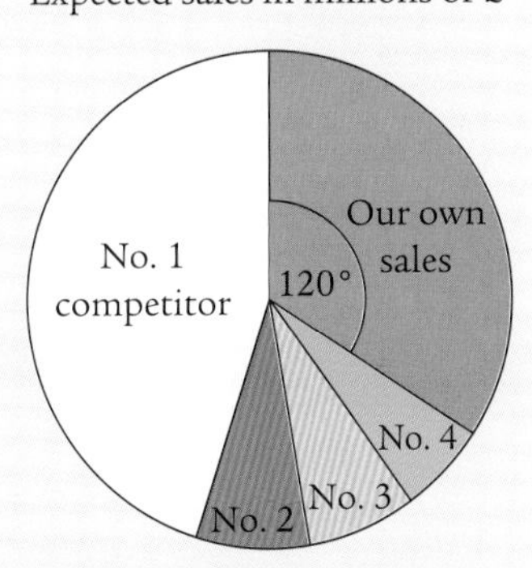

At the end of the meeting you have a decision to make. Surely you want your company to be the market leader and will decide to adopt Ralph's sales policy ... or will you?

What you need to know

1 This bar chart appears to show that sales rocketed during 2010 but is that really what happened?

Do you think that the sales really were better in 2010?

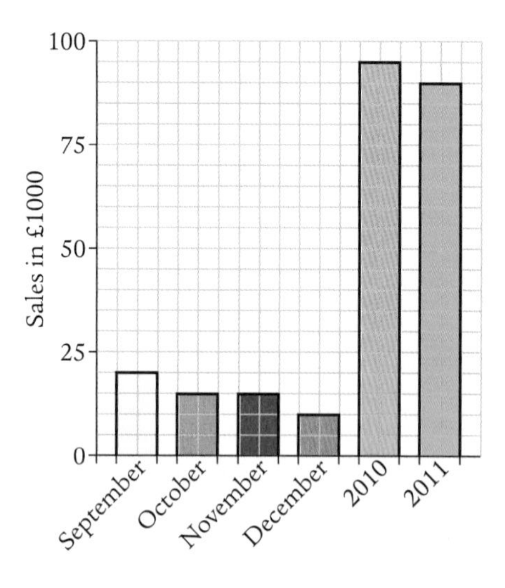

2 Draw angles of:

a 90° **b** 43° **c** 120° **d** 220°.

3 Express each angle as a fraction of a complete turn.

a 90° **b** 120° **c** 300°

6.1 Frequency tables with grouped data

The frequency table is a way of organising information.

Often they are used when the information is gathered verbally to make sense of a list of disorganised **data**.

The information gathered can then easily be displayed in a chart.

When the information is numerical and widely spread, it is sometimes useful to group it into **class intervals**.

Otherwise you could end up with a lot of entries, each with a frequency of 1. Then trends and patterns would be difficult to spot.

When entering unsorted data into a frequency table, it's best to make use of tally marks. Go through your data, one piece at a time, entering a tally mark for each piece as you come across it.

This is much better than scanning through the data looking for all the points that belong to the first class, then scan again for the points that belong to the second, etc. This wastes time and also offers a lot more opportunities to make mistakes.

Example 1

Response times for the Automobile Road Rescue have been noted over 20 calls.

The results, recorded to the nearest minute, are as follows:

25 30 76 62 27 42 61 62 57 47
45 38 41 42 42 52 36 44 41 53

a By using convenient class intervals, sort the data into groups.

b Hence give an estimate for the most likely time you would wait to be rescued.

a How big should the class intervals be?

The range of the data is $76 - 25 = 51$.

We would normally want to have about 5 or 6 intervals, suggesting that we place 10 numbers in each.

In this case, it is convenient to start at 20, so use intervals 20–29, 30–39, etc.

Class interval (min)	Tally	Frequency
20–29	||	2
30–39	|||	3
40–49	~~||||~~ |||	8
50–59	|||	3
60–69	|||	3
70–79	|	1

b Waiting times range from 25 min to 76 min, but the most likely time to wait is between 40 and 49 minutes.

Data: a set of individual pieces of information, which is to be collected, displayed or analysed.

Class interval: data can be put into groups to make it easier to handle. These groups are described by their class interval.

Example 2

Gemma works in market research. She has been working for a mobile phone company gathering information about text messages. She asked 35 people to keep a record of the number of text messages they sent in one week.

Here are the results:

210	232	209	151	243	35	83	123	192	476
427	349	246	221	81	72	212	352	137	251
184	312	245	216	168	273	139	328	406	225
197	163	213	185	179					

a Make a frequency table using suitable class intervals.

b Create a bar chart to illustrate the data, commenting on the features it shows up.

a The range of data is 476 − 35 = 441. So if we want 5 or 6 classes we should make the class size 100. A convenient place to start is 0–99.

Number of text messages	Tally	Frequency
0–99	||||	4
100–199	~~||||~~ ~~||||~~ |	11
200–299	~~||||~~ ~~||||~~ |||	13
300–399	||||	4
400–499	|||	3

b Most customers are sending between 200 and 299 messages, although a similar number sent 100–199 messages a week.

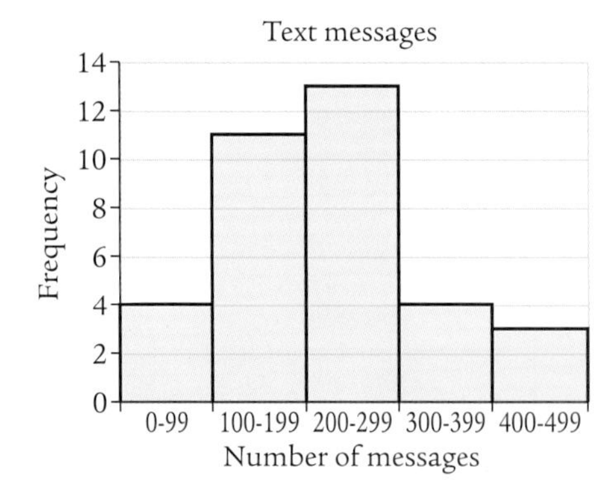

Exercise 6.1A

1 Enter this data set into a frequency table using class intervals of 30–39, 40–49, etc.

34 45 61 52 33 42 57 32 37 34 48 59

2 Construct and complete frequency tables for the following data sets using the class intervals suggested.

a 28 44 52 37 47 32 45 36 49 41 45 56

Class intervals: 20–29, 30–39, 40–49, 50–59.

b 4 6 14 31 38 24 27 9 25 22 38 28

Class intervals: 0–9, 10–19, 20–29, 30–39.

c 2·7 3·4 5·8 3·6 3·7 2·2 3·4 2·6 3·8 3·5 2·4 4·1

Class intervals: 2·0–2·9, 3·0–3·9, 4·0–4·9, 5·0–5·9.

d 435 620 528 474 564 523 541 615 543 422 483 611

Class intervals: 400–449, 450–499, 500–549, 550–599, 600–649.

3 Mr McPake's class sat a maths test. Here are the results:

73% 84% 71% 63% 59% 97% 75% 89% 47% 91%
68% 73% 51% 82% 86% 83% 72% 43% 73% 58%
62% 75% 77% 73% 81% 95% 86% 53% 61% 77%

a Make a frequency table using class interval sizes of 10, starting with 40.

b In which class interval did most students fall?

4 Here are the scores of 21 players in a golf tournament.

68 78 75 81 69 77 83
74 75 77 83 71 79 82
76 73 73 72 76 81 73

a Construct a frequency table of these results using class sizes of 5 starting with 65.

b If the golfers in the 65–69 class are described as A-grade golfers, in what grade are most of the golfers?

5 Here are the journey times, measured to the nearest minute, of 30 students on their way to school.

12 15 23 37 19 15 3 11 12 11
7 13 24 22 17 12 9 14 12 8
11 14 12 27 23 17 12 12 13 18

a Using convenient class intervals (start with 0–4), construct a frequency table of the data.

b If all 30 students leave home at 8.30 a.m., during what time spell do most students arrive?

Continuous data

When the data is continuous, say, weight, w, in kilograms, then describing classes 30–39, 40–49, etc. won't do because a weight of 39·5 kg would fall between the classes.

Instead we define the classes as $30 \leqslant w < 40$, $40 \leqslant w < 50$, etc. In that way, 39·5 definitely falls in the $30 \leqslant w < 40$ class.

Example 3

During an experiment with plant food, 30 plants have their heights measured in centimetres.

47·9	39·2	34·7	42·7	39·0	35·4
38·2	42·2	35·5	35·4	28·6	45·2
35·7	44·3	49·2	34·6	47·5	34·2
48·8	30·3	40·3	35·7	34·7	32·4
39·5	38·3	47·1	35·5	45·0	47·2

a Make a frequency table of the data using class intervals $25 \leqslant h < 30$, $30 \leqslant h < 35$, etc.

b What was the most common height for a plant in the experiment?

a

Height	Tally	Frequency
$25 \leqslant h < 30$	|	1
$30 \leqslant h < 35$	~~||||~~ |	6
$35 \leqslant h < 40$	~~||||~~ ~~||||~~ |	11
$40 \leqslant h < 45$	||||	4
$45 \leqslant h < 50$	~~||||~~ |||	8

Note that the 'score' of 45 falls within the final class, $45 \leqslant h < 50$.

b The most common height for a plant is between 35 cm and 40 cm.

Example 4

A speed camera is set up on a straight stretch of road.

The speed of 30 cars as they pass the camera is taken in miles per hour.

55·0	58·3	59·9	67·0	52·2	53·0
55·5	60·2	57·7	71·9	54·4	63·4
73·1	57·3	54·6	74·6	64·8	69·1
69·4	68·6	71·0	63·8	60·4	63·9
53·4	59·7	59·3	67·1	68·9	70·2

a Make a frequency table using suitable class intervals starting at 50 mph.

b How many cars drove at 70 mph or more?

a

Speed (mph)	Tally	Frequency
$50 \leqslant v < 55$	~~||||~~	5
$55 \leqslant v < 60$	~~||||~~ |||	8
$60 \leqslant v < 65$	~~||||~~ |	6
$65 \leqslant v < 70$	~~||||~~ |	6
$70 \leqslant v < 75$	~~||||~~	5

b Five cars were going at 70 mph or more.

Exercise 6.1B

1 Through a road tunnel the traffic is meant to travel at a minimum speed of 15 mph and a maximum of 30 mph.

A short sample was taken to give an idea of how many motorists were taking note of the restrictions.

These are the speeds taken halfway through the tunnel.

18·1	30·1	16·9	13·1	11·2	16·8
25·6	32·8	18	12·1	16·6	28
25·5	14·6	15·5	31·3	15·9	18·5
22·4	17·3	32·6	11·9	12·1	11·3
13·7	24·9	28·2	31·7	21·2	27·5

a Make a frequency table using $10 \leqslant v < 15$, etc. as class intervals.

b How many of the sample were observing the restrictions?

2 **a** One week at a slimming group all of the members lost weight. Here is the weight, in grams, that was lost by each member.

920	870	960	510	670	900	970	430	1120	750	720
1470	320	880	960	1370	420	730	810	430	520	1260

Using class intervals $300 \leqslant w < 500$, $500 \leqslant w < 700$, etc., construct a frequency table.

b The next week the slimmers all lost weight again. Here are their weight losses.

540	920	910	600	720	1450	1100	510	380	850	810
1510	1100	1200	1300	1220	530	850	920	910	620	1330

Using the same class intervals as in part **a**, construct a frequency table.

c Compare the two frequency tables.

Do you think that the slimmers did better in the first week or the second week?

3 Elaine, Karen, Fiona and Lee had heard that walking would improve their health.

They decided to give it a go and kept a record of how long they spent each week walking.

After four weeks here are the times, in minutes, that each person spent walking.

Elaine: 155, 145, 173, 167
Karen: 155, 168, 178, 164
Fiona: 186, 184, 183, 186
Lee: 183, 185, 172, 174

a Use class intervals (start with $140 \leqslant t < 150$) to construct one frequency table combining the walking times of all four ladies.

b It is recommended that 30 minutes of brisk walking 5 times a week provides significant health benefits.

Is the median time that the ladies spent walking likely to provide these benefits?

6.2 Bar charts and line graphs

Displaying the information

Two of the most common ways of displaying information are bar charts and line graphs.

In a **bar chart**, the x-axis does not have to be in any order. It doesn't even have to show numerical data.

The height of each bar is an indication of the frequency.

Here 59 students were asked to name a colour of the rainbow.

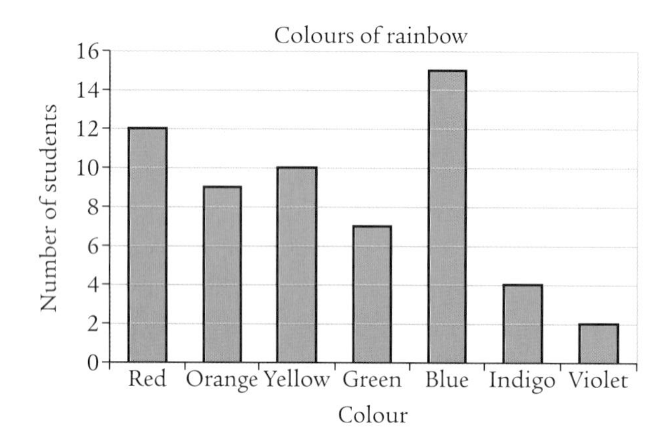

The bar chart lets us see that:

- blue was named the most, with a frequency of 15
- six times as many students picked red as violet.

Note that by being careful to include a title and to label the axes, no information goes missing. We would be able to reconstruct the table that the chart is based on.

In a **line graph** the order on the x-axis is important.

The line joins the points merely to emphasise a pattern or show up a trend.

However, values between the points need not make sense.

In this line graph we see there were 5 passengers on the bus at stop 2 and only 4 at stop 3.

We cannot think that when the bus is halfway between these stops there are 4·5 passengers as the line might suggest.

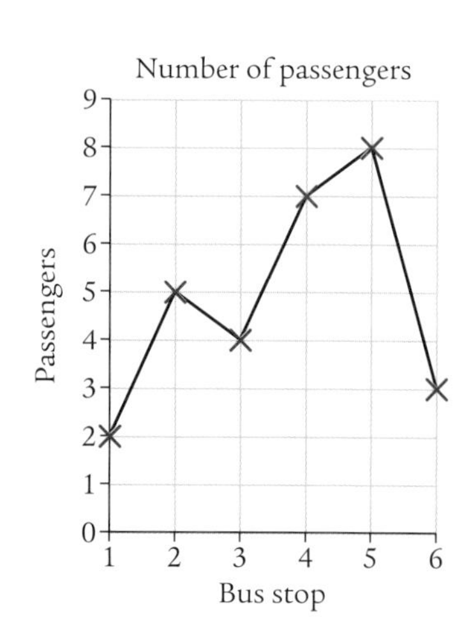

Example 1

Silverburn High School is holding an election for head boy.

The results are given to the head teacher in a table.

Candidate	William Gibson	Billy Menmuir	Robert Masterton	Eric Carruthers	James Cant
No. of votes	26	36	22	16	9

The head teacher wants the results displayed for the rest of the school to see.

a Draw a bar chart to illustrate the data.

b Identify who should be deputy head boy.

a

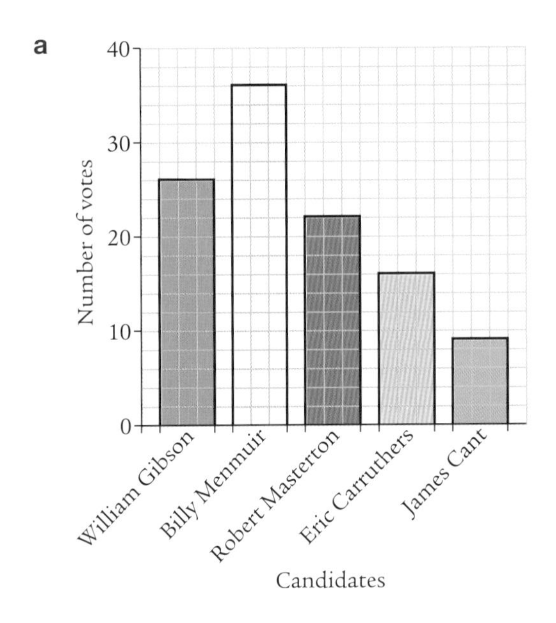

b The person who came in second in the vote, William Gibson, should probably be the deputy head boy.

Example 2

Graham is aware that he has put on weight. He decides to try to get his weight back down again. He weighs himself every Monday morning and keeps a record of his weight on a spreadsheet on his computer.

Date	Weight (kg)
2nd January	102
9th January	100
16th January	99
23rd January	98
30th January	98
6th February	97
13th February	95
20th February	96
27th February	96
5th March	94
12th March	93
19th March	93
26th March	92

a Draw a line graph to illustrate the data.

b Describe the trend in Graham's weight over the period recorded.

c When did the records go against the trend?

a

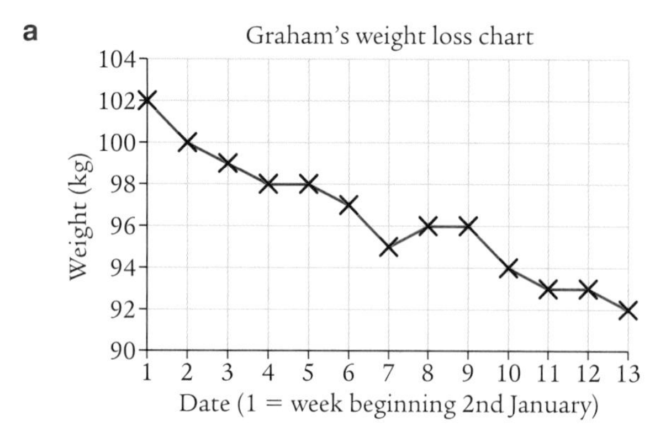

b In general Graham's weight is dropping. (This is called the **trend**.)

c In week 8 his weight increased, which was against the trend.

Exercise 6.2A

1 Class 4B carry out surveys. The results are shown in the tables below.

Make a bar chart for each one.

a

Favourite colour	Red	Blue	Green	Yellow	Pink	Orange
Frequency	3	7	7	2	9	1

b

Favourite flower	Rose	Daffodil	Sweet pea	Lobelia	Fuchsia
Frequency	8	6	10	2	3

2 Remember Mr McPake's Maths class from Exercise 6.1A?

The completed frequency table looked like this.

Maths mark (%)	Frequency
40–49	2
50–59	4
60–69	4
70–79	10
80–89	5
90–99	3

Make a bar chart using the information in the frequency table.

3 Lyle has had his height measured on his birthday every year since he was born.

Birthday	Birth	1	2	3	4	5	6	7	8	9	10
Height (cm)	48	73	87	95	103	109	115	122	128	134	139

On 2 mm graph paper draw a horizontal axis going from 0 (birth) to 10, letting 5 small squares represent one year. Draw a vertical axis going from 0 to 140. Let 5 small squares represent 10 cm.

Put scales on your axes as shown.

Label your axes. 'Age in years' for the horizontal axis and 'Height in centimetres' for the vertical axis.

Plot all the points from the table. Start with (0, 48) '0 years (birth), 48 centimetres'.

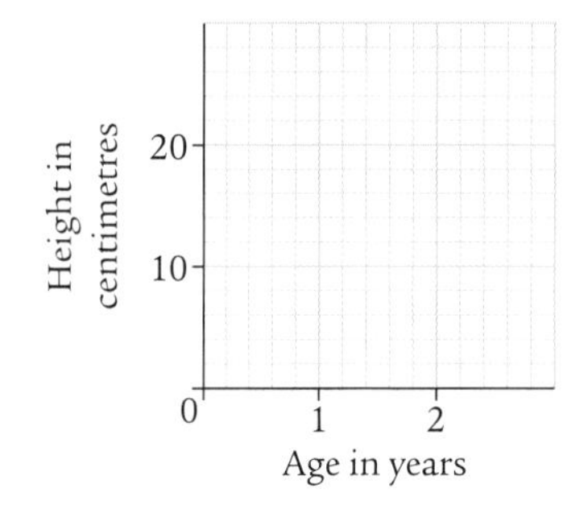

Join up all the points and complete your line graph of Lyle's height by giving it a title.

4 Make line graphs using the data in these tables.

Describe any trend you see.

a

Date	5 June	12 June	19 June	26 June	3 July	10 July	17 July	24 July	31 July
Money in bank account (£)	1800	1400	1000	600	2200	1700	1300	900	300

b

Month	Sept 2010	Oct 2010	Nov 2010	Dec 2010	Jan 2011	Feb 2011	March 2011	April 2011	May 2011
Percentage of people intending to vote for the Labour Party	35	36	36	38	37	41	42	43	43

c

Date	5 June	12 June	19 June	26 June	3 July	10 July	17 July	24 July	31 July
Height of plant (cm)	12	14	17	21	23	24	25	25	25

5 The graph of Lyle's height (see Question **3** above) will look something like this:

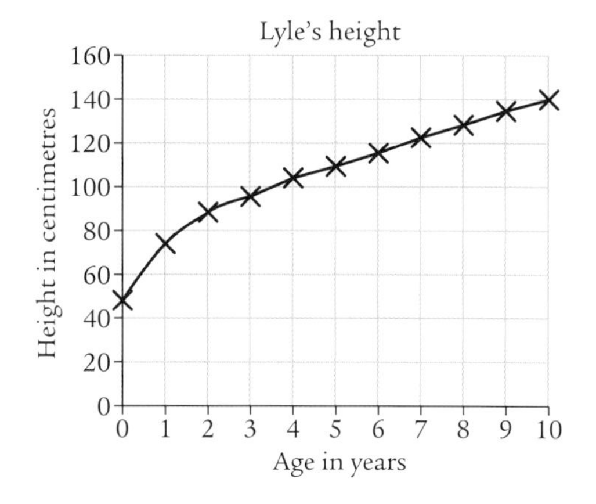

a Estimate what height Lyle was when he was:

i $7\frac{1}{2}$ years old **ii** $9\frac{1}{2}$ years old.

b At the age of $3\frac{1}{2}$, Lyle went to a theme park where you are only allowed on the log flume if you are taller than 1 metre. Was he allowed on the log flume?

6 The following graph shows the temperature of a patient who was admitted to hospital at 7 a.m. with a temperature of 40·4 °C.

The patient had his temperature recorded every 30 minutes.

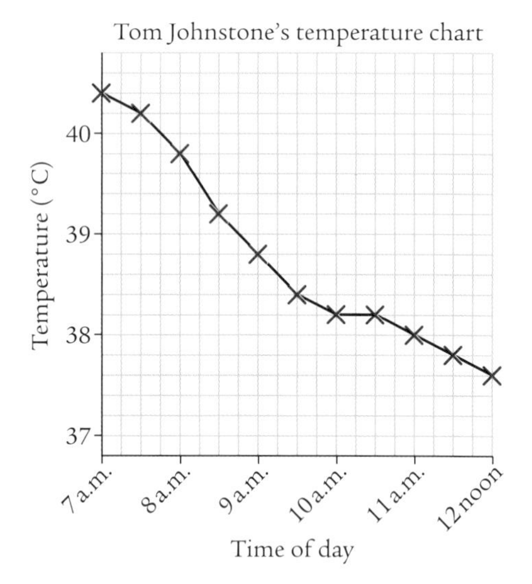

a What was the patient's temperature at 8 a.m.?

b By how much did the patient's temperature drop between 8 a.m. and 8:30 a.m.?

c Using the graph, estimate at what time the patient's temperature dropped below 39·0 °C.

d Once the temperature is below 39·0 °C the patient's condition is downgraded from 'serious' to 'worth observing'. At what time did the patient's condition get downgraded?

Comparing data sets

Graphs can be used when comparing data sets.

Example 3

The graph compares how two taxi firms charge for their time.

a What does each firm charge the instant they are hired?

b At what point does Terry Taxis become more expensive than City Cabs?

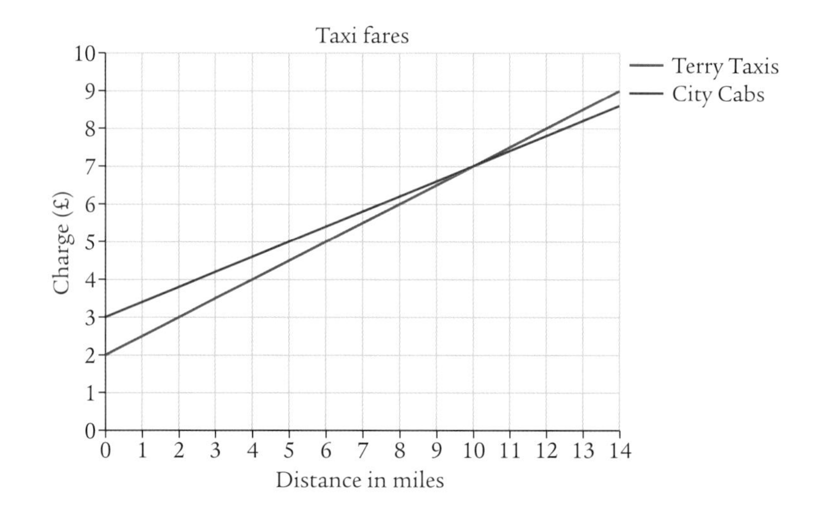

a For zero miles Terry Taxis charges £2 while City Cabs charges £3.

b At 10 miles they both charge the same namely £7. After that Terry Taxis is dearer.

Example 4

The chart compares the absentee figures of S1 and S4 for one week in term.

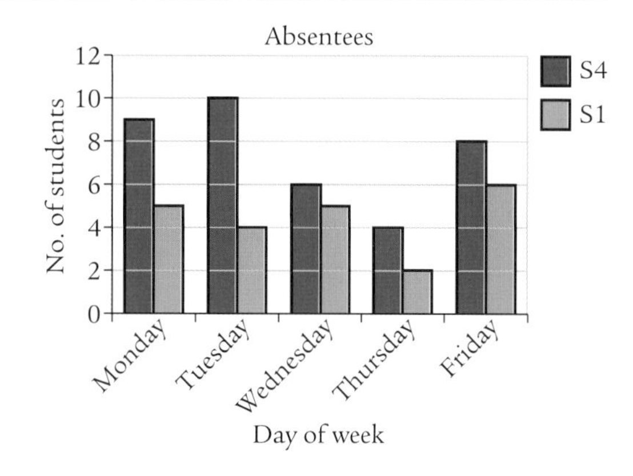

a When was the difference at its biggest?

b What was the difference on Friday?

a The biggest difference occurred on Tuesday. S4 were worse by 6 students.

b On Friday the difference was 2 students. S4 had the bigger figure.

Exercise 6.2B

1 This bar graph shows the result of a survey where people were asked, 'How many mobile phones have you had in your life?'

It compares boys' responses with girls' responses.

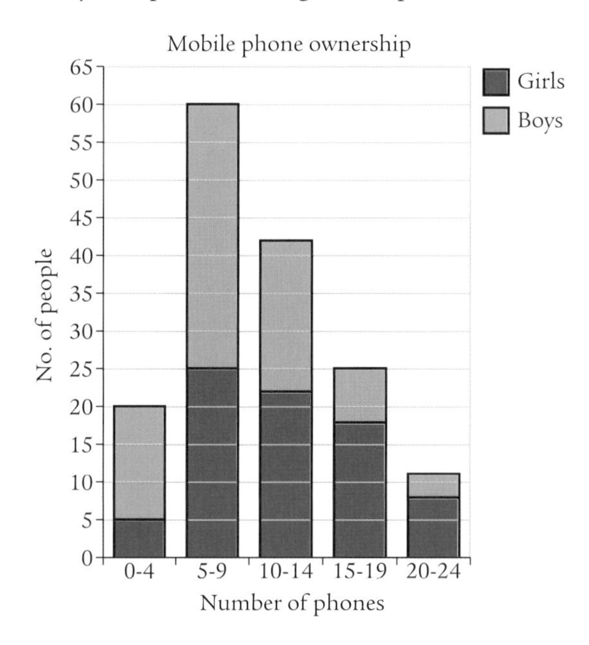

Column 1 tells us that 20 people had between 0 and 4 phones. Of these 5 were girls and 15 were boys.

a Describe the information held in column 2.

b Estimate how many more girls than boys claimed to have owned between 10 and 14 phones in their life.

c In how many categories did the girls outnumber the boys?

2 The Botany class mark off a small area between the meadow and the wood and keep a regular count once a month of the flowering harebells and buttercups they find there.

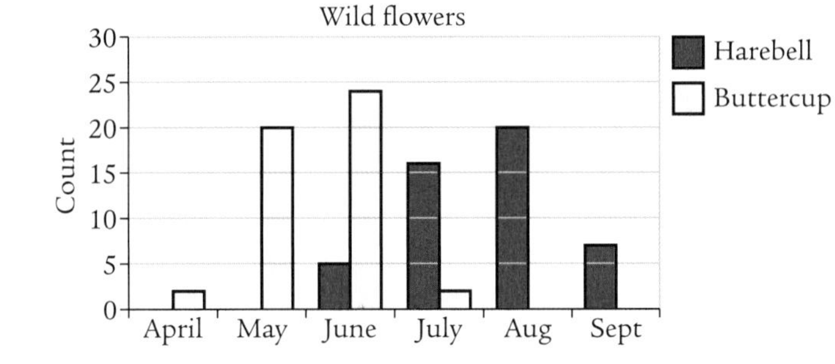

a In which months were there:

i harebells but no buttercups

ii buttercups but no harebells?

b What was the difference in counts between buttercups and harebells in June?

c What is the likely season for **i** buttercups **ii** harebells?

3 In a Media Studies project, students are asked to select one programme during a particular slot, Wednesday at 7.30 p.m. from a fixed set of channels.

Gender details are recorded.

Channel	BBC1	BBC2	STV	Ch 4	Vh 5
Programme	Rogue Cars	Top Chef	Emmerdown	News	Gadgets
Male	8	3	3	5	7
Female	2	7	7	5	3

a Draw a comparative bar graph to illustrate the data.

b Is there any gender bias in the preference for:

i Rogue Cars **ii** Top Chef **iii** News?

4 In a chemistry experiment, a beaker containing alcohol and water and another containing just water, are brought to the boil. Every minute the temperature is recorded.

Minute	0	1	2	3	4	5	6	7	8
Alcohol and water	21	41	61	72	74	74	74	74	74
Water only	21	42	64	76	86	94	98	100	100

Temperatures were recorded in degrees Celsius.

a Draw a comparative line graph to show what happened over the 8 minutes.

b What happened around the fourth minute? Mention both liquids.

c What happened in the seventh minute?

d Find out what you can about latent heat.

5 An advertising company wish to test the effectiveness of a particular advert.

They run the advert in only one neighbourhood.

They count the sales figures in that region and in one other where the advert did not run.

The advert was run in Week 4 of the survey.

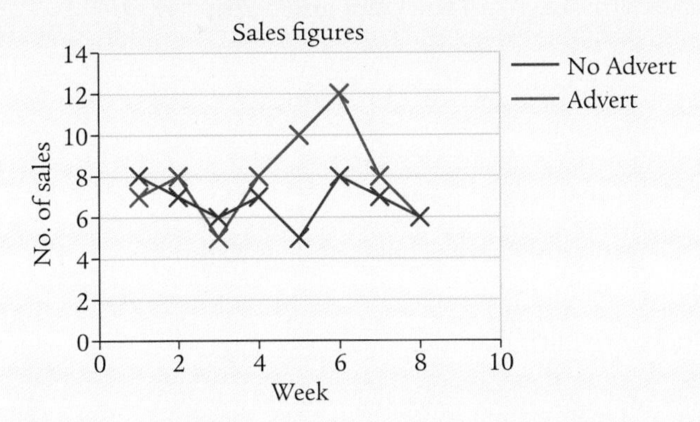

a Work out the average sales in the first four weeks for each region.

b Work out the average sales in each region after the advert had run.

c Compare the sales figures in the light of the advertising campaign.

6.3 Stem-and-leaf diagrams

We often wish to organise a large list of data into numerical order.

A quick and easy way of doing this is by means of a **stem-and-leaf diagram**.

Mike wrote down random numbers taken from the range 1 to 49 to study the lottery.

39	29	18	38	20
35	45	17	4	5
1	13	27	9	26
12	16	41	27	21
39	36	8	3	8
5	16	19	33	4

He wants to know if they really are random.

He puts the numbers in order as shown:

Random numbers (1–49)

Stem	Leaf								
0	1	3	4	4	5	5	8	8	9
1	2	3	6	6	7	8	9		
2	0	1	6	7	7	9			
3	3	5	6	8	9	9			
4	1	5							

$n = 30$ 1|2 represents the number 12

The diagram has:

- a title, explaining the context of the data
- a stem, in this case 0, 1, 2, 3, 4 which are the 'tens column' digits
- leaves, which are the 'units' digits that go with the stem
- a key that tells us how many pieces of data there are ($n = 30$)
- a key that tells us how to read the entries. (In 1|2, a 1 in the stem and a 2 in the leaves represents 12.)

Note that the red 7 represents 27 because it is in the 2-level of the stem.

Since the numbers are sorted into numerical order, we can also pick out the **quartiles**, the scores that split the sorted list into 4 equal parts.

There are 30 numbers and since $30 \div 4 = 7$ remainder 2, the list is split into 4 parts of 7 ...

The first part is separated from the second by the 8th number,

the second is separated by the third by the space between the 15th and 16th number,

the third part is separated from the fourth by the 23rd number, giving:

Lower quartile (8th number) = 8

Median (between the 15th and 16th) = $(18 + 19) \div 2 = 18{\cdot}5$

Upper quartile (23rd number) = 33.

Example 1

Stella is looking at the number of goals scored by teams in an amateur football league.

Here is a list of the numbers of goals they scored.

55	63	53	46	44	53	48	53	32	42	46
29	82	74	70	36	71	45	44	35	34	28
73	59	63	72	45	35	52	53	34	24	19

a Make a sorted stem-and-leaf diagram.

b Work out the median and quartiles of the data.

c Identify the mode of the data and calculate the range.

a A suitable stem would be 1, 2, 3, 4, 5, 6, 7, 8, representing the 10s digits.

Entering the data a row at a time produces an unsorted stem-and-leaf diagram.

1	9							
2	9	8	4					
3	2	6	5	4	5	4		
4	6	4	8	2	6	5	4	5
5	5	3	3	3	9	2	3	
6	3	3						
7	4	0	1	3	2			
8	2							

Sorting the leaves and adding the title and key gives:

Goals scored in the league

Stem	Leaf							
1	9							
2	4	8	9					
3	2	4	4	5	5	6		
4	2	4	4	5	5	6	6	8
5	2	3	3	3	3	5	9	
6	3	3						
7	0	1	2	3	4			
8	2							

$n = 33$ 2|4 represents 24 goals

b There are 33 pieces of data: $33 \div 4 = 8$ remainder 1.

So the list splits into 4 as follows:

1st to 8th 9th to 16th 17th 18th to 25th 26th to 33rd.

The lower quartile is between the 8th and 9th number: $(35 + 35) \div 2 = 35$ goals.

The median is the 17th number: 46 goals.

The upper quartile is between the 25th and 26th number: $(59 + 63) \div 2 = 61$ goals.

c The number 53 occurs 4 times, which is more often than any other number.

So mode is 53.

Range = highest − lowest = $82 - 19 = 63$ goals.

Exercise 6.3A

1 A company did a survey to find out what ages of people bought their product. Here are the ages of the people who bought their product in a superstore one Saturday morning:

25	28	74	36	42	72	22	28	31	74	50	69	37
25	41	65	18	75	68	72	37	26	71	63	37	29

a Construct a sorted stem-and-leaf diagram for this data.

b Identify the median and quartiles.

c Calculate the range.

2 The lottery balls that came out of the drum for five draws in May 2012 were:

3	4	18	24	44	49	17	1	19	39	40	42	46
22	5	6	21	41	48	49	1	5	18	21	23	33
20	37	2	7	8	10	22	23	46				

a Construct a stem-and-leaf diagram for this data.

b Is there one number that crops up more than others in the sample?

c Find the quartiles and median.

d In which quarter of the ordered list does the number 20 belong?

3 Fifteen people work in an office. Here are their journey to work times in minutes:

17 36 42 25 22 29 51 15 43 25 25
16 36 27 23

a Construct a stem-and-leaf diagram to illustrate this data.

b Write a short report describing the data, making use of highest and lowest times, and the quartiles.

Comparing data sets again

Miss Swan wanted to compare the results of her S3 class with the results from Mrs Steven's class. To do this she decided to use a **back-to-back** stem-and-leaf diagram.

Here we use a common stem and enter one set of marks to the right and the other to the left of the stem. In both cases the leaves get larger as they go away from the stem.

Marks in S3 test

Miss Swan's class		Mrs Steven's class
8 6	**1**	
6 4	**2**	9
9 9 7 7 7 4 4 4 2	**3**	7 7 7 9 9 9
7 5 5 2 2 2	**4**	5 5 7
8 5 0 0	**5**	0 3 5
	6	1 1 3 6
	7	4 4 4 6
7 2	**8**	2 7 9 9 9
	9	7 7 9
$n = 25$		$n = 29$

3|7 represents 37 marks

The diagram clearly shows that Miss Swan's class has only two marks above 60 while Mrs Steven's class has 16 marks above 60. We can say with the help of this back-to-back stem-and-leaf diagram that, in general, Mrs Steven's class has performed better than Miss Swan's class.

Example 2

A material is being invented to absorb shock so that people don't hurt themselves when they fall on it.

The material is tested by dropping ball bearings onto the surface and measuring how far they bounce away from where they initially drop. The smaller this measurement is, the better.

This bounce is measured in centimetres for two materials.

Material 1									
6	14	8	42	60	22	45	7	60	48
17	27	28	26	32	15	7	15	50	5
4	23	27	30	20	49	7	58	40	19

Material 2									
51	18	67	15	44	37	74	12	52	13
73	19	70	45	56	57	55	61	25	34
43	16	32	31	51	55	59	49	16	35

Make a back-to-back stem-and-leaf diagram to compare the two materials.

Which material is better for the purposes described?

Material 2		Material 1
	0	4 5 6 7 7 7 8
9 8 6 6 5 3 2	**1**	4 5 5 7 9
5	**2**	0 2 3 6 7 7 8
7 5 4 2 1	**3**	0 2
9 5 4 3	**4**	0 2 5 8 9
9 7 6 5 5 2 1 1	**5**	0 8
7 1	**6**	0 0
4 3 0	**7**	
$n = 30$		$n = 30$

1|4 represents 14 cm

We can see that Material 1 produces the smaller bounce:

lower quartile = 14 cm (for Material 2 this is 25 cm)

median = 24·5 cm (compared with 44·5 cm)

upper quartile = 42 cm (compared with 56 cm).

Exercise 6.3B

1 In diving competitions the judges give each dive a degree of difficulty. The degree of difficulty ranges from 1·2 for a really easy dive up to 4·5 for a really tricky one.

Here is a list of the 15 degrees of difficulty given in a competition:

3·2 3·5 4·2 2·8 1·9 3·4 1·7 3·8 4·1
2·8 3·3 3·8 2·9 3·1 3·8

a Make a stem-and-leaf diagram for this data.

b What is the median degree of difficulty?

c What is the modal degree of difficulty?

d What is the range of the degrees of difficulty?

2 Here is a stem-and-leaf diagram which shows the birth weights in kilograms of 20 babies born in a hospital.

Birth weight of babies

2	2 5 7 8 9
3	1 3 3 6 6 6 7 7 8 8 9 9
4	2 2 3

$n = 20$ 2|3 represents 2·3 kg

a How many of these babies weighed under 3·5 kg at birth?

b How many babies weighed over 3·8 kg at birth?

c What is the median birth weight of a baby from this group?

d What is the modal birth weight of these babies?

3 Stella decides to compare the goals conceded by teams in the 1st Division with the goals conceded by teams in the 3rd Division.

Here are the lists of goals conceded by teams in each division:

1st Division

25 29 41 35 33 59 49 49 61 78 70

3rd Division

63 31 40 67 46 43 50 73 85 83 84

a Make a back-to-back stem-and-leaf diagram with this data.

b Stella thinks that, in general, the teams in the 3rd Division will concede more goals than the teams in the 1st Division.

i Look at your stem-and-leaf diagram. Do you agree with Stella?

ii Find the median number of goals conceded in each division. Does this help to confirm what Stella thinks?

c Compare the ranges of the two divisions. Which division has the largest range of scores?

4 In a science experiment a weight is allowed to slide down a plane inclined at an angle of 30° to the horizontal.

The distance it goes, in centimetres, beyond the end of the incline is measured.

3·3	1·3	3·2	1·1	3·5	1·0
4·8	2·3	4·9	5·9	5·4	1·1
2·1	5·5	2·7	1·1	2·1	2·0
2·7	2·3	3·5	5·1	1·5	4·4

a Make a stem-and-leaf diagram to help sort the data

b Find the median and quartiles of the data.

The plane is inclined a further 5° and the experiment repeated.

3·7	5·9	6·1	4·2	4·8	6·7
2·5	6·6	4·6	4·1	4·2	4·0
4·5	3·0	3·7	4·4	6·0	3·6
3·5	5·1	4·1	4·3	3·5	6·0

c Make a back-to-back stem-and-leaf diagram and compare the results.

d What effect on the weight does increasing the angle of the plane have?

6.4 Pie charts

How much of the whole?

Pie charts allow us to illustrate the fraction of the data that falls into each category.

The area representing a category is proportional to the frequency of that category.

At a pottery, the quality controller classifies pots before they are put on sale.

Rejects are discarded.

The table shows the results of inspecting a batch of 180 pots.

Class	Frequency	Relative frequency	Angle
Very good	45	$\frac{45}{180} = \frac{1}{4}$	90°
Good	65	$\frac{65}{180} = \frac{13}{36}$	130°
Fair	60	$\frac{60}{180} = \frac{1}{3}$	120°
Reject	10	$\frac{10}{180} = \frac{1}{18}$	20°
Total	180	$\frac{180}{180} = 1$	360

The relative frequency is the fraction of the pots that are 'Very good', etc.

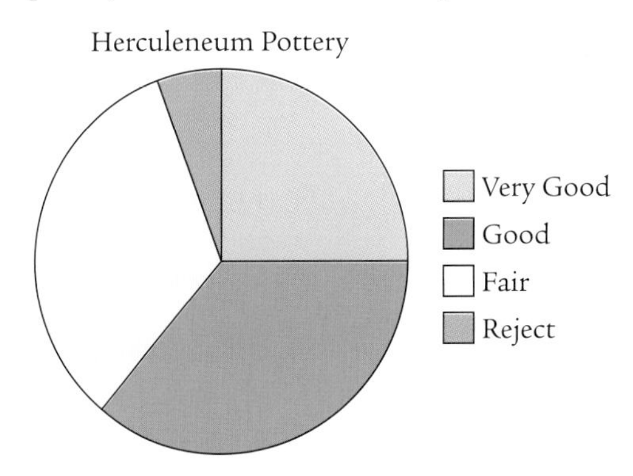

A key can be used to identify which sector represents which category.

Note that 'Very good' represents a quarter of the pots inspected: 90° is a quarter of 360°.

Example 1

James is a gardener and has to write a report for the estate he works for.

In one particular part, he has sampled the patch for weeds so that he can order a weedkiller.

Weed	Frequency
Buttercup	6
Daisy	9
Dandelion	12
Speedwell	9

a Copy the table and add columns for 'Relative frequency' and 'Angle at centre'.

b Draw a pie chart to illustrate the types of weeds in his patch.

a

Weed	Frequency	Relative frequency	Angle at centre
Buttercup	6	$6 \div 36 = \frac{1}{6}$	$= \frac{1}{6} \times 360° = 60°$
Daisy	9	$9 \div 36 = \frac{1}{4}$	$= \frac{1}{4} \times 360° = 90°$
Dandelion	12	$12 \div 36 = \frac{1}{3}$	$= \frac{1}{3} \times 360° = 120°$
Speedwell	9	$9 \div 36 = \frac{1}{4}$	$= \frac{1}{4} \times 360° = 90°$

b

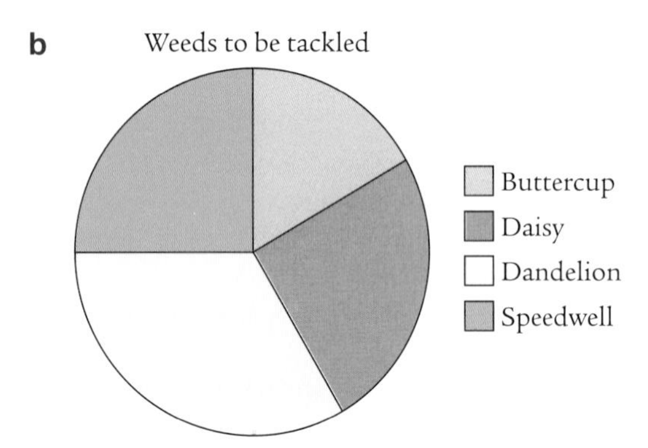

Example 2

Gemma was researching the number of text messages that people sent in one week.

Number of text messages	Frequency
0–99	4
100–199	11
200–299	14
300–399	4
400–499	3

a Draw a pie chart to illustrate the data.

b Suppose that next week the same question was asked of 500 people and the same pattern unfolded.

i How many people would you expect to say that they had sent between 300 and 399 text messages?

ii How many would say that they had sent fewer than 300?

a A total of 36 people were asked, so adding a 'Relative frequency' column and an 'Angle at centre' column to the table we get:

Number of text messages	Frequency	Relative frequency	Angle at centre
0–99	4	$4 \div 36 = \frac{1}{9}$	$\frac{1}{9} \times 360° = 40°$
100–199	11	$11 \div 36 = \frac{11}{36}$	$\frac{11}{36} \times 360° = 110°$
200–299	14	$14 \div 36 = \frac{7}{18}$	$\frac{7}{18} \times 360° = 140°$
300–399	4	$4 \div 36 = \frac{1}{9}$	$\frac{1}{9} \times 360° = 40°$
400–499	3	$3 \div 36 = \frac{1}{12}$	$\frac{1}{12} \times 360° = 30°$

This leads to the pie chart:

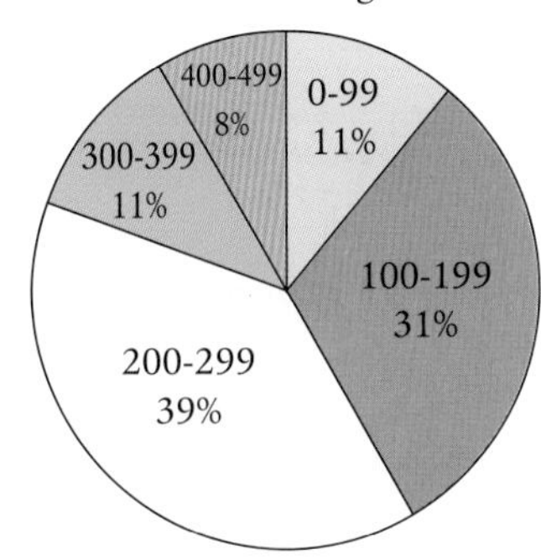

b **i** $\frac{1}{9}$ of the sample said they had sent 300–399 texts, so $\frac{1}{9}$ of $500 = 56$ people, to the nearest whole number.

ii $\frac{29}{36}$ of the sample said they sent fewer than 300 texts, so $\frac{29}{36}$ of $500 = 403$ people, to the nearest whole number.

Exercise 6.4A

1 A cat breeder kept a record of the number of kittens in a litter.
Here is the frequency table that he used:

Size of litter	Frequency	Relative frequency	Angle at centre
3	7		
4	14		
5	12		
6	6		
7	1		

Copy the frequency table and complete the 'Relative frequency' and 'Angle at centre' columns.

Draw a pie chart.

2 In one year a Mathematics department had £3000 to spend. They spent £1000 on textbooks, £750 on photocopying, £850 on stationery and £400 on calculators.

a What fraction of their budget did they spend on:

i textbooks **ii** photocopying **iii** stationery **iv** calculators?

You are going to draw a pie chart for the department's budget.

b Calculate the angle at the centre of the sectors which represent:

i textbooks **ii** photocopying **iii** stationery **iv** calculators.

c Draw the pie chart.

3 An opinion poll asked, 'If there was an election for the Scottish Parliament tomorrow, for whom would you vote?'

Of those asked, 38% said Labour, 36% said Scottish National Party, 12% said Conservative Party, 10% said Liberal Democrats and 4% said they would vote for other parties.

a By calculating 38% of 360°, work out the angle at the centre of the Labour Party sector.

b In the same way work out the angles at the centre of the other sectors.

c Draw the pie chart.

4 In a survey of where a group of people chose to go for their holidays, the following frequency table was obtained:

Holiday destination	Frequency	Relative frequency	Angle at centre
UK	17		
France	11		
Spain	12		
Italy	6		
Portugal	3		
USA	6		

Calculate the angles required for each sector and draw a pie chart to represent the holiday choices of this group of people.

Exercise 6.4B

1 A hotel wanted to investigate whether their short-stay guests stayed for 1, 2 or 3 nights.

They looked at their bookings over the last year and then made this pie chart.

The hotel had 3600 short-stay guests.

a How many guests stayed for 1 night?

b How many stayed for 2 nights?

c How many stayed for 3 nights?

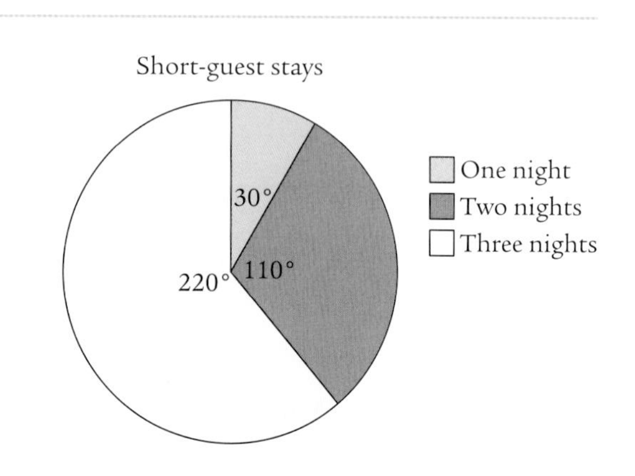

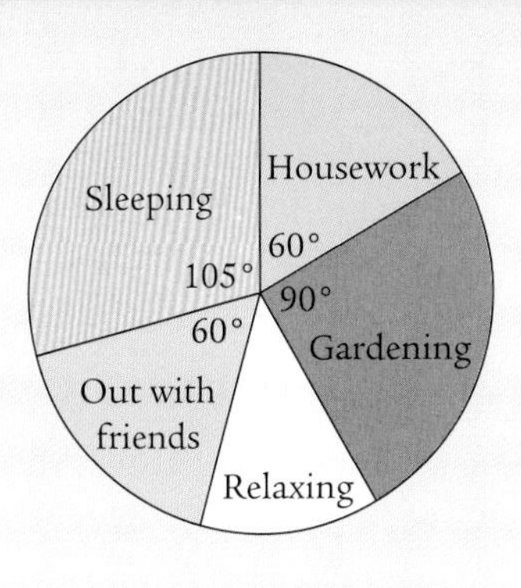

2 This pie chart shows how Mr Simpson spent his time one weekend, from midnight on Friday to midnight on Sunday.

a How many hours are there from midnight on Friday to midnight on Sunday?

b Calculate how long Mr Simpson spent:

i sleeping ii doing housework iii gardening iv out with friends.

c What size is the angle at the centre of the 'Relaxing' sector?

d How long did he spend relaxing?

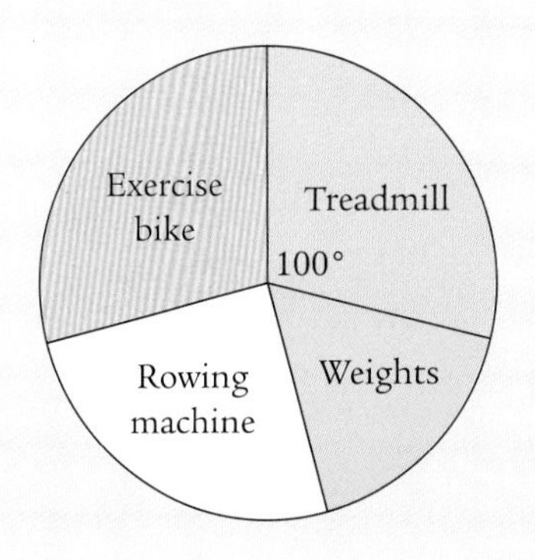

3 A fitness centre has to keep track of the usage of its equipment so that regular maintenance can take place.

This pie chart shows the time in hours each piece of equipment was used for.

During the period of the survey, the treadmill was used for 25 hours.

How long was the period of the survey?

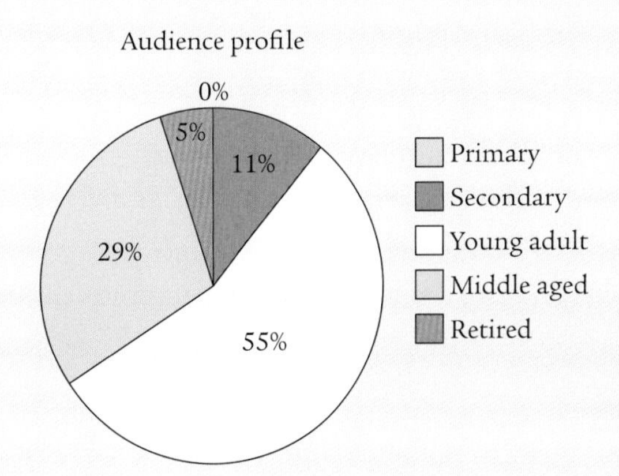

4 A TV company carries out a survey on who is watching a particular programme.

This will allow them to plan which adverts to show while the programme is running.

a Which category of the audience was the most represented in the sample?

b If 25 000 people were surveyed, how many were:

i at primary school

ii at secondary school

iii retired people?

c How many more young adults than middle-aged people were asked?

d Which advert would be the most suitable to air during this programme?

i Cutie Dolls

ii pop festival

iii pension scheme

iv SAGA Holidays

Preparation for assessment

1 At a coffee morning there is a £100 prize for guessing how many sweeties are in a jar.

Here are the guesses of the first 20 people:

170	215	185	219	180	190	210	200	183	175
201	212	211	210	215	206	199	190	205	203

Group the data into five class intervals and make a frequency table.

Draw a pie chart to show how the first 20 people made their guess.

2 A car dealership sold 2400 'Siesta' cars last year.

The cars were available in five different colours and the pie chart shows how the buyers of the cars chose their colour.

How many of the 'Siesta' cars sold last year were Colorado Red?

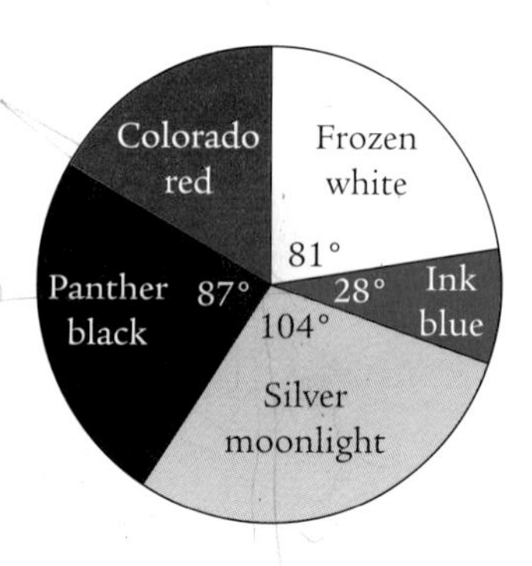

3 A consumer magazine conducts a survey to compare two types of shaving gel, 'Smoothglide' and 'Progel'.

Twenty people trialled both gels, giving each a mark out of 100. Here is a back-to-back stem-and-leaf diagram to illustrate the results.

Shaving gel survey

Smoothglide		Progel
7 5 0 0	**1**	
3	**2**	
3 2 1 1	**3**	7 7 8 9 9
8 7 7 7 6 4	**4**	2 2 2 2 3 7
2 2 1 0 0	**5**	5 8
	6	3 7
	7	2
	8	2
	9	5 5
$n = 20$		$n = 20$

2|5 represents 25

a Find the median mark given to each shaving gel.

b Which gel performed better in the survey?

4 Here is a line graph which was shown in a newspaper to illustrate how the points scored by the Scotland rugby team increased during a game.

The red crosses show when Scotland scored points.
For example, after 15 minutes they scored a penalty, which scores 3 points. After 30 minutes they scored another penalty so their score went up to 6 points.

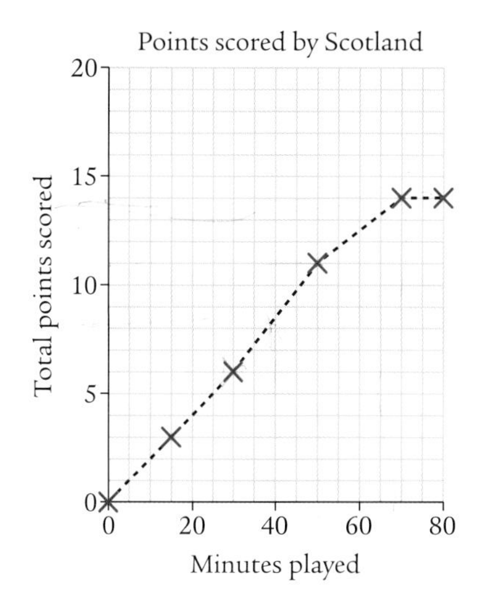

a How many points did Scotland have at the end of the game?

b How many points did Scotland have after 50 minutes?

c By how many points did Scotland's score increase between 50 minutes and 70 minutes? What happened during this time?

d Half-time comes after 40 minutes. How many points does the graph suggest that Scotland had at half-time? Is this possible?

e Explain how this graph could be misleading.

5 Remember the choice that you had to make between Ralph's sales policy and Masoud's sales policy?

Here is Ralph's bar chart.

a What are the expected sales for our company?

b Find the value of the total expected sales of all five companies.

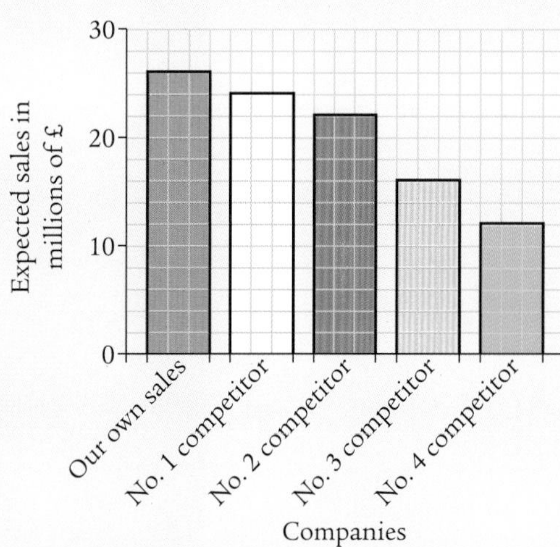

Now look at Masoud's pie chart. Notice that the angle at the centre for our own sales is 120°.

Expected sales in millions of £

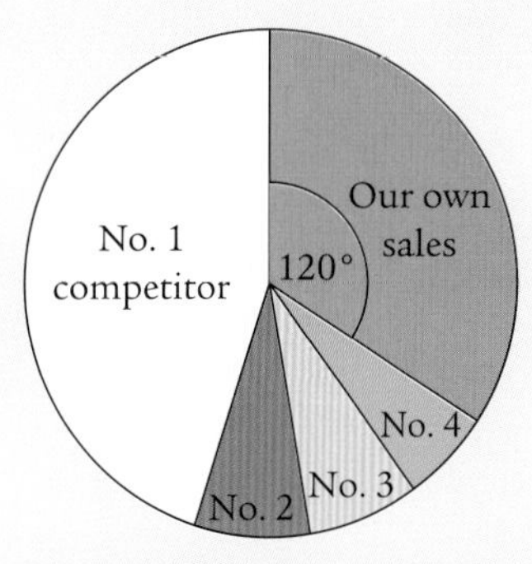

Remembering that your answer to part b gives the total expected sales for all 5 companies, calculate the expected sales for our company if we follow Masoud's policy.

As managing director of the company you must make decisions which maximise the profit for the shareholders. This means that you must choose the policy which expects the biggest sales. Which policy do you choose?

7 Equations and formulae

Before we start...

George is starting his career in property development.

He has just bought his first flat for £65 000. He also paid £25 000 to renovate the flat.

He is now going to rent the flat out to a tenant.

He has to decide on how much rent he will charge.

He remembers from a college course that he should aim for a yield of 0·065 (or 6·5%).

The formula to calculate yield is $Y = \frac{R}{C}$, where Y is yield expressed as a decimal fraction, £R is the annual rent and £C is the total cost of the flat and renovations.

To George it seems that the formula is no good since it is a formula for working out yield, but he wants a formula for working out rent.

He gets advice from his two friends, Alan and Stevie.

Alan tells him that he will need to rearrange the formula so that he can obtain one that works out the rent.

Stevie, on the other hand, tells him that he can actually use the formula as it stands to calculate the rent.

Who should George believe?

How much rent should he charge?

What you need to know

1 Simplify these expressions by collecting like terms.

a $2x + 3y + 4x + 5y$ **b** $7x + 9y - 5x - 3y$

c $8x + 3y - 5x - 7$ **d** $14x - 3y + x + 5y$

2 Multiply out the brackets in these expressions.

a $3(a + b)$ **b** $5(2m + 3n)$ **c** $10(2x - 7)$ **d** $4(3 - 2y)$

3 Multiply out these brackets and then collect like terms.

a $7(2x + 3) - 8$ **b** $5(3p + 1) - 4p$

c $4(5z - 2) + 7(z + 1)$ **d** $6(y - 3) + 7$

4 Adding and subtracting are **inverse** operations: one of the operations 'undoes' the other.

For example, adding 8 to a number undoes the action of subtracting 8 from the number.

What would you do to undo each of these actions:

a adding 2 **b** subtracting 7 **c** adding 15

d adding (-3) **e** subtracting $(-1{\cdot}5)$?

5 Multiplying and dividing are also inverse operations of one another.

What would you do to undo these:

a multiplying by 4 **b** dividing by 3

c multiplying by (-12) **d** dividing by (-10)?

6 Calculate:

a $-1 + 2$ **b** $-3 - 7$ **c** $-7 + (-11)$ **d** $-6 - (-5)$.

7.1 Linear equations

One-step equations

An equation is a mathematical statement that tells us that two expressions are equal.

The algebraic equation contains numbers and letters.

When a letter can be replaced by any number, it is referred to as a **variable**.

Some replacements will make the statement false.

Some replacements make the statement true: these are known as **solutions** of the equation.

We say such replacements **satisfy** the equation.

To find solutions, we can imagine the equation is a set of scales balanced on the equal sign.

If we do the same thing to both sides of an equation, we don't disturb the balance.

By carefully choosing what to do to both sides we can simplify the equation until we are left with a statement of the form 'letter = number'.

Example 1

Solve each equation.

a $x + 7 = 34$

b $y - 5 = 2$

a $x + 7 = 34$ — We're adding 7 to x, the inverse is subtracting 7

$\Rightarrow x + 7 - 7 = 34 - 7$ — Subtracting 7 from both sides to isolate the term in x

$\Rightarrow x = 27$

b $y - 5 = 2$ — We're subtracting 5 from y, the inverse is adding 5

$\Rightarrow y - 5 + 5 = 2 + 5$ — Adding 5 to both sides to isolate the term in y

$\Rightarrow y = 7$

Example 2

Solve each equation.

a $3x = 18$

b $\frac{z}{7} = 12$ $\left(\text{Remember } \frac{z}{7} = z \div 7\right)$

Letters can represent both constants and **variables**. If the value of a constant is unknown, this doesn't make it a variable.

a
$$3x = 18$$ We're multiplying x by 3, the inverse is dividing by 3
$$\Rightarrow \quad 3x \div 3 = 18 \div 3$$ Divide both sides by 3 to isolate the term in x
$$\Rightarrow \quad x = 6$$

b
$$\frac{z}{7} = 12$$ We're dividing x by 7, the inverse is multiplying by 7
$$\Rightarrow \quad \frac{z}{7} \times 7 = 12 \times 7$$
$$\Rightarrow \quad z = 84$$

Exercise 7.1A

1 Solve each equation, showing each step clearly.

a $x + 1 = 7$ **b** $x - 3 = 1$ **c** $a + 6 = 9$ **d** $b - 5 = 3$ **e** $c + 2 = 8$

2 Find the value of the variable which satisfies each of these equations:

a $5x = 15$ **b** $3x = 24$ **c** $\frac{x}{12} = 2$ **d** $\frac{x}{3} = 1$ **e** $9x = 63$

f $6m = 66$ **g** $\frac{n}{10} = 6$ **h** $4x = 76$ **i** $\frac{y}{9} = 11$ **j** $2z = 76$.

3 Solve these equations:

a $x + 7 = 15$ **b** $3x = 27$ **c** $y - 4 = 39$ **d** $z + 18 = 21$ **e** $x - 14 = 15$

f $\frac{z}{4} = 22$ **g** $b - 23 = 26$ **h** $c + 43 = 52$ **i** $\frac{k}{5} = 20$ **j** $y - 28 = 24$.

4 John knows that a rectangular carpet is 7 metres long and has an area of 28 m^2.

He wants to know the breadth of the carpet.

Let x metres represent the breadth of the rectangle.

a Why do we know that $7x = 28$?

b Solve this equation to find the breadth of the carpet.

5 Sally sells Christmas trees.

She started with 25 trees and now has 7 left.

Let t trees stand for the number of trees sold.

a Explain why $t + 7 = 25$.

b Solve this equation to find out how many trees she's sold.

c Sally thinks of a different way to work out how many trees she's sold.

Each tree costs £10 and she's already collected £180.

i Explain why $10t = 180$.

ii Solve this equation to confirm the number of trees sold.

6 In World War II the allied invasion of Normandy was meticulously planned.

The actual date of the invasion depended on the weather conditions. It was a variable.

During planning they called the day of the invasion D-Day.

The day after was called $D + 1$, the next $D + 2$, and so on.

We know that $D + 5$ was the 11th June.

a Make up an equation in D.

b Solve it to find out the date of D-Day.

Rearrangements

Sometimes we are given the equation in a slightly different form.

It helps to remember that:

i $a + b = b + a$ e.g. $3 + 4 = 4 + 3$

ii if $a = b$, then $b = a$ e.g. $3 + 4 = 7$ can be written as $7 = 3 + 4$

iii $a - b = -b + a$ e.g. $8 - 3 = -3 + 8$

If an equation appears in a slightly different form we can always make it look familiar by rearrangement.

Example 3

Solve each equation for x.

a $3 + x = 17$ **b** $25 = x + 12$ **c** $28 = 7x$ **d** $-14 + x = 3$

a

$3 + x = 17$ — $3 + x = x + 3$ ... we're adding 3 to x, the inverse is subtracting 3

$\Rightarrow 3 + x - 3 = 17 - 3$

$\Rightarrow x = 14$

b

$25 = x + 12$

$\Rightarrow x + 12 = 25$ — Adding 12 to x, the inverse is subtracting 12

$\Rightarrow x + 12 - 12 = 25 - 12$

$\Rightarrow x = 13$

c

$28 = 7x$

$\Rightarrow 7x = 28$ — We're multiplying x by 7, the inverse is dividing by 7

$\Rightarrow 7x \div 7 = 28 \div 7$

$\Rightarrow x = 4$

d

$-14 + x = 3$ — $a - b = -b + a$

$\Rightarrow x - 14 = 3$ — We're subtracting 14 from x, the inverse is adding 14

$\Rightarrow x - 14 + 14 = 3 + 14$

$\Rightarrow x = 17$

Exercise 7.1B

1 Solve each equation.

a $5 + x = 8$ b $18 = a - 6$ c $45 = 9b$ d $12 + c = 20$ e $28 = m + 22$

f $38 = 12 + n$ g $p + 25 = 100$ h $47 + y = 50$ i $27 = y - 24$ j $58 = 8y$

2 Find the value of the variable.

a $m - 9 = 4$ b $x + 7 = 3$ c $9 + x = 7$ d $x - 8 = -12$

e $18 = y + 22$ f $n + 17 = 95$ g $y - 34 = -40$ h $-50 = a - 27$

i $b + 63 = 26$ j $c - 100 = -150$

3 Solve these equations giving your answer as a common fraction in its simplest form.

a $2x = 15$ b $34 = 4y$ c $10m = 55$ d $21 = 14z$ e $8n = 44$

f $54 = \frac{p}{2}$ g $\frac{g}{5} = 16$ h $10 = \frac{20}{p}$ i $14 = \frac{42}{h}$ j $18 = \frac{3}{m}$

4 The temperature at midday was 5 °C. It had risen by 7 °C since 6 a.m.

What was the temperature at 6 a.m.?

Let x stand for the temperature at 6 a.m.

a Form an equation in x starting with '5 = ...'.

b Solve it to find the temperature at 6 a.m.

5 The total amount of money paid into an account is £75.
This was done over three equal instalments.

How much was each instalment?

a Let £y be the amount of money that makes up an instalment.

Construct an equation to model this situation.

b Solve to find the size of one instalment.

6 Kerr is doing well in a Fantasy Football League competition.

He had a total of 73 points after the first three weeks.

In the fourth week his total score changes to 92.
How many points were gained in week 4?

a Let x be the number of points gained in week 4.

Explain why the equation $73 = 92 - x$ models the situation.

b Solve this equation for x.

c After the fifth week his score is 81 points.

i Form a similar equation that could be used to deduce the number of points gained in week 5.

ii Solve this equation to work out his score from week 5.

7 A manufacturer makes square rugs. Each rug is edged with a trim that costs £5 per metre.

a Explain why the cost, £C, to edge a rug of side x m can be calculated from the formula $C = 20x$.

b If the cost were £300, form an equation in x and solve it to find the length of the rug.

c i Explain why $x = \frac{P}{4}$ might be a suitable formula for calculating the length of a side when you know the perimeter, P m, of the square.

ii Set the value of x at 18 and solve the equation to calculate P.

8 In Britain, electricity is supplied to the home at 240 volts.

In Science they will tell you that the current, I amps, needed to run an appliance of resistance R ohms can be calculated from the formula $I = \frac{240}{R}$.

a For $I = 100$, form an equation and solve it for R.

b Calculate the resistance when the current is 50 amps.

7.2 More linear equations

Two-step equations

As you learned in Chapter 3, equations of the form $ax + b = c$ are called linear equations.

Solving these still involves isolating the variable.

Although this may take several steps, we still aim for 'letter = number'.

Example 1

Solve these equations.

a $4x + 5 = 41$ b $6y - 15 = 33$

a
$$4x + 5 = 41$$
Isolate the term in x ... 5 is being added, the inverse is subtracting 5
$$\Rightarrow 4x + 5 - 5 = 41 - 5$$
$$\Rightarrow 4x = 36$$
Isolate x ... it is being multiplied by 4, the inverse is dividing by 4
$$\Rightarrow 4x \div 4 = 36 \div 4$$
$$\Rightarrow x = 9$$

b
$$6y - 15 = 33$$
$$\Rightarrow 6y - 15 + 15 = 33 + 15$$
$$\Rightarrow 6y = 48$$
$$\Rightarrow 6y \div 6 = 48 \div 6$$
$$\Rightarrow y = 8$$

Exercise 7.2A

1 Solve:

a $2x + 7 = 13$ b $3x - 5 = 7$ c $7x + 6 = 55$ d $4x - 21 = 3$

e $6x - 1 = 11$ f $9x + 4 = 40$ g $5x + 3 = 13$ h $8x - 5 = 19$.

2 Find the value of the variable that satisfies each equation.

a $5m + 13 = 33$ b $2n - 17 = 15$ c $5y - 12 = 73$ d $10z + 17 = 47$

e $8p - 3 = 93$ f $7q + 2 = 100$ g $3r - 1 = 98$ h $200t - 23 = 377$

3 A carpenter is making a cuboid storage box.

It is made from five squares of plywood and a lid of Perspex.

He charges £19, saying the Perspex lid cost £4.

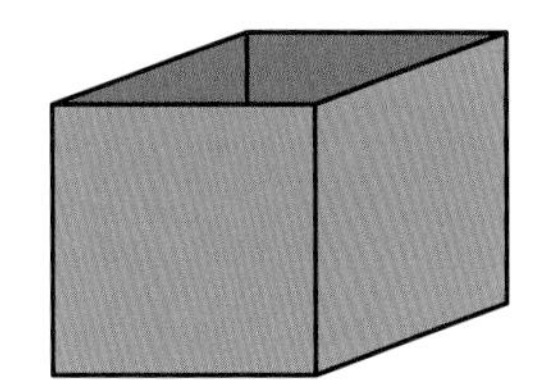

a Let £c be the cost of a square of plywood.
Form an equation in c for this situation.

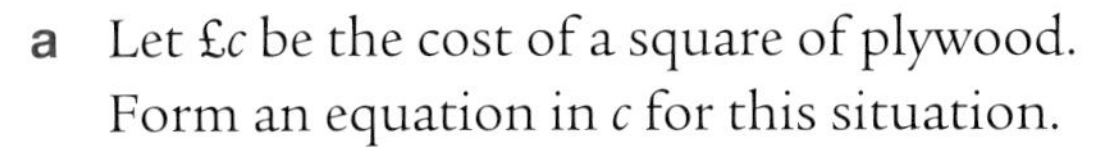

b Solve it to find the cost of a square of plywood.

4 In Chemistry Suzie learns that to find the number of hydrogen atoms in a particular type of molecule you double the number of carbon atoms and add 2.

a Let C stand for the number of carbon atoms in the molecule.

Write down an expression for the number of hydrogen atoms in the molecule.

b Let the number of hydrogen atoms in the molecule be 20.
Form an equation in C.

c Solve this equation to find the number of carbon atoms in the molecule.

5 A set of points (x, y) all lie on the same straight line.

To find the y-coordinate of any point on this line, we multiply the x-coordinate by 3 and add 2.

a Make up an expression in x for the y-coordinate.

b Form an equation in x for when the y-coordinate is 23.

c Solve it to find the x-coordinate of the point on the line that has a y-coordinate of 23.

Exercise 7.2B

1 Solve these equations.

a $2x + 7 = 8$ b $6x - 5 = 19$ c $x + 12 = 46$ d $8x - 7 = 49$

e $36 + 5y = 41$ f $23 + 9y = 50$ g $3 + 4y = 47$ h $35 + 7y = 39$

2 Three friends regularly went out for a meal.

They had £10 left in the 'kitty' from previous meals.

One evening when the bill came, they used this as part-payment for the meal. The remainder of the cost was shared equally by the friends.

Let £x be the amount that each person paid.

a Write an expression in x for the total amount they paid.

b Write down an expression for the figure written on the bill.

c If the figure written on the bill was £76 form an equation in x.

d Solve this equation for x to find the amount paid by each person.

3 A straight line has the equation $y = 2x - 3$.

a Solve the equation $11 = 2x - 3$ to find the point with a y-coordinate of 11.

b **i** Solve the equation $0 = 2x - 3$ to work out where the line crosses an axis.

ii Which axis?

c By considering both $5 = 2x - 3$ and $-5 = 2x - 3$, find the points where the line is 5 units away from the x-axis.

4 Mrs Parkes gave her Home Economics class homework.

She asked them to research, on the internet, safe cooking times for turkey.

On one internet site the advice given was '2 hours plus 15 minutes for every kilogram'.

a Let w kg represent the weight of the turkey. Write an expression for the amount of time, in minutes, required to cook a turkey.

b A particular turkey takes four and a half hours (remember to convert this to minutes) to cook.

i Form an equation in w to describe this situation.

ii Solve the equation to find the weight of the turkey.

5 In Sport Science, the following equation was used to model how the record for the mile altered over the years: $R = 17{\cdot}27 - 0{\cdot}0068y$, where R is the record in minutes for the mile and y is the year that the record was reached.

a Let $R = 4$ and solve the resulting equation to estimate the year that the record was 4 minutes.

b The first sub-4-minute mile was run by Roger Bannister in a time of 3·98 minutes. Form an equation and solve it to find the year he did it.

c The formula was created using the data from the years 1913 to 1980. Do you think it would be useful for finding out when the 3-minute mile will be broken?

7.3 Equations with the variables on both sides

Equations may have variables on both sides.
Perform whatever action is necessary to get 'letters = number' and then solve as before.

Example 1

Solve: $6x + 5 = 3x + 23$.

$$6x + 5 = 3x + 23$$
$\Rightarrow \quad 6x + 5 - 3x = 3x + 23 - 3x$ Action to lose the $3x$ from the right-hand side
$\Rightarrow \quad 3x + 5 = 23$
$\Rightarrow \quad 3x + 5 - 5 = 23 - 5$ Action to lose the 5 from the left-hand side
$\Rightarrow \quad 3x = 18$
$\Rightarrow \quad 3x \div 3 = 18 \div 3$ Action to isolate x
$\Rightarrow \quad x = 6$

Exercise 7.3A

1 Solve:

a $5x + 7 = 3x + 17$ **b** $6x + 11 = 4x + 27$ **c** $7x + 5 = x + 35$
d $8y + 15 = 5y + 42$ **e** $8y + 15 = 4y + 51$ **f** $11y + 37 = y + 97$
g $5p - 14 = 3p + 12$ **h** $6q - 17 = q + 28$ **i** $3m - 46 = 2m + 3$

2 **a** The Mercury Mail delivery company takes contracts to deliver in bulk. They ask for a standing charge of £15 plus £5 per parcel.
Let p stand for the number of parcels they agree to deliver. Write down an expression in p for the total cost of the contract.

b The Hermes Handling company has a similar scheme. Their standing charge is £25 and they charge £3 per parcel.
Write down an expression for the cost of delivering p parcels with Hermes.

c **i** Equate the two expressions.
ii Solve the resultant equation to find out the number of parcels for which both companies would charge the same.

3 At the car factory, camper vans are parked against a wall as they come off the assembly line.
It was noticed that 5 camper vans made a queue 4 metres short of the wall.
However, 7 vans made a queue 8 m longer than the wall.

a Let x m stand for the length of a van. Use the statements to form two expressions for the length of the wall.

b Equate the expressions and solve the resultant equation to find the length of a van.

c Find the length of the wall.

4 Spike is in a woodwork class and needs six strips of beading of equal length to put an edge round his coffee table.

He goes to the store cupboard and gets a long piece of beading. From this he cuts five of the strips he needs. There is a 40 cm piece left over.

a Use this information to help you write down an expression for the length of beading he took from the store cupboard. (Let x cm be the length of one strip that he cut off.)

b He goes back to the store and gets another length of beading the same size as before. This time he cuts one strip from it and there is a 248 cm piece left over.

Write down another expression for the length of beading he took from the cupboard.

c **i** Make an equation in x.

ii Solve it to find out the length of a strip.

iii What is the length of the original pieces he got from the store?

iv What was the perimeter of his coffee table?

5 How things are priced can be quite important.

A company had been offering tablet computers for sale for a deposit of £100 plus 10 equal monthly instalments.

a Write down an expression for the total cost of the tablet computer. (Let £x be one instalment.)

b Sales were low, so the company changed the terms of the scheme. It made the deposit £40 and asked for 12 monthly instalments of the same size as in the original deal.

Write down another expression for the cost of the tablet computer.

c Assuming that the actual cost of the tablet hadn't changed,

i form an equation in x

ii solve it to find the size of an instalment

iii work out what the tablet computer cost.

6 During an experiment, one beaker held 420 ml of red dye and another held 70 ml of blue dye.

The person carrying out the experiment had a small measure, which held an unknown volume. (Treat it as x ml.)

He discovered that he could equalise the contents of the beakers by removing two measures from the red dye and adding five measures to the blue dye.

a Write an expression for the contents, after the adjustment, of:

i the beaker with the red dye

ii the beaker with the blue dye.

b Form an equation in x and solve it to find the size of the measure.

c How much was in each beaker after the adjustment?

d Suppose the beakers now have the same volume of contents.

Three measures of red are removed and mixed with the blue.

Three measures of the blue/red mixture are removed and mixed with the red.

Is there more red dye in the blue beaker or more blue dye in the red beaker?

Example 2

On the quayside they are shifting a delivery of coal.

When the heap is used to fill 9 skips there are 5 tonnes left over.

When it is used to fill 5 skips there are 25 tonnes left over.

a What weight does a skip hold?

b How much coal is on the quayside?

(If you are not given a letter to use as a variable, pick your own ... but state what it stands for.)

a Let x tonnes be the weight in one skip.

The first statement tells us the heap weighs $9x + 5$ tonnes.

The second statement tells us that the heap weighs $5x + 25$ tonnes.

Since it is the same heap we can form an equation:

$$9x + 5 = 5x + 25$$

$$\Rightarrow \quad 9x + 5 - 5x - 5 = 5x + 25 - 5x - 5$$

$$\Rightarrow \quad 4x = 20$$

$$\Rightarrow \quad 4x \div 4 = 20 \div 4$$

$$\Rightarrow \quad x = 5$$

Note that $x = 5$ is not what was asked for, so the answer is: a skip holds 5 tonnes of coal.

b The heap weighs $9x + 5$ tonnes.

$\Rightarrow \quad 9 . 5 + 5 = 50$ Note that $9 . 5$ means 9×5

The heap on the quayside weighs 50 tonnes.

Check: we could work out the heap another way, viz.
$5x + 25 = 5 . 5 + 25 = 50$ ✓

Example 3

Where do the lines $y = 5x - 2$ and $y = 2x + 7$ intersect?

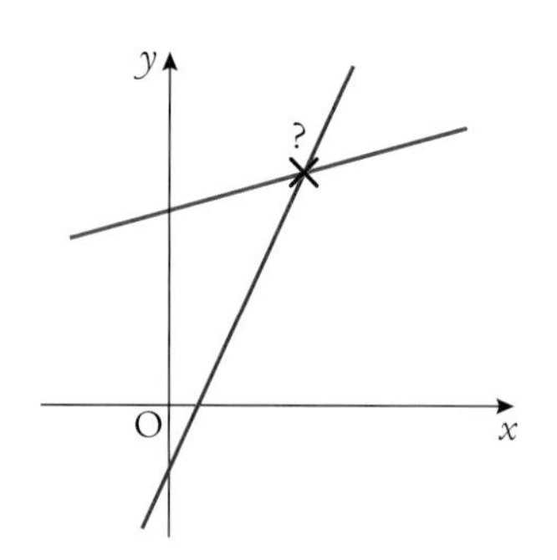

Remember that the equation of a line gives you an expression for y in terms of x.

When the lines intersect, the y-coordinates must be equal ... obviously.

So $5x - 2 = 2x + 7$

$\Rightarrow \quad 5x - 2 + 2 - 2x = 2x + 7 + 2 - 2x$

$\Rightarrow \quad 3x = 9$

$\Rightarrow \quad 3x \div 3 = 9 \div 3$

$\Rightarrow \quad x = 3$ — Note that $x = 3$ is only part of what was asked for

$\Rightarrow \quad y = 5x - 2 = 5.3 - 2 = 13$ — Note that 5.3 means 5×3

The lines intersect at (3, 13).

Exercise 7.3B

1 Solve:

a $4x - 13 = 2x - 3$ **b** $5y - 15 = 2y - 9$ **c** $6z - 35 = 2z - 3$

d $3p - 22 = p - 6$ **e** $7q - 57 = 2q + 3$ **f** $12r - 4 = 5r - 11$.

2 Find the value of the variable which satisfies each equation.

a $6a - 27 = 23 - 4a$ (Hint: add $4a$ to both sides.) **b** $8a + 16 = 38 - 3a$

c $5 - 2b = 4b - 13$ **d** $24 - 7b = 2b + 33$ **e** $c - 42 = 18 - 5c$

3 Where do the following pairs of lines intersect?

a $y = 4x + 7$ and $y = 10x - 41$ **b** $y = 6x + 23$ and $y = 7x + 17$

c $y = 3x + 28$ and $y = 7x + 4$ **d** $y = x + 3$ and $y = -2x + 12$

4 Colin is a plumber. For any job he charges a call-out fee of £50. He then adds £15 per hour for the time that he spends on the job.

Another plumber, Lawrie, has a call-out fee of only £20 but then adds £30 per hour.

a Form an equation using the two expressions and solve it to find the length of time for which both plumbers would charge the same.

b Which plumber would you choose for a job lasting 1 hour?

c Which plumber would you choose for a job lasting 3 hours?

5 In Physics, Hilary is carrying out an experiment using two toy racing cars, one red and one blue, rolling down an inclined plane.

The velocity at any time, t seconds after the timer is started, can be calculated using the formula:

$$v = u + at$$

where u m s^{-1} is the initial velocity, a m s^{-2} is the acceleration and t is the time in seconds. (m s^{-1} means the same as m/s.)

a The red car has an acceleration of 0·2 m s^{-2} and an initial velocity of 1 m s^{-1}.

Write an expression for the velocity of the red car at time t.

b The blue car has an acceleration of 0·4 m s^{-2} and an initial velocity of 0·5 m s^{-1}.

Find an expression for the velocity of this car at time t.

c After how many seconds will the cars have the same velocity?

6 **a** Find where the lines with equations $y = 3x - 4$ and $y = x + 2$ intersect.

b Where does the line $y = x$ cut $y = 3x - 4$?

c Does the line $y = x$ cut $y = x + 2$?
Form an equation and try to solve it. Comment.

7.4 More equations: simplify first

Brackets and like terms

In Chapter 1 you learned how to multiply out brackets and collect like terms to simplify expressions.

This will come in handy here.

Example 1

Solve $6(3x - 1) - 8x = 34$.

$$6(3x - 1) - 8x = 34$$
$$\Rightarrow \quad 18x - 6 - 8x = 34 \quad \text{Multiplying out the brackets}$$
$$\Rightarrow \quad 10x - 6 = 34 \quad \text{Collecting like terms}$$
$$\Rightarrow \quad 10x - 6 + 6 = 34 + 6$$
$$\Rightarrow \quad 10x = 40$$
$$\Rightarrow \quad 10x \div 10 = 40 \div 10$$
$$\Rightarrow \quad x = 4$$

Example 2

The Ardossan–Brodick Ferry crosses the Clyde between Arran and the mainland.

Let its calm-water cruising speed be x metres/minute.

Crossing with the tide adds another 40 m/min to its speed and it can complete the crossing in 50 minutes.

Crossing against the tide slows it down by 40 m/min and the crossing takes 60 minutes.

a Calculate the ferry's calm-water cruising speed.

b What is the distance, D, between Ardrossan and Brodick?

a Distance = speed × time

With the tide: $D = (x + 40) \times 50 = 50(x + 40)$.

Against the tide: $D = (x - 40) \times 60 = 60(x - 40)$.

Since both distances are the same:

$$\begin{aligned}
& 50(x + 40) = 60(x - 40) \\
\Rightarrow\quad & 50x + 2000 = 60x - 2400 \\
\Rightarrow\quad & 50x + 2000 - 50x + 2400 = 60x - 2400 - 50x + 2400 \\
\Rightarrow\quad & 10x = 4400 \\
\Rightarrow\quad & 10x \div 10 = 4400 \div 10 \\
\Rightarrow\quad & x = 440
\end{aligned}$$

The ferry's calm-water cruising speed is 440 metres per minute.

b We use either expression to calculate the distance and we can use the other expression to confirm our answer.

$$D = 50(x + 40) = 50(440 + 40) = 50 \times 480 = 24\,000\text{ m} = 24\text{ km}.$$

The distance between Ardrossan and Brodick is 24 kilometres.

Check: we can use the other expression to confirm our answer.

$$D = 60(x - 40) = 60(440 - 40) = 60 \times 400 = 24\,000\text{ m} \checkmark$$

Exercise 7.4A

1 Solve:

a $2(x + 3) = 8$ **b** $3(y - 1) = 6$ **c** $4(z - 5) = 8$

d $7(2m + 5) = 77$ **e** $15 = 3(4n + 1)$ **f** $4(3p + 2) = 56$

g $2(5q - 12) = 36$ **h** $10(3x - 4) = 20$ **i** $3(y - 2) - y = 2$

j $5(2z + 7) - 4z = 41$ **k** $8m + 7(2 - m) = 20$ **l** $4(n + 5) + 3 = 31.$

2 Alan is a gardener. He is going to extend a flower bed in a park.

The bed currently measures 4 metres by 2 metres.

He wants the bed to have an area of $12{\cdot}5\ \text{m}^2$.

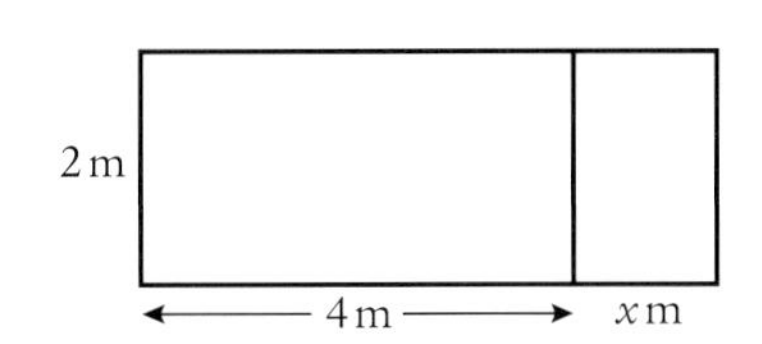

a Write down an expression for the length of the extended flower bed.

b Write down an expression (using brackets) for the area of the extended flower bed.

c Form an equation and solve it to find by how much he needs to extend the bed.

3

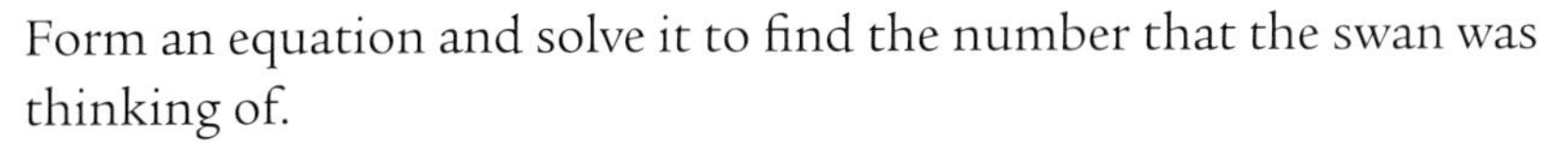
Form an equation and solve it to find the number that the swan was thinking of.

4 Lewis Carroll wrote a nonsense poem called *The Hunting of the Snark*. In the poem two characters, the Beaver and the Butcher, try to do a problem and get to the answer 3.

Taking x as the subject to reason about –

A convenient number to state –

We add Seven, and Ten, and then multiply out

By One Thousand diminished by Eight.

The result we proceed to divide, as you see,

By Nine Hundred and Ninety Two:

Then subtract Seventeen, and the answer must be

Exactly and perfectly true.

a Build up an expression in x just as it is described in the poem.
(Hint: 'One Thousand diminished by Eight' would look like $(1000 - 8)$.)

b What value would you give x if you wanted the expression to equal 3?

c What does this have to do with inverse operations?

5 Thomas Maxwell could row a boat at a speed of x metres per minute.

The river ran with a current of 10 metres per minute. So when Thomas rowed upstream his actual speed was $x - 10$ metres per minute.

a He rowed upstream for 15 minutes. Write an expression in x for the distance he travelled.

b He rowed downstream, with the current, and had to row for 12 minutes to get back to where he started.

i Form an expression for the distance he rowed downstream.

ii Since both distances are the same, form an equation in x.

iii Solve the equation to find Thomas' rowing speed in still water.

c What was the total distance he rowed?

6 Joe commutes between Glasgow and Edinburgh.

His car has a comfortable average cruising speed.

However, conditions are such that when going to Edinburgh he can add 5 miles per hour on to his average speed. It takes him an hour to get there.

On the way back to Glasgow things are different, and because of congestion he can take 15 mph off his average cruising speed. It takes him 1·4 hours for the journey.

a Let x stand for his average cruising speed.
Form two expressions for the distance between Glasgow and Edinburgh in terms of x.

b Form an equation and solve it to find his average cruising speed.

c Hence find the distance between Edinburgh and Glasgow.

Once you are comfortable with the manipulations, you need not show every step ... though you should be ready to explain them.

Example 3

Solve: $8(x + 3) + 5(3 - x) = 45 + 7(x - 1)$.

	$8(x + 3) + 5(3 - x) = 45 + 7(x - 1)$	
$\Rightarrow$	$8x + 24 + 15 - 5x = 45 + 7x - 7$	Removing brackets
$\Rightarrow$	$8x - 5x - 7x = 45 - 7 - 24 - 15$	'Letters = numbers'
$\Rightarrow$	$-4x = -1$	Combining like terms
$\Rightarrow$	$x = \frac{1}{4}$	Isolating x

Example 4

In Physics, they examine how weights on a beam will balance.
The theory says that for weights w_1 and w_2 to balance, the beam has to be suspended from a point P (see the diagram) so that $w_1d_1 = w_2d_2$.

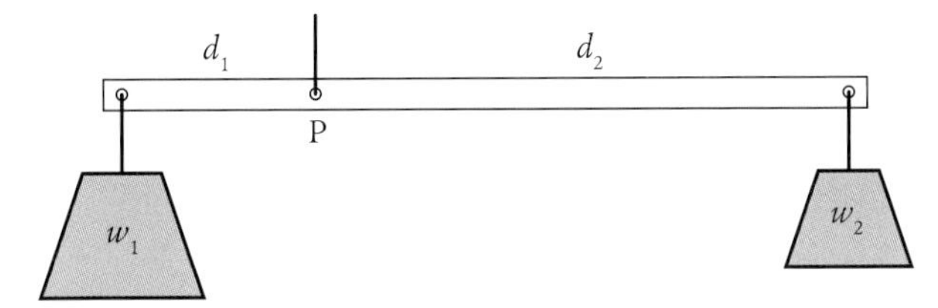

If the beam is 100 cm long, where should the point P be located so that a 24 kg weight will counterbalance a weight of 72 kg?
We wish to know how far from the end of the beam to suspend the 72 kg weight.
Let this distance be x cm.
This means the other weight is $(100 - x)$ cm from the other end.
Using the theory:

$$w_1d_1 = w_2d_2$$
$$\Rightarrow \quad 72x = 24(100 - x)$$
$$\Rightarrow \quad 72x = 2400 - 24x$$
$$\Rightarrow \quad 72x + 24x = 2400$$
$$\Rightarrow \quad 96x = 2400$$
$$\Rightarrow \quad x = 25$$

So P should be positioned so that it is 25 cm from the 72 kg weight and 75 cm from the 24 kg weight.

Exercise 7.4B

1 Solve:

a $3(y + 7) + 2(y - 2) = 28$ **b** $6(3y - 4) + 2(y - 3) = 70$ **c** $5(2y - 1) + 6(y + 2) = 23$

d $3(4x + 7) = 5(x + 7)$ **e** $4(5x - 2) = 5(2x + 3) + 2$ **f** $7(x + 5) = 2(1 - 2x)$.

2 Rab and Wayne are laying a driveway.

The lorry delivering the boxes of bricks has left them some distance away. They have to carry the boxes to where they are going to be used.

The boxes are too heavy, so Rab takes out five bricks from each box that he carries. Wayne takes three bricks out of each box that he carries.

a If there were b bricks in each box initially, write down an expression for the number of bricks in one of the boxes that Rab carries.

b Write down an expression for the number of bricks in one of the boxes that Wayne carries.

An hour later they stop for a break and Rab has shifted six boxes while Wayne has shifted five boxes.

c Write down an expression (using brackets) for the number of bricks:

i that Rab has carried in total

ii that Wayne has carried in total.

d They have both carried the same amount of bricks.

i By equating the two expressions find a value for b.

ii Hence find the number of bricks that were initially in a box.

3 Jean makes two patchwork quilts.
The measurements are in centimetres.

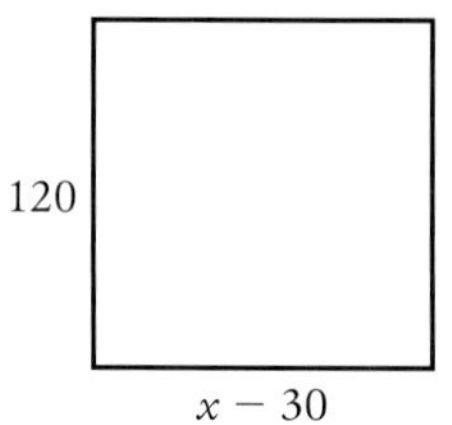

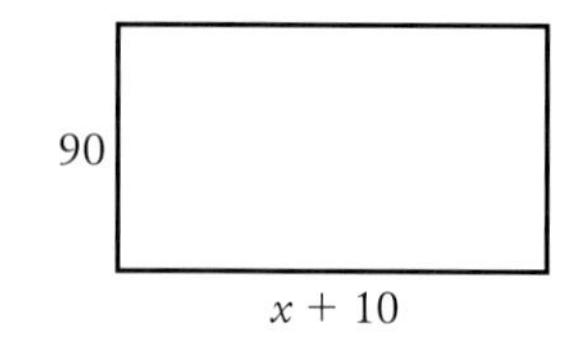

a If the two quilts have equal area, construct an equation and solve it to find the value of x.

b If, instead, Jean wanted the area and the perimeter of the second quilt to have the same numerical value calculate the value of x. Comment.

c Explore the possibility of the first quilt having the same property.

4 The Music Maestro Ministore sells CDs all at the same price.

They are reviewing their pricing policy.

If they put £1 on the price of a CD, I could just buy five CDs.

On the other hand, if they were to drop the price by £2, I could afford seven.

a By finding two expressions for how much money I've got, find the current cost of a CD.

b How much money do I have?

5 A man is 38 and his son is 6 years old.

How long will it be before the father is twice as old as his son?

(Hint: let the number of years till this occurs be x. Find an expression for their ages then. Knowing the father is then twice as old as the boy will let you form an equation to solve.)

6 An employer has a workforce of 18 people.
His total wage bill for a day is £870.
Some get paid £40 per day and the rest £65.

How many workers does he employ at each rate?

7 A barrier at a level crossing is 3·5 metres long, swinging on a pivot at P and counterbalanced by a weight, w_1.

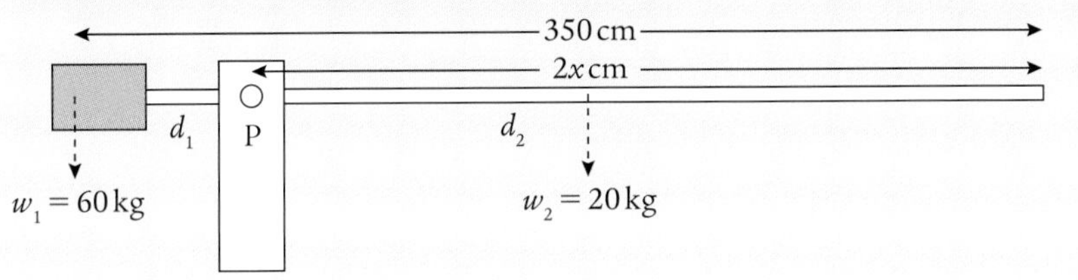

The barrier weighs 20 kg, which can be thought of as acting halfway along the arm.
The counter-balance weighs 60 kg.

The theory is that for the barrier to work smoothly $w_1d_1 = w_2d_2$.

Let the arm be $2x$ cm long.

a Form an equation in x and solve it to find the length of the arm.

b How far is the counterweight from the pivot point?

7.5 Formulae

Substitution and evaluation

When we are given values for variables we can **substitute** these values into expressions and formulae ... and **evaluate** them.

When we write down our answer it is safer to make the substitution and then evaluate rather than trying to do both together.

Example 1

Given that $x = 3, y = 5$ and $z = -1$, evaluate these expressions:

a $3x + 2y - z$ **b** $3y^2 + 2z$ **c** $\sqrt{y^2 - x^2}$ **d** $x(3z + 4) - 2y$.

a $3x + 2y - z$

$= 3.3 + 2.5 - (-1)$ Note the use of the dot for multiplication ... it helps clarity

$= 9 + 10 + 1$

$= 20$

b $3y^2 + 2z$

$= 3.5^2 + 2.(-1)$ Note: we're not tempted to evaluate while substituting (-1) for z

$= 75 - 2$

$= 73$

c $\sqrt{y^2 - x^2}$

$= \sqrt{5^2 - 3^2}$

$= \sqrt{25 - 9}$ Note: one step at a time

$= \sqrt{16}$

$= 4$

d $x(3z + 4) - 2y$

$= 3.[3.(-1) + 4] - 2.5$

$= 3.(-3 + 4) - 10$

$= 3.1 - 10$

$= -7$

Example 2

In Science, we know that $V = IR$, where V volts is the voltage, I amps is the current and R ohms is the resistance.

Calculate V, when $I = 2$ and $R = 6$.

$V = IR$

Given $I = 2$ and $R = 6 \Rightarrow V = 2.6 = 12$.

The voltage is 12 volts.

Example 3

A washer in a machine has to take some pressure.

To work this out we need its area.

This we get from the formula $A = \pi(R^2 - r^2)$.

What is the area of the washer with outside radius $(R) = 1{\cdot}2$ cm and inside radius $(r) = 0{\cdot}9$ cm?

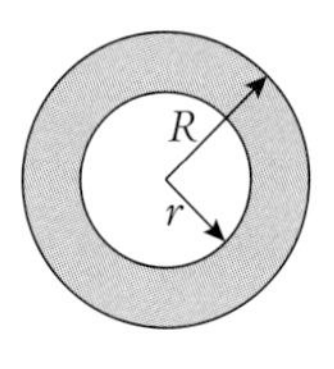

$A = \pi(R^2 - r^2)$

Given $R = 1{\cdot}2$ and $r = 0{\cdot}9$.

$\Rightarrow \quad A = \pi(1{\cdot}2^2 - 0{\cdot}9^2)$

$= \pi . 0{\cdot}63$

$= 1{\cdot}98$ (to 3 s.f.)

The area of the washer is $1{\cdot}98\ \text{cm}^2$.

Exercise 7.5A

1 Given that $a = 2$, $b = 3$ and $c = -5$, evaluate the following expressions.

a $a + b + c$ **b** $4a$ **c** $3b^2$ **d** $5c^2$

e $a^2 - 1$ **f** $b + 2c$ **g** $6b^2 + 3a^2$ **h** $\sqrt{10a - 4}$

i $\sqrt{c^2 - b^2}$ **j** $2ab + ac$ **k** $-5ac$ **l** a^3b^2

m $\sqrt{4c^2} - 4a$ **n** $3ca + ac$ **o** $\dfrac{3b}{5a}$ **p** $\dfrac{a}{b} + \dfrac{1}{3}$

q $\dfrac{18}{b} + \dfrac{14}{a}$ **r** $\dfrac{1}{a} + \dfrac{b}{6} + \dfrac{c}{10}$ **s** $\sqrt{\dfrac{27b}{8a}}$

2 Calculate the value of the formula using the values of the variables given.

a $N = 2M - 3$ (given $M = 14$) **b** $V = u - 5$ (given $u = 17$)

c $l = \dfrac{P}{6}$ (given $P = 78$) **d** $s = \sqrt{A}$ (given $A = 81$)

e $C = x + y$ (given $x = 24$, $y = 7{\cdot}2$) **f** $Q = PR$ (given $P = 27$, $R = 5$)

g $L = \dfrac{M}{N}$ (given $M = 28{\cdot}5$, $N = 3$) **h** $y = 3x + c$ (given $x = 6$, $c = -2$)

i $l = \sqrt{A} + 2b$ (given $A = 64$, $b = 3{\cdot}5$) **j** $A = l^2 + ab$ (given $l = 7$, $a = 3$, $b = 4$)

3 The number of points, P, that a football team has in a league depends on its performance.

It is calculated using the formula:

$P = 3w + d$

where w is the number of games won and d is the number of games drawn.

No points are awarded for losing a game.

a **i** Letting $w = 1$ and $d = 0$, calculate how many points a team gets for a win.

ii Letting $w = 0$ and $d = 1$, calculate what they get for a draw.

b Use the formula to calculate the points awarded to a team who have:

i won 7 and drawn 3

ii won 6 and drawn 6.

4 To estimate the heating bills of different factory spaces, an engineer first works out the volume of the space using the formula $V = lbh$, where l is length, b is breadth and h is height.

Use the formula to calculate the volume of these spaces.

a

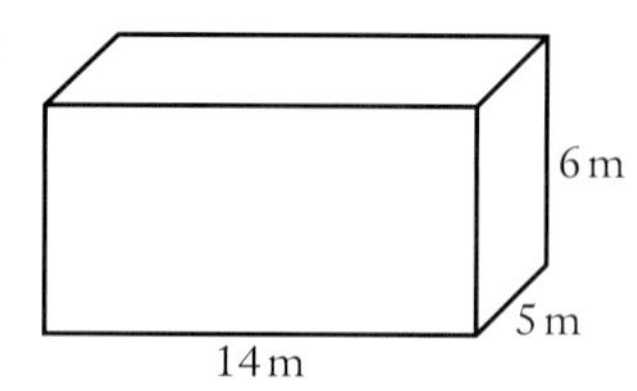

b

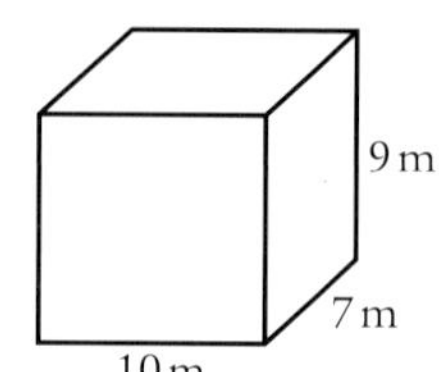

5 Claire works with the police. She studies the data from average speed cameras.

She uses the formula $S = \frac{D}{T}$, where S mph is average speed, D miles is distance and T hours is time.

The speed limit is 60 miles per hour (mph).

Which of these vehicles has an average speed of over 60 mph:

a a Vauxhall Corsa, which travelled 110 miles in 2 hours

b a Ford Focus, which travelled 182 miles in 3 hours

c a Renault Laguna, which travelled 35 miles in half-an-hour?

6 **a** For a sky diver, there are two important formulae. The first is:

$$v = u + 9{\cdot}8t$$

where v metres per second is the speed of heading towards the Earth, u metres per second is his speed when he started the stopwatch, and t seconds is the time that has elapsed since starting his watch.

i A sky diver started his watch when he was travelling at 5 m/s. How fast was he going 3 seconds later?

ii How fast was he travelling after 7 seconds?

b The second formula is:

$$s = ut + 4{\cdot}9t^2$$

where s metres is the distance he's travelled since the watch was started and u and t are as described above.

i If $u = 1$ m/s, calculate how far he has fallen in 6 seconds.

ii How far has he fallen in 7 seconds?

iii How far did he fall during the 7th second?

iv How far did he fall during the 10th second?

7 When studying the motion of the planets in 1619, Kepler discovered the connection between the time it takes a planet to go round the Sun, T years, and the distance the planet was from the Sun, R astronomical units (AU).

Using his law, $T = \sqrt{R^3}$, answer the following questions.

Also, when working with spreadsheets we can't get the computer to do these manipulations.

If we want it to work out V, then we have to give it a formula for V ... not a formula for L that contains V.

The method used to change the subject of a formula is very similar to the method used to solve an equation.

Example 1

Make x the subject of the formula $G = x + a$.

Treat this as an equation where we are trying to find out the value of x ... trying to isolate x.

$$G = x + a$$

$$\Rightarrow \quad G - a = x + a - a$$

$$\Rightarrow \quad G - a = x$$

$$\Rightarrow \quad x = G - a$$ x is now the subject of the formula

Example 2

Make D the subject of the formula $S = \frac{D}{T}$.

$$S = \frac{D}{T}$$ D is being divided by T, the inverse is to multiply by T

$$\Rightarrow \quad S \times T = \frac{D}{T} \times T$$

$$\Rightarrow \quad ST = D$$

$$\Rightarrow \quad D = ST$$ D is now the subject of the formula

Example 3

Change the subject of the formula, $E = 3w - k$, to w.

Our focus is to isolate the w.

$$E = 3w - k$$

$$\Rightarrow \quad E + k = 3w - k + k$$ To isolate the term that contains w

$$\Rightarrow \quad E + k = 3w$$

$$\Rightarrow \quad (E + k) \div 3 = 3w \div 3$$ To isolate w

$$\Rightarrow \quad \frac{E + k}{3} = w$$

$$\Rightarrow \quad w = \frac{E + k}{3}$$ w is now the subject of the formula

Exercise 7.6A

1 Make the coloured variable the subject of the formula in each case.

a $a = b + c$ **b** $q = p + r$ **c** $s = t - u$

d $V = IR$ **e** $D = ST$ **f** $h = \frac{v}{n}$

g $d = \frac{C}{\pi}$ **h** $C = 5H + S$ **i** $F = 15H - D$

2 **a** **i** Make t the subject of the formula $v = t - u$.

ii Hence calculate t when $u = 12$ and $v = 7$.

b **i** Make m the subject of the formula $d = \frac{m}{v}$.

ii Hence calculate m when $v = 2$ and $d = 8940$.

3 Martin is a taxi driver who charges 80p per mile and 20p per minute.

He makes a spreadsheet to calculate the charge, P pence, in terms of the distance d miles and t minutes, using the formula $P = 80d + 20t$.

	A	B	C
1	Miles	Minutes	Charge
2	10	5	900
3	8	6	760
4	12	7	1100
5	20	8	1760
6	20	14	1880

Column A holds the miles, column B the minutes and column C holds the formula for working out P, the charge in pence ... =80*A2+20*B2.

a Make d the subject of the formula $P = 80d + 20t$.

b If a new spreadsheet is made to calculate values of d with P values in column A and t values in column B, write out a formula for d in column C.

4 Charlie is doing an experiment in Physics.

He is testing Ohm's law.

Ohm's law is a formula $V = IR$, where V volts is the voltage, I amps is the current and R ohms is the resistance.

He varies the voltage and the resistance, then measures the current that is flowing.

It would be more convenient if I was subject of the formula.

a Make I the subject of the formula $V = IR$.

b On a spreadsheet he records values of V and R.
The computer evaluates the corresponding values of I in the third row.

Setting	i	ii	iii	iv
V volts	12	9	6	18
R ohms	3	6	3	4.5
I amps				

Use the formula that you have for I to calculate these values.

5 Frank was formerly a famous footballer. Now he earns a living by making after-dinner speeches about his experiences. He charges a fee of £400 for each dinner that he attends plus £50 for every hour that he is there.

This can be described by the formula $F = 50H + 400$, where £F is his fee and H hours is the length of time he is there.

Sometimes the organisers of the dinner will tell him how much they are willing to pay, £F.

Frank then has to work out the number of hours he has to work.

To do this he needs to change the subject of the formula to H.

a Express H in terms of F.

b An amateur football club are willing to pay Frank a fee of £600 to speak at their annual dinner. Use your formula from **a** to work out how long he will stay.

6 Alanna is buying a new carpet for her flat.

The total cost to Alanna is given by the formula $C = AP$, where £C is the total cost, A m^2 is the area of the room to be carpeted and £P is the cost of a square metre of carpet.

When she goes to the warehouse, all the carpets have their cost per square metre displayed.

a Make P the subject of the formula $C = AP$.

b The room in Alanna's flat has an area of 8 m^2.
The total cost that she can afford is £250.

Work out the maximum cost per square metre that Alanna can afford.

Example 4

The volume of a cone is given by the formula $V = \frac{1}{3}\pi r^2 h$, where r is the radius of the base and h is the height of the apex of the cone.

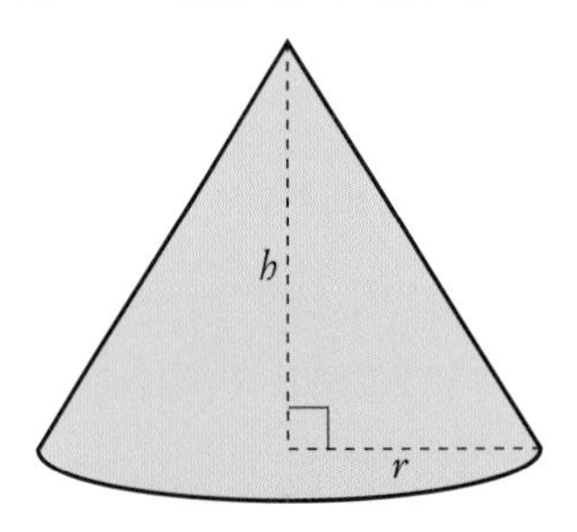

Make r the subject of the formula.

$V = \frac{1}{3}\pi r^2 h$

$\Rightarrow \quad 3V = \pi r^2 h$ Multiplying by 3 ... inverse of finding $\frac{1}{3}$

$\Rightarrow \quad \frac{3V}{\pi h} = r^2$ Dividing by πh ... inverse of multiplying by πh

$\Rightarrow \quad r^2 = \frac{3V}{\pi h}$

$\Rightarrow \quad r = \sqrt{\frac{3V}{\pi h}}$ Taking the square root ... inverse of squaring

Example 5

The length of the hypotenuse h cm, of a right-angled triangle with shorter sides, a cm and b cm can be found from the formula:

$$h = \sqrt{a^2 + b^2}.$$

Make b the subject of the formula.

$$h = \sqrt{a^2 + b^2}$$

$\Rightarrow \quad h^2 = a^2 + b^2$ Squaring ... inverse of taking the square root

$\Rightarrow \quad b^2 = h^2 - a^2$ Isolating the term in b

$\Rightarrow \quad b = \sqrt{h^2 - a^2}$ Taking the square root ... inverse of squaring

Exercise 7.6B

1 The formula for the area of a circle is $A = \pi r^2$.

- **a** Make r the subject.
- **b** Hence calculate the radius of a circle with an area of 153·86 cm^2. (Use $\pi = 3{\cdot}14$.)
- **c** Calculate the **diameter** of a circle with an area of 28·26 cm^2.

2 In Physics, the formula $P = I^2R$ is used to calculate the power, P joules, when I amps is the current and R ohms is the resistance.

- **a** **i** Make R the subject of the formula.
  - **ii** Hence calculate the resistance when the power is 2025 joules and the current is 9 amps.
- **b** **i** Change the subject of the formula to I.
  - **ii** Hence calculate the current given that the resistance is 16 ohms when the power is 2500 joules.

3 A lifestyle adviser likes to give her clients a target weight.

She knows the formula for body mass index, $B = \frac{w}{h^2}$, where w kilograms is the weight of the person and h metres is their height.

Since she is usually looking to calculate a target weight, she would prefer that w was the subject of the formula.

- **a** Rearrange the formula so that w is its subject.
- **b** The ideal BMI of a man is 23. Use the new formula to work out the ideal weight for a man of height 1·8 metres.
- **c** The ideal BMI of a woman is 21. Use the new formula to calculate the ideal weight for a woman who is 1·6 metres tall.

4 The area of a trapezium can be calculated from the formula $A = \frac{h}{2}(a + b)$, where a cm and b cm are the lengths of the parallel sides and h cm is the perpendicular distance between them.

a Make a the subject of the formula.

b If the area of a trapezium is 48 cm^2 and the parallels are 6 cm apart, what is the length of a when b is i 1 cm ii 2 cm iii 3 cm?

5 The population, P, of herons on a river has been modelled by $P = \frac{(4L - 5y)}{5}$, where L is the population last year and y is the number of years since the record started.

a Make L the subject of the formula.

b Calculate last year's population when this year's is 50 herons and we are in year 6 of the survey.

c Has the population increased or decreased?

6 The mean weight of a group of N boys is X_{old} kg.

One more boy of weight x kg joins the group.

The new mean weight can be calculated using the formula $X_{\text{new}} = \frac{NX_{\text{old}} + x}{N + 1}$.

a Make X_{old} the subject of the formula.

b **Challenge:** Make N the subject of the formula.

Preparation for assessment

1 Make x the subject of the formula $y = 3(2x + d)$.

2 Solve for k:

a $6k - 11 = 3k + 4$

b $5(k + 3) = 5$.

3 Martha is buying a new television by hire purchase.

The overall cost of the TV is £949·99.

She pays a deposit of £109·99 and then 24 equal monthly instalments.

a Let £x be the cost of one of the monthly instalments.
Write down an expression in x for the total of all the instalments.

b Write down an expression for the total cost of the TV (including the deposit).

c Form an equation in x and solve it to find the cost of each instalment.

4 Mary has £700 in a bank account and £30 in a building society.

She deposits £10 in each account every week.

How long will it be before the building society account is half the size of the bank account?

(Hint: let x be the number of weeks and form expressions for the amount in each account after x weeks.)

5 In chemistry the alkanes form a group of hydrocarbons.

Molecules of the alkanes are made up of only carbon and hydrogen atoms.

```
     H             H   H              H   H   H
     |             |   |              |   |   |
  H—C—H        H—C—C—H          H—C—C—C—H
     |             |   |              |   |   |
     H             H   H              H   H   H
```

CH_4 Methane

C_2H_6 Ethane

C_3H_8 Propane

	Methane	Ethane	Propane
Number of carbon atoms, C	1	2	3
Number of hydrogen atoms, H	4	6	8

a The next molecule in the sequence is butane, which has 4 carbon atoms.
How many hydrogen atoms will it have?

b Write down a formula for finding H (the number of hydrogen atoms) when you know C (the number of carbon atoms).

c Use this formula to work out how many hydrogen atoms there are in a molecule of hexane, which has 6 carbon atoms.

d How many hydrogen atoms are there in a molecule of an alkane which has 20 carbon atoms?

e Calculate how many carbon atoms there are in an alkane molecule which has 60 hydrogen molecules.

6 Remember George, who was starting his career in property development?

He had spent £65 000 on a flat and a further £25 000 to renovate it: £90 000 in total.

He wanted to charge a rent so that the yield was 0.065 (or 6·5%).

The formula to calculate the yield was $Y = \frac{R}{C}$, where Y is yield, £R is the annual rent and £C is the total cost of the flat and renovations.

He gets advice from his two friends, Alan and Stevie.

Alan tells him that he will need to rearrange the formula so that he can obtain one that works out the rent.

Stevie, on the other hand, tells him that he can actually use the formula as it stands to calculate the rent.

Who should George believe?

How much rent should he charge?

8 Pythagoras' theorem

Before we start...

James wants to study the Finnieston Crane.

He took a photograph and placed it on a document on the computer.

Using the drawing tools he drew the line AB.

Double-clicking this line, he was given the information that its height is 1·18 cm and its width is 7·62 cm.

Is this enough information for James to deduce the length of AB?

What you need to know

1 Given the length of the side of a square, can you calculate its area?

Calculate the area of each of these squares:

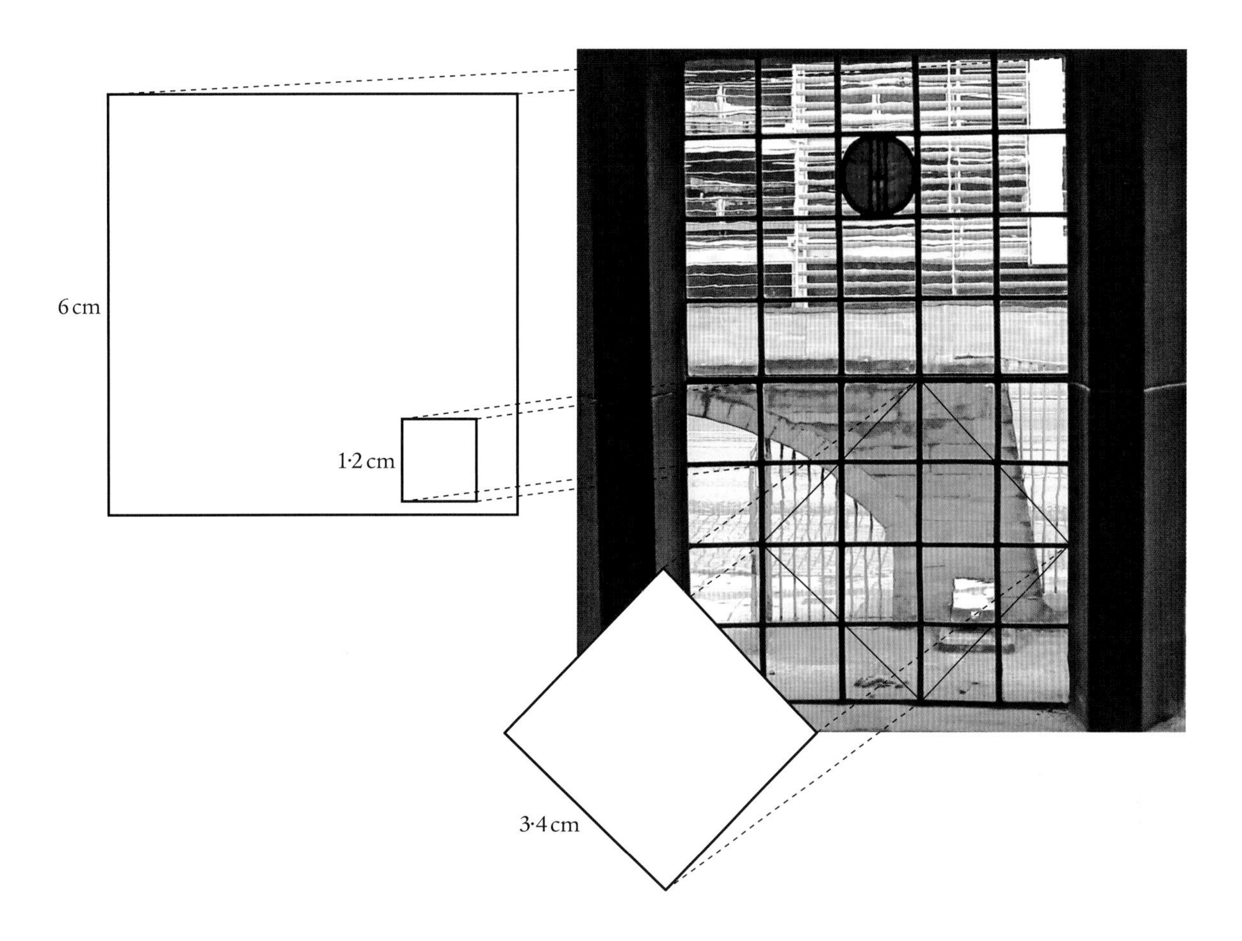

2 Given the area of a square, can you calculate the length of its side?

What is the length of a square of area:

a $49\ \text{cm}^2$ **b** $1{\cdot}96\ \text{m}^2$ **c** $7{\cdot}29\ \text{km}^2$?

3 Enough slabs to make a path 72 m long by 2 m wide were used, instead, in the making of a square patio.

What is the length of the side of the patio?

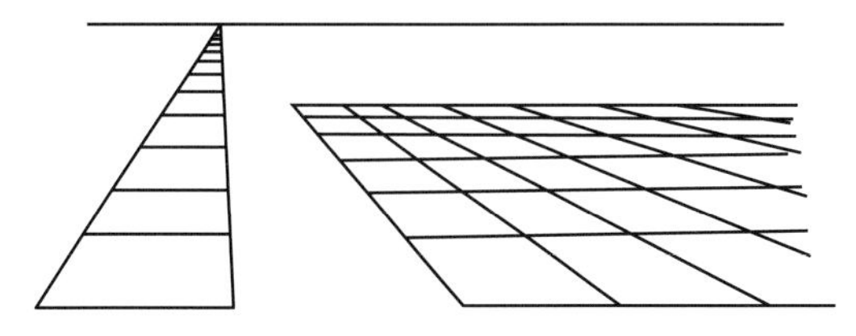

8.1 Pythagoras' theorem

A useful relation

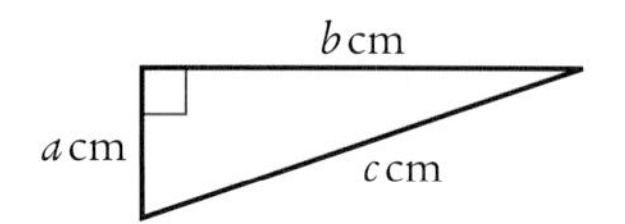

Take four congruent right-angled triangles with sides of length a cm, b cm and c cm.

Let the longest side, called the hypotenuse, be the one of length c cm.

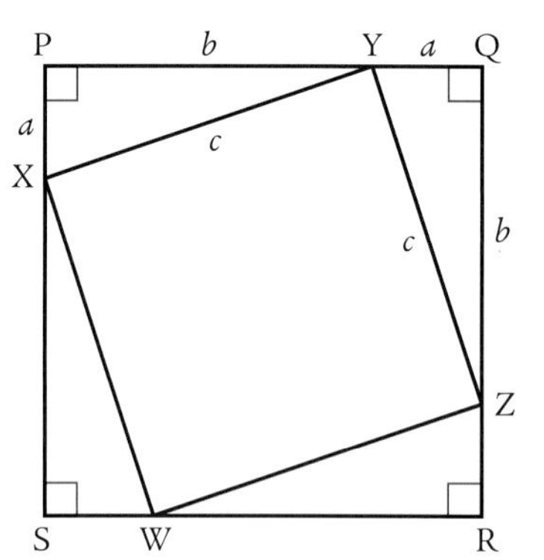

1 Arrange them to form a square PQRS as shown.

- **a** Can you prove that WXYZ is a square?
  - **i** Given PQ is a straight line, calculate the size of $\angle XYZ$.
  - **ii** Check all sides are equal.
  - **iii** Check all angles are right angles.
- **b** Was it important that the triangles were right-angled?

In this layout we can see the area of PQRS = 4 triangles + c^2. ①

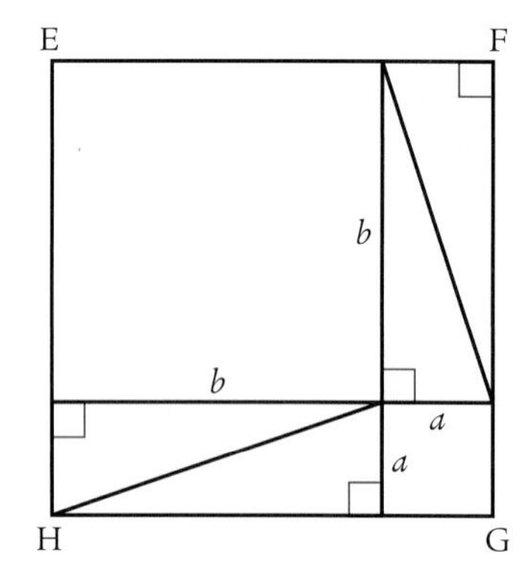

2 Rearrange the triangles to define the square EFGH as shown.

- **a** Prove that EFGH is a square with the same area as PQRS.
  - **i** What is the length of a side?
  - **ii** What is the size of each of its angles?
- **b** Was it important that the triangles were right-angled?

In this layout we can see the area of EFGH = 4 triangles + $a^2 + b^2$. ②

Combining ① and ② we get PQRS = EFGH:

$$\Rightarrow \quad 4 \text{ triangles} + c^2 = 4 \text{ triangles} + a^2 + b^2$$

$$\Rightarrow \quad c^2 = a^2 + b^2$$

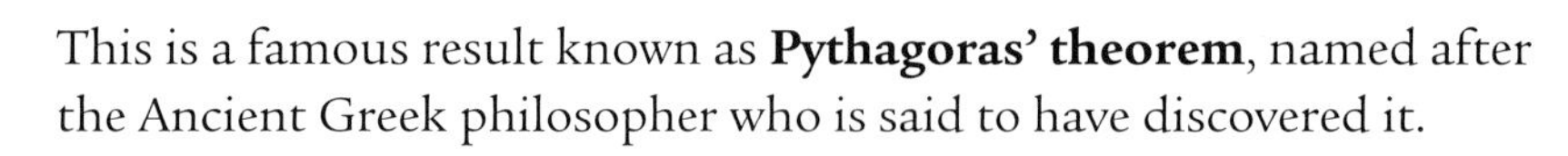

This is a famous result known as **Pythagoras' theorem**, named after the Ancient Greek philosopher who is said to have discovered it.

In words it says,

In any right-angled triangle the square on the hypotenuse is equal to the sum of the squares on the other two sides.

Hypotenuse: the word 'hypotenuse' comes from an ancient Greek term meaning 'stretched under'.

Using the result

Example 1

In 1955 the Greeks honoured their philosopher Pythagoras by presenting his theorem on a postage stamp.

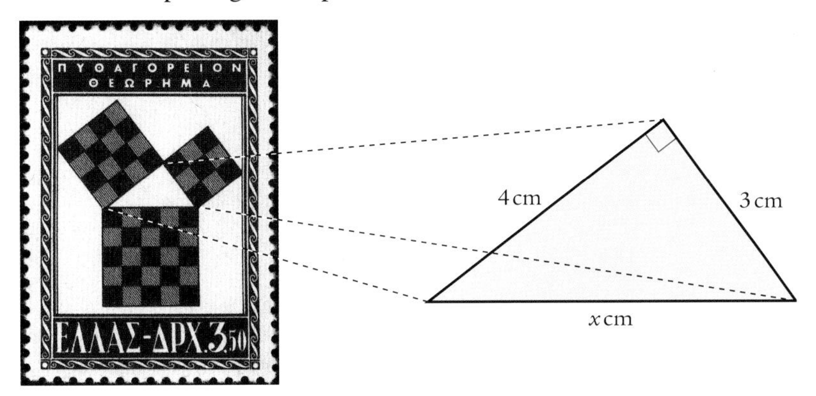

Can you see what it is trying to show?

The shorter sides of the triangle are of lengths 3 cm and 4 cm.

Let the length of the hypotenuse be x cm.

The triangle is right-angled ... so we can use Pythagoras' theorem:

$\Rightarrow \quad x^2 = 3^2 + 4^2$

$\Rightarrow \quad x^2 = 9 + 16$

$\Rightarrow \quad x^2 = 25$

$\Rightarrow \quad x = \sqrt{25}$

$\Rightarrow \quad x = 5$

So the hypotenuse is 5 cm long. (Have another look at the stamp.)

Example 2

The central span of the Bell's Bridge in Glasgow is suspended on a pylon by several steel cables.

One cable is fixed to the bridge 40 m from the foot of the pylon to a point 19·5 m up the pylon.

What is the length of the cable?

The cable forms the hypotenuse of a right-angled triangle.

$\Rightarrow \quad x^2 = 40^2 + 19{\cdot}5^2$

$\Rightarrow \quad x^2 = 1600 + 380{\cdot}25$

$\Rightarrow \quad x^2 = 1980{\cdot}25$

$\Rightarrow \quad x = \sqrt{1980{\cdot}25}$

$\Rightarrow \quad x = 44{\cdot}5$

The cable is 44·5 metres long.

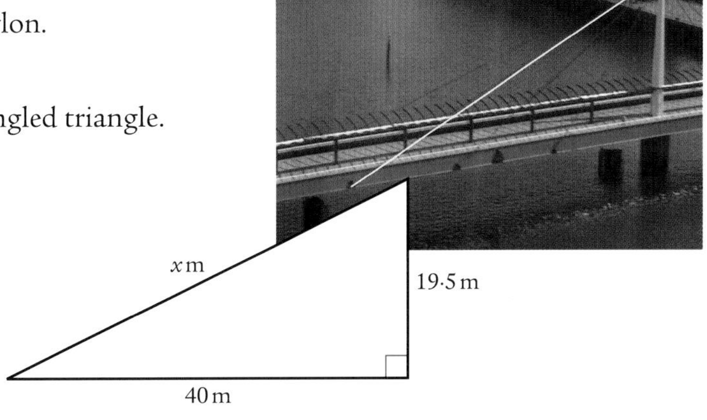

Exercise 8.1A

1 Calculate the hypotenuse in each triangle.

a

b

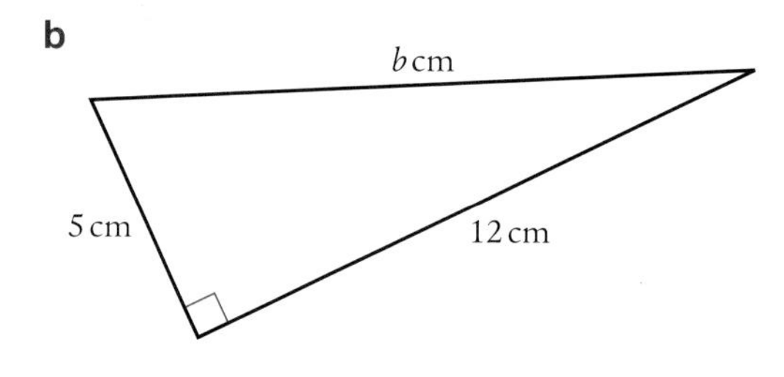

c

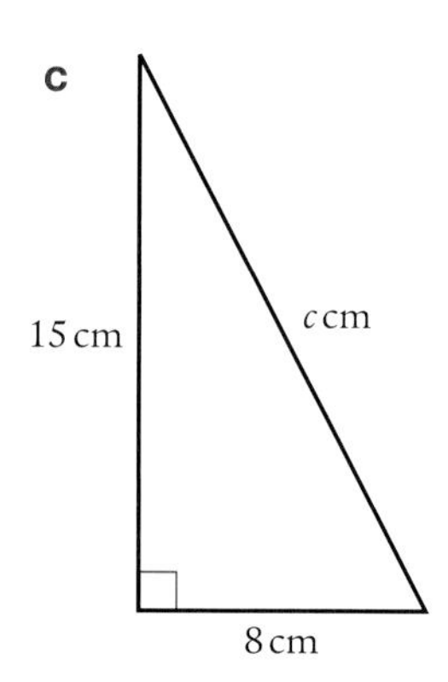

2 Use Pythagoras' theorem **where possible** to calculate the length of the longest side in each case.

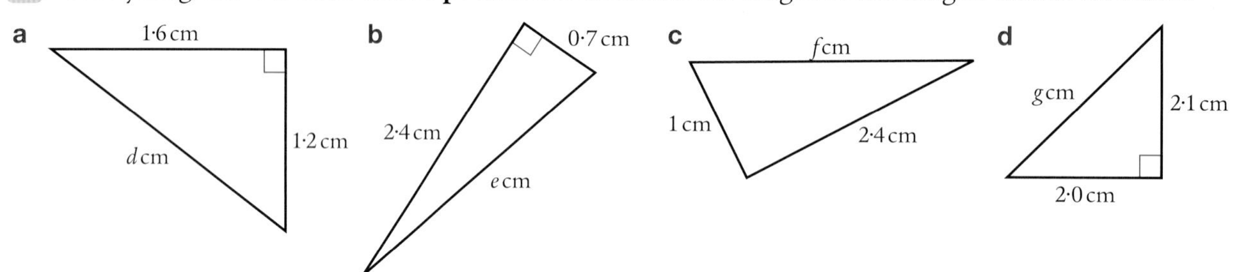

3 Calculate the length of the hypotenuse, correct to 1 decimal place, of a right-angled triangle whose shorter sides are:

a 39 cm and 10 cm

b 8·1 cm and 7·6 cm

c 12 cm and 12 cm.

4 Kincardine is 20·8 miles east and 14·5 miles north of the centre of Glasgow.

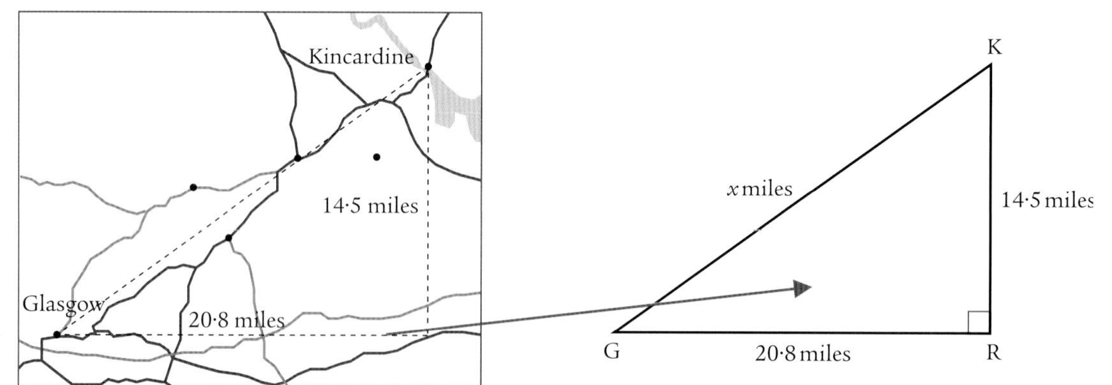

a How far apart are Kincardine and Glasgow?
Give your answer correct to 1 decimal place.

b My satnav gives the distance as 27·5 miles.
Explain the difference in values.

5 Flora McDonald helped Bonnie Prince Charlie escape to Skye after the Battle of Culloden. Her monument in Skye, though made of stone, once blew over in a gale. To prevent this happening again a steel brace was attached to it 7 m off the ground and anchored on the ground 2 m from the foot of the monument.

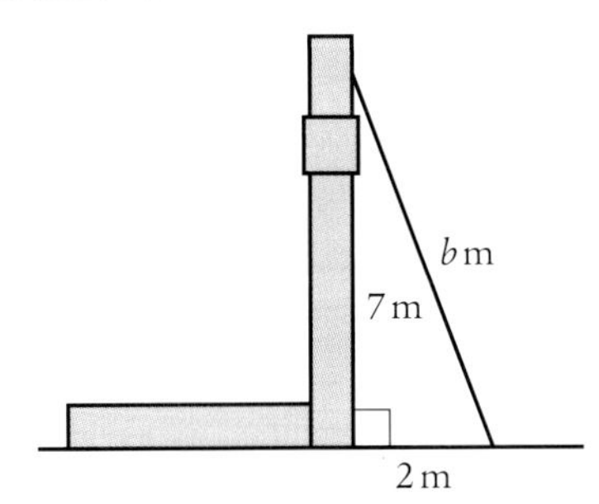

a How long is the steel brace in metres, correct to 1 decimal place?

b Why are triangles used to brace, or fix the position of, things?

6 A street lamp is designed as shown. The post is 8 m tall.

A metre from the top an arm CB goes out at right angles to the post. It is 1·4 m long.

A strut AB supports the weight of the globe.

a How long is the strut AB?

b What is the distance CD?

c How far is B from D?

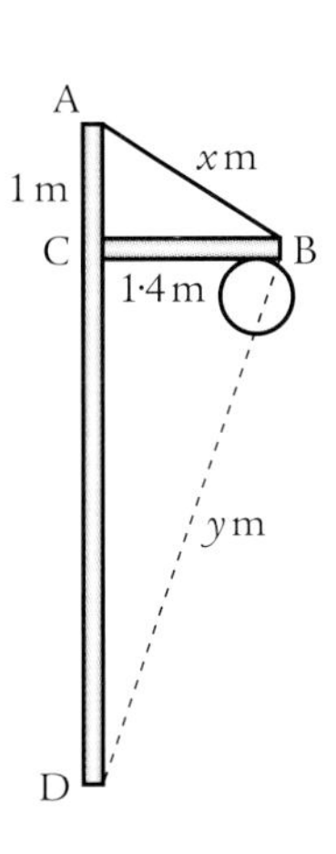

7 A road junction has two pedestrian crossings.

If a pedestrian wants to go from A to B he should use both crossings.

However, when the 'green man' shows, people tend to quickly cross on the diagonal without using the crossings.

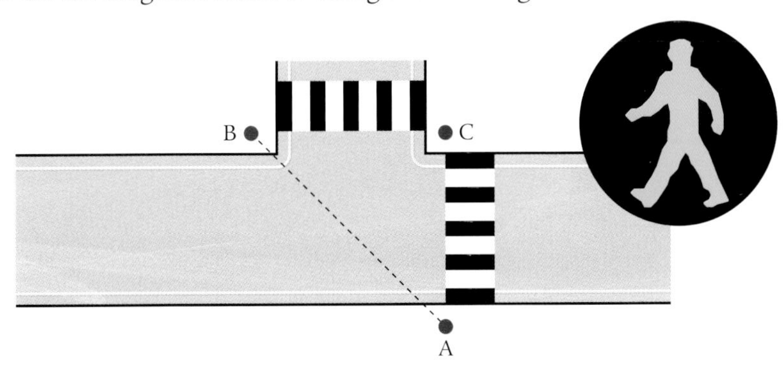

a Given that A to C is 11 m and C to B is 12 m, what length should a person walk to get from A to B?

b What distance is saved by crossing on the diagonal?

8 In AD 72 the Romans laid siege to a fortress in Judea called Masada.
They built a huge ramp in order to reach the defences of the fortress.
The ramp was 114 m high and 315 m along its base.

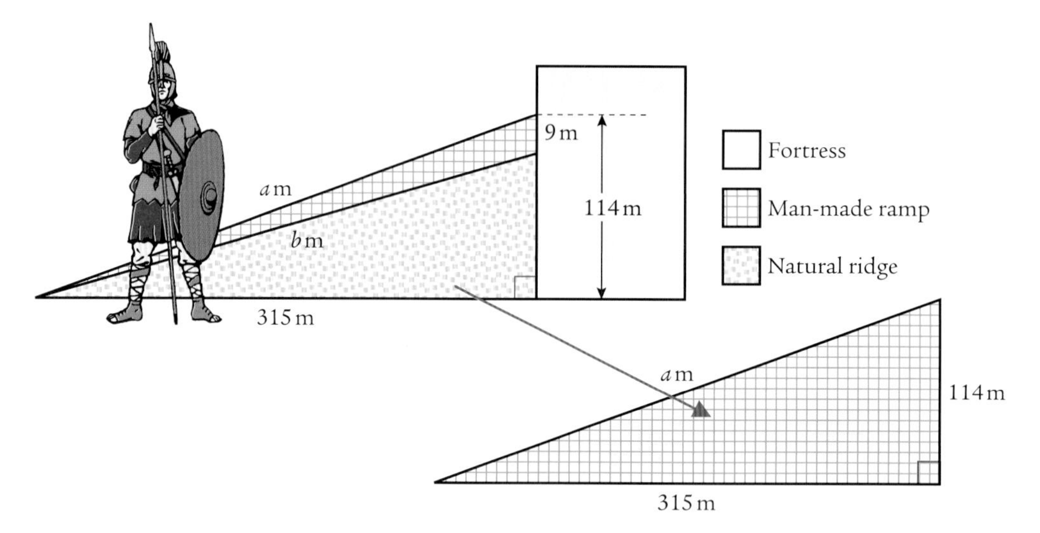

a How long was the slope of the ramp?

b Archaeologists say that the ramp was actually built upon a natural rock ridge and that only the last 9 m had to be built up.

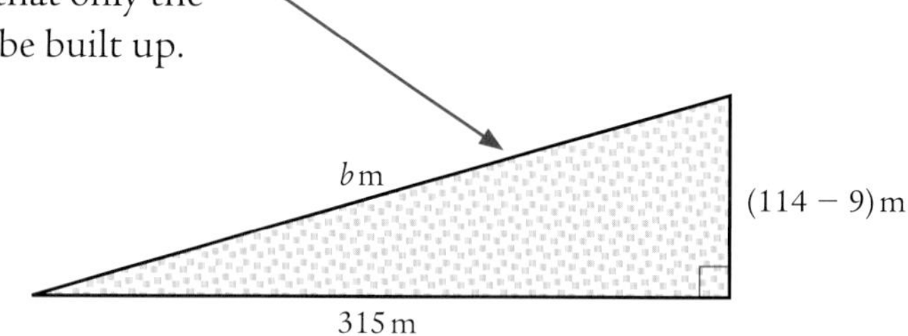

How long was the ridge?

Exercise 8.1B

1 Calculate the perimeter of each triangle.

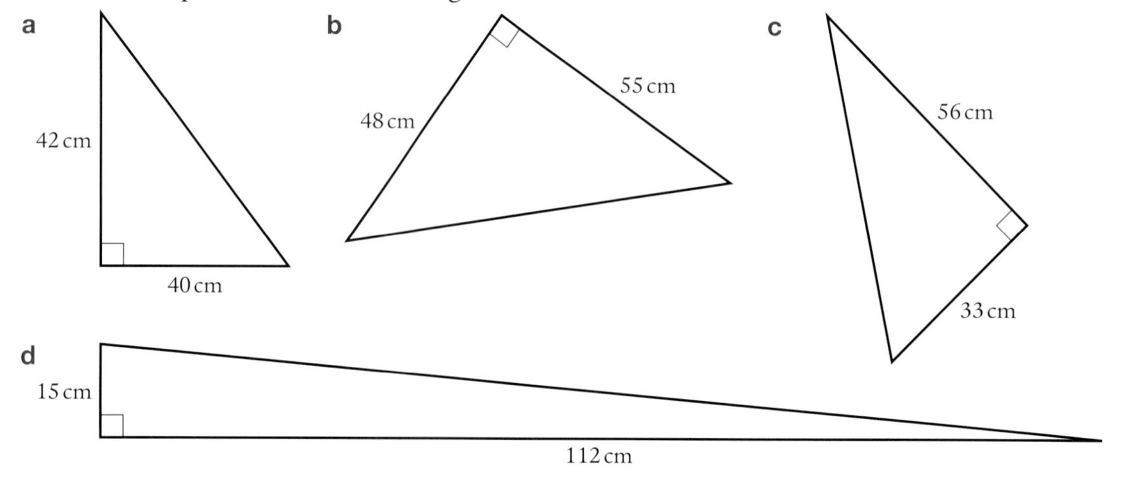

2 In 1774 Charles Mason climbed the mountain Schiehallion in Tayside to do experiments that enabled him to calculate the 'weight' of the Earth.

It is a well-climbed mountain.

At the car park at the start of the walk there is a map from which was noted the following information.

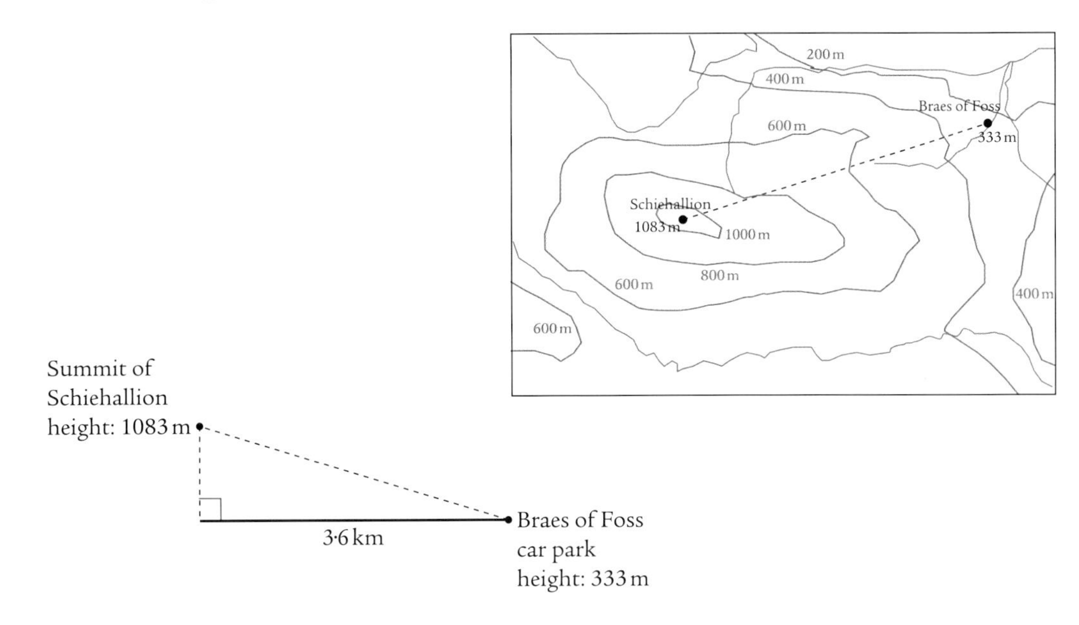

Note that from a map we read horizontal distances.

Heights we get from contours and spots.

a What is the difference in height between the summit and the car park?

b Calculate the distance between the car park and the summit, along the 'line-of-sight' to the nearest metre.

c On the map, we are told that the path to the summit is 9·5 km.

Why is it more than twice the distance calculated in part **b**?

3 A farmer's field is a right-angled triangle caught between two roads and a river. The farmer wishes to put up a perimeter fence.

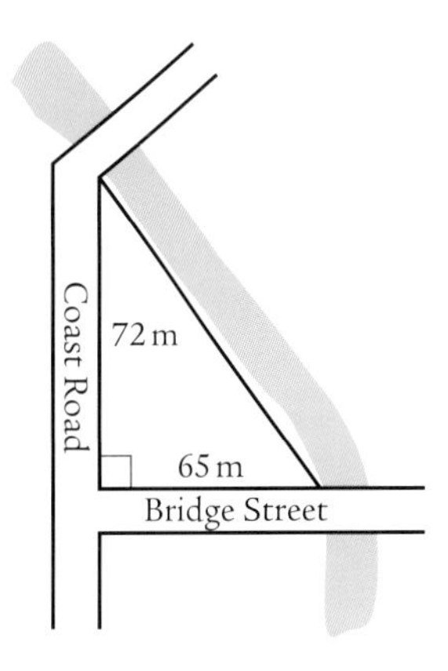

a What length of fence will he need along the river bank?

b What is the perimeter of the field?

c He believes he can get the job done for £2·50 a metre plus VAT (20%).

What does he think this will cost him?

4 In the village of Rye Tangled, at the summer fête, the tradition is to have a race.

The course is in the shape of a right-angled triangle whose shorter sides are 40 m and 96 m.

There are two teams, the Uppies and the Doonies.

The race starts at some point on the hypotenuse.

The Uppies run clockwise round the course and the Doonies anticlockwise.

The race finishes at the right angle.

Where should the race start to be fair to both teams?

5 Triangle A is right-angled and has shorter sides 10 cm and 7·5 cm.

Triangle B is also right-angled and has shorter sides 5 cm and 12 cm.

- **a** Calculate the sum of the shorter sides for each triangle.
- **b** Guess which triangle will have the larger perimeter.
- **c** Calculate the perimeter of each triangle.
- **d** Was your guess correct?

6 A science experiment is set up to show the benefits of an inclined plane.

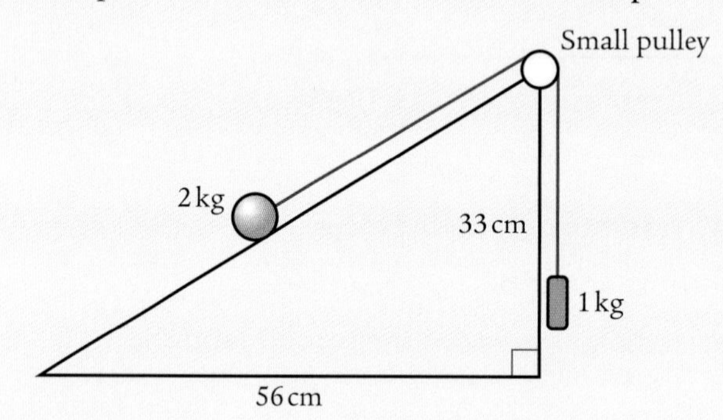

It shows how a 1 kg weight can pull a 2 kg weight up the plane.

- **a** What is the length of the slope?
- **b** If the string is the same length as the slope, how far up the plane will the 2 kg weight be when the 1 kg weight hits the ground?
- **c** How far from the pulley will it be?

7 The inner tube of a kitchen roll has a circumference of 12 cm. Its length is 27 cm.

A seam spirals up the tube making three turns from one end to the other.

How long is the seam?

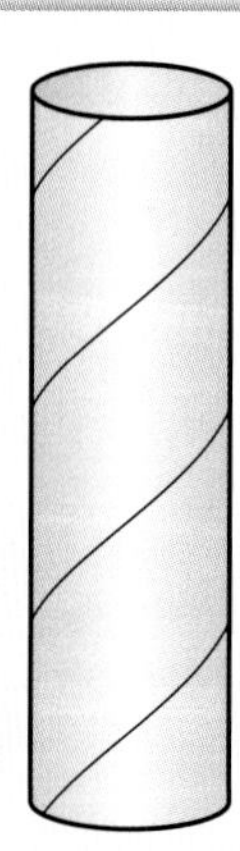

8.2 Finding a shorter side

Example 1

The hypotenuse of a right-angled triangle is 26 cm long.

One of the shorter sides is of length 10 cm.

What is the length of the third side of the triangle?

A quick sketch shows the situation:

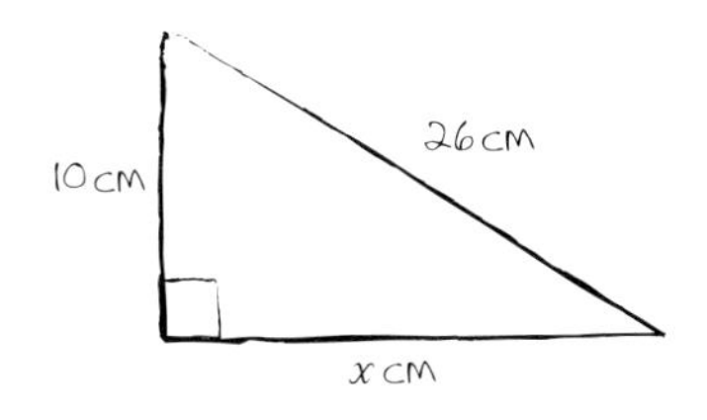

The triangle is right-angled, so Pythagoras' theorem can be used.

$\Rightarrow \quad 26^2 = x^2 + 10^2$ Notice that the hypotenuse is known

$\Rightarrow \quad 676 = x^2 + 100$

$\Rightarrow \quad x^2 = 676 - 100$

$\Rightarrow \quad x^2 = 576$

$\Rightarrow \quad x = \sqrt{576}$

$\Rightarrow \quad x = 24$

The third side of the triangle is 24 cm long.

Example 2

A buoy is tethered to the sea floor by a cable of length 20 m.

The current keeps the cable taut at a point 16 m from the point where it is tethered.

How deep is the sea at this point?

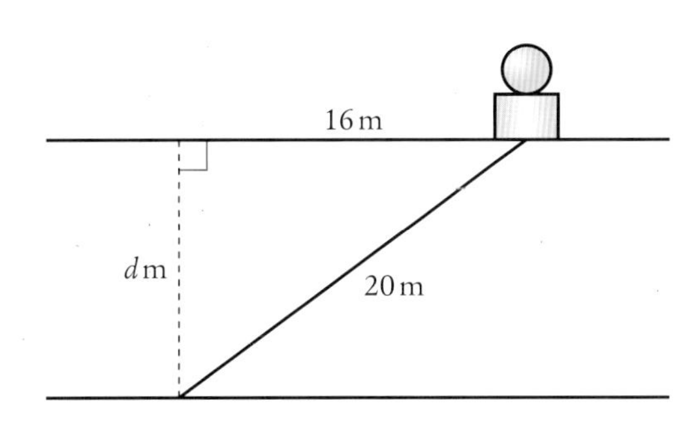

The triangle is right angled, so Pythagoras' theorem can be used.

$\Rightarrow \quad 20^2 = d^2 + 16^2$ Notice that the hypotenuse is known

$\Rightarrow \quad 400 = d^2 + 256$

$\Rightarrow \quad d^2 = 400 - 256$

$\Rightarrow \quad d^2 = 144$

$\Rightarrow \quad d = \sqrt{144}$

$\Rightarrow \quad d = 12$

The buoy is tethered in 12 metres of water.

Exercise 8.2A

1 Calculate the size of the third side of each triangle.

a

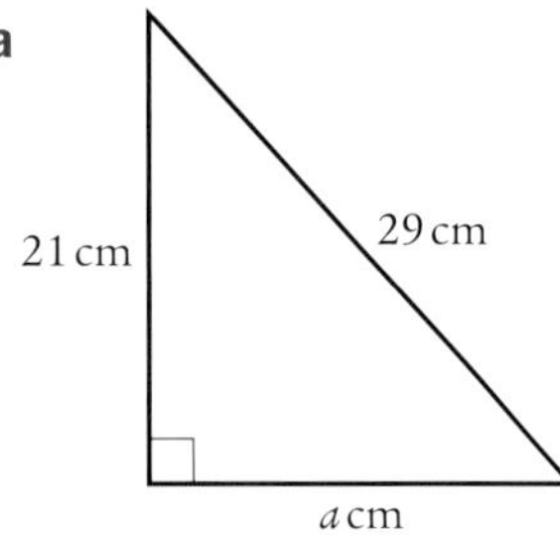

b

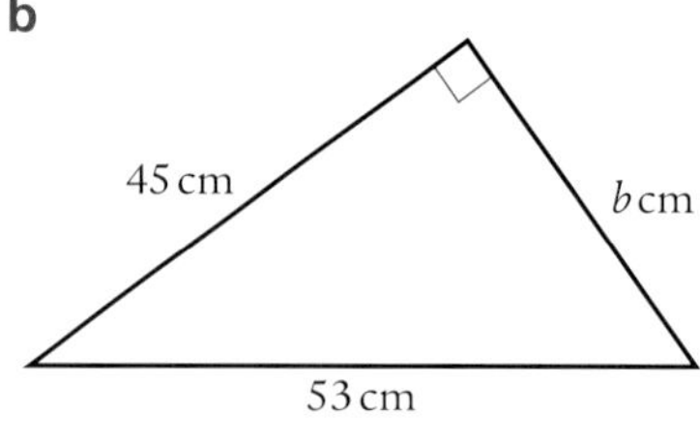

c

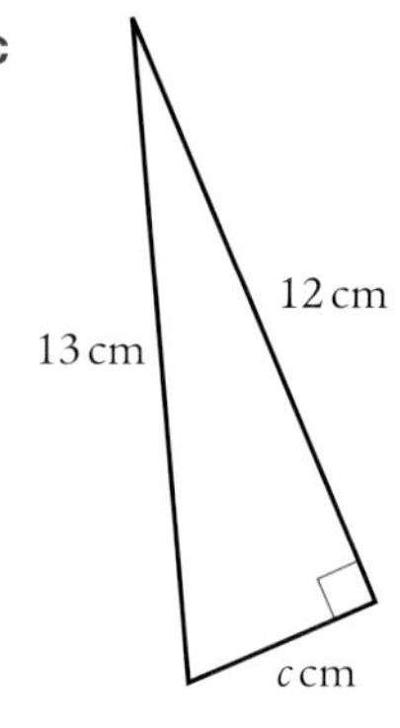

2 Find the exact length of the third side of each triangle.

a

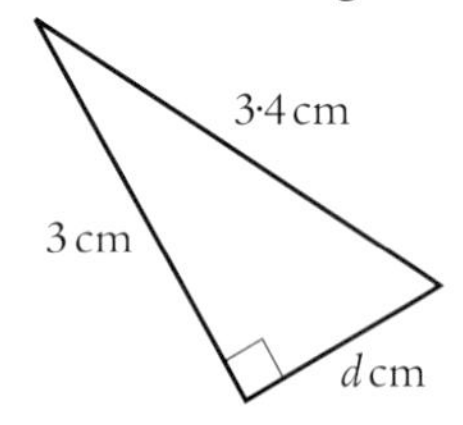

b

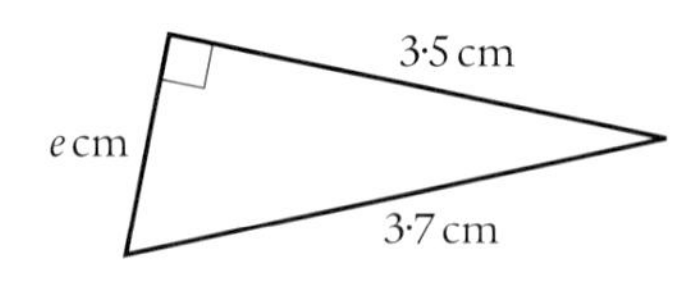

c

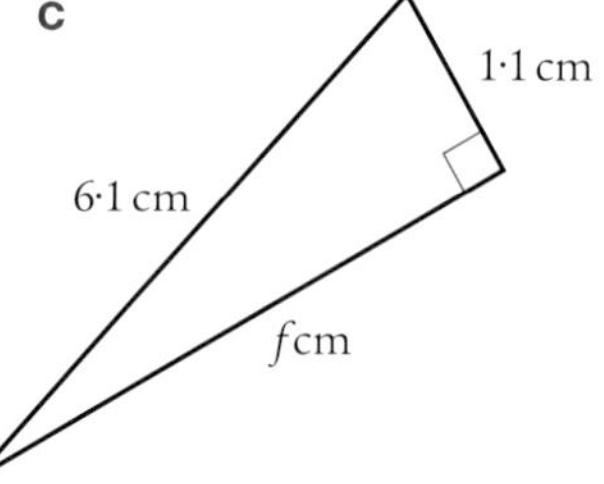

3 Find the length of the third side of a right-angled triangle correct to 1 decimal place when:

a the hypotenuse is 10 cm and one of the shorter sides is 2·7 cm

b the hypotenuse is 25 cm and one of the shorter sides is 17 cm

c one of the shorter sides is 7·9 cm and the hypotenuse is 22·3 cm.

4 The man who came to fix the TV aerial leant a 12·5 m ladder against the house.

The foot of the ladder was 3·5 m from the house.

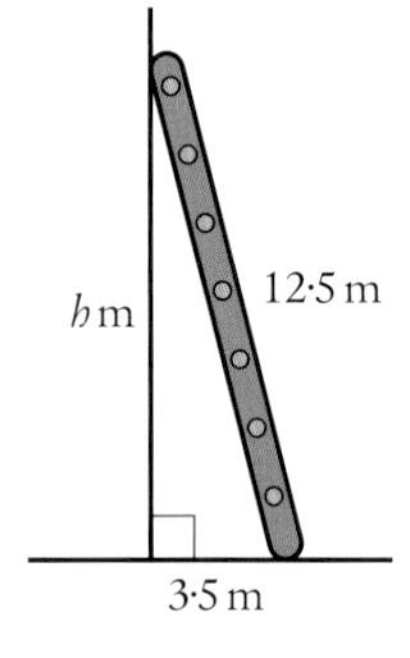

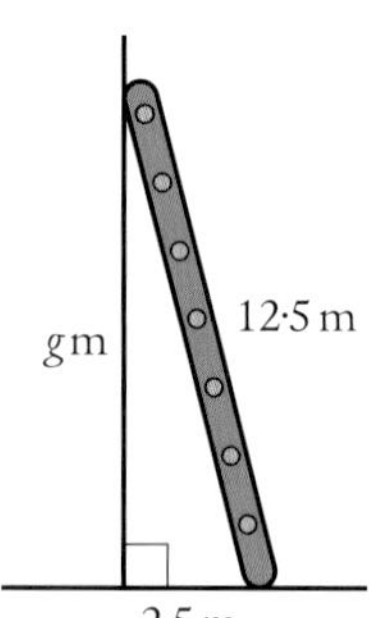

a How far up the side of the house did the ladder reach?

b Why might it be reasonable to assume the angle between the ground and the house is a right angle?

c If we put the foot of the ladder 1 m closer to the house, would the top of the ladder go 1 metre higher?

Exercise 8.2B

1 Suppose you are told that the two shorter sides of a right-angled triangle are the same length and that the hypotenuse is of length 10 cm.

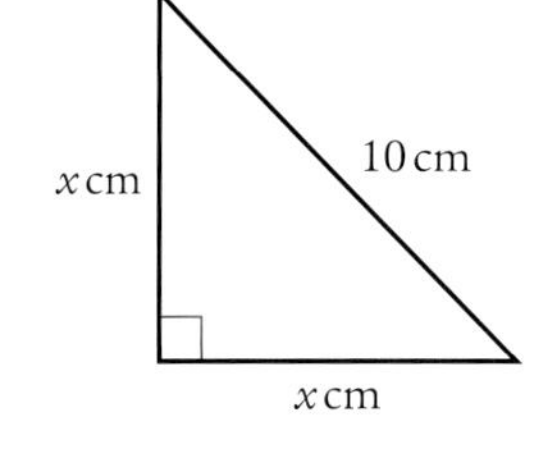

Copy and complete the calculation to find the lengths of the shorter sides.

The triangle is right-angled, so we can use Pythagoras' theorem.

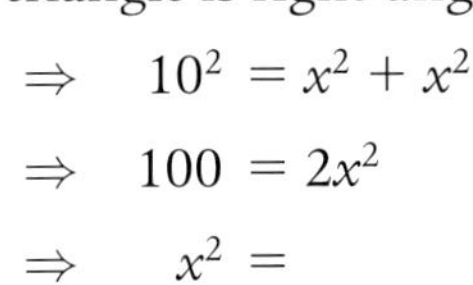

$\Rightarrow \quad 10^2 = x^2 + x^2$

$\Rightarrow \quad 100 = 2x^2$

$\Rightarrow \quad x^2 =$

$\Rightarrow \quad x =$

2 Calculate the shorter sides in an isosceles right-angled triangle that has a hypotenuse of length:

a 16 cm **b** 37 cm **c** $\sqrt{2}$ cm.

3 Calculate the perimeter of an isosceles right-angled triangle, which has a hypotenuse of length 8·4 m.

4 The ferry route from Lochranza to Claonaig is 7·9 km.

Claonaig is 5·2 km north of Lochranza.

a How far west of Lochranza is Claonaig?

b How are we sure that we are working with a right-angled triangle?

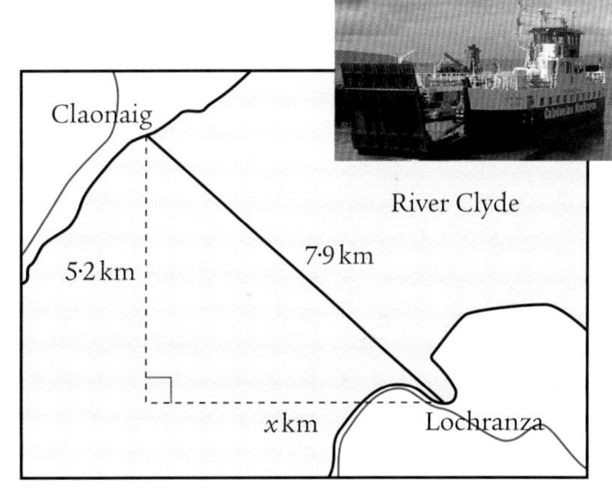

Lochranza

5 Jeff wants to create a temporary playpen in his garden.

He places a 6 m long piece of fencing against a corner wall.

From above it looks like this.

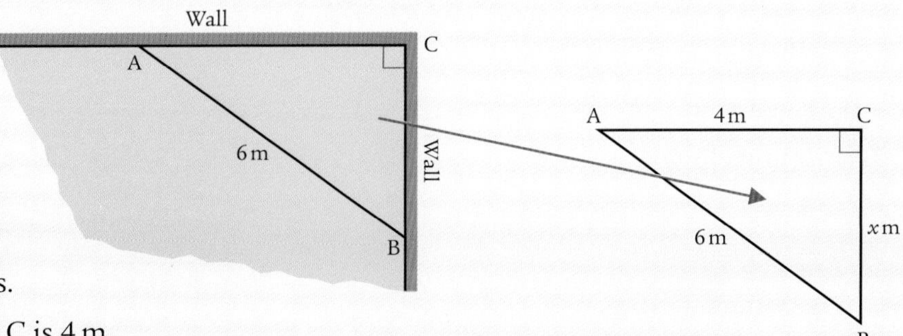

a If the distance of A from C is 4 m,

i what is the perimeter of the triangle created

ii what area is created in square metres?

b Work out the new values when the board is shifted so that AC is 4·25 m.

6 When we are looking at a half-moon, we are seeing the moment when the Earth, Moon and Sun form a right-angled triangle with the right angle at the Moon

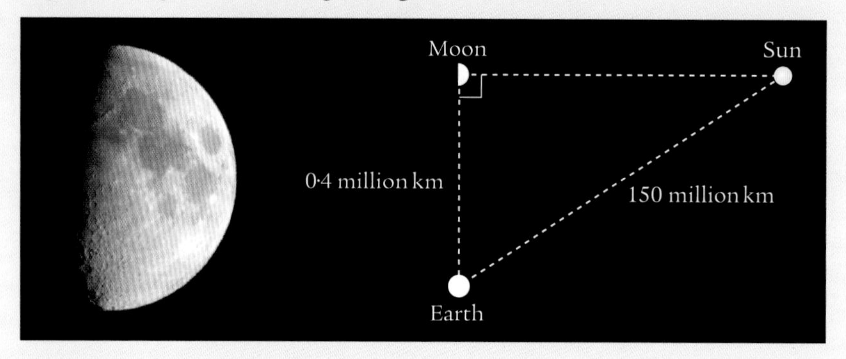

a Calculate the distance from the Moon to the Sun at this moment correct to 1 decimal place.

b Comment on your answer.

8.3 Point-to-point

The theorem of Pythagoras can be used very effectively with coordinates.

Example 1

What is the distance between the points A(2, 1) and B(8, 9)?

Plotting both points, we can pick out a right-angled triangle using the grid lines, with AB as the hypotenuse.

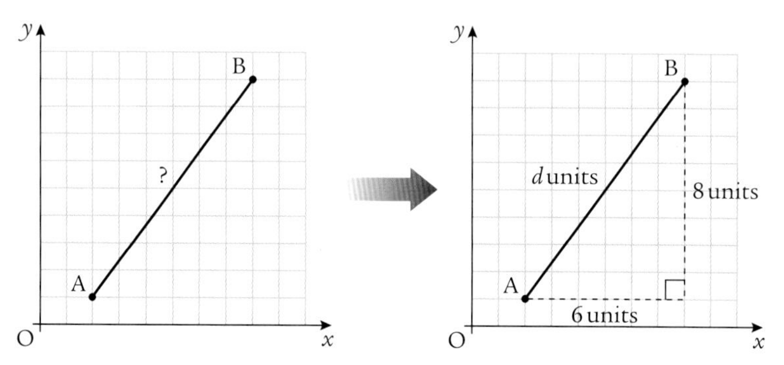

The triangle is right-angled, so we can use Pythagoras' theorem:

$\Rightarrow \quad d^2 = 6^2 + 8^2$

$\Rightarrow \quad d^2 = 36 + 64$

$\Rightarrow \quad d^2 = 100$

$\Rightarrow \quad d = \sqrt{100}$

$\Rightarrow \quad d = 10$

AB is 10 units long.

Example 2

If the coordinates are so big that it is not practical to draw the coordinate grid exactly, a sketch is enough to let you visualise the triangle required.

What is the distance between A(2, 1) and C(22, 100)?

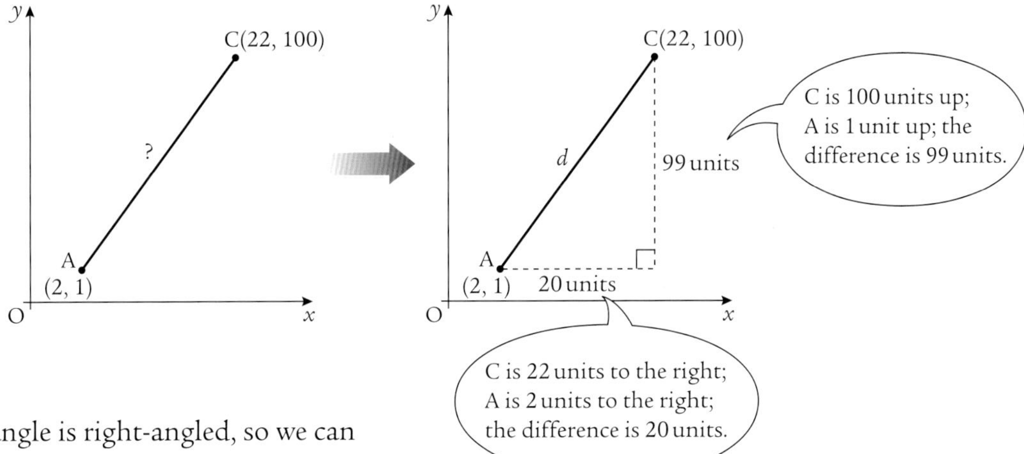

The triangle is right-angled, so we can use Pythagoras' theorem:

$\Rightarrow \quad d^2 = 20^2 + 99^2$

$\Rightarrow \quad d^2 = 400 + 9801$

$\Rightarrow \quad d^2 = 10\,201$

$\Rightarrow \quad d = \sqrt{10\,201}$

$\Rightarrow \quad d = 101$

AC is 101 units long.

Exercise 8.3A

1 Calculate the distance between

a A(2, 3) and B(5, 7)

b C(1, 1) and D(6, 13)

c O(0, 0) and (24, 7).

2 How far apart are P(−3, 4) and Q(4, −2)?

Give your answer correct to 1 decimal place.

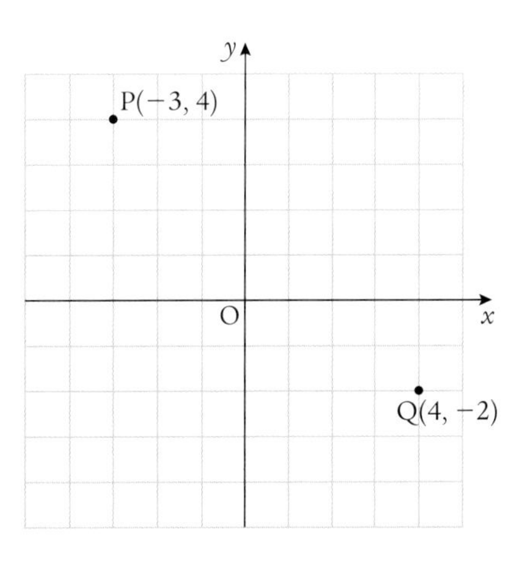

3 A(2, 7), B(16, 55) and C(32, 47) are three points.

a Calculate the distance between A and B.

b Find the length of the line BC.

c Show that A, B, and C are the vertices of an isosceles triangle.

4 For the purposes of reference, a map is divided into squares.

It is known that the direct distance between Caddonfoot and Galashiels is 4·5 km.

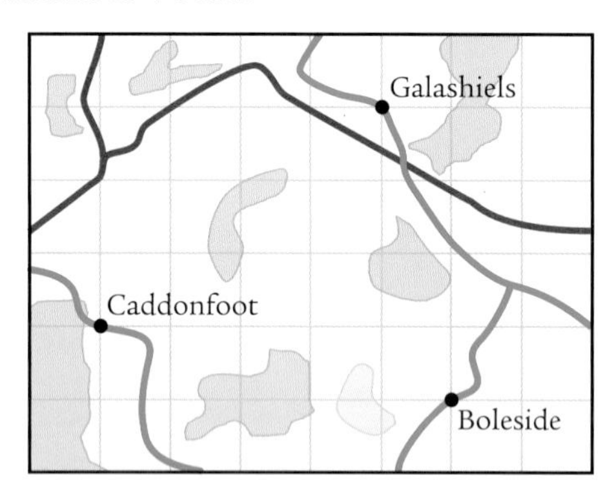

a What is the length of the edge of one square?

(Hint: work out the distance between Galashiels and Caddonfoot in 'squares'.)

b Calculate the direct distance between Caddonfoot and Boleside.

Exercise 8.3B

1 **a** Calculate the length of AB.

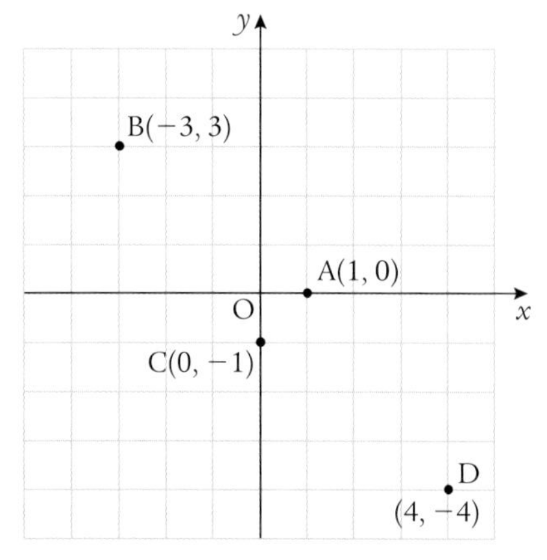

b Prove that A, B, C, and D are the vertices of a rhombus.

c The area of a rhombus is half the product of its diagonals.

$$\text{Area}_{\text{rhombus}} = \tfrac{1}{2}d_1d_2$$

Calculate the area of ABCD.

2 When Meg sends a text-message to her pal Helen, it goes to the nearest receiver and from there it gets relayed to Helen by the network.

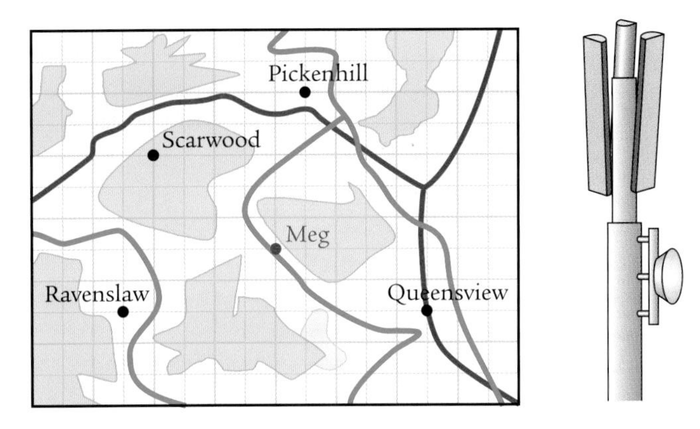

a Find the distance in units between Meg and

i Scarwood **ii** Pickenhill.

b Which receiver, Pickenhill, Queensview, Ravenslaw or Scarwood, is closest to Meg?

c If each square has a side of 0·5 km, what is the distance between Meg and the nearest receiver?

3 In Geography, you learn that the whole of the UK is subdivided by Ordnance Survey into **sheets**.

Each sheet is further divided into squares.

Places can be described by giving references.

If we give a four-figure reference then the squares of the grid are 1 km long.

The four-figure reference for the mountain Goatfell is NR 99 42.

A mathematician would say this is the point (99, 42) on the sheet called NR.

a The four-figure reference for the Arran Outdoor Centre is NR 94 50.

Write this as a mathematician would.

b Calculate the distance between Goatfell and the Arran Outdoor Centre.

c If we want to be more accurate we would use six-figure references.

The reference for Goatfell becomes NR 991 415, which means the point (99·1, 41·5).

The six-figure reference for the Arran Outdoor Centre is NR 938 502.

Find the distance between Goatfell and the Arran Outdoor Centre using these more accurate references.

d What is the distance between the centre of Glasgow (NS 590 655) and Paisley (NS 485 635)?

8.4 Spotting right-angled triangles

Many geometric shapes are based on the right-angled triangle.

This can be exploited when solving problems.

Example 1

An isosceles triangle has a base of 12 cm and an altitude of 8 cm.

Calculate:

a the lengths of the equal sides

b the perimeter of the triangle.

Within the shape of the isosceles triangle we can pick out a right-angled triangle.

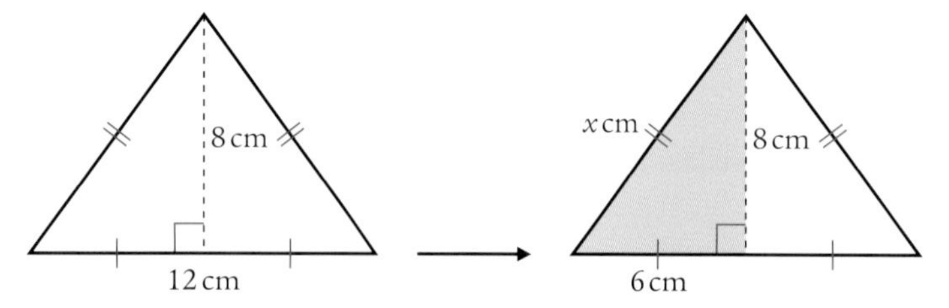

The triangle is right-angled, so we can use Pythagoras' theorem:

$\Rightarrow \quad x^2 = 6^2 + 8^2$

$\Rightarrow \quad x^2 = 36 + 64$

$\Rightarrow \quad x^2 = 100$

$\Rightarrow \quad x = \sqrt{100}$

$\Rightarrow \quad x = 10$

a The equal sides of the isosceles triangle each have length 10 cm.

b Perimeter of triangle = 10 + 10 + 12 = 32 cm

Example 2

A 50-inch wide-screen TV is 43·6 inches wide.

The size of a TV screen is described in shops by giving the length of its diagonal in inches.

a What is the height of the screen?

b What is its perimeter?

Within the shape of the rectangle that represents the screen we can pick out a right-angled triangle.

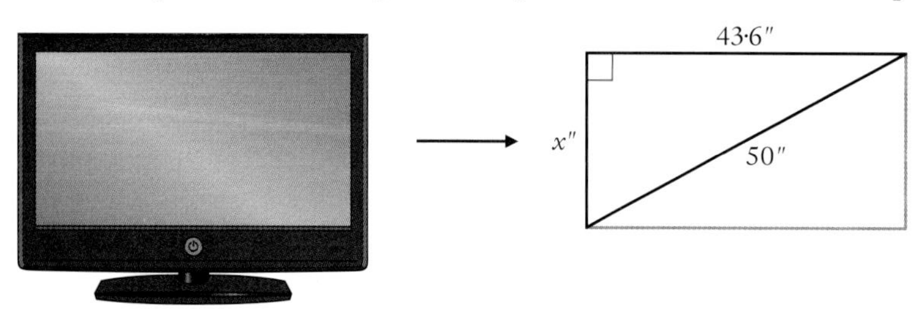

a The triangle is right-angled, so Pythagoras' theorem can be used:

$\Rightarrow \quad 50^2 = x^2 + 43{\cdot}6^2$ Notice that the hypotenuse is known

$\Rightarrow \quad 2500 = x^2 + 1900{\cdot}96$

$\Rightarrow \quad x^2 = 2500 - 1900{\cdot}96$

$\Rightarrow \quad x^2 = 599{\cdot}04$

$\Rightarrow \quad x = \sqrt{599{\cdot}04}$

$\Rightarrow \quad x = 24{\cdot}5$ (to 1 d.p.)

The height of a 50-inch wide-screen TV is 24·5 inches.

b Perimeter of screen = 24·5 + 24·5 + 43·6 + 43·6 = 136·2 inches

Exercise 8.4A

1 A rectangle is 60 cm by 11 cm. What is the length of its diagonals?

2 A square has a diagonal of side 12 cm. Find its perimeter.

3 The sides of a triangle are 80 mm, 80 mm and 36 mm.

Calculate its altitude using the 36 mm side as the base.

4 Calculate the altitude of an equilateral triangle of side 24 cm.

5 A rhombus has diagonals of length 4 cm and 9·6 cm.

Calculate:

a the length of one side of the rhombus

b its perimeter.

6 A kite is as shown.

Its sides are of lengths 58 mm and 104 mm.

The diagonal that crosses the axis of symmetry is 80 mm long.

Calculate the length of the diagonal which lies along the axis of symmetry.

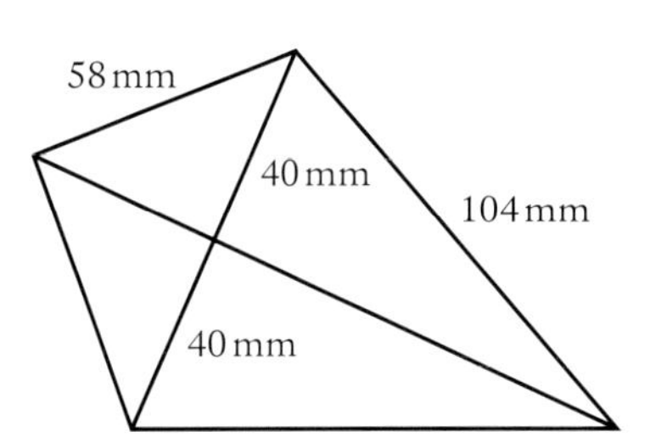

7 A symmetrical trapezium has parallel sides of length 21 cm and 10 cm.

They are 4·8 cm apart.

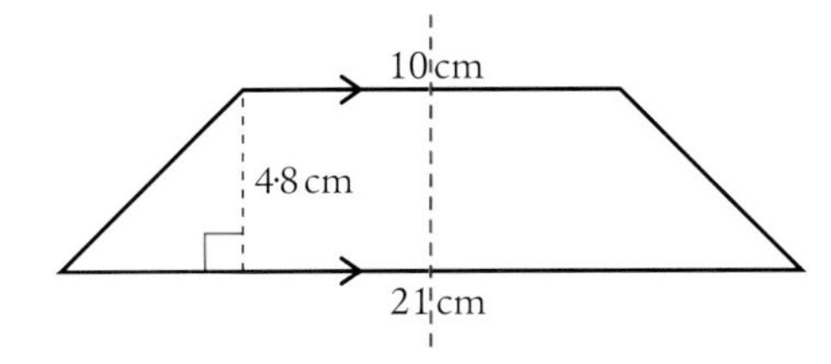

a Calculate the lengths of the other sides.

b Hence calculate the perimeter of the trapezium.

Exercise 8.4B

1 The size of a TV screen is described in adverts by giving the length of its diagonal in inches.

a On a 40-inch screen standard TV the picture area is 32 inches wide.

How high is the picture area?

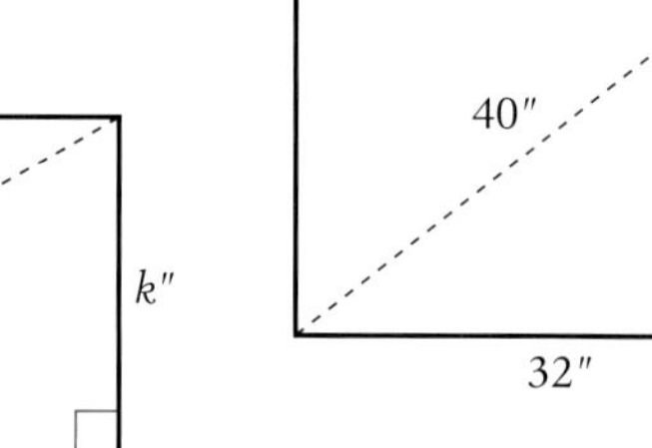

b On a widescreen TV of the same 'size', 40 inches, the picture area is 34·9 inches wide.

How high is the picture area?

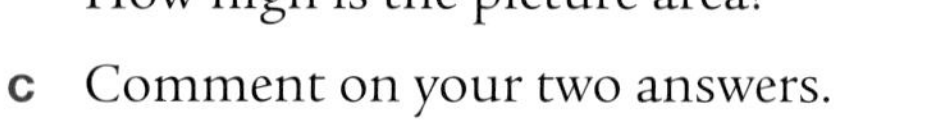

c Comment on your two answers.

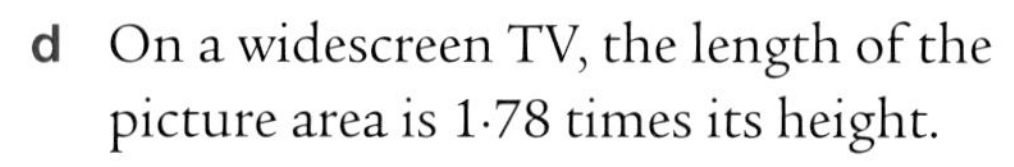

d On a widescreen TV, the length of the picture area is 1·78 times its height.

What size of widescreen TV would you need to buy to get a picture area as high as the one you get on a standard 40-inch TV?

2 An upholstery manufacturer wishes to mass-produce cushions. These are made from squares of cloth with a frill round the perimeter.

a If these cushions have a 30 cm diagonal, what length of frill is required for one cushion (i.e. the perimeter of the square)?

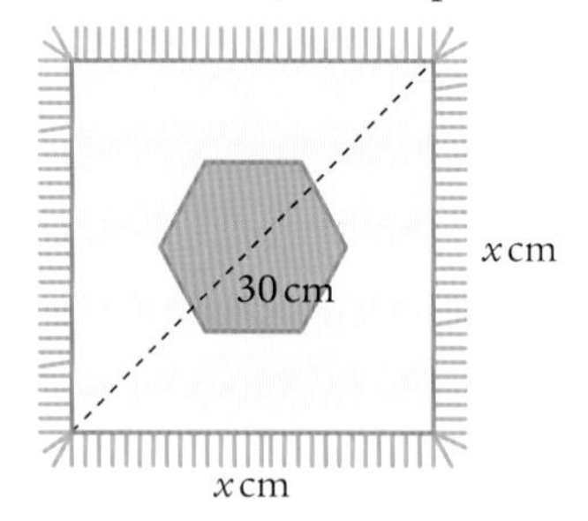

b If, instead, the manufacturer decides to make the diagonal 31 cm long, how much more frill will be needed when making 1000 cushions?

3 A circle of radius 10·9 cm and centre C has a chord AB that is 12 cm long.

What is the shortest distance from C to AB?

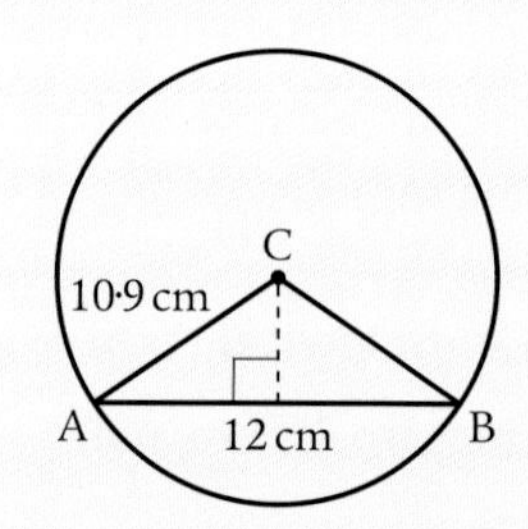

4 The sides of a triangle are 80 mm, 80 mm and 36 mm (see Exercise 8.4A, Question 3).

a Calculate its altitude using the 36 mm side as the base.

b Calculate its area.

c Calculate its altitude using an 80 mm side as the base.

5 The London Eye is a large Ferris wheel on the Thames Embankment.

It has a radius of 60 m.

There are 32 capsules evenly spaced round it that carry people up to get a view of London.

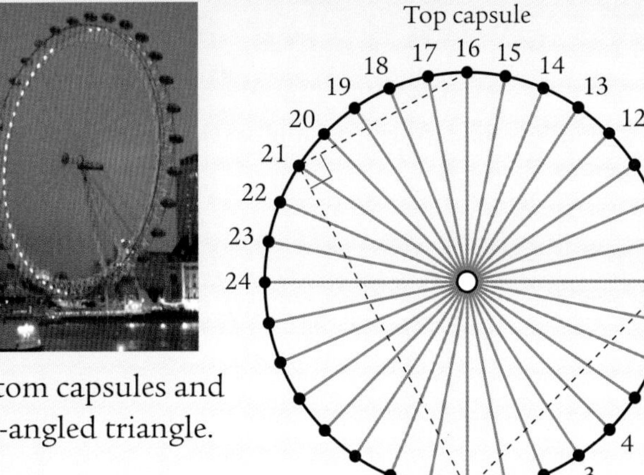

a How far from the bottom capsule is capsule 8?

b If you form a triangle with the top and bottom capsules and any one other, it is guaranteed to be a right-angled triangle.

Now there's a thing!

Capsules 16 and 21 are 28·3 m apart.

How far is capsule 21 from the bottom capsule?

6 In science we learn about machines.

One such machine is the scissor jack.

It allows one person to manually lift a car.

The scissor jack can be modelled by a rhombus.

As it opens and closes, the length of its sides stay the same.

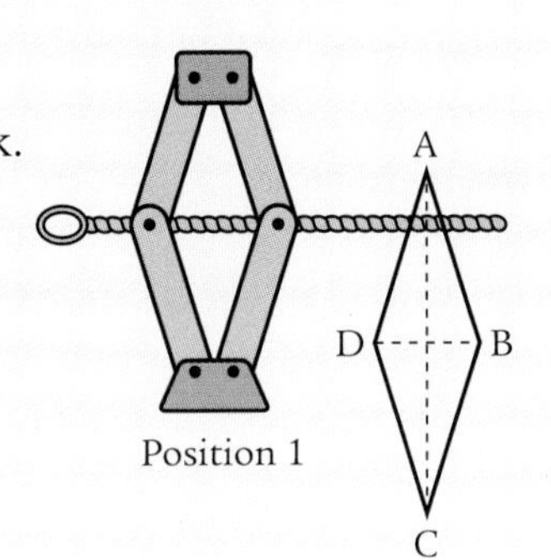

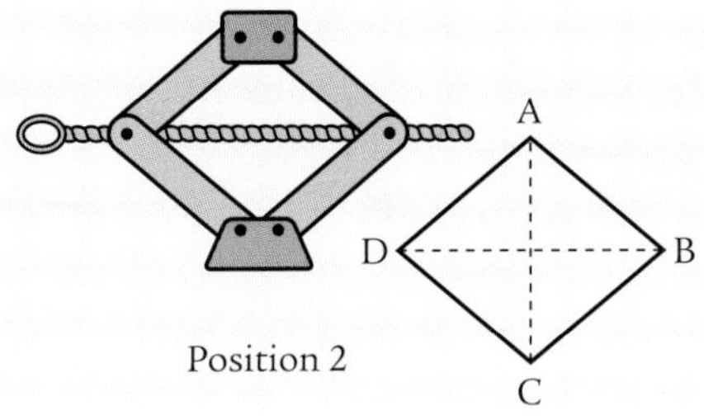

a In position 1, the diagonals are of length 11 cm and 60 cm. What is the length of one side of the rhombus?

b In position 2, the height AC is 20 cm. Calculate the width BD, correct to 1 decimal place.

7 The Eiffel Tower in Paris can be modelled by 4 symmetrical trapezia and an isosceles triangle.

Eiffel Tower	Height (m)	Width (m)
Ground	0	100
Stage 1	57	66
Stage 2	116	37
Stage 3	196	12
Top level	276	6
Top	300	0

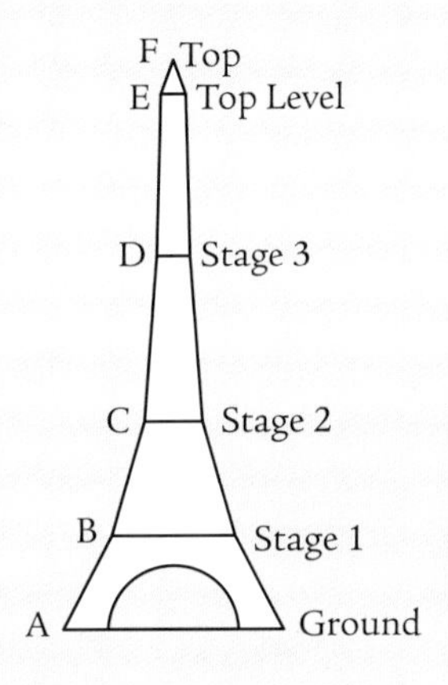

Using the data from the table, the first two stages look like:

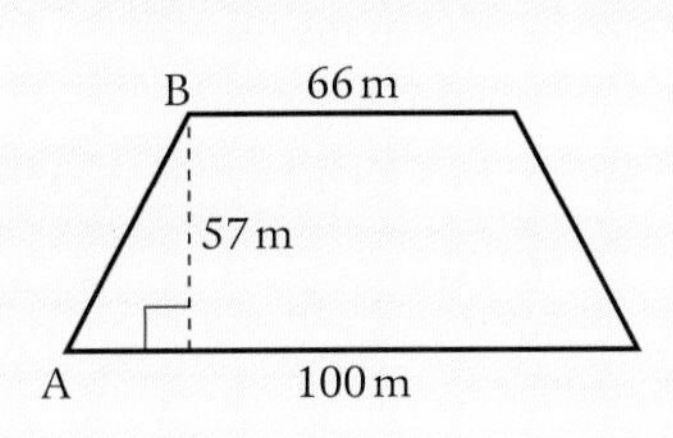

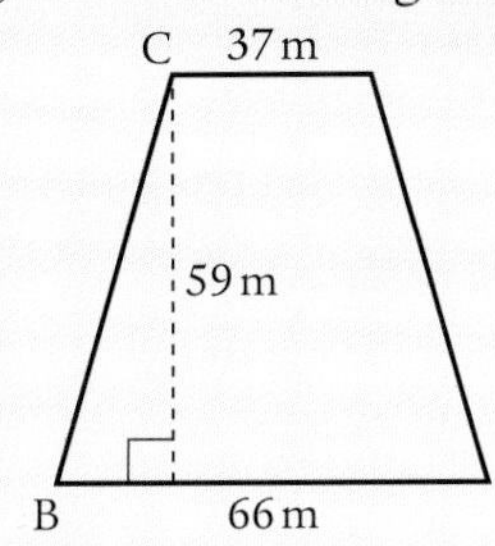

a Calculate the length of: **i** AB **ii** BC.

b Where did the figure of 59 m come from in the second diagram?

c Calculate the length of: **i** CD **ii** EF.

Preparation for assessment

1 Calculate the length of each marked side.

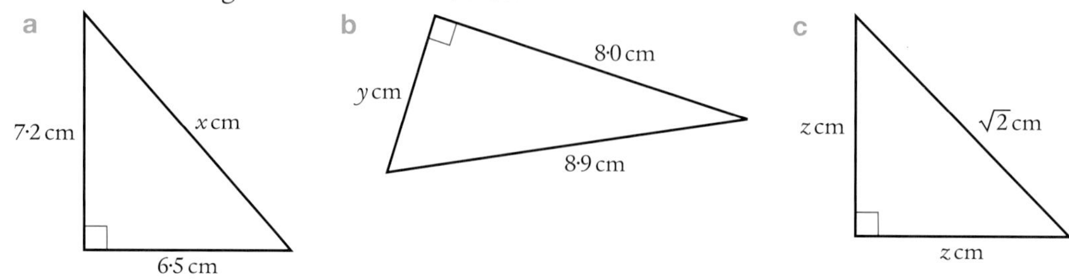

2 a Calculate the length of the hypotenuse when the shorter sides of a right-angled triangle are 44 cm and 117 cm.

b Calculate the length of the third side of a right-angled triangle when one side is of length 10·5 cm and the hypotenuse is of length 13·7 cm.

c Jennifer said that two sides of a right-angled triangle were 1 cm and 2 cm long. What might be the length of the third side?

3 a Calculate the distance between P and Q in the diagram.

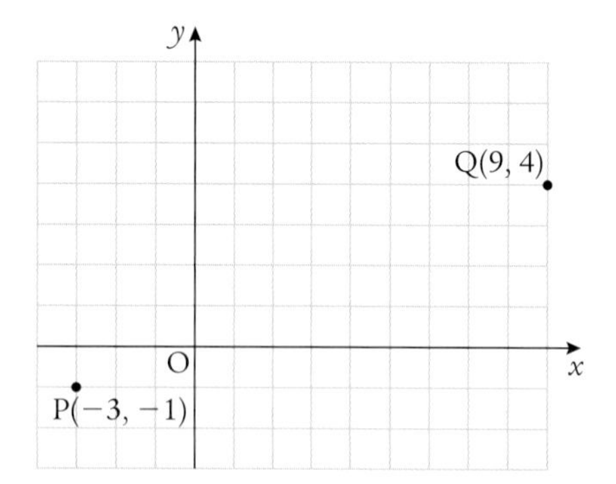

b How far apart are the points (3, 7) and (63, 18)?

4 The Tradeston, or Squiggly, Bridge has an 'S'-shaped walkway across the Clyde. This was a technology solution to the problem of making the bridge high enough for traffic to sail under it but keeping the slope easy enough for pedestrians.

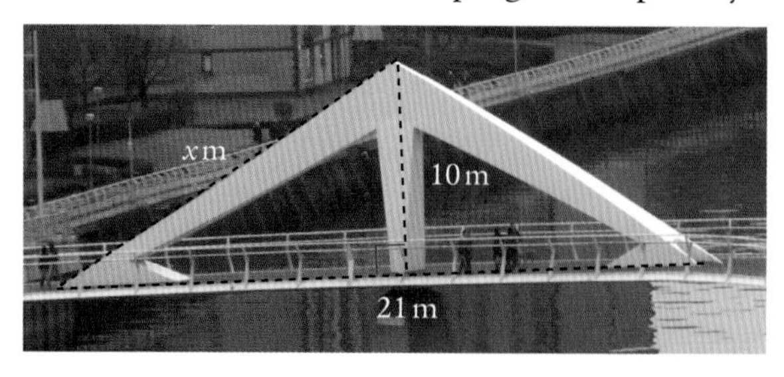

It is held up by two structures in the shape of isosceles triangles.

The base of one of the isosceles triangles is 21 metres.

The height of the triangle is 10 metres.

What is the length of the sloping edge of the triangle?

5 ABCD is a square of side 1 cm.

ABED is a kite. C is the midpoint of AE.

a Calculate the length of AC.

b What is the length of BE?

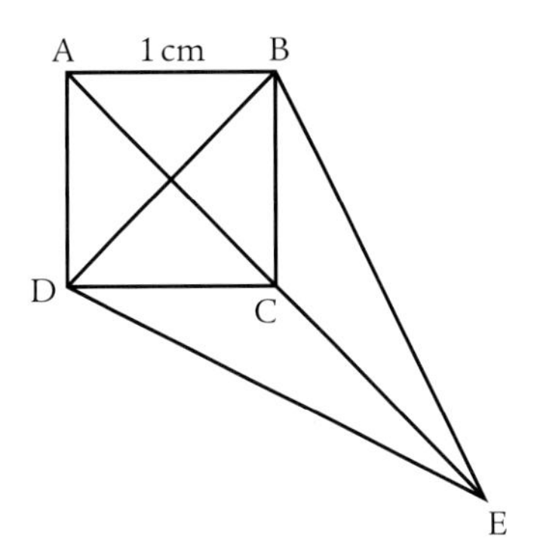

6 Remember the Finnieston Crane?

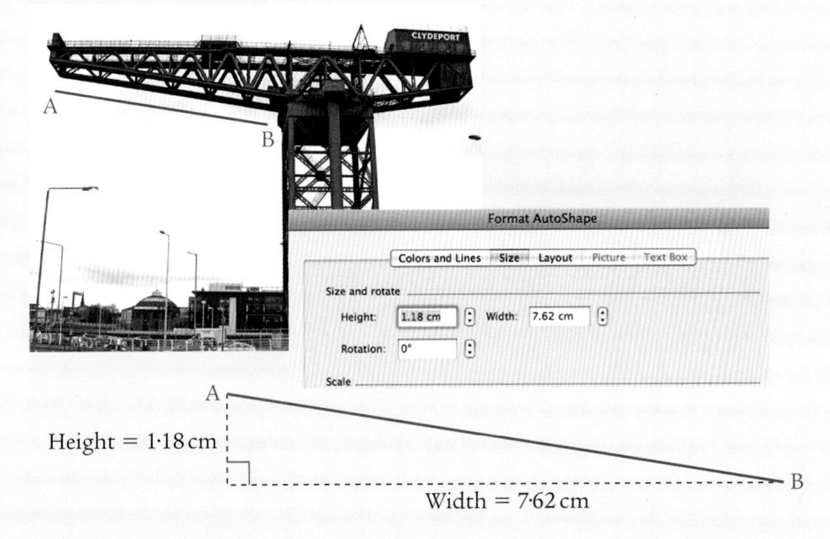

What is the length of the line AB?

9 Related angles

Before we start...

A jackdaw sits on the hand-rail over two angles that are related ... labelled A and B.

I'm sure you know the relationship.

But do you know how C and B are related?

Angles D and E only really exist in the photograph.

How are D and E related?

The window panes are rectangular ... but what shape do they appear in the photo?

The hand-rail divides the picture into two quadrilaterals of the same type. What type?

The rail is one-third of the way up the picture on the left but halfway up on the right.

What fraction of the picture is above the rail?

What you need to know

1 ABC is a straight line with BD perpendicular to it.

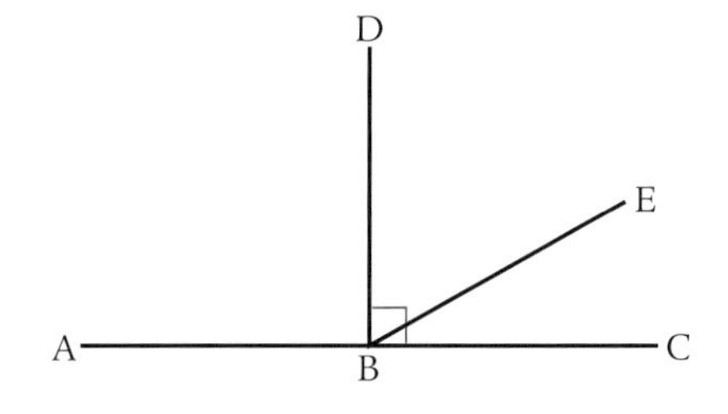

a Name a pair of complementary angles.

b Name a pair of supplementary angles.

c Name an obtuse angle.

2 State the value, giving reasons, of:

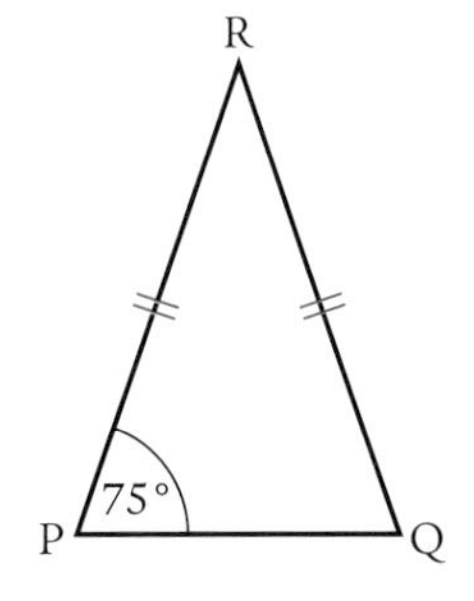

a ∠PQR

b ∠PRQ.

3 Every set of congruent quadrilaterals will tile.

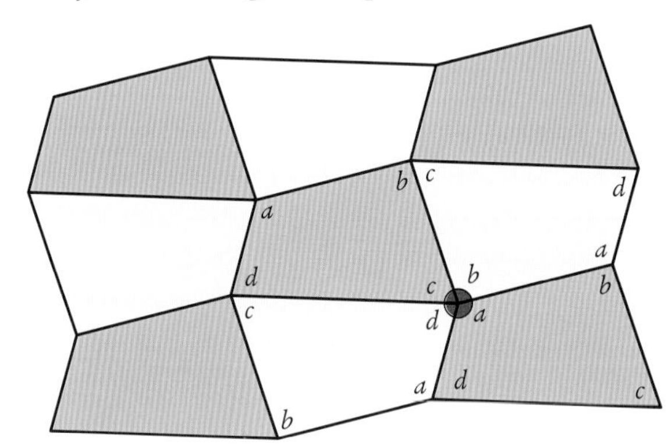

a What is the sum of the angles round the red dot?

b What is the sum of the angles of a quadrilateral?

4 Calculate the labelled angles.

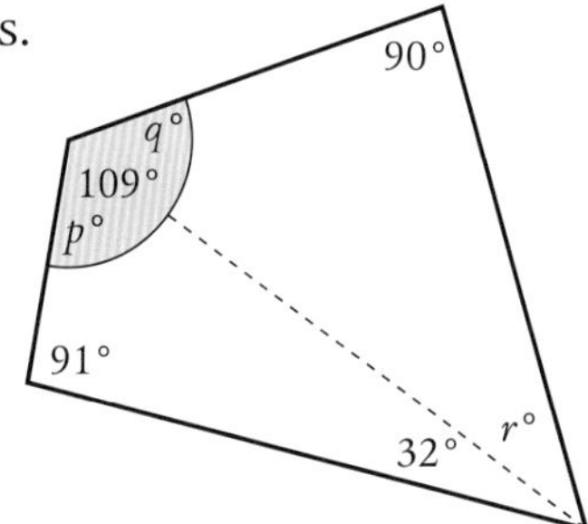

9.1 Using angle facts

You should always make your thinking clear by giving reasons for your steps.

Example 1

Write an equation in x and solve it to find $\angle DBC$.

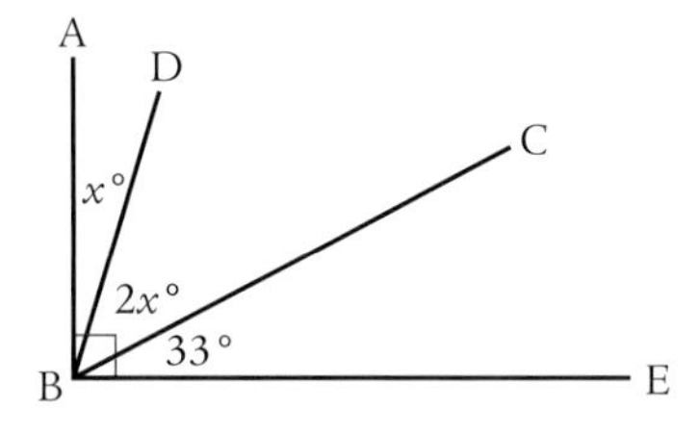

$x + 2x + 33 = 90$ All the parts of a right angle

$\Rightarrow \quad 3x = 90 - 33$

$\Rightarrow \quad 3x = 57$

$\Rightarrow \quad x = 57 \div 3$

$\Rightarrow \quad x = 19$

So since $\angle DBC = 2x°$

then $\angle DBC = 38°$.

Example 2

Write an equation in a, and solve it to find the angles in the triangle.

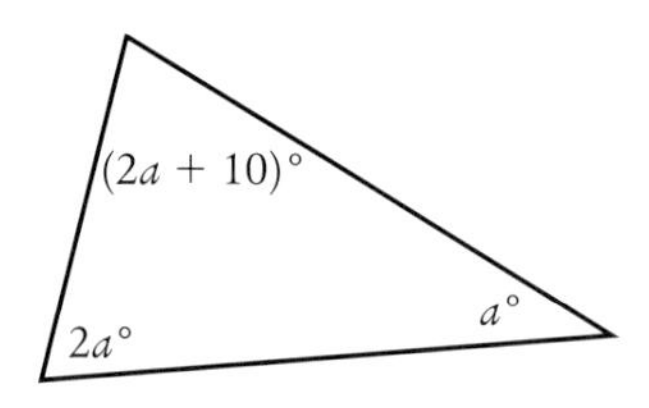

$(2a + 10) + 2a + a = 180$ The three angles of a triangle

$\Rightarrow \quad 5a = 170$

$\Rightarrow \quad a = 170 \div 5$

$\Rightarrow \quad a = 34$

So the angles of the triangle are $2 \times 34° = 68°$, $(2 \times 34 + 10)° = 78°$, and $34°$.

Check it: $68° + 78° + 34° = 180°$ ✓

Exercise 9.1A

1 Calculate the value of each letter.

a

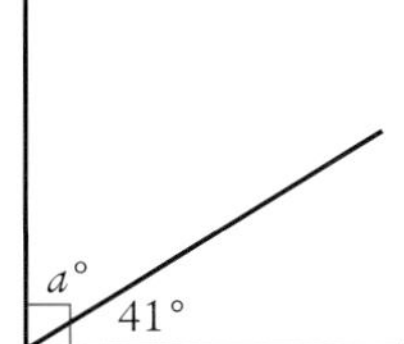

b

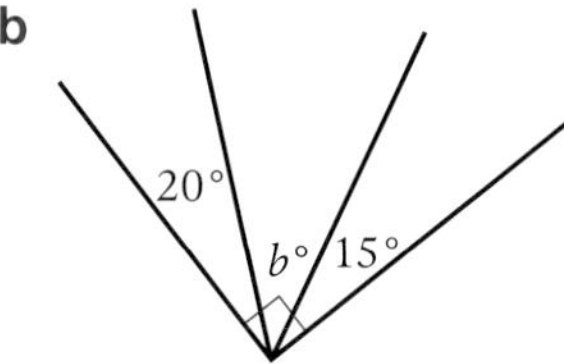

c

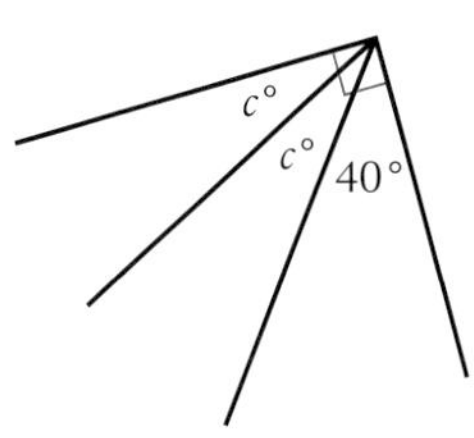

2 Form an equation that connects *d* and *e*.

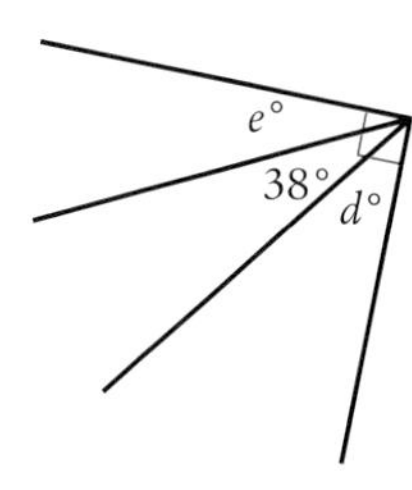

3 Explain why there is something wrong with this diagram.

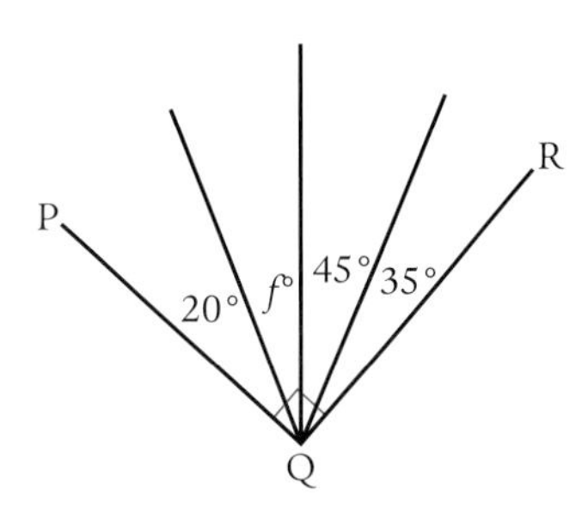

4 In each diagram, AOB is a straight line.

Calculate the values of *g* and *h*.

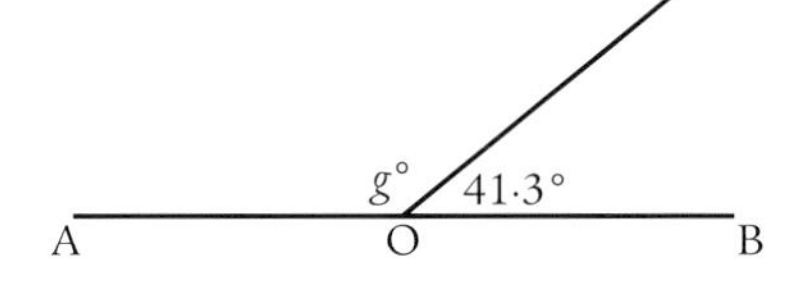

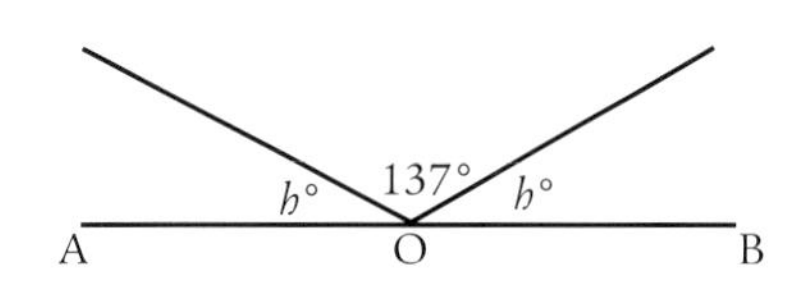

5 AOB is a straight line.

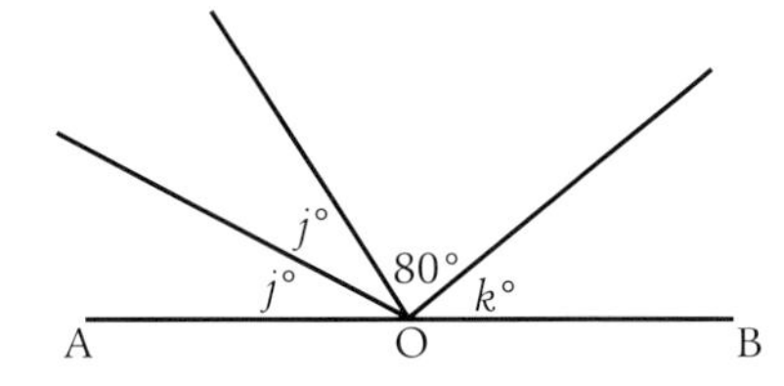

a Form an equation that connects *j* and *k*.

b Calculate their values when *k* is twice the size of *j*.

6 AOB is a straight line. ∠COD is a right angle.

If you are told that ∠COB is 120°, explain why $q = r$.

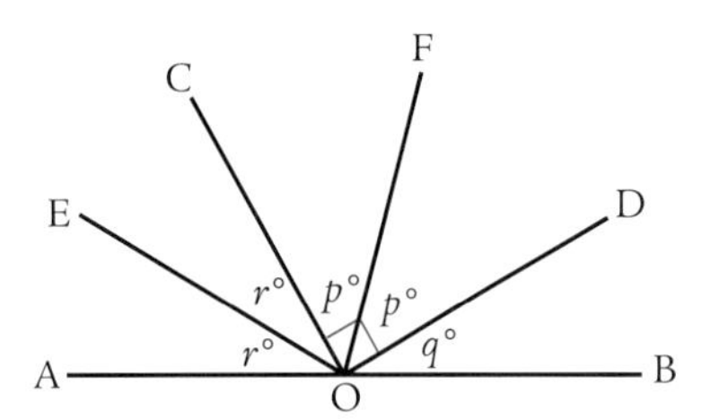

7 A joiner has a rectangular plank of wood ABCD.

He makes a diagonal cut, EF, through it where F is the midpoint of DC.

He rotates the two pieces to form a new shape AEGB as shown.

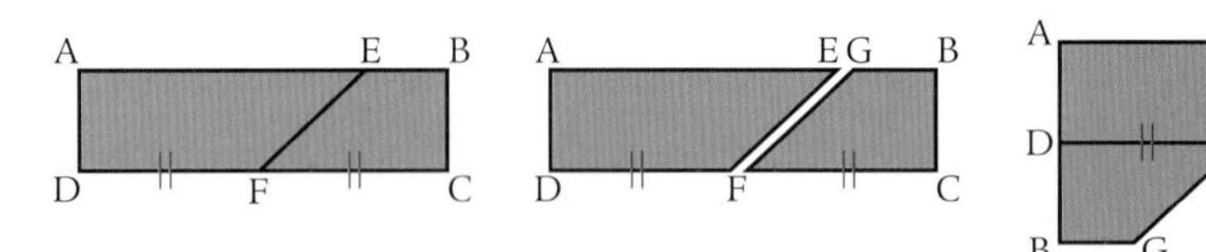

Referring to the new shape, explain why we know that:

a ADB is a straight line

b EFG is a straight line

c ∠AEG and ∠BGE in the trapezium are supplementary.

8 Three roads intersect to form a triangle.

The middle-sized angle is 20° more than the smallest one.

It is also 20° less than the largest angle.

a Find the size of all three angles in the triangle.

b Sketch the diagram and work out the sizes of all the other angles in the diagram.

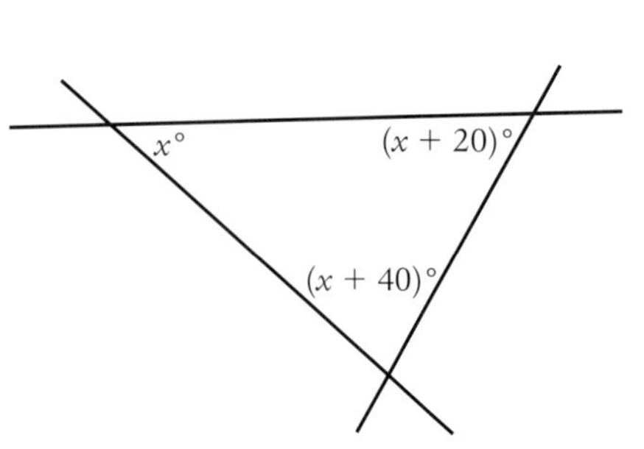

9 In a textbook on astronomy the following diagram is used to explain an eclipse of the Moon. The diagram is symmetrical.

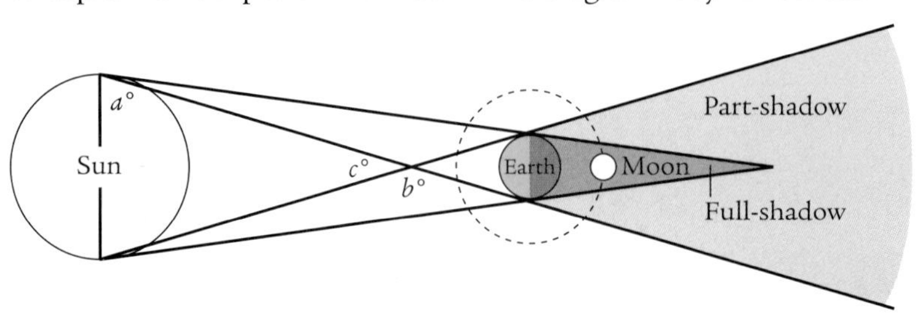

What is the relationship between:

a b and c

b a and c

c a and b?

Exercise 9.1B

1 The walls of certain buildings require support.

Here, the wall is being supported by a buttress.

a Sketch the diagram.

b Find the size of as many angles in the diagram as you can.

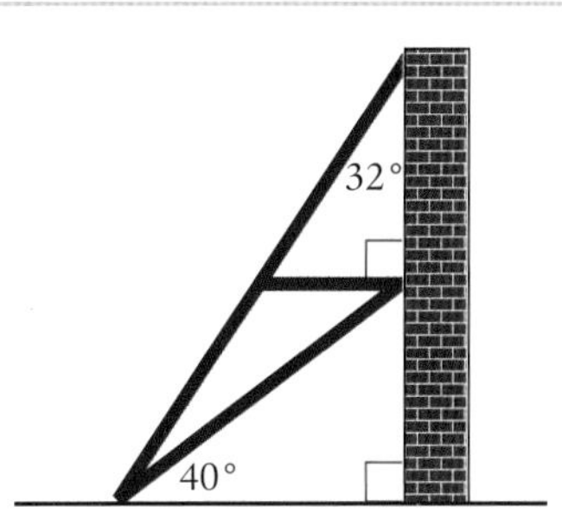

2 When a beam of light is reflected by a mirror, the angle at which the beam strikes the mirror (the angle of incidence) will always be equal to the angle at which the beam leaves the mirror (the angle of reflection).

This angle is usually measured from the **normal**. (The normal is a line perpendicular to the mirror.)

In the diagram, the mirror has been inclined at an angle of $a°$ to the horizontal.

Explain what will happen when $a° = (90 - x)°$.

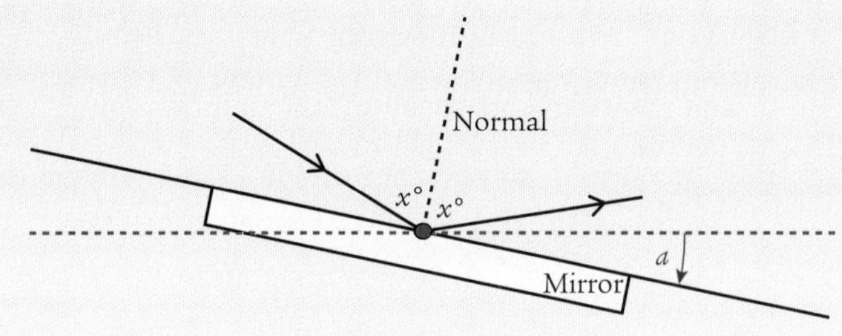

3 Stephen Higgins was snookered on the blue ball.

He thinks the white ball will need to come off the right-hand cushion at an angle of 32° if he is to strike the blue.

At what angle should he strike the left-hand cushion?

(Assume the angle of incidence is equal to the angle of reflection, each time the cue ball strikes the cushion.)

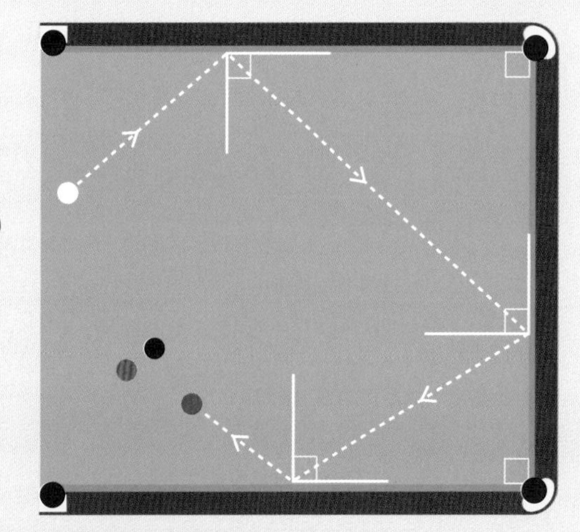

9.2 The FXZ of angles

Class discussion

1 When two lines cross, the point of intersection is known as a **vertex**.

A pair of angles across from each other on the vertex are called **vertically opposite angles**.

In the diagram below, ∠AVC and ∠BVD are vertically opposite angles. ∠CVB and ∠AVD are another pair.

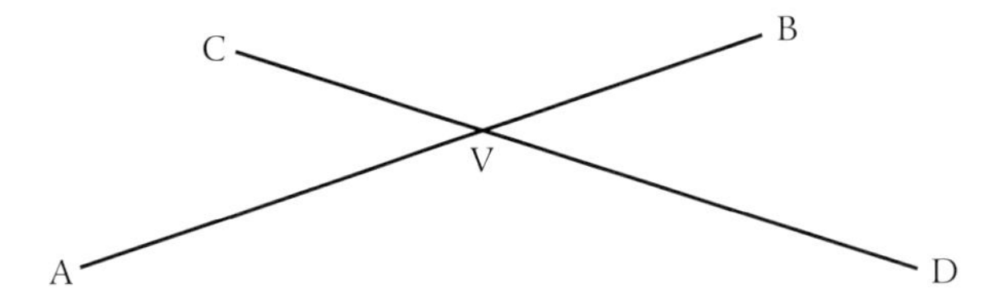

a What can you say about the pair of angles ∠AVC and ∠CVB?

b What can you say about the pair of angles ∠CVB and ∠BVD?

c So, what can you say about the pair of angles ∠AVC and ∠BVD?

d Use a similar stepped argument to show that the vertically opposite angles ∠CVB and ∠AVD are equal.
These are often called 'X angles' because the angles in a letter X are vertically opposite.

2 Two lines can be defined as being parallel if they both cross a third line at the same angle. I know AB is parallel to CD because I'm told they both cross EF at an angle of $x°$.

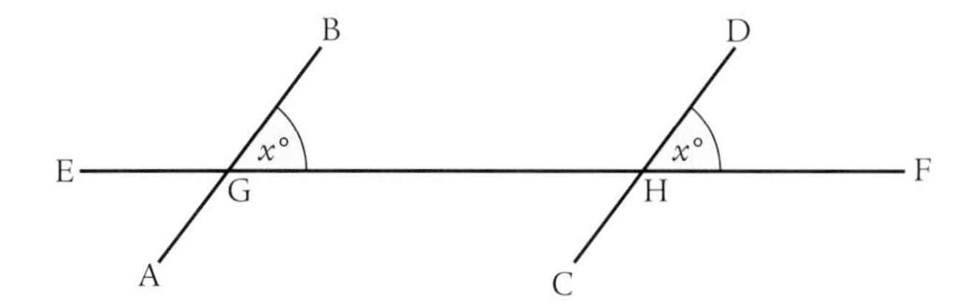

We can now find all the other angles round the two intersections G and H because they form the pattern of vertically opposite angles ...
... and the pattern round G is identical to the pattern round H.

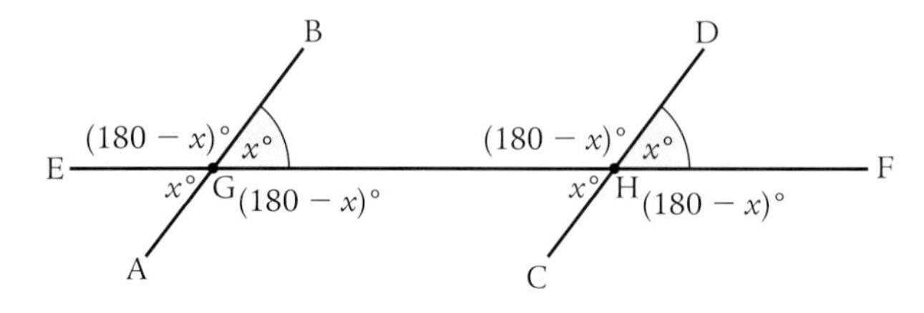

Explain why each angle has been marked as it is.

3 Within the above pattern of lines, we can focus on particular parts. Mathematicians have given these special names:

a **Corresponding angles**

These are angles that take up corresponding positions at the two intersections.

They are both above the **transversal** and to the right of the parallel.

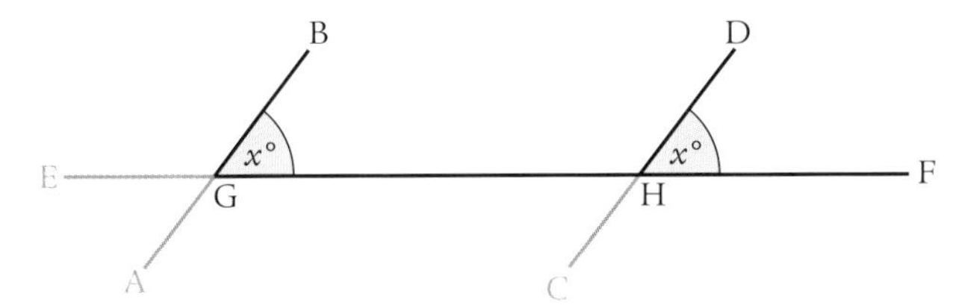

i There are four pairs to find. Identify them.

ii What can be said about corresponding angles?

These are often called 'F angles' because the angles in a letter F are corresponding.

b **Alternate angles**

These are angles that alternate from one side of the transversal to the other: in this case, above or below the transversal.

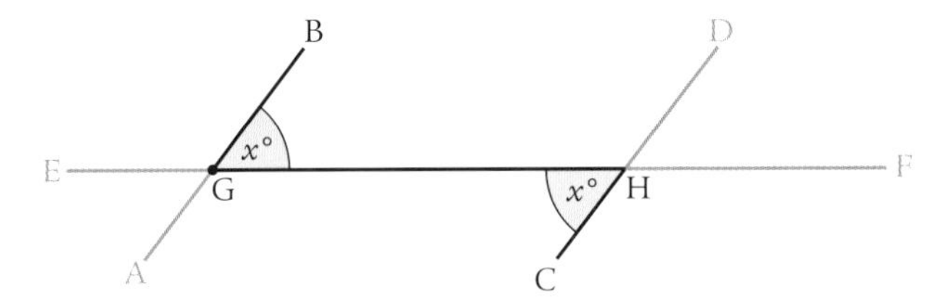

i There is another pair of alternate angles to find. Identify it.

ii What can be said about alternate angles?

These are often called 'Z angles' because the angles in a letter Z are alternate.

c **Co-interior angles**

These are angles trapped together between the parallels on the same side of the transversal.

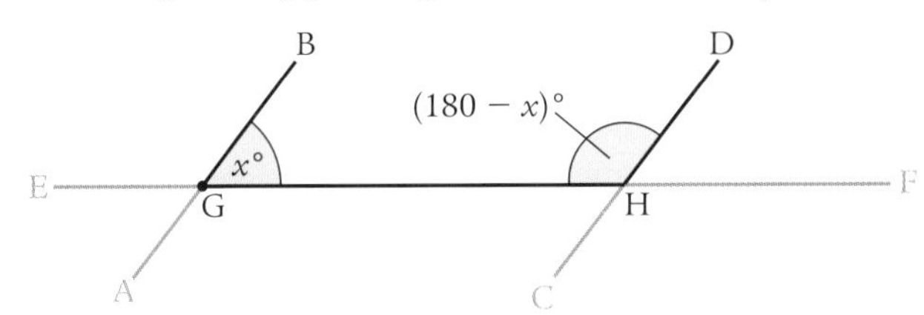

i There is another pair of co-interior angles to find. Identify it.

ii What can be said about co-interior angles?

These are often called 'U angles' because the angles in a squared-off letter U are co-interior.

This relationship is particularly handy when working with bearings.

Always explain which relationship you are using when putting forward an argument.

Transversal: a line that cuts across two or more lines.

Example 1

Two pairs of parallel lines cross as shown.

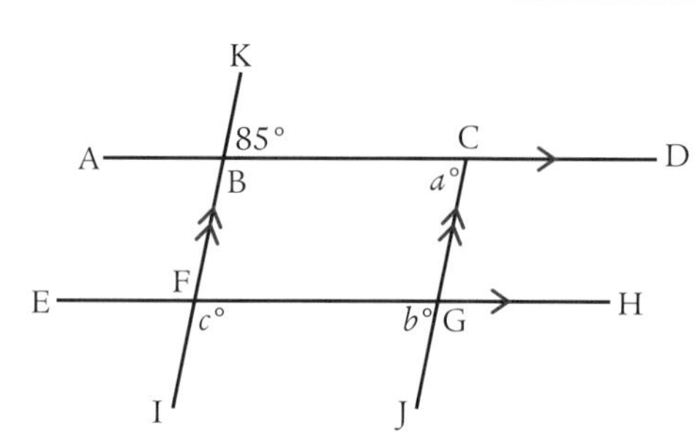

Calculate the values of $a°$, $b°$ and $c°$ in that order, giving reasons for your decisions.

$a° = 85°$ ∠KBC and ∠BCG are alternate angles and so are equal

$b° = 85°$ ∠BCG and ∠FGJ are corresponding angles and so are equal

$c° = 95°$ ∠FGJ and ∠GFI are co-interior angles and so are supplementary (180 − 85 = 95)

Example 2

The *Sapphire* and the *Ruby* are in a yacht race.

The *Sapphire* (S) is on a bearing of 080° from the *Ruby* (R).

What is the bearing of the *Ruby* from the *Sapphire*?

(N and T are both due North.)

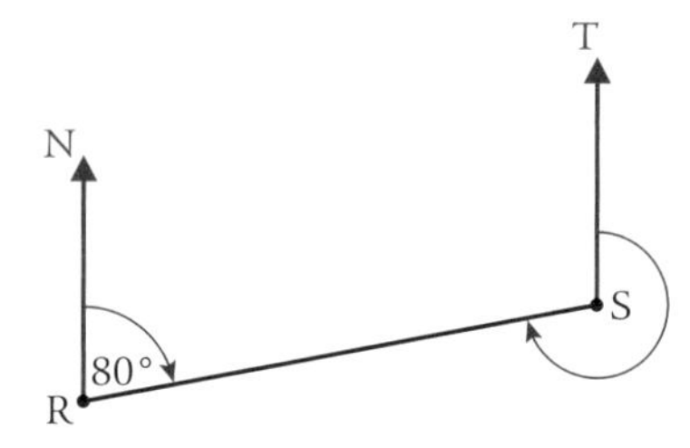

RN and ST are parallel (both pointing due North).

⇒ ∠NRS and ∠TSR are co-interior ... so they are supplementary.

⇒ ∠TSR = 180° − ∠NRS = 180° − 80° = 100°.

⇒ Bearing of *Ruby* from *Sapphire* = 360° − 100° = 260° Angles round a point add to 360°

Exercise 9.2A

1 In this diagram AB and CD are parallel lines both being cut by the transversal EF.

Calculate the size of each labelled angle.

Give a reason for your answer by relating each to the angle marked 43°.

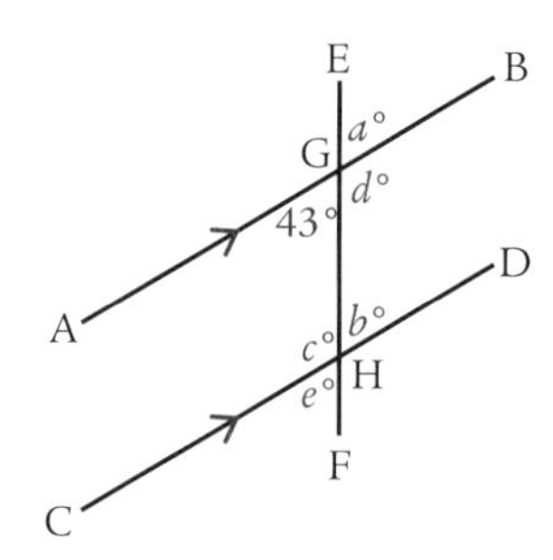

2 A metal frame in the shape of a right-angled triangle is shown.

a Find the size of each named angle, giving reasons for your answers.

i ∠CDE **ii** ∠DEB **iii** ∠ACB

b Looking at the trapezium ABED, what do you notice about:

i the pair of angles ∠BAD and ∠EDA

ii the four angles?

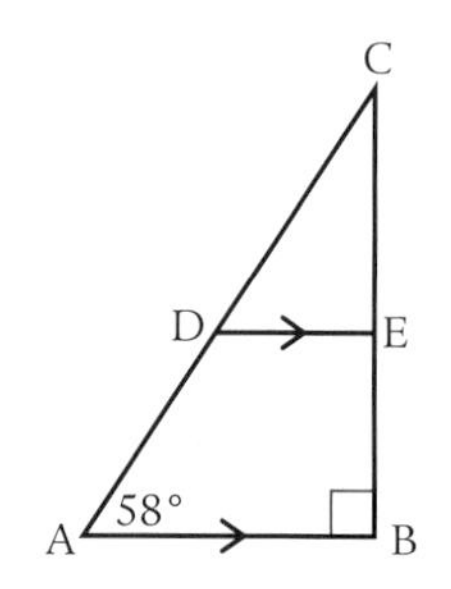

3 The side-frame to support a child's swing is in the shape of an isosceles triangle (ΔPRT) with tie bars, QS and QT to stiffen the frame.

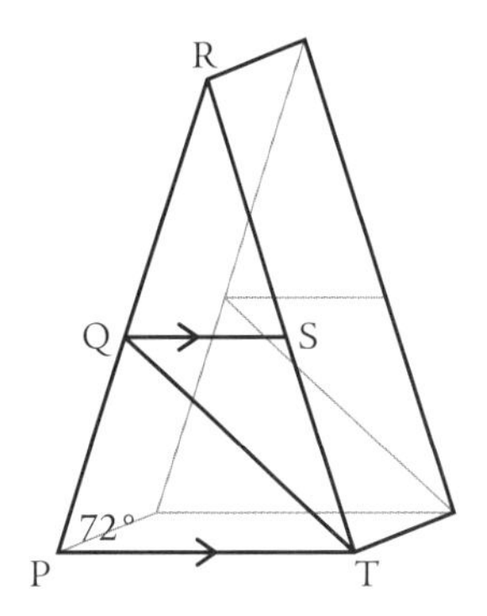

a Calculate the size of:

i $\angle RQS$

ii $\angle RSQ$

iii $\angle QRS$.

b Name an angle equal to $\angle SQT$.

c Name an angle supplementary to $\angle TPQ$.

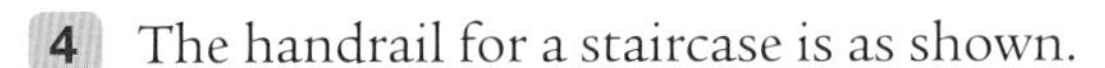

4 The handrail for a staircase is as shown.

The upright supports are all vertical.

The inclined runners are all parallel to the handrail.

The acute angle $\angle BAD$ between the upright and the parallel runner is 68°.

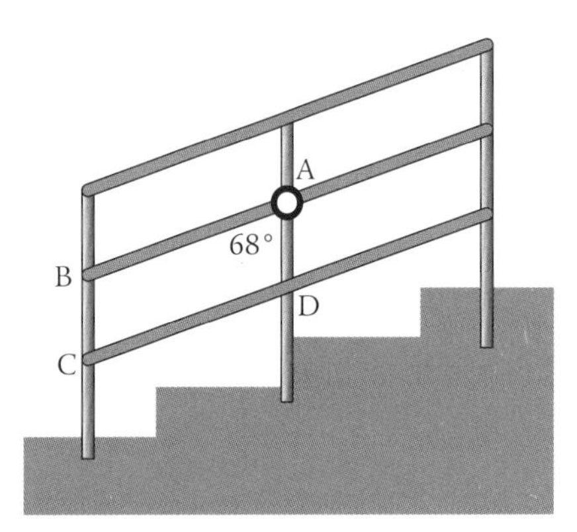

a By making use of co-interior angles, find the size of each angle in the parallelogram ABCD.

b What must be true about the opposite angles of any parallelogram?

c What must be true about the adjacent angles of any parallelogram?

Exercise 9.2B

1 The technology class investigated the design of ironing boards. When the board is in use, the legs form isosceles triangles.

The board has two settings: 'normal' and 'low'.

The diagram shows that one of the angles is 64° when the board is in the low setting.

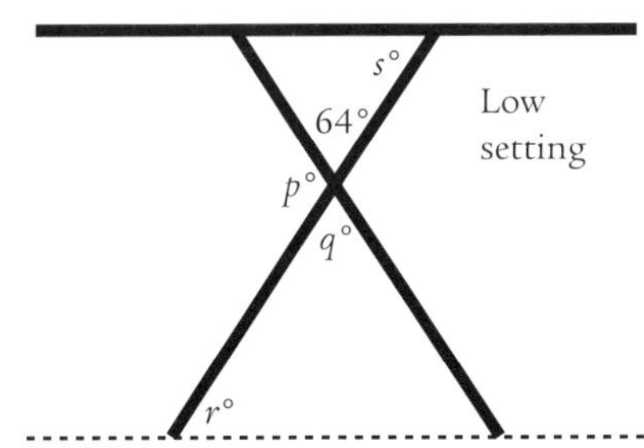

a Calculate the size of each labelled angle.

b When the board is at the normal setting, the angle is 8° smaller.

Calculate the size of p, q, r and s when the board is in the normal setting.

c In the normal setting the board is higher up. Explain why the 64° angle needs to become smaller for the board to rise.

2 A deck chair design incorporates different geometric shapes.

The wooden frame is constructed as shown in the diagram.

The designer says that $\angle BCE$ is to be $71°$.

What size will the following angles be?

a $\angle DCE$

b $\angle CEF$

c $\angle CDF$

d $\angle EFG$

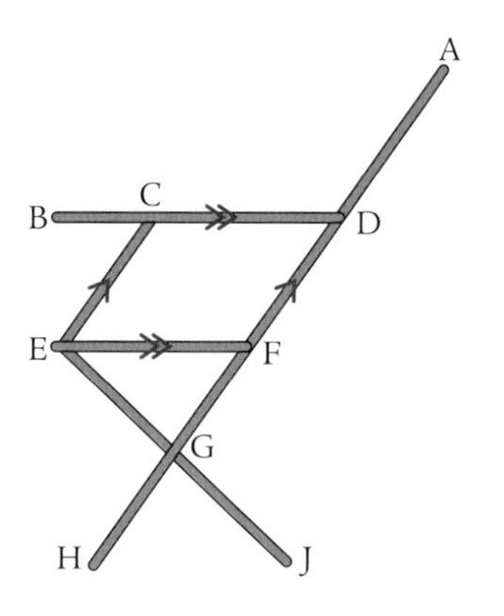

3 The weight of the roof in many houses is carried by roof trusses.

The diagram shows what a typical roof truss looks like.

The trusses are symmetrically positioned.

What is the size of the angle at the apex of the roof?

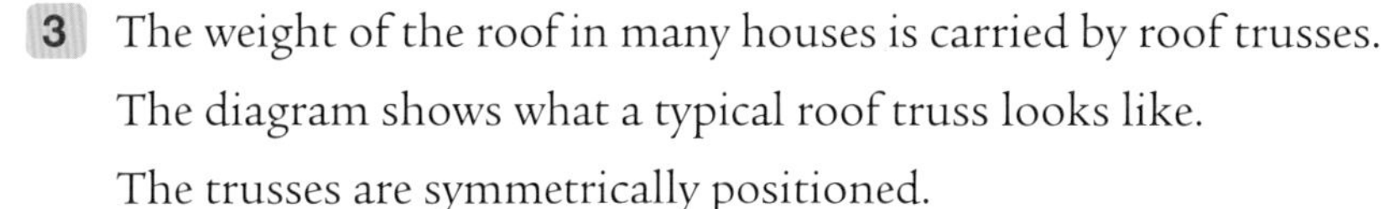

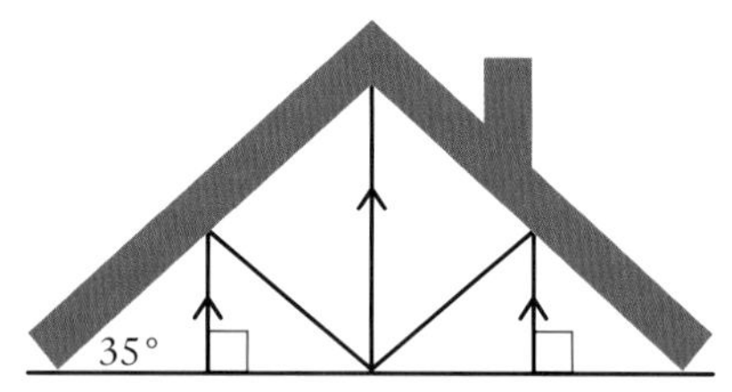

4 ABCD is a trapezium.

When the non-parallel sides are extended, they intersect at E.

$\angle AEB = 40°$ and $\angle EAB = a°$

a Express the angles in the trapezium in terms of a.

b Verify that the four angles of the trapezium add up to $360°$.

c By looking at the case where $\angle AEB = x°$ and $\angle EAB = a°$, prove that the angles of any trapezium add up to $360°$.

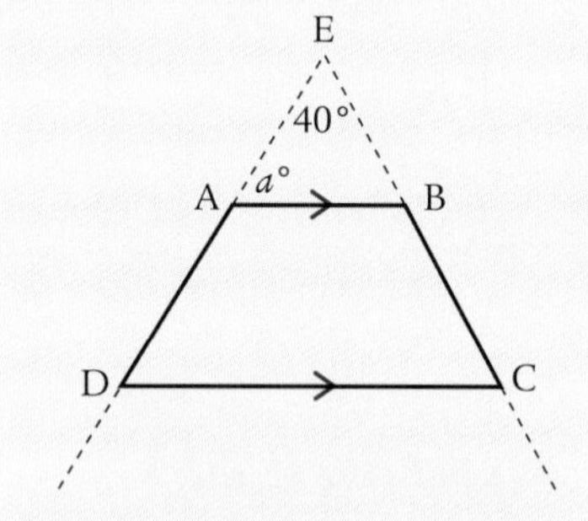

5 The lifeboat house at Dougerie has an unusual gable-end.

The roof ridge PQ makes an angle of $61°$ with the upper part of the roof.

The guttering, RS is parallel to PQ and makes an angle of $154°$ with the lower end.

What is the size of $\angle PTR$?

Draw a convenient line parallel to PQ.

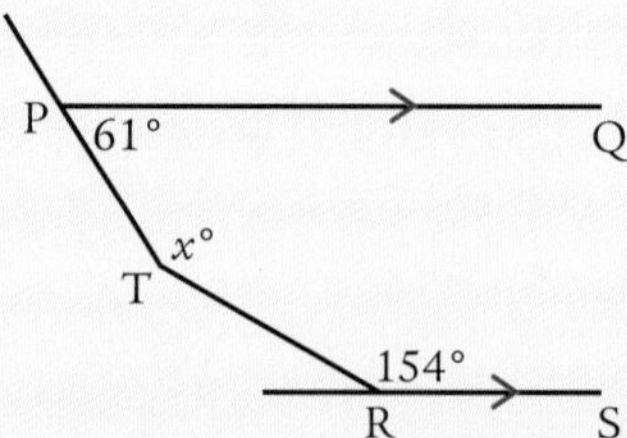

6 In the model of a rift valley studied in Geography, the beds of rock are parallel.

a Calculate the values of a, b and c.

b The down-thrust block is shaped like a trapezium.

Calculate the size of each angle in the trapezium.

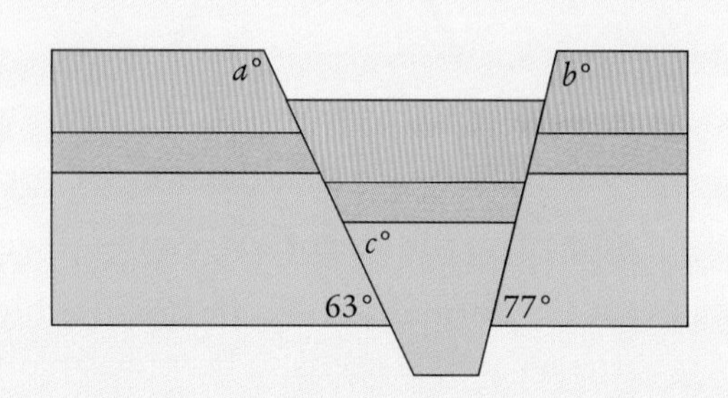

9.3 Bearings

Example 1

The harbour master records the direction to various ships.

Using 3-figure bearings,
the *Amber* is on a bearing of 037°, the *Beryl* is on a bearing of 123°,
the *Calypso* is on a bearing of 255°, the *Deirdre* is on a bearing of 310°.

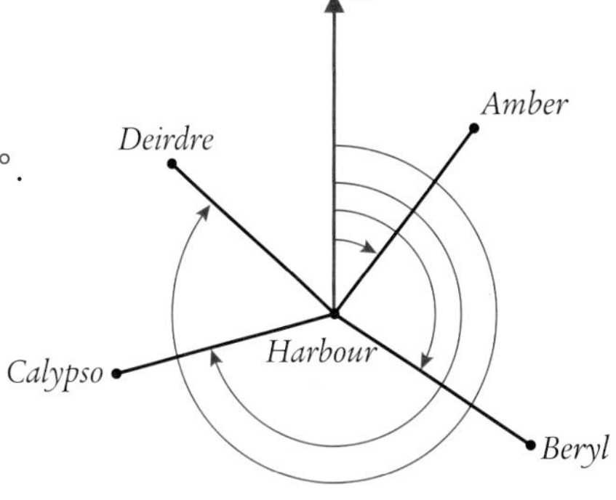

Through what angle must the harbour master turn to:

a look from the *Amber* to the *Beryl* clockwise

b look from the *Amber* to the *Calypso* clockwise

c look from the *Deirdre* to the *Calypso* anticlockwise?

All bearings are measured clockwise, so:

a $123° - 37° = 86°$ **b** $255° - 37° = 218°$ **c** $310° - 255° = 55°$.

Example 2

The radar system at Air Traffic Control (A) shows an aircraft at B on a bearing of 127°.

What is the bearing of Air Traffic Control from the aircraft?

(This is often referred to as a **back-bearing**.)

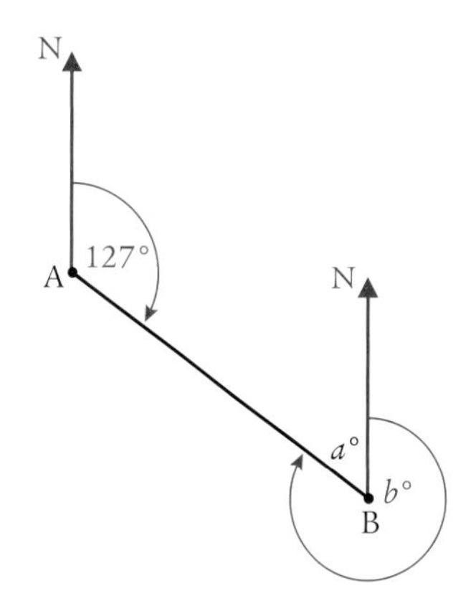

$127° + a° = 180°$ Co-interior angles

$a° = 53°$

So the bearing of A from B is $360° - 53° = 307°$.

Exercise 9.3A

1 A ferry sails from Brodick (B) to Ardrossan (A) on a bearing of 071°.

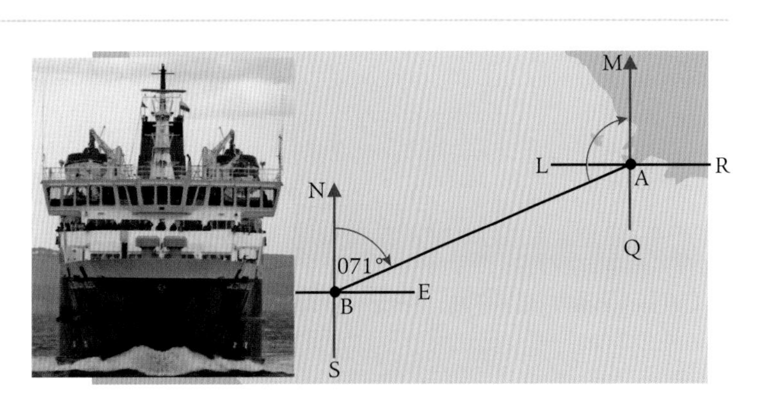

a Name an angle complementary to ∠NBA.

b Name an angle equal to ∠NBA because it is an alternate angle.

c Name two other pairs of alternate angles.

d Calculate the size of ∠BAM, giving a reason for your answer.

e Calculate the bearing of Brodick from Ardrossan.

2 A fishing boat sails from Peterhead (P) on a bearing of 056° to the fishing grounds at A.

It then changes bearing to 141°, following the shoals of fish, and sails to point B.

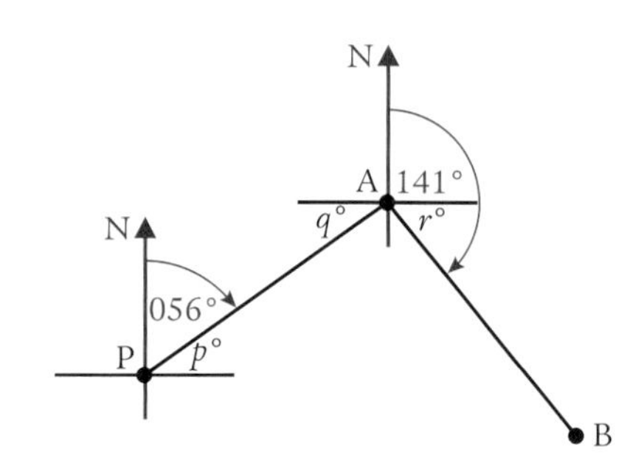

a Write down the sizes of the angles marked $p°$, $q°$ and $r°$.

b What size is ∠PAB?

c Through how many degrees did the boat turn when it changed direction at A?

d How is your answer to **c** linked to the two bearings given in the question?

3 An Airbus A320 flies out of Edinburgh International Airport (E) on a bearing of 255°.

After a few minutes, it is instructed to change bearing to 028°.

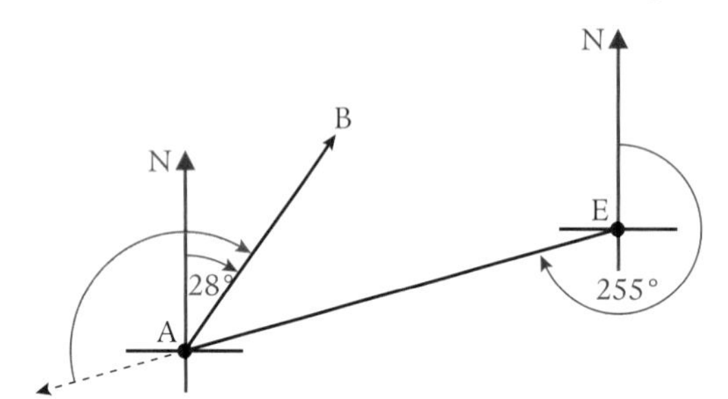

a What size is ∠EAB?

b Through how many degrees clockwise does the pilot turn the plane at A to head for point B? (The angle is marked by the blue arc.)

4 The *Amethyst* and the *Crystal* are in a race.

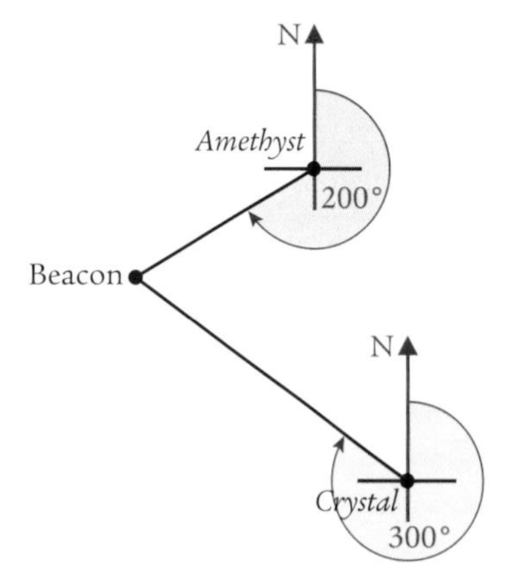

They both take their bearings on an on-shore beacon.

From the *Amethyst* the beacon is on a bearing of 200°.

From the *Crystal* the beacon is on a bearing of 300°.

Through what angle must an observer at the beacon turn to look from the *Amethyst* to the *Crystal*?

Exercise 9.3B

1 The first priority of anyone working in air traffic control is to ensure the safety of passengers. The greatest disaster is when two planes collide in mid-air.

Air traffic controller A is tracking a Boeing 737, which is north of him flying on a bearing of 057°.

At the same time, air traffic controller B, who is 50 km due east of A, is tracking an Airbus A310, which is overhead, flying on a bearing of 237°.

a If the planes stay on their present bearings, why are both controllers confident about this situation?

b Is there any situation when a plane flying on a bearing of 057° will be a danger to a plane flying on a bearing of 237°?

2 Helicopters are used to supply off-shore oil rigs in the North Sea.

A large number are based in Aberdeen.

One such helicopter flew from Aberdeen on a bearing of 048° for a distance of 62 km.

It then changed to a bearing of 167° for a distance of 45 km.

a Make a scale drawing of the helicopter's route. (Use a scale of 1 : 100 000.)

b Work out how far it now is from its base in Aberdeen.

c What bearing should the pilot fly to return to base?

3 A cabin cruiser is sailing on a bearing of $(abc)°$, where a, b and c are the digits of a number.

The captain is sure that the general direction of the cruiser is within 2° of South-East.

a Which digit is a most likely to represent ... 0, 1 or 2?

b Which digit is b most likely to represent ... 3, 8 or 9?

c Which digit is c most likely to represent ... 0, 6 or 9?

4 The *Amethyst* and the *Crystal* are in a race.

They both take their bearings on an on-shore beacon.

From the *Amethyst* the beacon is on a bearing of $x°$.

From the *Crystal* the beacon is on a bearing of $y°$.

Through what angle must an observer at the beacon turn to look from the *Amethyst* to the *Crystal*?

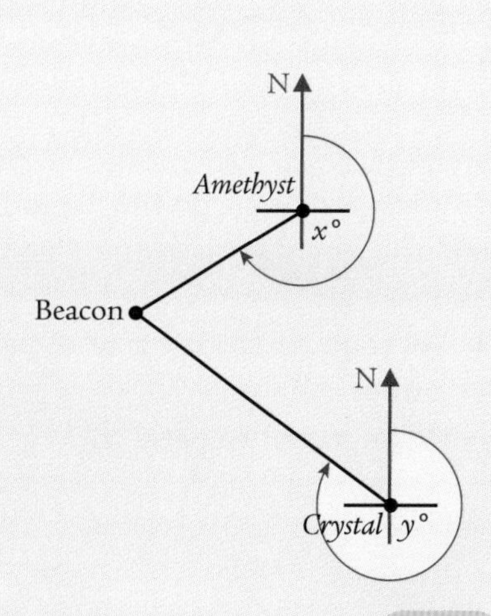

9.4 Angles in quadrilaterals

Class discussion

Within the quadrilaterals, and their diagonals, we can find various sets of related angles.

These depend mainly on the properties of the shape concerned.

In each case discuss the relation suggested by the given property:

1 A kite has one axis of symmetry. Congruent kites tile.

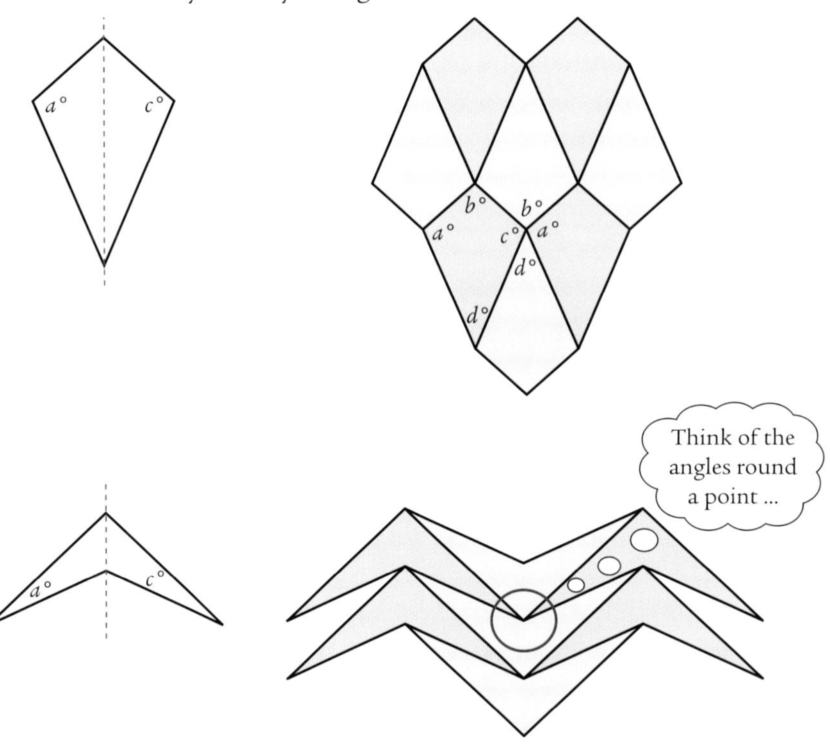

2 A rhombus has two axes of symmetry. Congruent rhombuses tile.

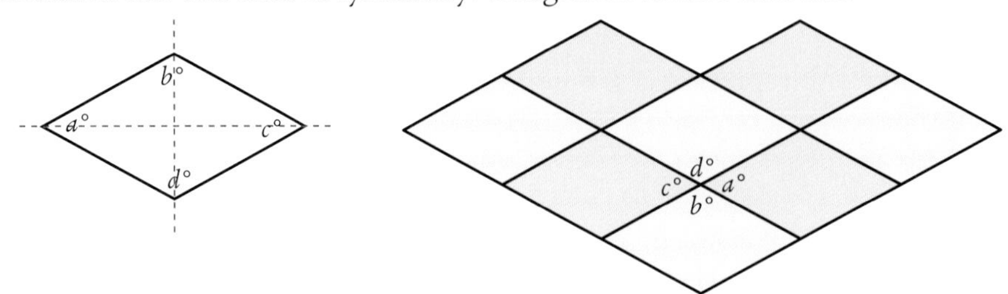

At what angle do the diagonals intersect in both the kite and the rhombus? Justify your answer.

3 A trapezium has a pair of parallel sides.
Congruent trapezia tile.

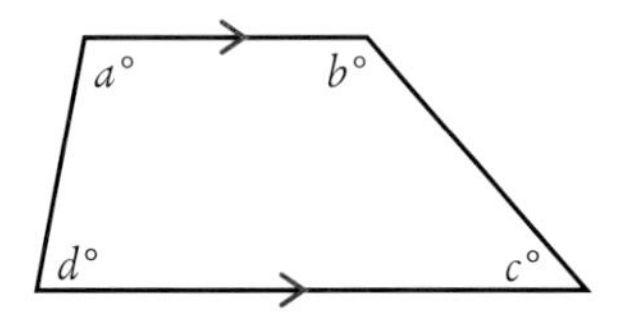

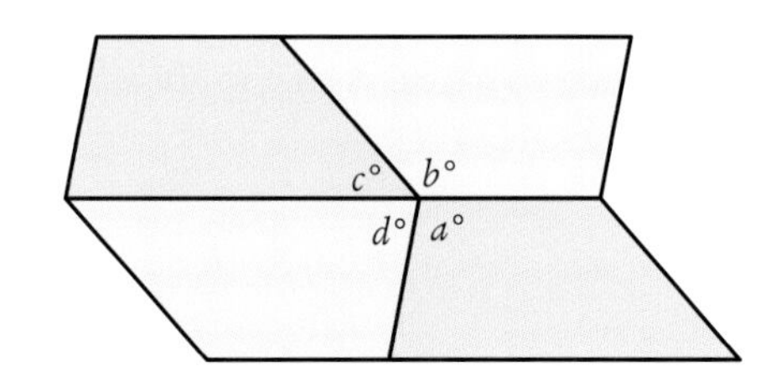

When we draw the diagonals, how many pairs of related angles can you find?

Look for alternate angles, vertically opposite angles, co-interior angles.

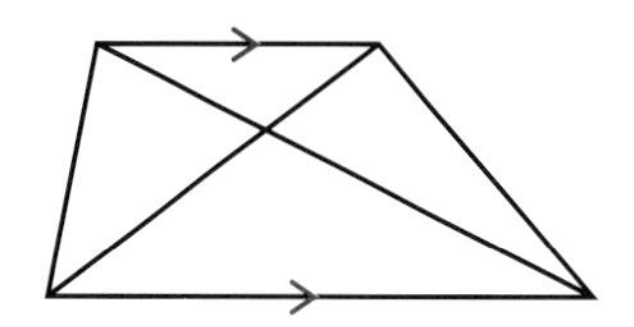

4 A parallelogram has two pair of parallel sides.
Congruent parallelograms tile.

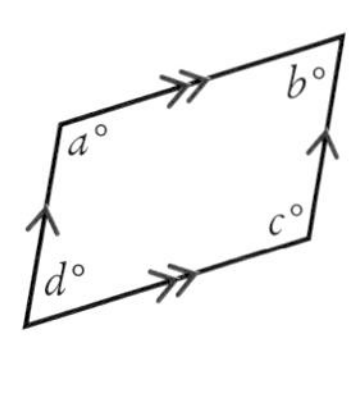

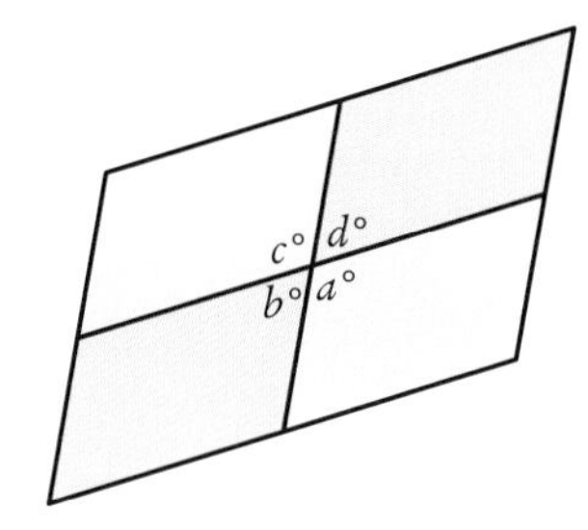

i When we draw its diagonals, how many pairs of related angles can you find?

ii What do we mean when we say that a parallelogram has a centre of symmetry?

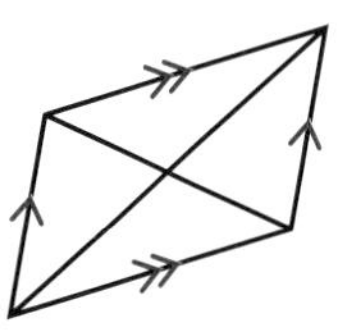

5 Any set of congruent quadrilaterals tile.

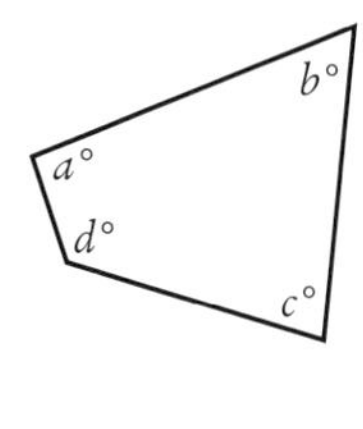

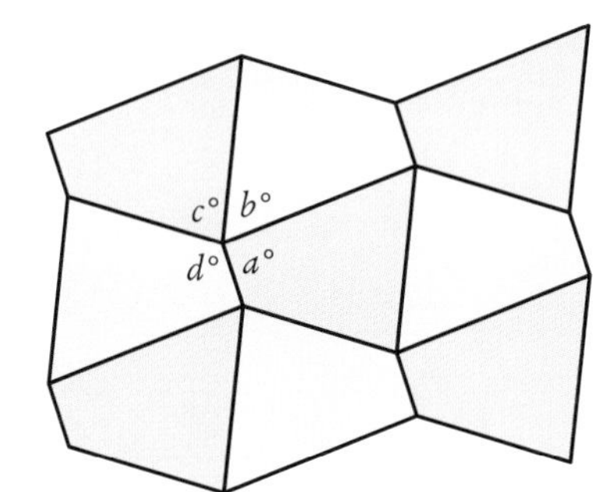

Example 1

ABCD is a kite with diagonal AC drawn.

∠BCA = 37°.

a Calculate the size of ∠ABC.

b If ∠BCD = 95°, calculate the size of ∠CDA.

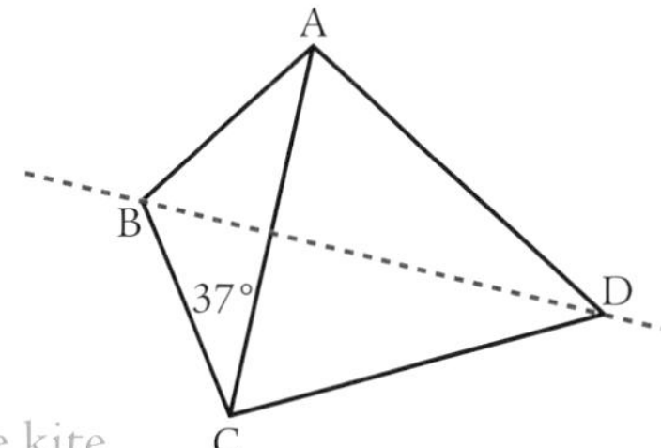

a ∠BAC = 37° — By symmetry of the kite

∠ABC = 180° − 37° − 37° — Angles in a triangle add up to 180°

= 106°

b ∠BCD = 95° ⇒ ∠BAD = 95° — By symmetry of the kite

∠ADC = 360° − 95° − 95° − 106° — Angles in a kite add up to 360°

= 64°

Example 2

PQRS is a rhombus with ∠PQR = 37°.

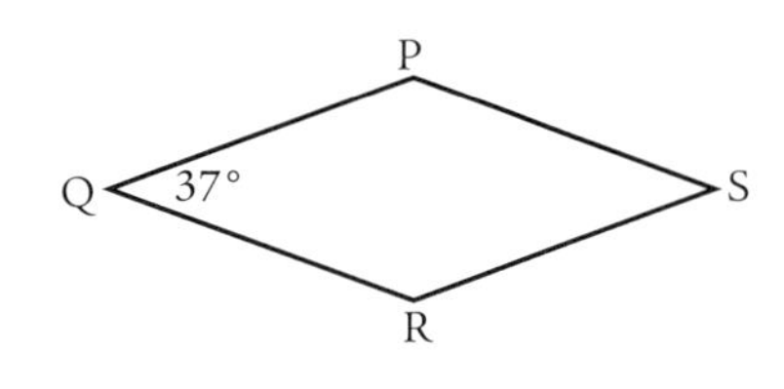

a Calculate the size of each of the other angles of the rhombus.

b When the diagonals are drawn, they intersect at E.

Calculate the size of: **i** ∠PQE **ii** ∠QPE.

a ∠PSR = ∠PQR = 37° — Symmetry of the rhombus

⇒ ∠QPS + ∠QRS = 360° − 37° − 37° — Four angles add up to 360°

⇒ ∠QPS + ∠QRS = 286°

⇒ ∠QPS = 143° — ∠QPS = ∠QRS from the symmetry of the rhombus

⇒ ∠QRS = 143°

b ∠PQE = $\frac{1}{2}$ of ∠PQR = $\frac{1}{2}$ of 37° = 18·5°

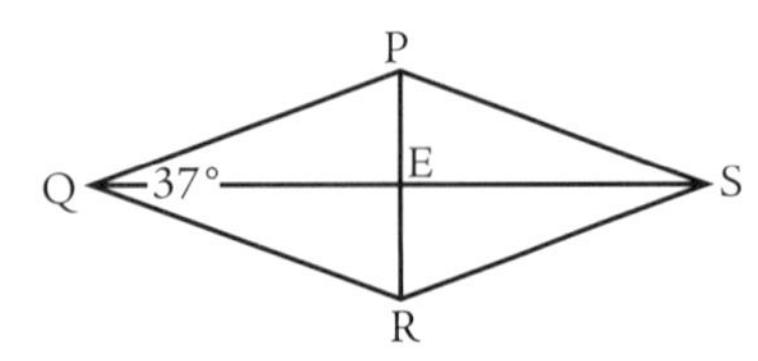

∠PEQ = 90° — Diagonals of a rhombus intersect at right angles

⇒ ∠QPE = 90° − 18·5° = 71·5° — Complementary angles

Example 3

PQRS is a parallelogram with ∠SPQ = 72°.

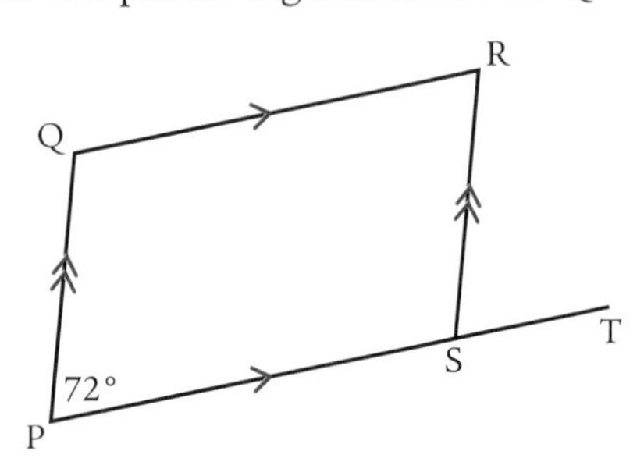

PS is extended to T forming the exterior angle ∠RST.

a What is the size of ∠QRS?

b What is the size of ∠PQR?

c Calculate the size of the exterior angle ∠RST.

a ∠QRS = 72° — Opposite angles are equal because of centre of symmetry

b ∠PQR = 180° − 72° = 108° — Co-interior angles ∠SPQ and ∠PQR are supplementary

c ∠RST = 72° — Corresponding angle to ∠SPQ

Exercise 9.4A

1 **i** Identify each shape. Broken lines are axes of symmetry.

ii Find the size of each angle indicated.

a

b

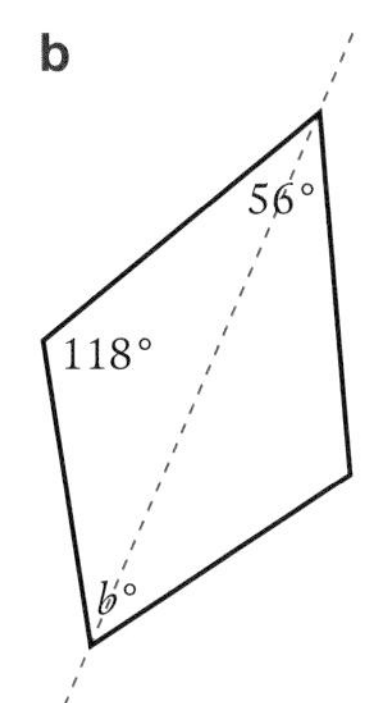

c

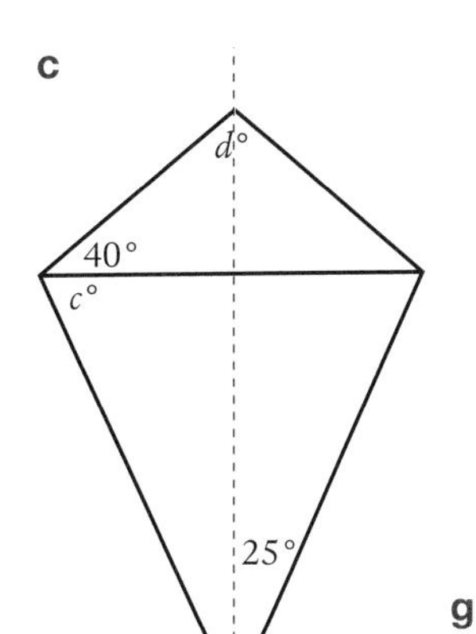

d

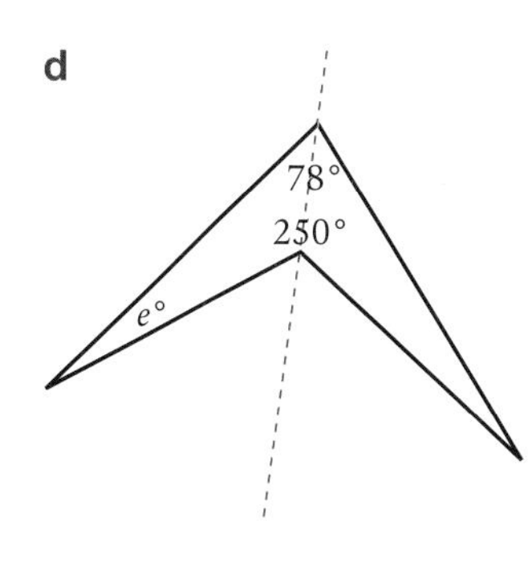

e

f

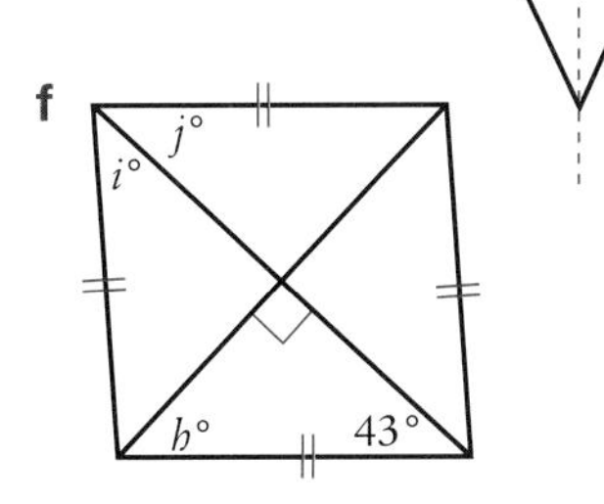

g

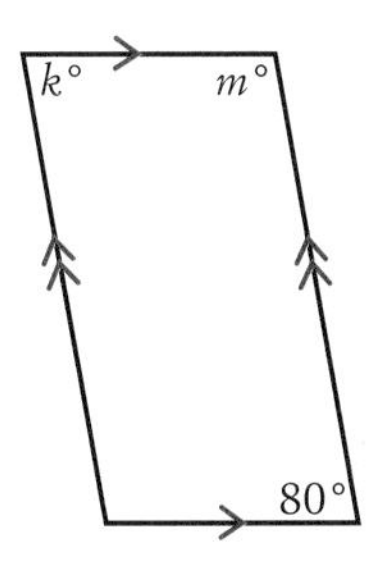

h

i

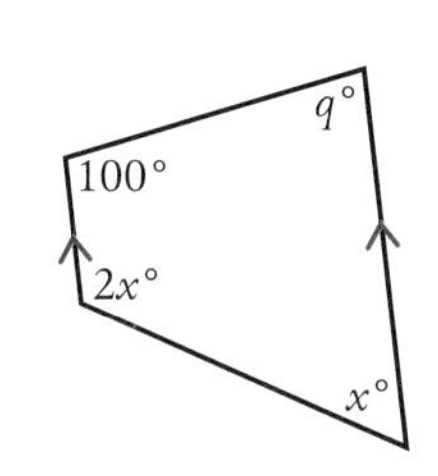

j

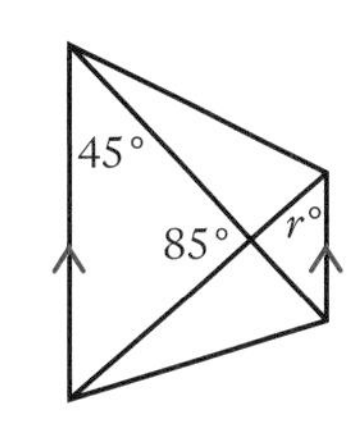

2 For each diagram either find a relationship between m and n or find their values.

a

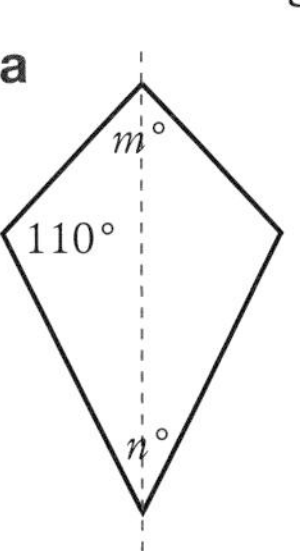

b

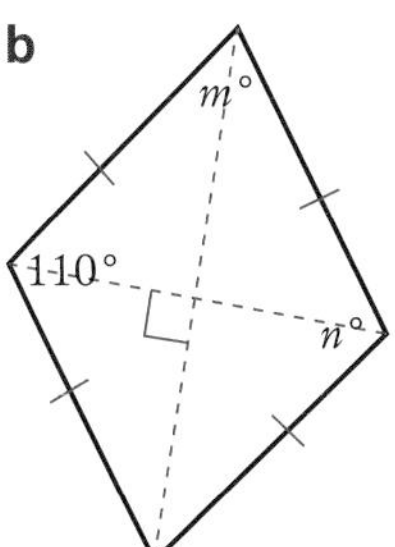

c

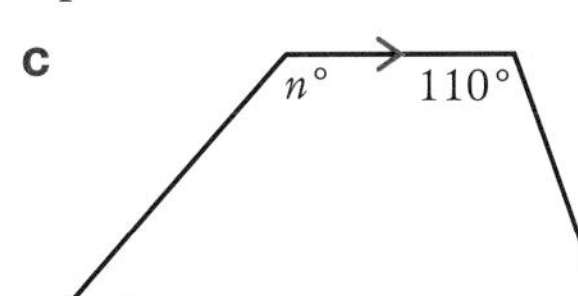

d

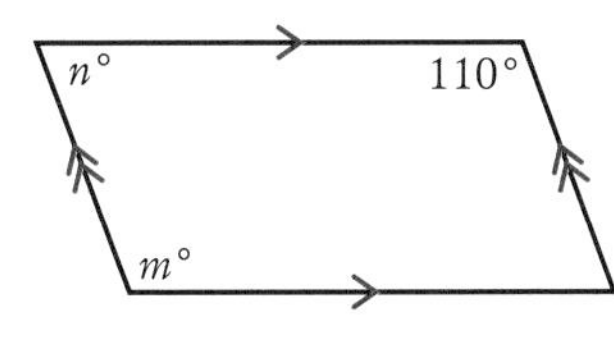

3 The Rocketeers were experimenting with rocket design.

For the sake of stability they considered flight stabilisers shaped as rhombuses.

These were attached to the body of the rocket at an angle of 136°.

Calculate the size of each angle in the rhombus.

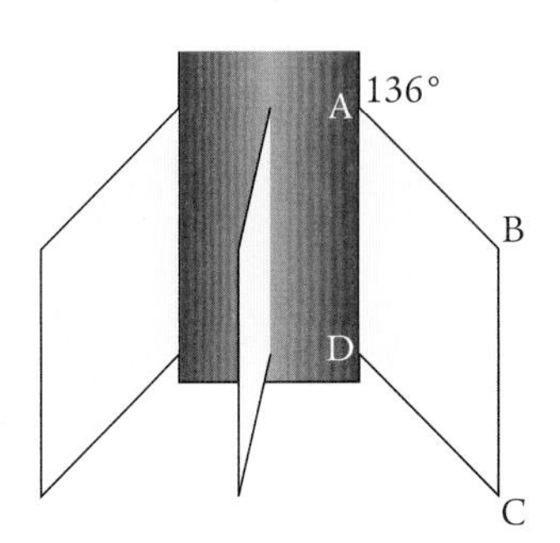

4 Sameena makes a square from four pieces of rod, all of the same length.

They are fixed by screws at each corner.

She loosens the screws and pushes the verticals 29° forward.

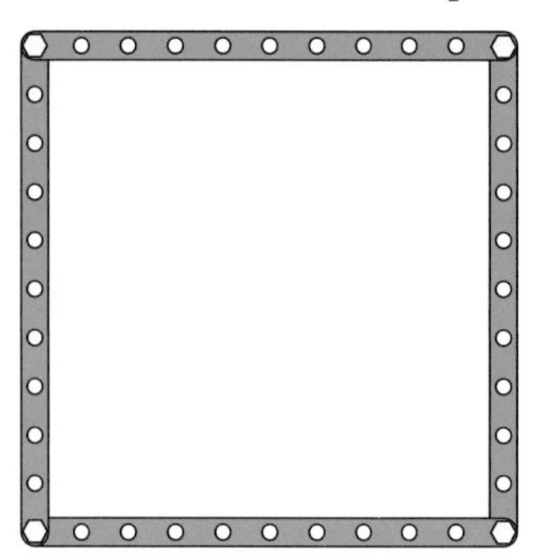

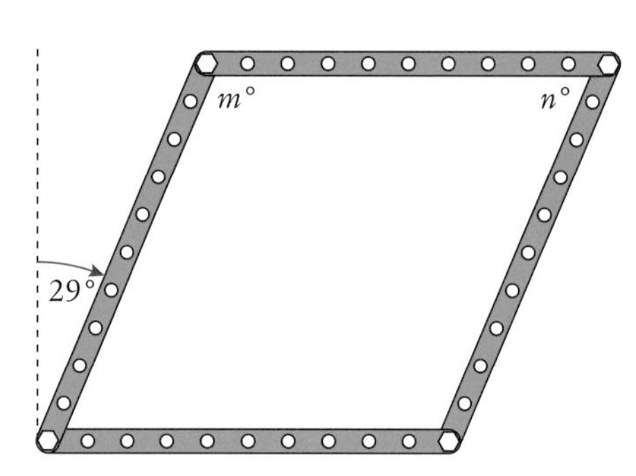

a What is the new shape she has created?

b What is the size of each of the four interior angles?

c **i** What is the relationship between m and n?

ii Find the value of n, when $m = 127{\cdot}8$.

iii How do m and n relate to the angle the vertical is pushed through?

5 An architect uses the same technique to explore the structure of a building.

He constructs a rectangle then distorts it off the vertical.

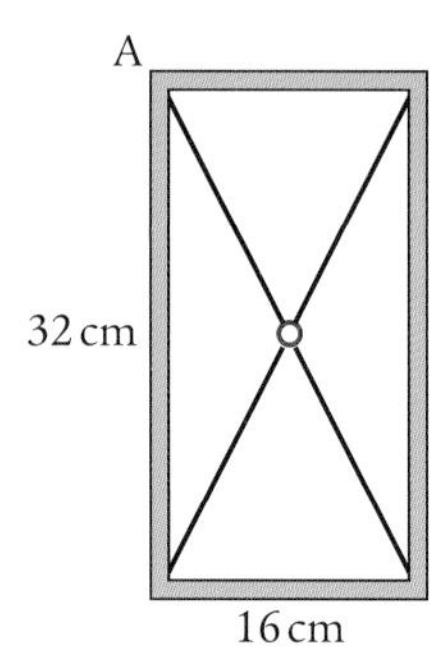

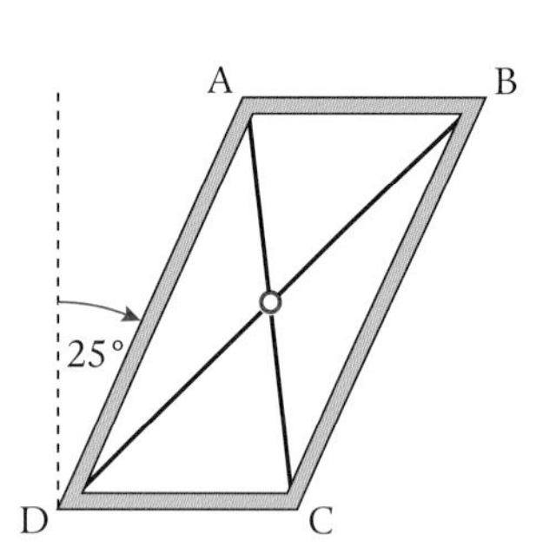

a Into what shape does the rectangle distort?

b Calculate the size of each internal angle when the structure has been offset by 25°.

c The architect is interested in the path the point of intersection of the diagonals takes. If it goes to the right of C, the structure becomes unstable.

i What path does the point A draw out as it moves?

ii Explore the point when the structure becomes unstable.

6 WXYZ is a parallelogram with diagonals intersecting at M. Angles are given to the nearest degree.

a What is the size of ∠YXW?

b What is the size of ∠XWZ?

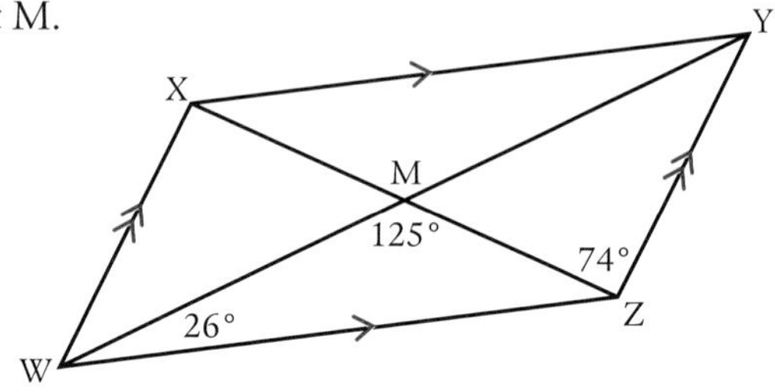

7 DEFG is a trapezium.

DG = GF and angles ∠DEF and ∠GFD are 52° and 27° respectively.

a What size is ∠FDE?

b What size is ∠DFE?

c What size is ∠DGF?

d How can you check you have the correct answers?

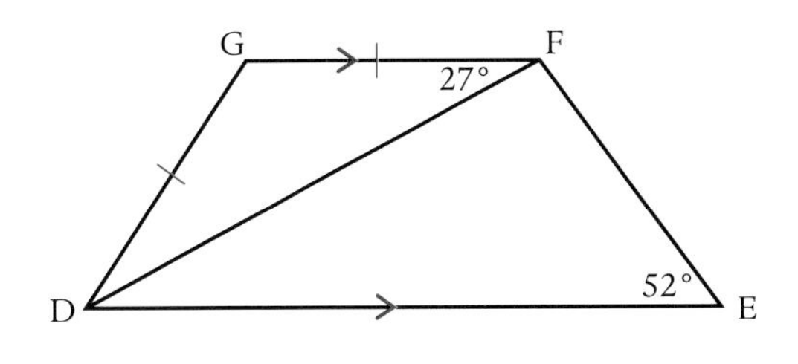

8 The trapezium JKLM is symmetrical with JM parallel to KL.

Side LM has been extended to N. Angle ∠JMN = 77°.

a What is the size of ∠KLM?

b What is the size of ∠KJM?

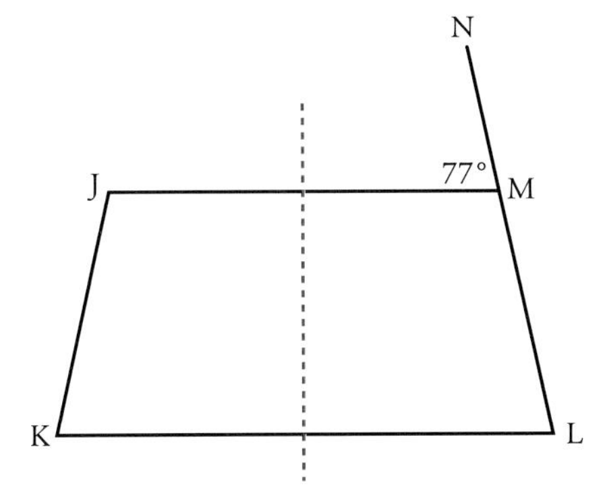

9 The side of a chimney stack on a roof is a trapezium.

Its sides are vertical and its top is horizontal.

The pitch of the roof is 25°.

a Calculate ∠ADC.

b What is the size of ∠DAB?

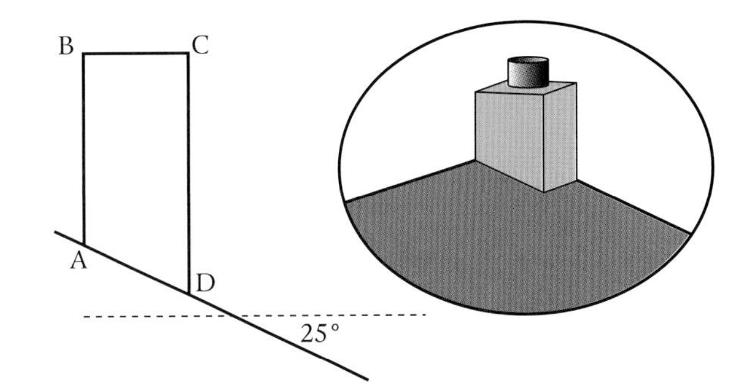

Exercise 9.4B

1 An apprentice joiner was asked to make a wooden frame in the shape of the kite as shown.

She was given four pieces of wood to make the frame.

They have to be cut in the shape of four trapezia.

What are the sizes of the angles in each of the trapezia?

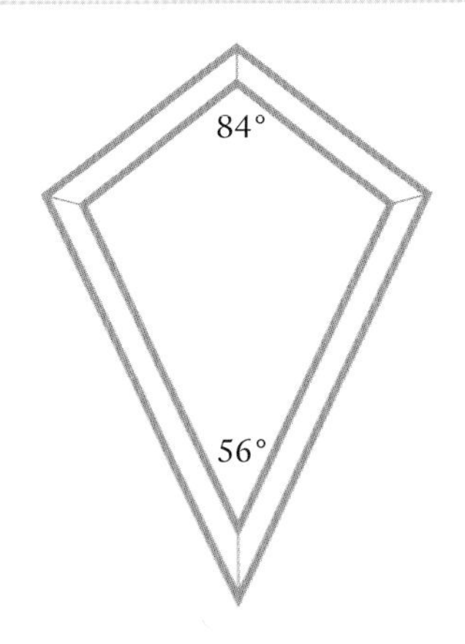

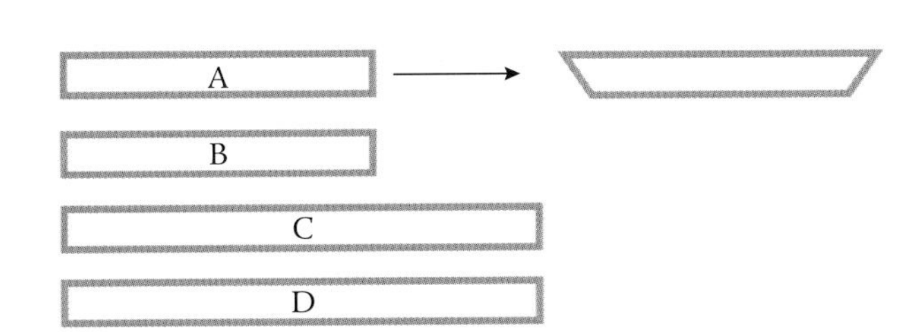

2 The aerial mast for the TV transmitter needs wires connected to ensure it remains stable in high winds and other challenging weather.

The wires are symmetrically placed around the mast.

The aerial mast is perpendicular to the ground.

a What are the sizes of the angles in kite ABCD?

b If a further support AC is put in place, what would be the size of ∠BAC?

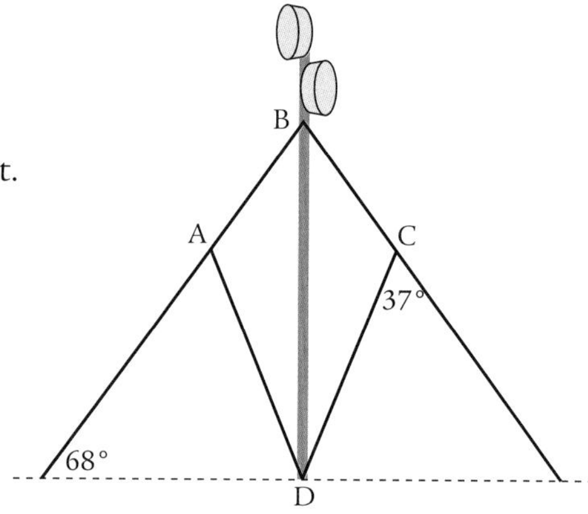

3 Kite PQRS has diagonals that intersect at M.

∠MSR = 30° and ∠QPM = a°.

a Copy the diagram and find an expression or value for each acute angle.

b Find an expression for the value of each of the four angles of the kite.

c The diagonals split the kite into four triangles.

Explain why the sum of the four angles of the kite is not 4 × 180°.

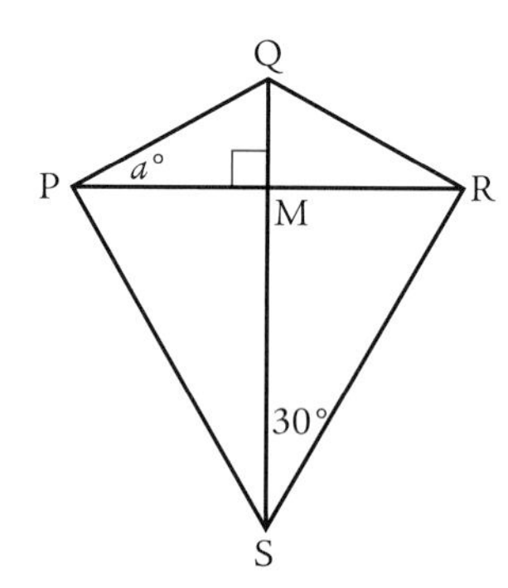

4 **a** A rail bridge will be constructed using two rhombuses joined end to end.

To stiffen the structure, the two rhombuses will be tied using a top tie.

The angle between the tie and the rhombus is 28°.

What are the sizes of the angles in each rhombus?

b A stretch of the Forth Bridge is based on six trapezia which share parallel sides.

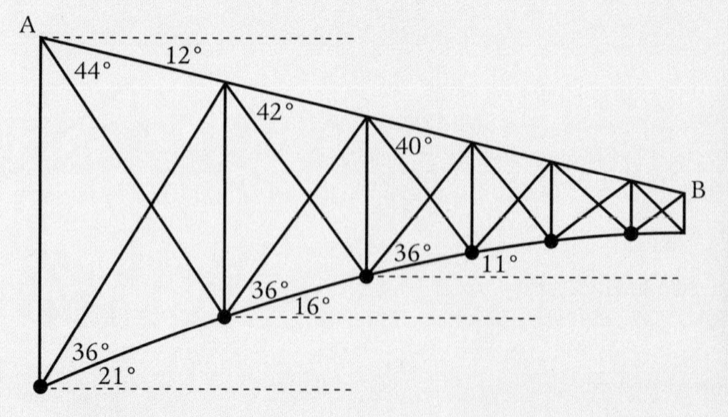

AB is a straight line at 12° to the horizontal.

Some very striking but simple angle patterns have been added to the sketch.

i Sketch the first trapezium and enter the value of each angle in it.

ii Repeat this with the fourth trapezium.

5 A jeweller designs a brooch that will be made up of rubies and emeralds.

The four rubies are identical and each is cut in the shape of a rhombus with an obtuse angle of 130°.

The rubies are set symmetrically round a centre.

Each of the four emeralds is also to be cut in the shape of a rhombus.

The emeralds are to fit exactly into the gaps between the rubies.

What size of angles will the gem cutter use when cutting the emeralds?

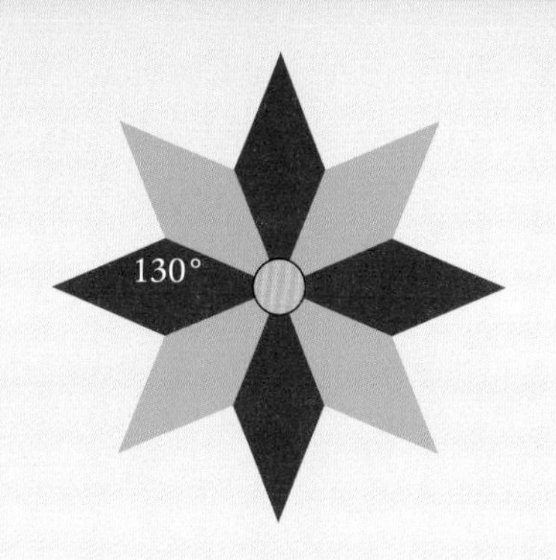

6 In 1828, a Scottish physicist named William Nicol carried out experiments with light rays.

He used a calcite crystal, referred to as 'the Nicol prism'.

The faces of a calcite crystal are in the shape of parallelograms.

Passing the light through the crystal he tracked the ray.

The light enters the crystal at 110°. It is reflected off the diagonal at 34°.

If ∠PQS = 36°, what size are the following?

a ∠UTS **b** ∠TSU **c** ∠QSR

d ∠TSR **e** ∠QPS

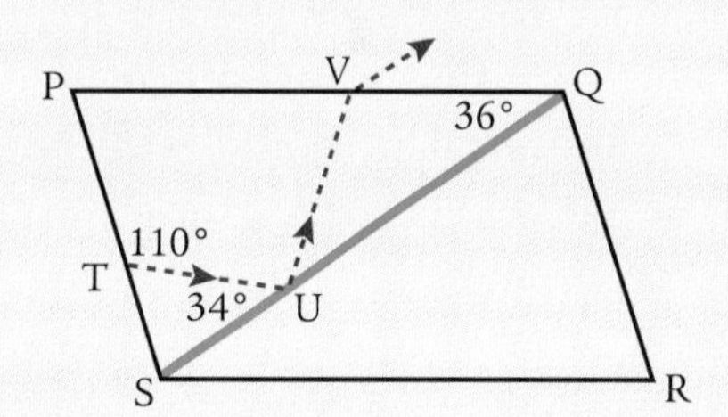

7 The designer of mountain bike frames uses a parallelogram as the main part of the structure with an isosceles triangle added for strength.

In the diagram, ABCD is the parallelogram and CDE the isosceles triangle.

If ∠ECD is 24°, ∠EDA is 146°, and ∠DAC is 57°, what size are the following?

a ∠CDE **b** ∠CDA

c ∠CBA **d** ∠ACB

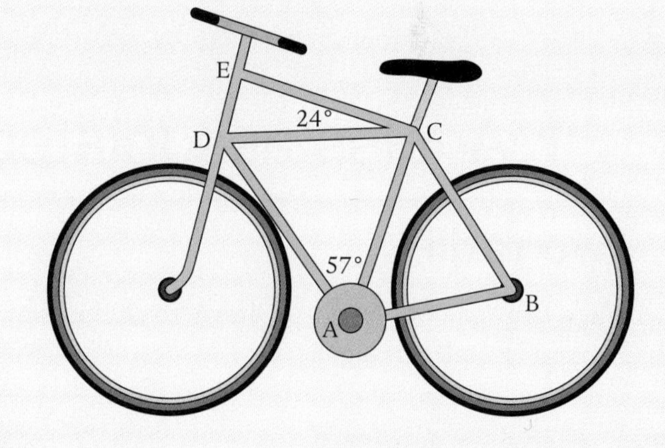

8 The Gallery of Modern Art has stained glass windows in the shape of symmetrical trapezia.

Each trapezium is to be made of four triangles. The diagonals and edges of the trapezium form the boundaries of the pieces.

The designer specifies two angles, as shown, to the glass cutter.

Copy the diagram and fill in all the other angles to help the glass cutter make templates for the four pieces.

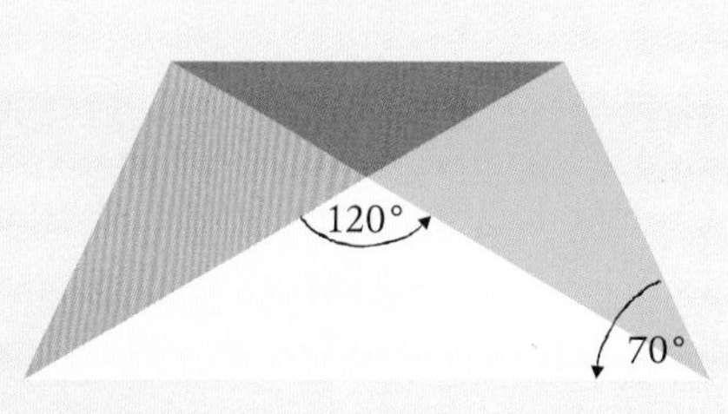

9 The 'Paper Cup Company' found that their cups were stable, and not liable to topple, as long as the obtuse angle at the bottom was not more than one-and-a-half times the acute angle at the top.

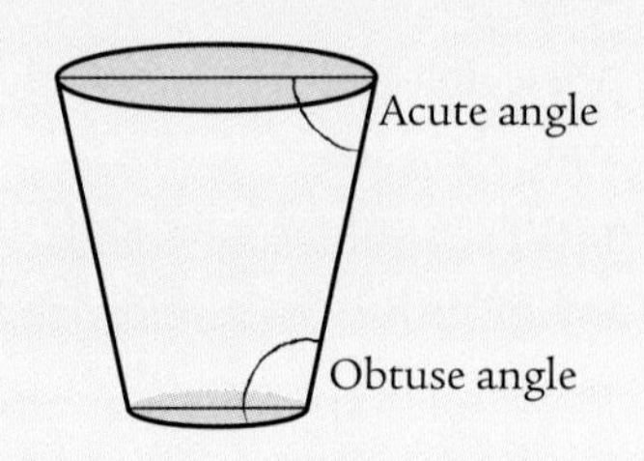

a State whether cups A, B and C are stable or unstable and give reasons for your decisions.

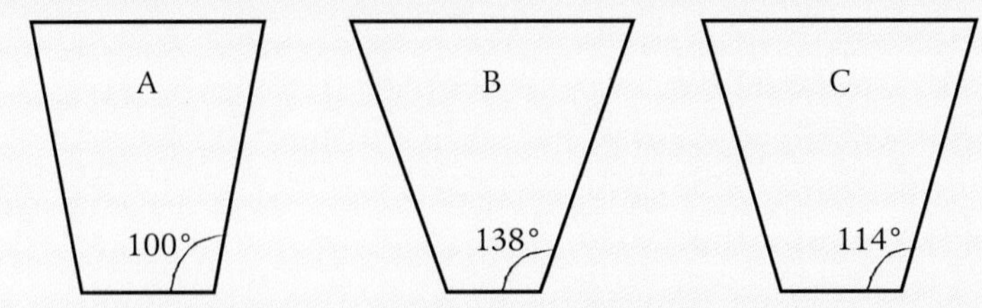

b What is the biggest angle they could use at the bottom of the cup before it becomes unstable?

10 A graphic designer is to design a logo that combines a number of familiar geometric shapes.

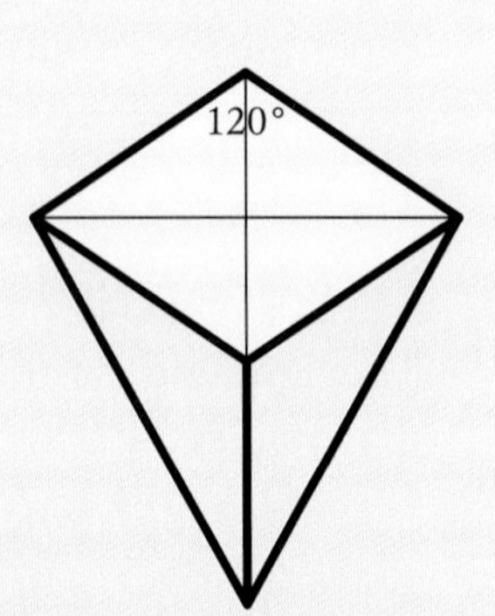

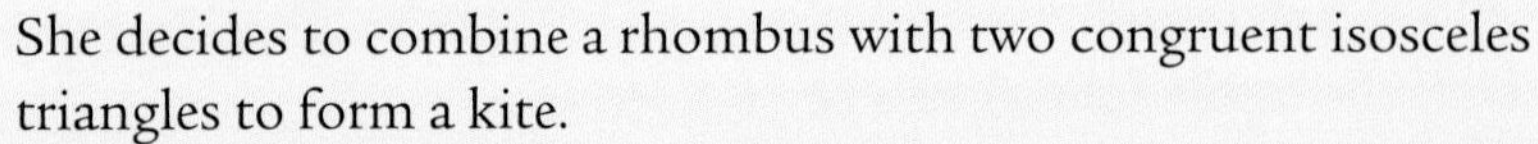

She decides to combine a rhombus with two congruent isosceles triangles to form a kite.

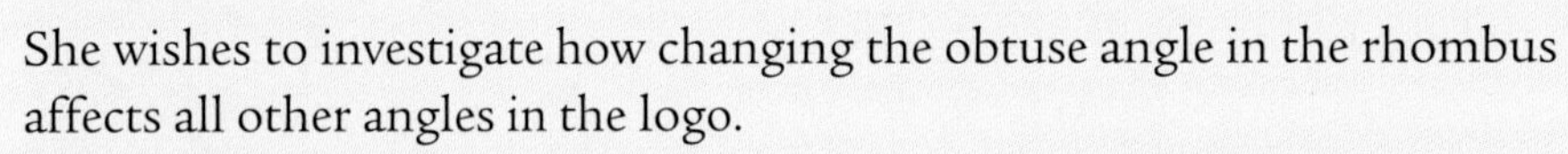

She wishes to investigate how changing the obtuse angle in the rhombus affects all other angles in the logo.

Start with an angle of 120° and complete the diagram, calculating all other angles.

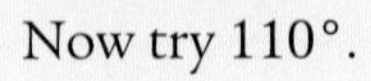

Now try 110°.

Investigate for different sizes of angles.

Can you spot different patterns when the obtuse angle in the rhombus changes?

9.5 Angles associated with the circle

Two radii

When two radii of a circle are drawn to the ends of a chord they form an isosceles triangle.

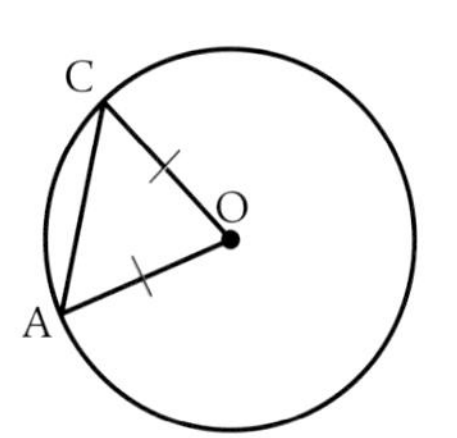

Angle drawn in a semicircle

Consider a circle, centre O and diameter AB.

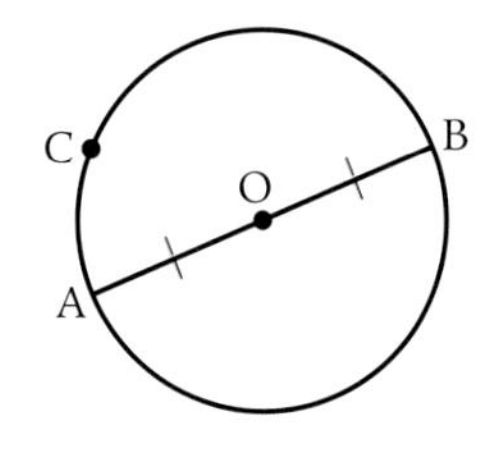

Suppose you are given any point, C, on the circumference to one side of AB.

You can always draw a diameter CD.

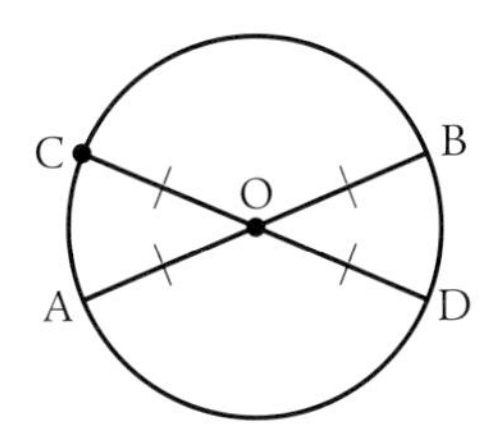

Note that AB = CD and that they bisect each other at O.

If we think of AB and CD as the diagonals of the quadrilateral ACBD, it would make the quadrilateral ACBD a rectangle.

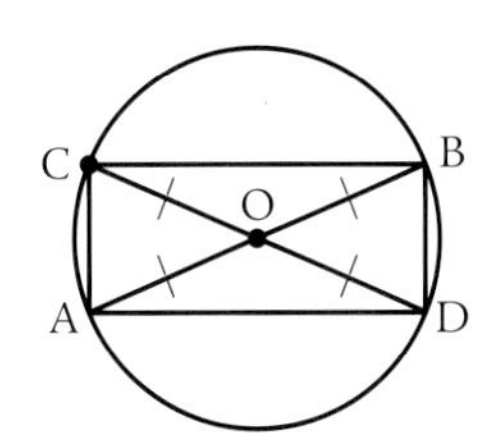

It would make the angle ∠ACB a right angle.

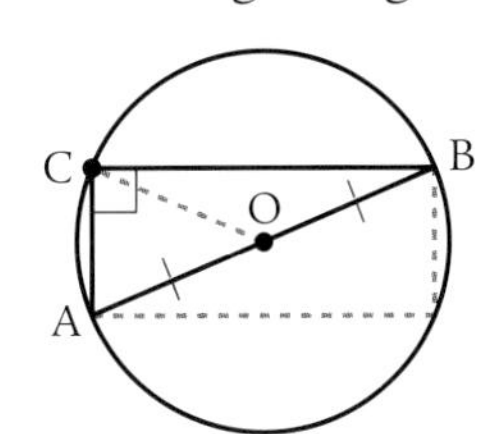

So any angle drawn to a point on the circumference from the ends of a diameter is a right angle ...

... *any* point!

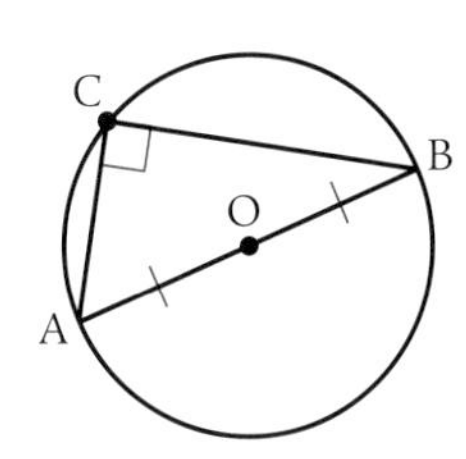

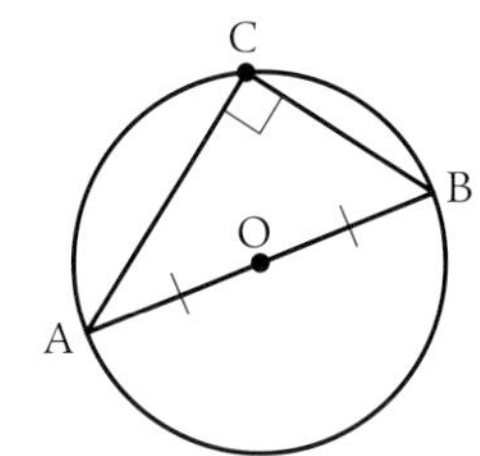

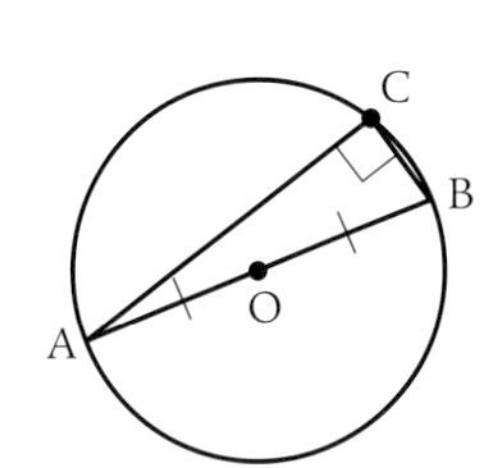

For obvious reasons this is often referred to as the **angle in a semicircle**.

Example 1

An astronomer photographs the Moon and wants to test that it is circular. PQ is easily identified as a diameter. Angle ∠PQR = 68°.

If the Moon's outline is indeed circular:

a what should be the size of ∠PRQ

b what should be the size of ∠QPR?

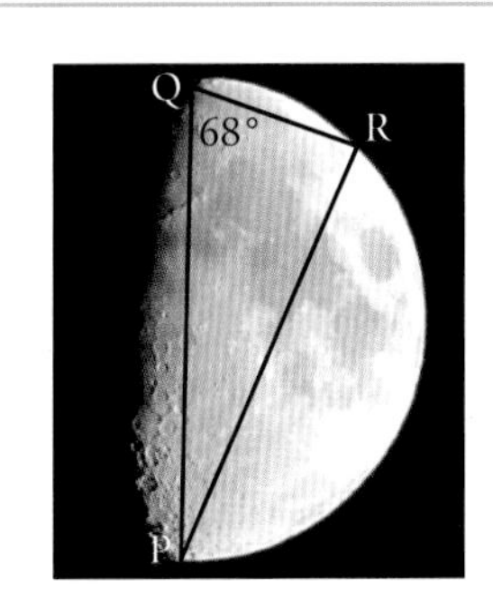

a ∠PRQ = 90° — Angle in a semicircle

b ∠QPR = 180° − 90° − 68° = 22° — Third angle in a triangle

Example 2

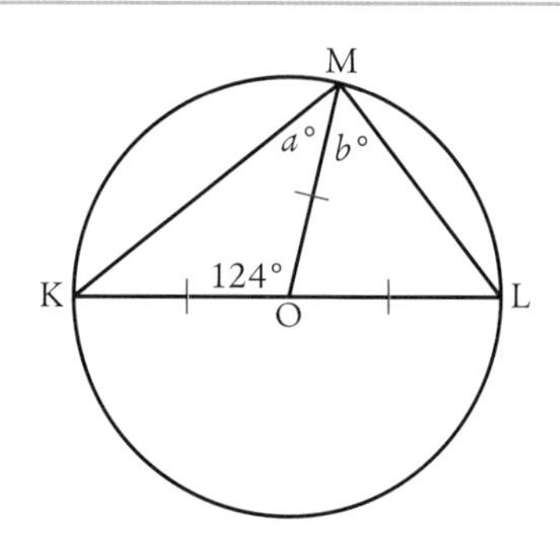

KOL is a diameter and OM is a radius.

a What is the size of the angle marked $a°$?

b What is the size of the angle marked $b°$?

c How can you check your answers by considering angle ∠KML?

a $\angle MKO = a°$ — Triangle KOM is isosceles

$a° + a° + 124° = 180°$ — Angles in a triangle add up to 180°

$\Rightarrow \quad 2a + 124 = 180$

$\Rightarrow \quad 2a = 56$

$\Rightarrow \quad a = 28$

b $\angle MOL = 56°$ — The supplement of 124°

$b° + b° + 56° = 180°$ — Angles in an isosceles triangle add up to 180°

$\Rightarrow \quad 2b + 56 = 180$

$\Rightarrow \quad 2b = 124$

$\Rightarrow \quad b = 62$

c $\angle KML = a° + b° = 28° + 62° = 90°$

We know angle ∠KML is 90° since it is the angle in a semicircle.

This agrees with our answers.

Exercise 9.5A

1 In each of the following diagrams, POQ is a diameter and O is the centre of the circle.
Identify a right angle and then calculate the size of each marked angle.

a

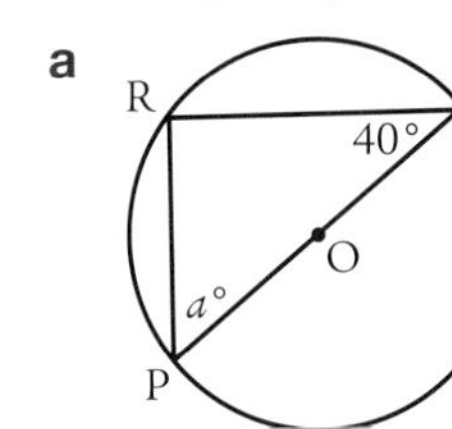

b

c

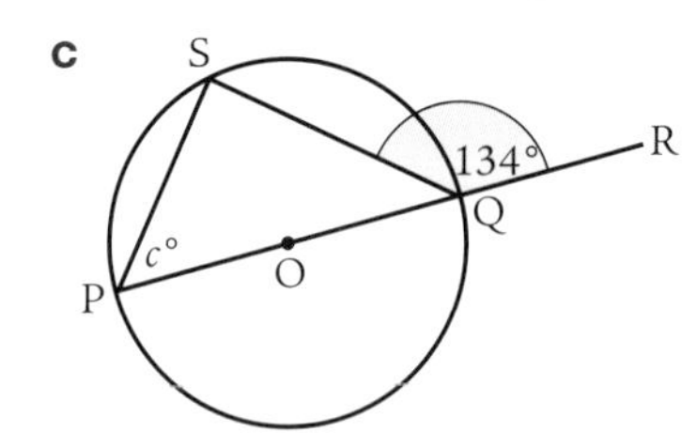

d

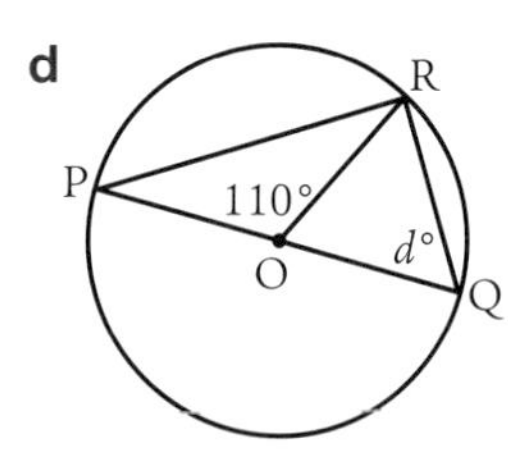

2 In the diagram, AC is a diameter and O is the centre of the circle.

a Name two right angles in the diagram.

b Make an equation relating p and q.

c What size is q if $p = 78{\cdot}2$?

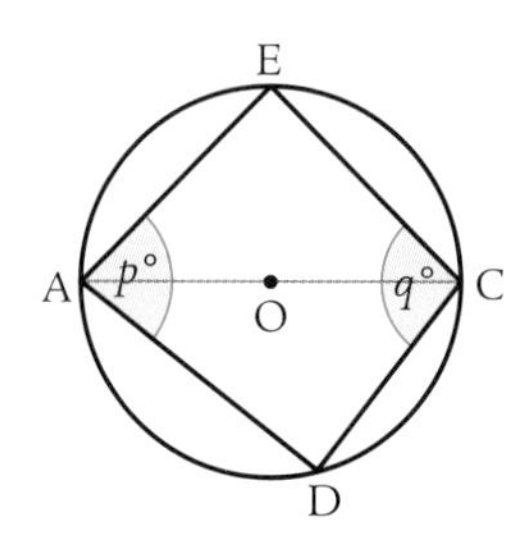

3 A wall clock is suspended from a cord to the wall.

To keep it secure it is attached to the back of the clock in three places: at B and D, which are diametrically opposite each other, and at C, which is at the bottom of the clock.

If $\angle BAD = 48°$, what size is $\angle CDA$?

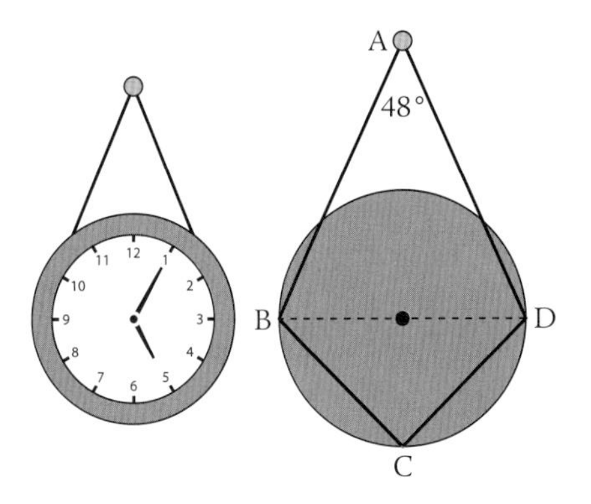

4 When two circles touch, the line joining their centres passes through the point of contact.

This fact is used a lot in the design of gears and other machinery.

So, ABC is a straight line.

A chord, AE is drawn at 62° to AC.

The line EB is drawn and cuts the small circle at D.

a Calculate the size of $\angle CBD$.

b What can be said about AE and DC? (Hint: think of alternate angles.)

5  When studying the geometry of the clock face, Gordon considered two angles:

i the centre angle, $\angle TCH$

ii the 6 o'clock angle, $\angle TSH$.

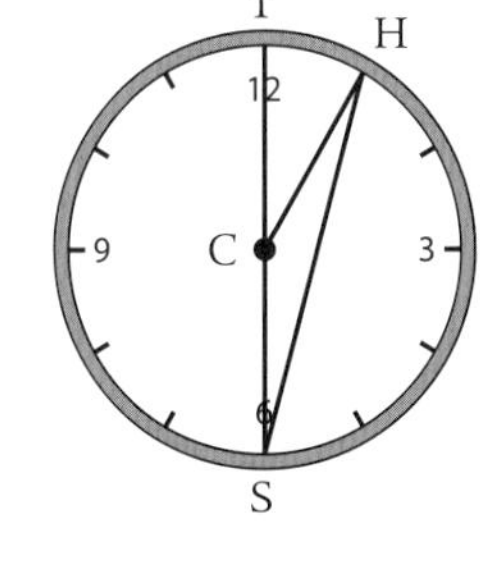

The diagram shows these angles at 1 o'clock.

For each of the following times, what is **i**, the centre angle, and **ii**, the 6 o'clock angle?

a 1 o'clock **b** 2 o'clock **c** 3 o'clock

d 4 o'clock **e** Investigate around the clock.

Exercise 9.5B

1 The Hill Walking Club got together with the Mountaineering Club and designed their logo.

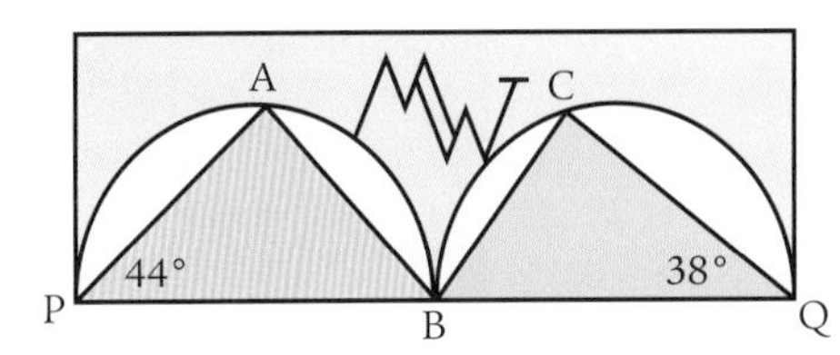

It has two semicircles with two triangles inside.

PB = BQ and both are diameters.

a What is the size of $\angle ABC$ in the logo?

b They decided they wanted $\angle ABC$ to be a right angle.
If they kept $\angle APB$ at 44°, what would be the size of $\angle CQB$?

2 During a football training session, five players were spaced around the centre circle as shown.

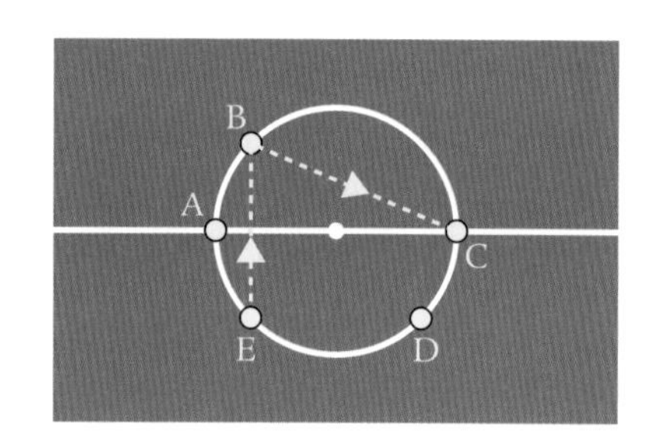

The coach had explained to them that a 'right-angle pass' was when the path of the ball made an angle of 90° as it passed from one player to a second and then onto a third.

He explained that E to B to C would not be a 'right-angle pass' because EBC was not 90°.

a There are six definite right-angle passes. Can you describe them?

b There are a further six right-angle passes possible, but only if a certain thing is true. What would need to be true?

3 The 'great hall' of a Tudor mansion was built with a semicircular ceiling.

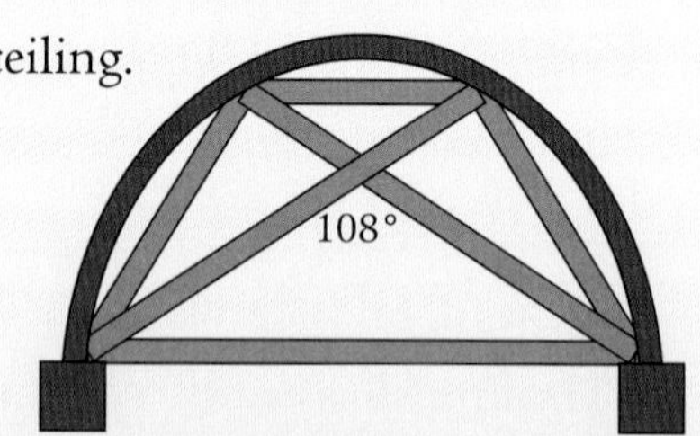

Through time the ceiling weakened and it was agreed that support frames should be added.

Each frame was a symmetrical trapezium with cross-members that met at 108°.

Make a sketch of the frame and find as many angles as you can.

4 Euclid was an Ancient Greek geometer. He noticed something curious about any quadrilateral whose vertices were lying on a circle.

A circle centre O and a quadrilateral whose vertices, P, Q, R and S lie on the circle, are drawn.

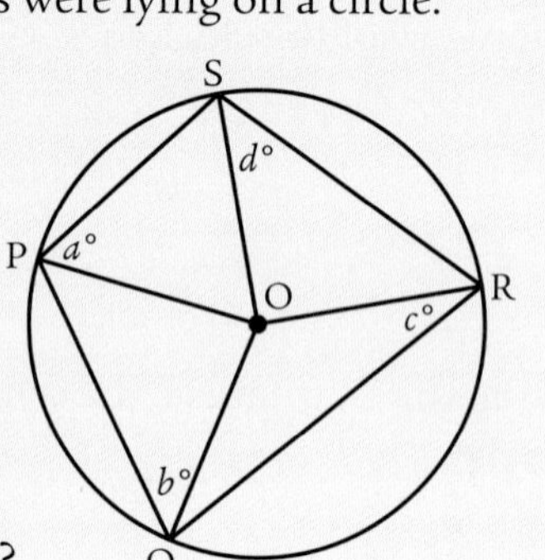

The radii to these vertices are also drawn.

Some angles have been labelled.

a Make a sketch and express each of the other angles in terms of a, b, c and d.

b What is the value of the sum $a + b + c + d$?

c What is the value of ∠SPQ + ∠SRQ?

d What is the value of ∠PQR + ∠PSR?

e What curious fact did Euclid discover?

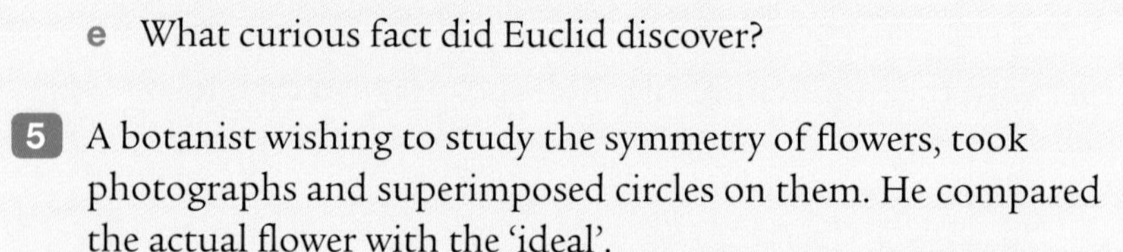

5 A botanist wishing to study the symmetry of flowers, took photographs and superimposed circles on them. He compared the actual flower with the 'ideal'.

a If this flower has perfect symmetry, what is the size of:

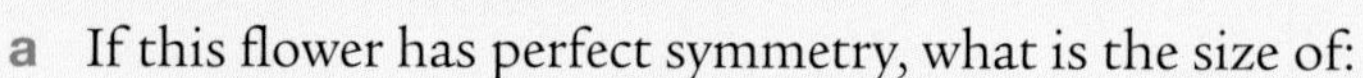

i ∠ROP ii ∠ORP iii ∠QRP iv ∠RQP?

b What fraction of ∠ROP is ∠RQP?

c If this flower has perfect symmetry, what is the size of:

i ∠SOP ii ∠OSP iii ∠QSP iv ∠SQP?

d i What fraction of ∠SOP is ∠SQP?

ii What fraction of ∠SOR is ∠SQR?

6 The same botanist, in the forest, came across a patch of the toadstool, Fly Agaric.

They have perfectly round tops and the botanist wanted to pinpoint the centre to make measurements.

First he held his notebook, as in Figure 1, making slight marks on the top at A and B. He then lightly marked A to B.

Figure 1

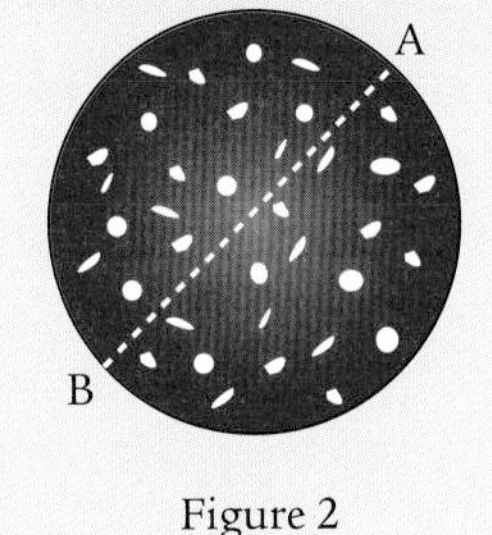

Figure 2

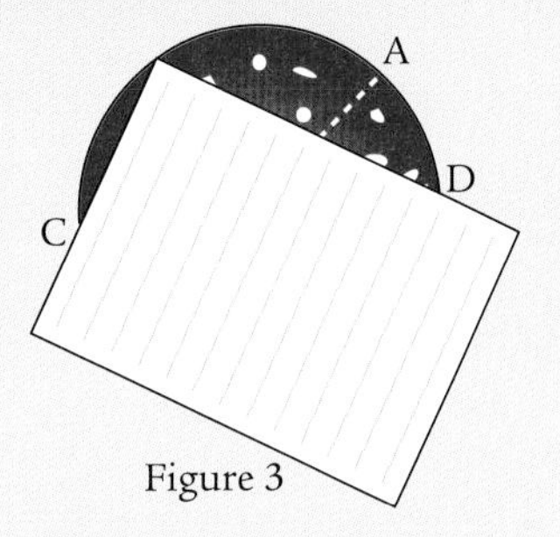

Figure 3

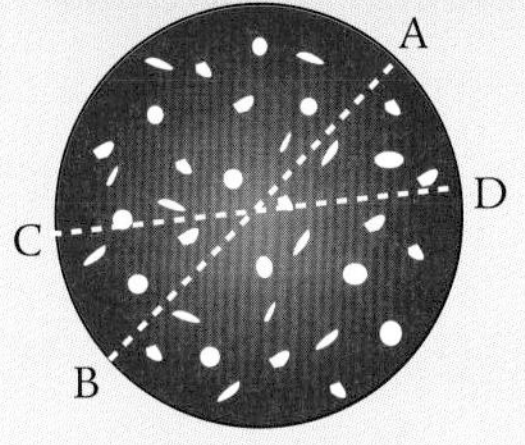

Figure 4

He repositioned his notebook and repeated the markings, ending up with two lines drawn ... and the centre of the mushroom identified.

Explain how the method works.

9.6 Circles, chords, tangents and radii

A **tangent** is a line that cuts a circle at only one point ... where it touches it.

The Latin for 'I touch' is 'tango'.

The point where the tangent touches the circle is called a 'point of tangency' or 'tangent point'.

Class discussion

Peter noticed that every time the assistant in the pizza shop cut the pizza, she always started with the four tangent points. He also noticed that the cuts passed through the centre of the pizza. He concluded that the cuts and the sides of the box were perpendicular. He further noticed that most pizza shops also cut diagonally. He wondered what shape of box could be used so that all eight points touched the sides of the pizza box.

a Discuss the possible shape of such a box and the pros and cons of making it.

On further investigation, Peter thought that the tangent to a circle will always be at right angles to the radius at the tangent point.

b Is it possible to make a situation where this does not happen?

By reversing his logic, Peter thought that a line drawn at right angles to a tangent, from the tangent point, will always pass through the centre of the circle.

c Is he allowed to reverse his logic like this?

d Would this be a useful thing to know?

Chords and symmetry

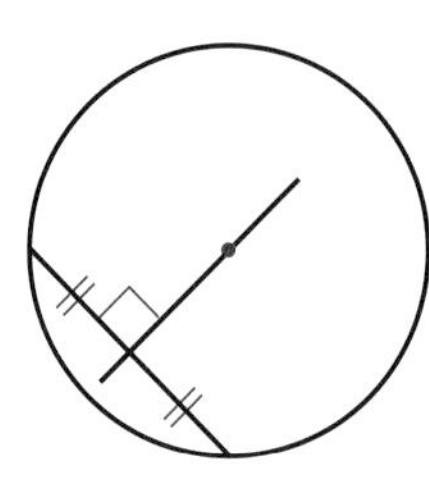

Because of symmetry we can see that:

i the perpendicular bisector of a chord **must** pass through the centre of a circle

ii the bisector of a chord passing through the centre of a circle **must** be perpendicular to the chord

iii the perpendicular to a chord passing through the centre of a circle **must** bisect the chord.

Given two of these facts, the third comes for free!

Tangent and radius

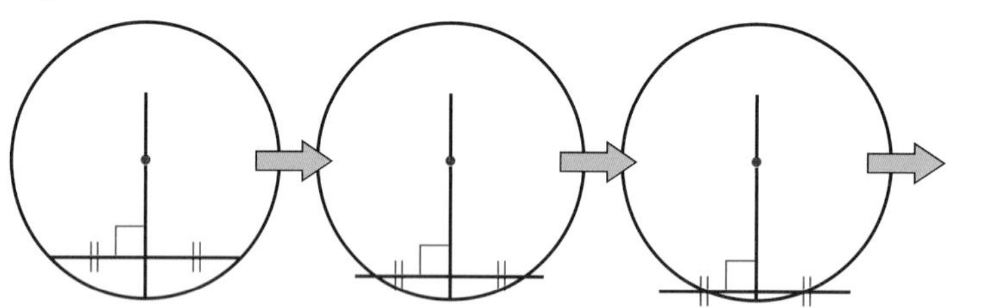

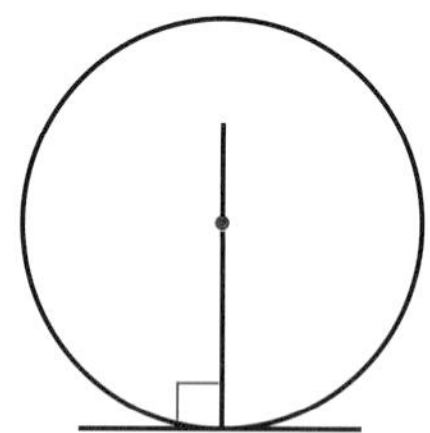

Imagine the chord slipping down until it becomes a tangent ...

At this point it is still at right angles to the radius at the point of contact.

Example 1

AB is the chord of a circle with centre O. M is the midpoint of the chord.

$\angle$OAM is 32°.

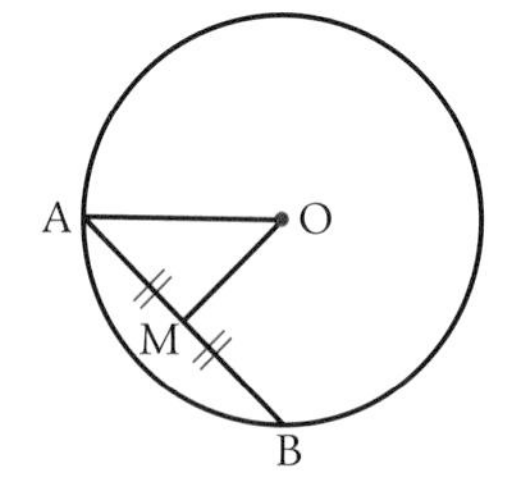

a State the size of $\angle$AMO.

b Calculate the size of $\angle$AOM.

a The line MO bisects the chord and passes through the centre of the circle. Therefore it is at right angles to the chord.

So $\angle AMO = 90°$.

b $\angle AOM = 90° - 32° = 58°$ Third angle in a right-angled triangle

Example 2

OA and OB are radii in a circle, at 137° to each other.

Tangents are draw from A and B that intersect at C.

OACB is known as a tangent kite.

What is the size of the angle marked $y°$?

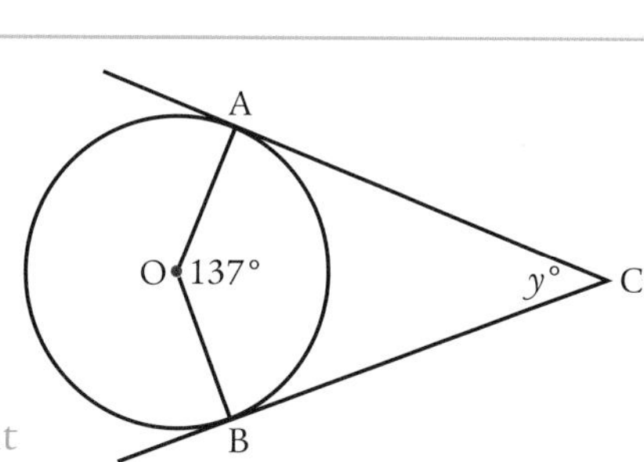

$\angle OAC = \angle OBC = 90°$ Tangent and radius are perpendicular at tangent point

$\Rightarrow \quad y° + 90° + 90° + 137° = 360°$ Angles in any quadrilateral add up to 360°

$\Rightarrow \quad y° + 317° = 360°$

$\Rightarrow \quad y° = 43°$

Example 3

POQ is a diameter in a circle and OT is a radius.

RS is a tangent to the circle at T.

If $\angle TOQ = 72°$, what size is $\angle PTR$?

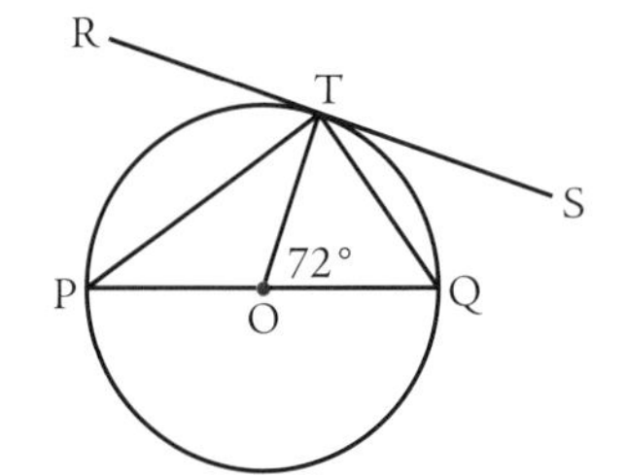

$\angle POT = 108°$ — Supplement of 72°

$\Rightarrow \quad \angle OTP + \angle OPT + 108° = 180°$ — Angles in a triangle

$\Rightarrow \quad 2\angle OTP + 108° = 180°$ — $\angle OTP = \angle OPT$ ... base angles of isosceles triangle

$\Rightarrow \quad 2\angle OTP = 72°$

$\Rightarrow \quad \angle OTP = 36°$

Now $\angle OTR = 90°$ — Angle between tangent and radius

$\Rightarrow \quad \angle PTR = 90° - 36°$

$\Rightarrow \quad \angle PTR = 54°$

Exercise 9.6A

1 AB is a chord of a circle with centre O.

A line passing through O meets AB at right angles at the point T.

AT is 6 cm long. $\angle TAO = 37°$.

Calculate:

a the size of $\angle AOT$ **b** the length of TB.

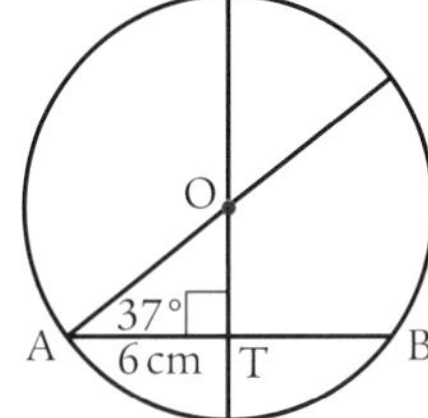

2 AT is the chord of a circle whose centre is not given.

A point P lies on the circumference closer to A than T.

The perpendicular bisector of AP cuts AT at K.

Funnily enough the perpendicular bisector of TP also cuts AT at K.

After reading this description, Margaret drew the following diagram.

If $\angle TAP = 50°$, calculate the sizes of the other two angles in triangle TAP.

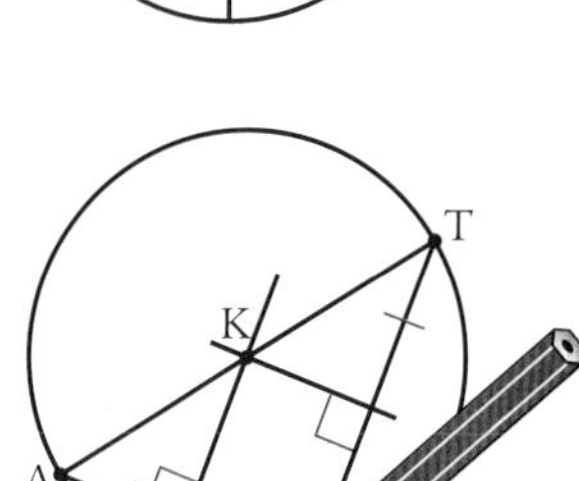

3 Each diagram shows a circle with centre O and radius OT.

A tangent is drawn touching the circle at T.

Calculate the values of the angles marked.

a

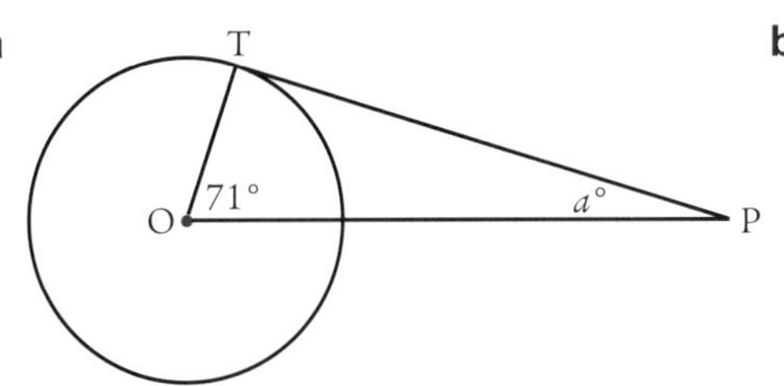

b

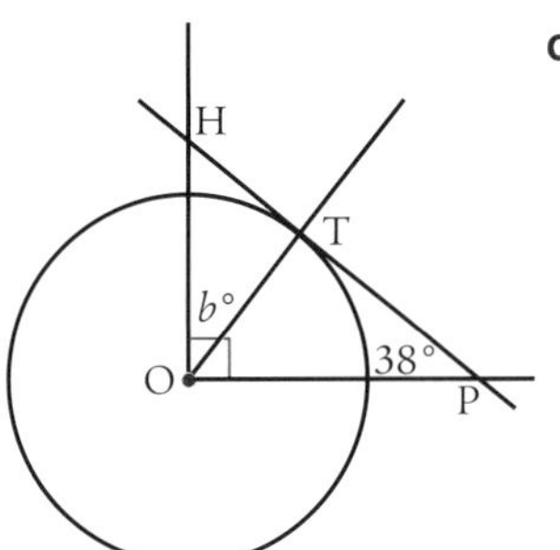

c

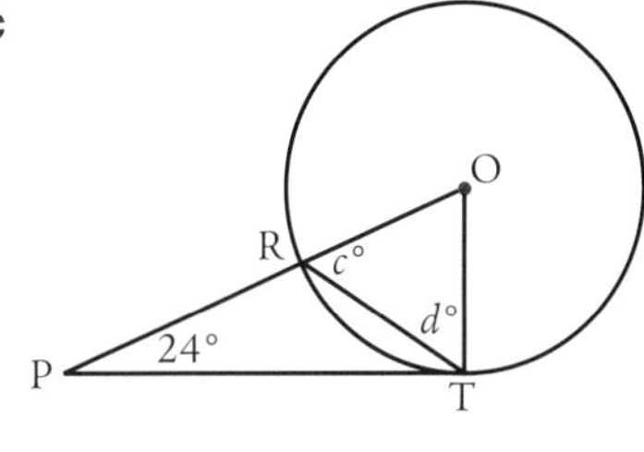

4 This is a sculpture outside a well-known hotel in Edinburgh.

It is a hemisphere. When it needs cleaning a ladder is propped up against it.

Wherever the ladder is placed the geometry is the same.

The diagram shows the cross-section.

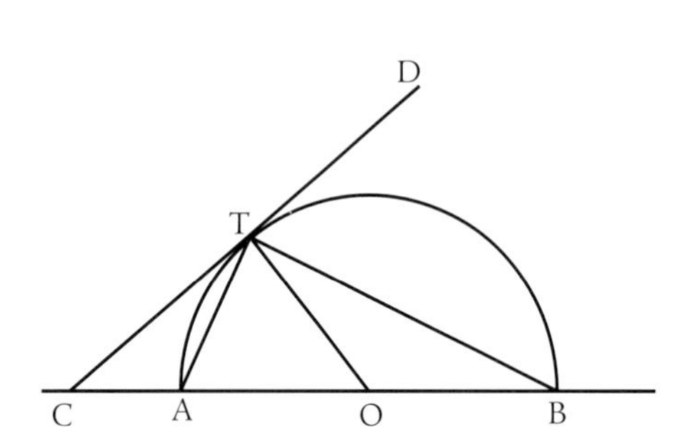

AOB is the diameter of a semicircle.

CD represents the ladder touching the hemisphere at T.

To do various calculations, the right angles in the diagram have to be recognised.

Name the three right angles in the diagram, giving reasons for your choices.

5 Creating computer graphics takes a lot of maths. To just create the shadow of a ball, we have to make use of the tangent kite.

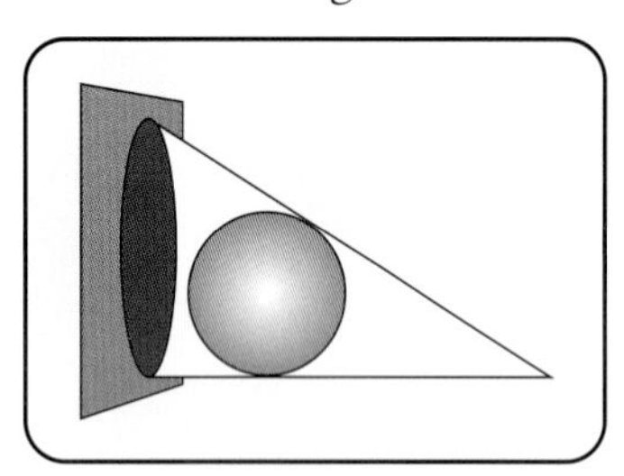

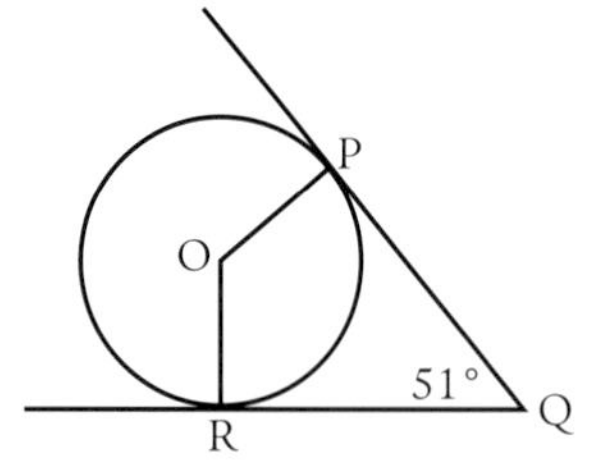

In this diagram, Q is a light source. The part of the wall that doesn't get light is the part between the tangents.

The circle, centre O, represents the ball. QP and QR are the tangents that represent the 'last' rays to get past the ball. OP and OR are radii of the circle.

a If $\angle PQR = 51°$, what is the size of $\angle POR$?

b As Q moves away from the ball, $\angle PQR$ varies. If $\angle PQR = x°$, what is the size of $\angle POR$?

Exercise 9.6B

1 A painter and decorator uses a sheepskin roller to paint a ceiling.

The painter thinks that when the angle between the horizontal and the handle is $58°$, the paint is applied more efficiently.

Let the end of the handle be H and the centre of the roller be C. The roller touches the ceiling at T.

What is the size of $\angle HCT$?

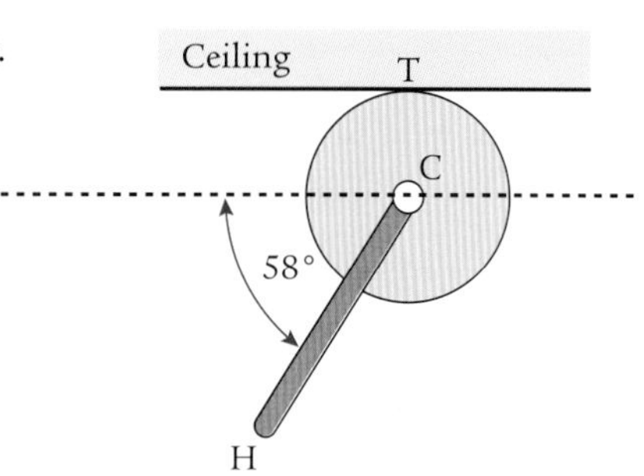

2 Ergonomics is the science of designing equipment.

In designing a wheelbarrow, it's been found that the most comfortable working position is when the handle makes an angle of 127° with the vertical.

What angle does the handle of the barrow make with the ground when in this position?

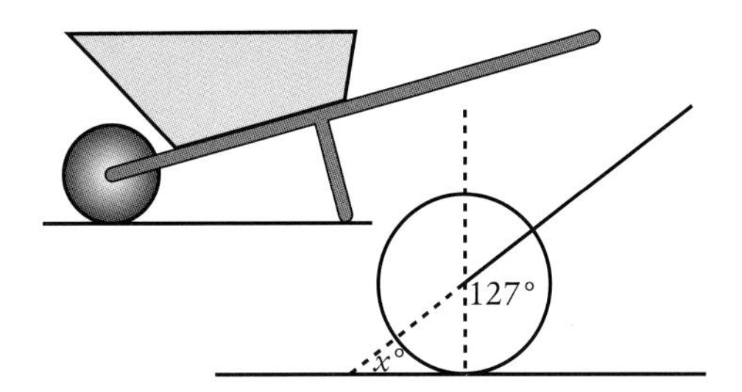

3 AB is a diameter of a circle. C is a point on the circumference. DE is a tangent to the circle, touching it at B.

a If ∠CBE = 30°, calculate the value of ∠BAC.

b If ∠CBE = x°, find ∠BAC.

c Make a general comment about the angle between a tangent and a chord.

4 a While model-making, Geoff makes wheels by drawing round circular objects like plates.

Of course by using this method he doesn't know where the centre of the circle is.

He needs this information fairly exactly as the wheels will look wobbly if the axles are fitted the least off-centre.

Though he doesn't have a set of compasses, he does have a ruler to measure length, but only to the nearest centimetre.

The ruler is also see-through, which allows him to use the markings to draw a perpendicular to any given line.

Describe how this ruler might be used to find the centre with pin-point precision.

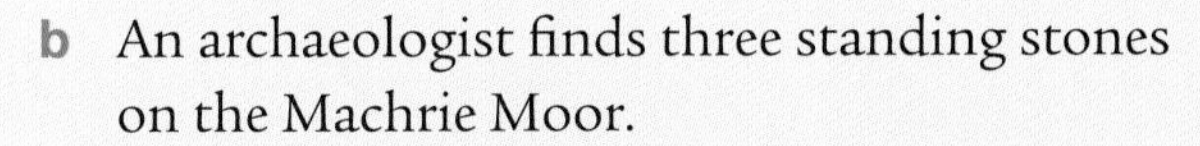

b An archaeologist finds three standing stones on the Machrie Moor.

He's fairly confident they are three of a set of stones that form a circle.

Unfortunately over the past 3000 years most of the stones are missing, having been used to build walls by farmers.

The archaeologist feels that evidence of the 'post holes' where these stones once stood could be found if the centre and radius of the original circle could be worked out.

He has placed a stick where he has worked out the centre.

Explain how he might have done this with only three stones being visible.

5 A boy is inside a large bubble walking on water. This is modelled in Figure 1.

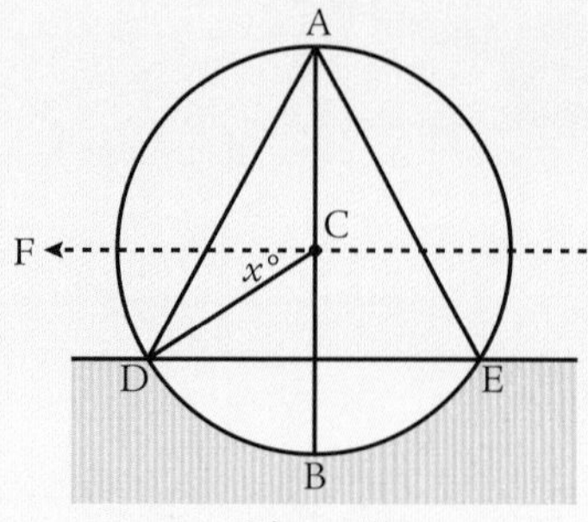

Figure 1

A P E Q R D B B

Figure 2

DE is the water-line. B represents the point where his feet are. C, the centre of the circle, is where he tries to keep his hands (and in line with his head) to keep himself stable.

F is straight ahead and CD is where he fixes his gaze to stop himself getting disoriented.

Let FCD be $x°$.

a Express the following in terms of $x°$, giving reasons:

i ∠CDE **ii** ∠CDA **iii** ∠DAC **iv** the reflex angle ∠DCE.

b As the boy walks, triangle ADE moves on and is replaced by PQR.
What can be said about ∠QPR and ∠DAE?

Preparation for assessment

1 BDF is an isosceles triangle. BE lies on its axis of symmetry.

BE is perpendicular to AC.

∠FDB is 67°.

What size is:

a ∠EBF **b** ∠DBA?

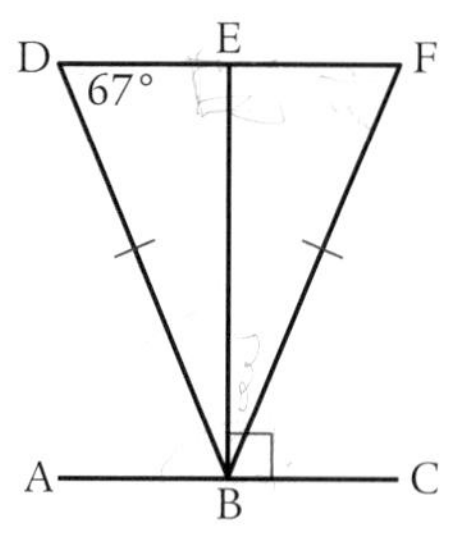

2 PQRS is a rhombus with ∠PQR = 56°.

PTRS is a kite with ∠TRP = 24°.

a What size is ∠TPQ?

b What size is ∠PTR?

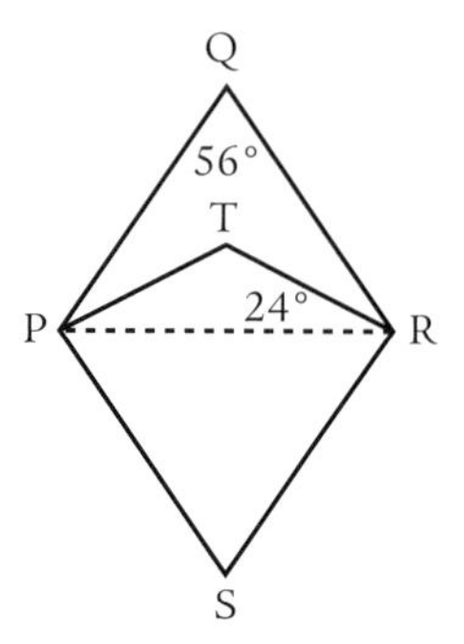

3 ABCD is a parallelogram with $\angle BAD = 64°$.

An isosceles triangle, DCE, is added to the parallelogram to complete a symmetrical trapezium.

What size is:

a $\angle ABC$ **b** $\angle CDE$ **c** $\angle DCE$?

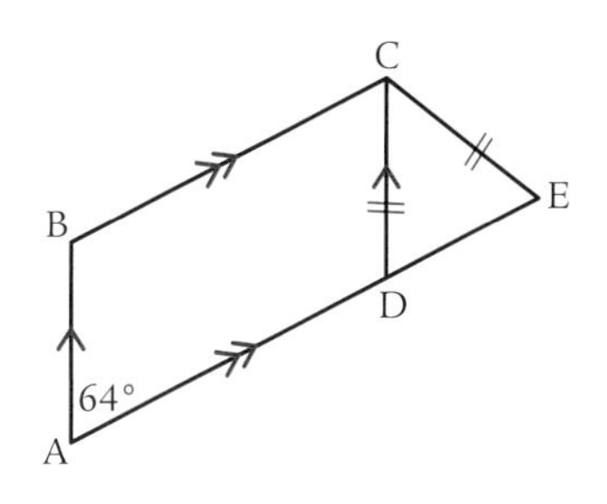

4 A ship sails from A to B on a bearing of 333° for a distance of 6 km.

At B it changes bearing to 243° and sails for a further 8 km to C.

a Make an accurate scale drawing of its journey.

b Calculate the size of $\angle ABC$ (i.e. don't measure).

c Calculate also the distance from A to C.

d Check your answers to **b** and **c** by measurement.

5 AB is the diameter of a circle, centre O, and C is a point on its circumference.

DE is a tangent to the circle touching it at A.

$\angle ABC = 58°$.

a What size is $\angle DAC$?

b BE is drawn parallel to AC, cutting the circle at F.

Calculate the value of:

i $\angle AEB$ **ii** $\angle ABE$ **iii** $\angle AFE$.

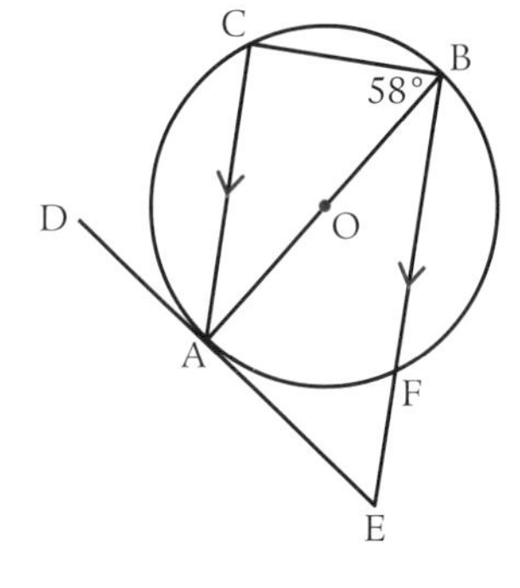

6 After the First World War, women who had lost their husband in the trenches were sent a large bronze plaque struck with their husband's name on it.

It became known as the 'Widow's Penny'.

It was posted in a strong cardboard envelope as shown.

A, C, I, and G would all fold over to meet at E.

a **i** Name four tangent kites.

ii Say what is special about them.

b The plaque had a diameter of 120 mm.

Use Pythagoras' theorem to calculate the length of AC.

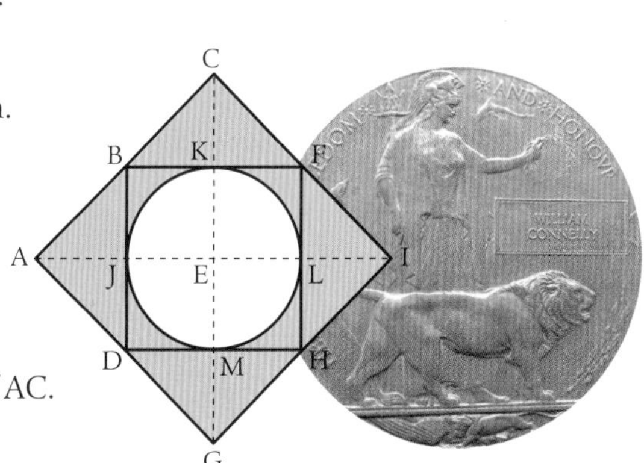

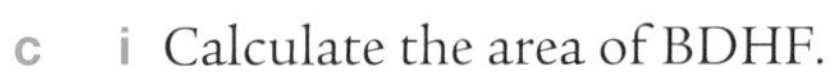

c **i** Calculate the area of BDHF.

ii Hence calculate the area of ACIG.

iii Hence find the length of AC.

d Compare methods used in finding AC.

7 Industrial archaeologists study a mill wheel before taking it apart for renovation.

The photo only shows one half of the wheel as the other is in the lade.

However, using rotational symmetry, the design of the whole wheel can be deduced.

a Looking at the photo,

i how many large spokes does the wheel have

ii how many slats does it have on its rim?

b In this mathematical model of the wheel, $\angle LOA$ has been measured as $48°$.

What is the size of:

i $\angle AOC$ ii $\angle DOR$ iii $\angle AEL$?

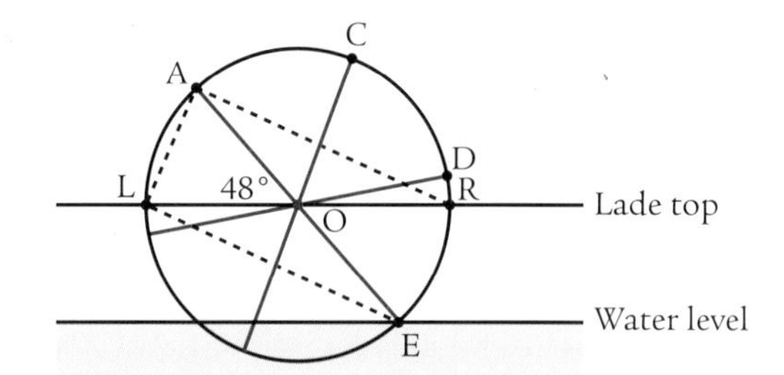

8 From an historical/engineering point of view, engineers model the 'Penny Farthing' bicycle.

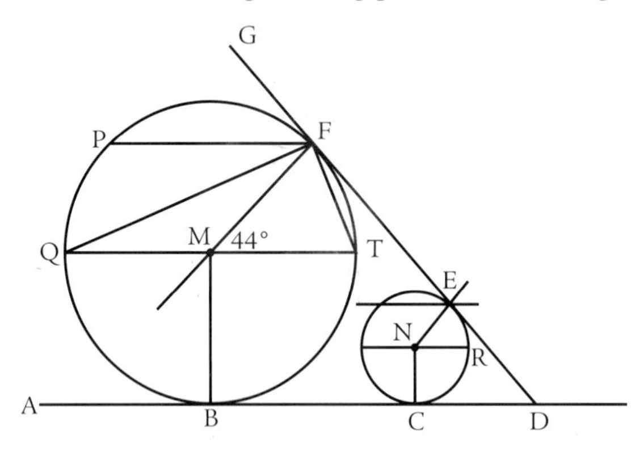

The ground AD is an obvious common tangent touching the larger circle at B and the smaller circle at C.

GD is a second common tangent with E and F being the points of contact.

a Name three right angles at F, giving reasons.

b $\angle FMT = 44°$. Calculate the value of:

i $\angle MTF$ ii $\angle ENR$.

c Sketch the circles and suggest where there might be two other common tangents.

9 Remember the jackdaw?

Go back to the start of the chapter and see if you can now answer the questions posed.

10 Trigonometry

Before we start...

From a map, I learned I was 300 m from the Wallace Monument which was sitting on the Abbey Craig.

Using a clinometer, I measured the angle of elevation of the top of the tower from where I stood.
It was 24°.

Is this enough information to let me calculate how high above me the top of the monument is
... without using scale drawings?

What you need to know

1 Maureen was designing a postage stamp. She drew it out as a rectangle.

The ratio of its length to its breadth was 3 : 1.

- **a** How many times larger than the breadth is the length?
- **b** What fraction of the length is the breadth?
- **c** If the length is 12 cm, how long is the breadth?
- **d** If the breadth is 8·1 cm, what is the length?
- **e** If the perimeter is 72 cm, what are the rectangle's dimensions?
- **f** **Difficult:** If the area is 243 cm^2 what are its dimensions?

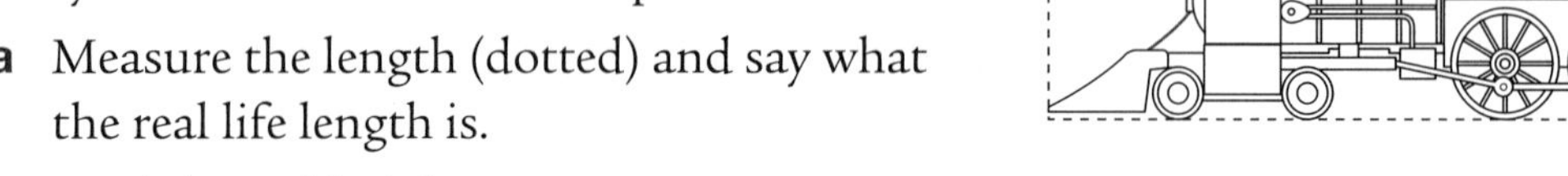

2 A model maker has plans for an old locomotive. They are drawn so that 1 cm represents 2 m.

- **a** Measure the length (dotted) and say what the real life length is.
- **b** Find the real height.
- **c** If the ratio of height to length on the model is 5 : 16, what is the ratio of height to length on the real thing?
- **d** If a metal plate in the model is a triangle with angles 30°, 60° and 90°, what will be the size of the angles in the real thing?
- **e** What things don't change when you scale things up or down?

3 One right-angled triangle is an enlargement of the other.

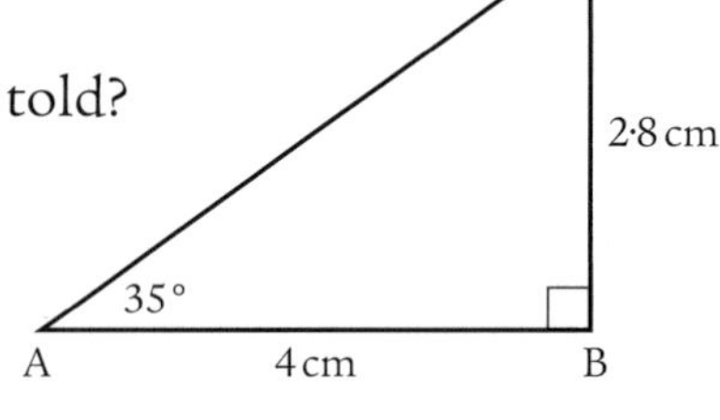

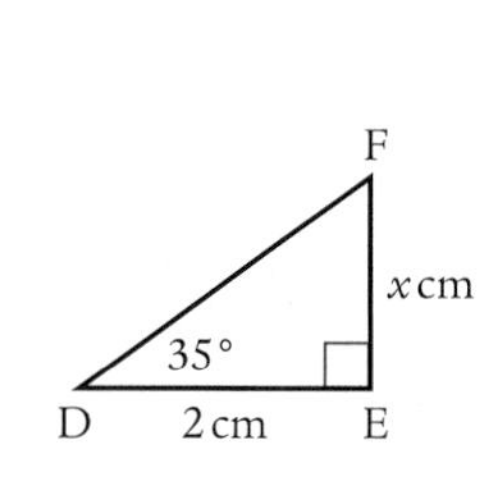

- **a** How could you tell, without being told?
- **b** Calculate the size of EF.
- **c** What is the ratio of:

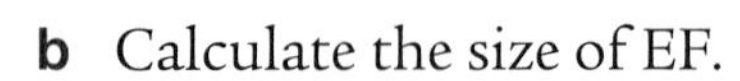

  i BC : AB **ii** EF : DE?
- **d** Name the side that is opposite: **i** ∠A **ii** ∠C **iii** ∠D **iv** ∠F.
- **e** What do we call a side opposite the right angle?
- **f** The angle A has two arms, viz. AC and AB. AC is the hypotenuse. We say that AB is **adjacent** to ∠A ...

  Name the side adjacent to: **i** ∠C **ii** ∠D **iii** ∠F.

Adjacent: 'neighbouring' or 'next to'.

10.1 The tangent of an angle

The ancients realised that, for a given angle in a right-angled triangle, because of the idea of enlargement, the ratio of the side opposite the angle to the side adjacent to the angle was fixed.

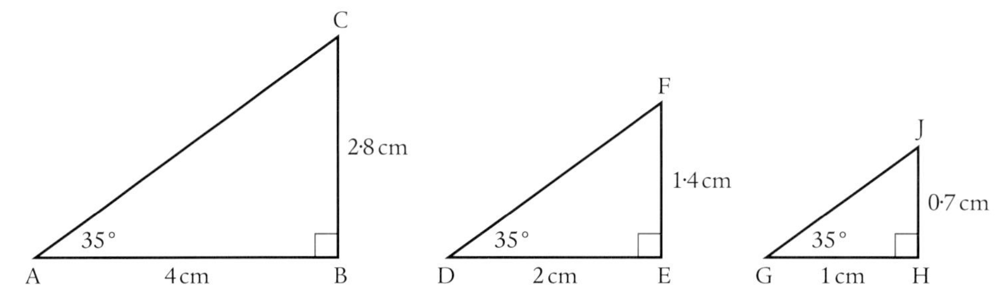

No matter the size of the triangle, if the angle is 35° then $\frac{\text{opposite side}}{\text{adjacent side}} = \frac{2·8}{4} = 0·700$.

Knowing this can save a lot of scale drawing.

The Ancient Greeks created tables of this ratio for every acute angle.

They gave the name of **tangent** to this ratio.

The tangent of 35° is 0·700 ... we write: tan 35° = 0·700.

Nowadays these tables are built into the calculator.

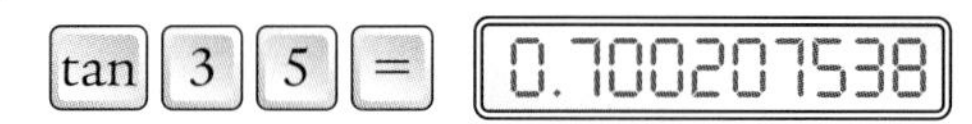

We usually round to 3 significant figures.

If we know the tangent of an angle and one side of the ratio we can work out the other side.

Example 1

A surveyor was 60 m from a tree.

He measured the angle of elevation to the top of the tree.

It was 20°.

How high is the tree?

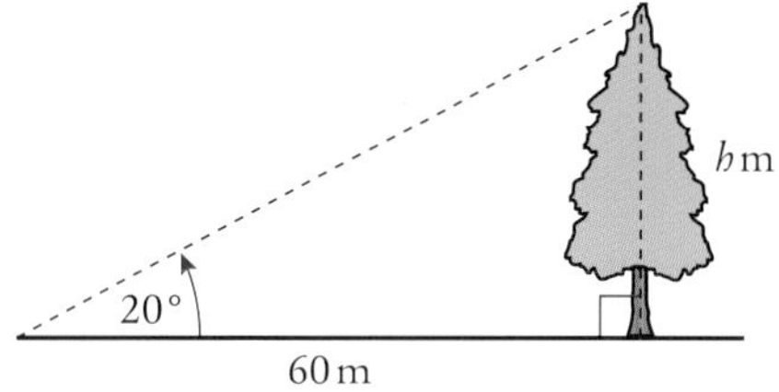

From the sketch we see that the tree is opposite the known angle and the distance from the tree is adjacent to the angle, so ...

$$\tan 20° = \frac{\text{opposite}}{\text{adjacent}}$$

$$= \frac{h}{60}$$

$\Rightarrow \quad 0·364 = \frac{h}{60}$ [tan] The calculator gives tan 20° = 0·364 (to 3 s.f.)

$\Rightarrow \quad 0·364 \times 60 = h$ Multiplying both sides of the equation by 60

$\Rightarrow \quad h = 21·84$

$\Rightarrow$ The height of the tree is 21·8 m (to 3 s.f.).

Angle of elevation: the angle your head turns through as you look up at an object.

Example 2

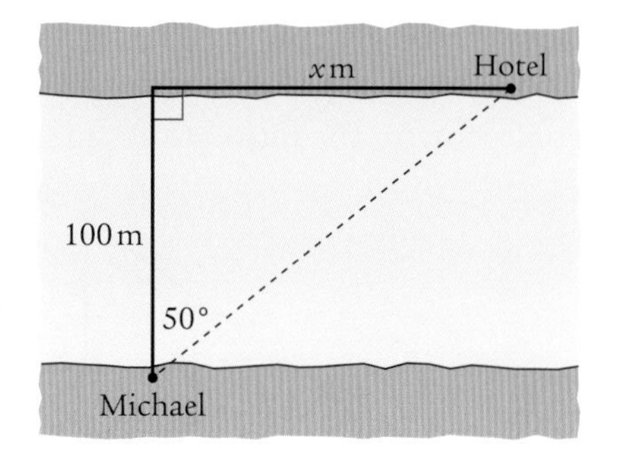

Michael knew the river was 100 m wide where he was standing.
He had to turn through 50° to face the hotel on the other side.
How far along the riverside from Michael is the hotel?

From the sketch we see that the distance we want is opposite the known angle and the width of the river is adjacent to the angle, so ...

$$\tan 50° = \frac{\text{opposite}}{\text{adjacent}}$$

$$= \frac{x}{100}$$

$\Rightarrow \quad 1{\cdot}19 = \frac{x}{100}$ tan — The calculator gives us $\tan 50° = 1{\cdot}191753593$

$\Rightarrow \quad 1{\cdot}19 \times 100 = x$ — Multiplying both sides of the equation by 100

$\Rightarrow \quad x = 119$

$\Rightarrow$ The hotel is 119 m along the riverside, correct to 3 s.f.

Note: in practice you would multiply the calculator value for tan 50° by 100 and then round.

Exercise 10.1A

1 **i** Identify the opposite side and adjacent side.

ii Write down the tangent of the marked angle.

a

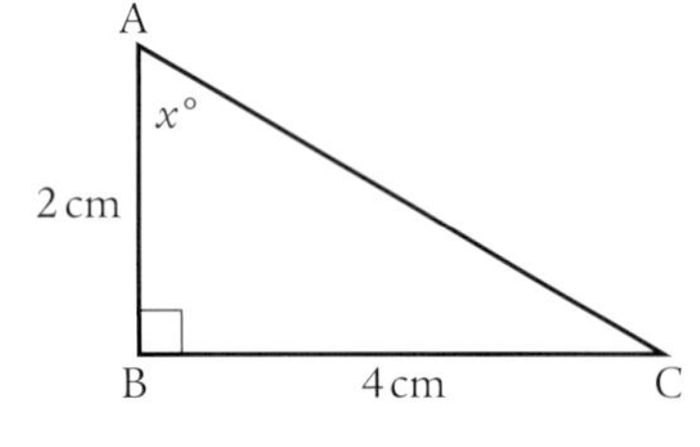

b

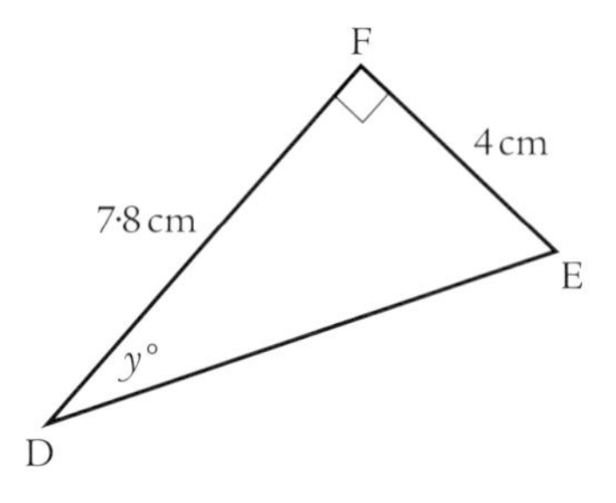

c

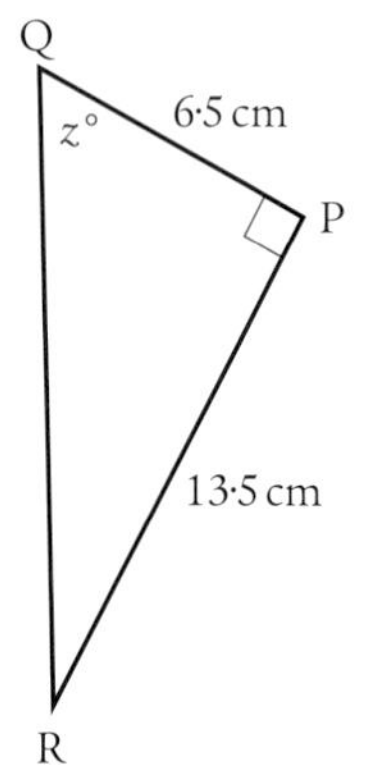

2 In each triangle calculate the marked side using the tangent ratio.

a

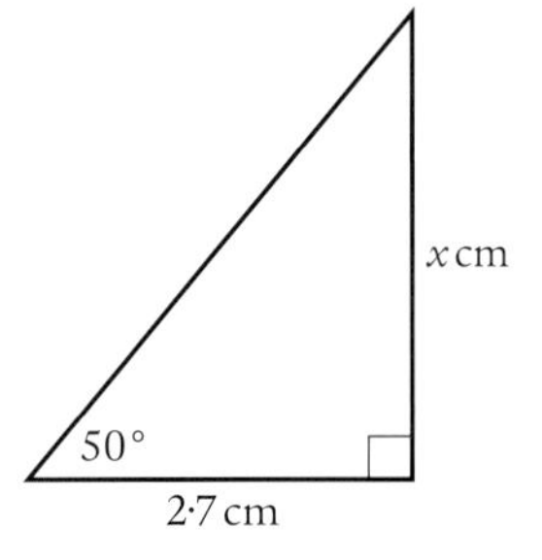

b

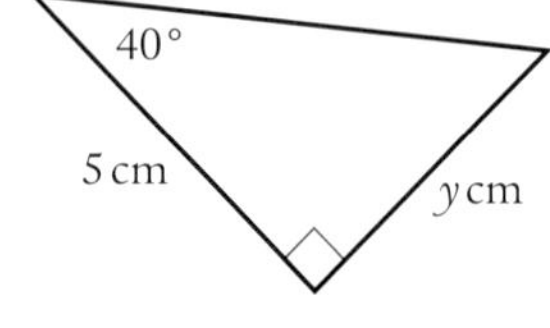

c

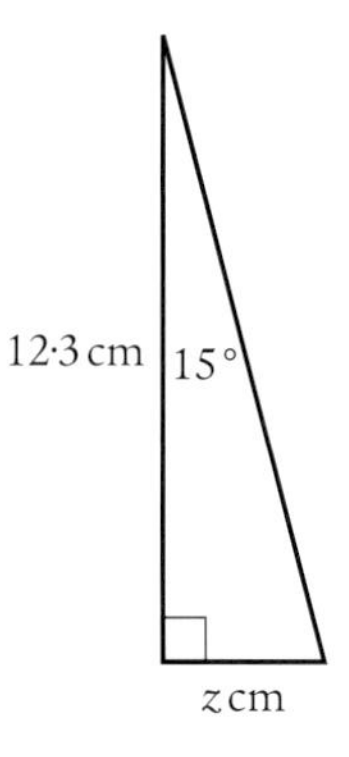

3 An archaeologist is surveying a standing stone circle on the moors.

From a distance of 8 metres she measures the angle to the top of a stone as 27°.

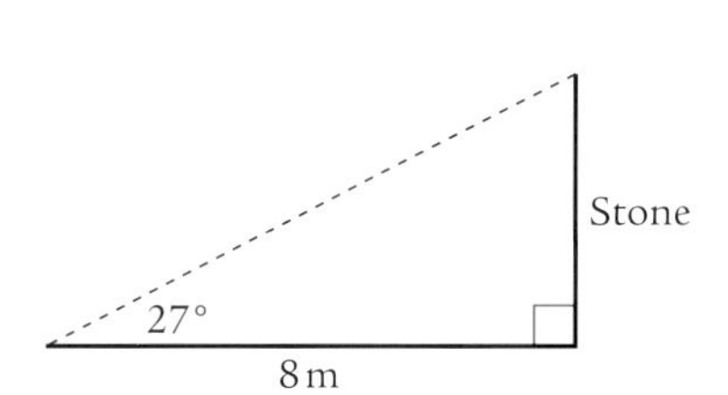

Calculate the height of the stone correct to 3 significant figures.

4 A yacht sails 40 metres from the shore when a photographer takes a snap.

The angle to the top of the mast from the shore is 24°.

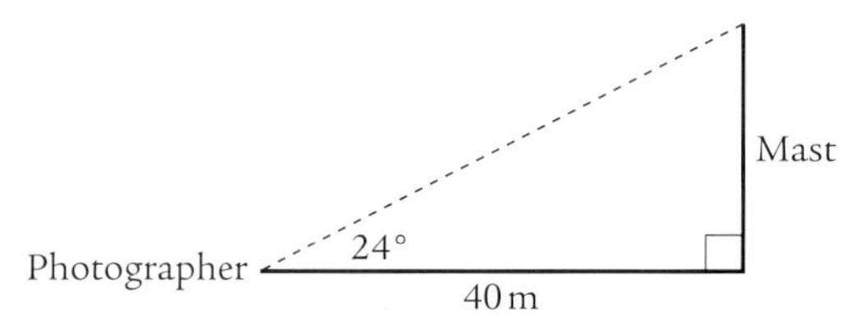

a Calculate the height of the mast correct to 3 significant figures.

b Using the photo, the angle the guy rope makes with the mast is 14°.

How far from the foot of the mast is the guy rope attached?

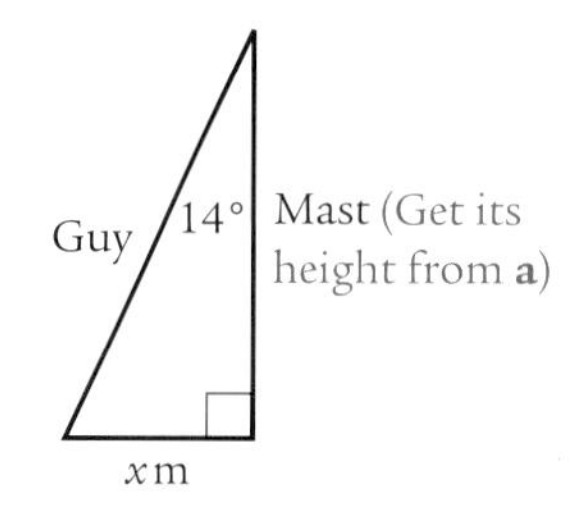

c The angle at the top of the sail is 19°.
It is 16·5 m high.

How long is the boom?

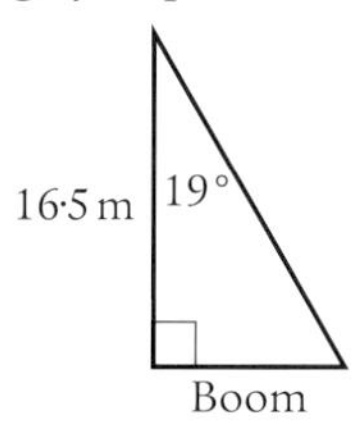

5 The gradient of a hill is often given as the tangent of the angle the hill makes with the horizontal. The result is often multiplied by 100 and given as a percentage on road signs.

So a 10% hill means a tangent of 0·1.

Calculate what would appear on the road-sign for the following hills:

a

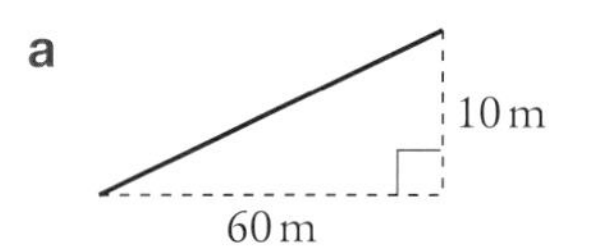

b

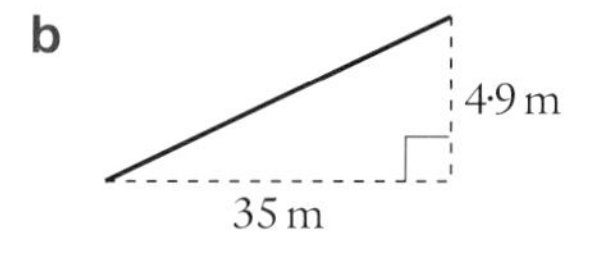

c

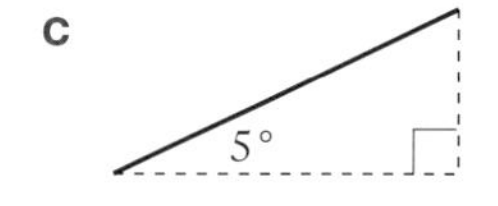

6 The yacht *Aurora* was in a position 8 km west of the Brodick Beacon.

The bearing of the beacon from the *Aurora* was 058°.

a What is the size of ∠BAC?

b How many kilometres north of the *Aurora* is the beacon?

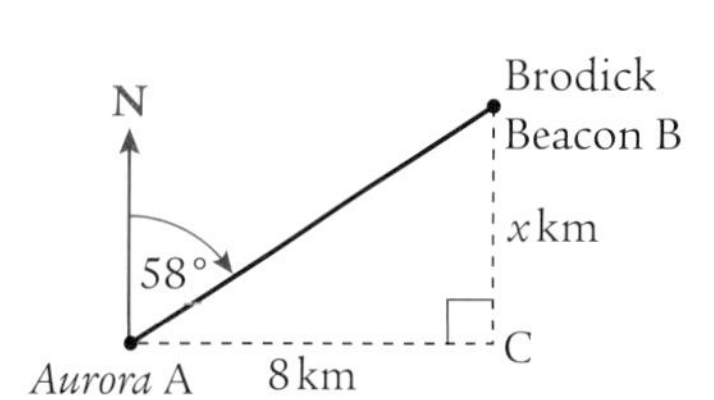

7 In rectangle ABCD, AB is 18 cm long.

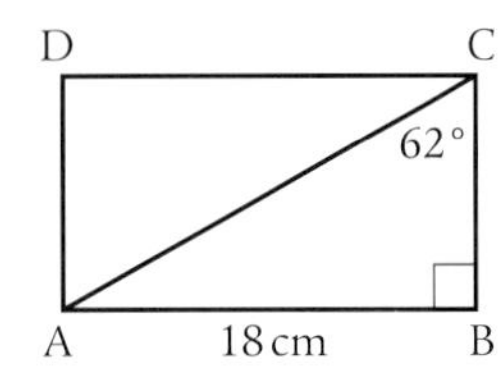

The diagonal AC makes an angle of 62° with CB.

a What is the size of ∠CAB?

b Use the tangent of ∠CAB to find the length of CB.

c Use Pythagoras' theorem to find the length of the diagonal.

8 An electricity cable is held aloft by poles.

Where the cable crosses a footpath, a sign states the maximum safe clearance is 4·8 metres.

Ten metres from the pole the angle of elevation to its top is 50°.

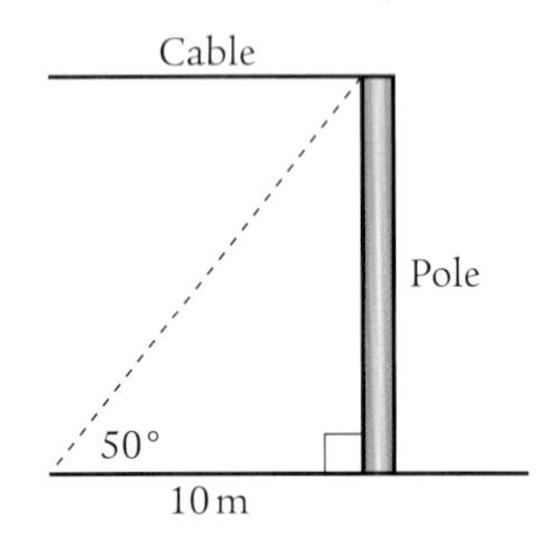

a Calculate the height of the pole.

b How much higher than 4·8 m is the cable?

This is called the safety margin.

c Express the safety margin as a percentage of the given safe clearance.

Example 3

The Science Tower in Glasgow is 122 m tall.

From a point directly across the river the angle of elevation is measured as 51°.

How wide is the river?

$$\tan 51° = \frac{\text{opposite}}{\text{adjacent}}$$

$$= \frac{122}{x}$$

$$\Rightarrow x \tan 51° = 122$$

$$\Rightarrow \quad x = \frac{122}{\tan 51°} = 98{\cdot}8$$

At this point the river is 98·8 m wide.

A memory aid

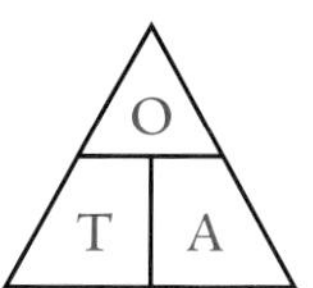

To minimise the amount of algebra used, you can memorise a triangle that helps:

If you want to find the opposite side, cover the O with your finger ...

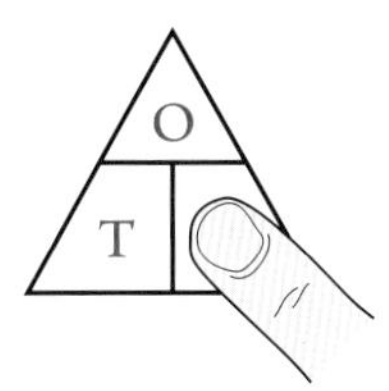

... a formula O = TA is 'revealed'.

The opposite side = the tangent times the adjacent side.

If you want to find the adjacent side, cover the A with your finger ...

... a formula $A = \frac{O}{T}$ is 'revealed'.

The adjacent side = the opposite side divided by the tangent.

Example 4

Goat Fell is 874 m above sea level.

A map maker taking measurements from Brodick Pier finds the angle of elevation to the top of the mountain is 8·0°.

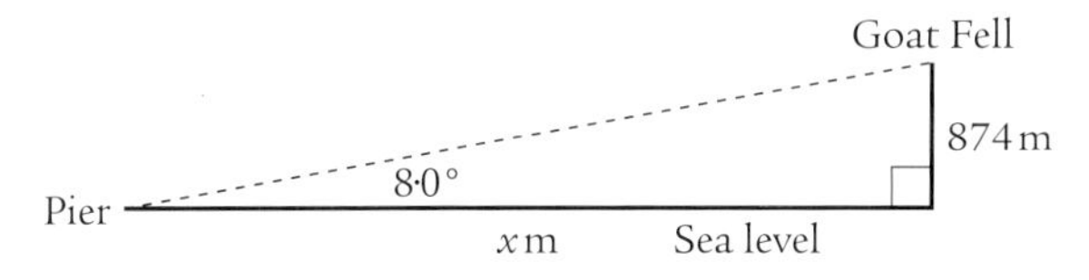

How far is the mountain from the pier?

In the right-angled triangle we want the side adjacent to the known angle.

$$A = \frac{O}{T}$$

$$x = \frac{874}{\tan 8{\cdot}0°}$$

$$x = 6220 \text{ (to 3 s.f.)}$$

The mountain is 6220 m from the pier.

Exercise 10.1B

1 In each case find the size of the side adjacent to the given angle.

a

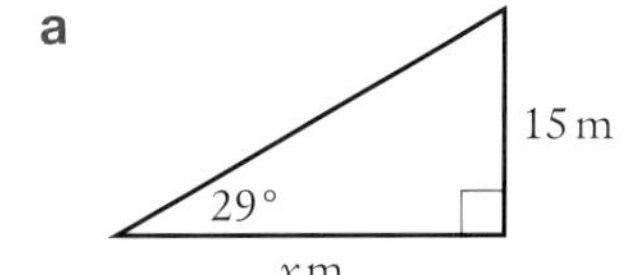

b

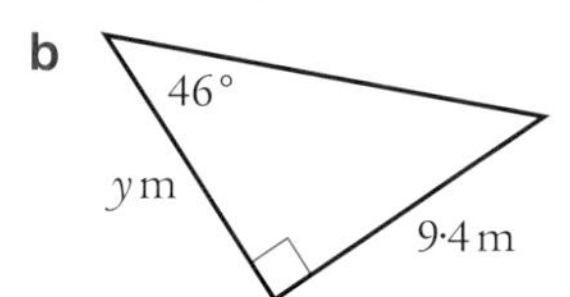

c

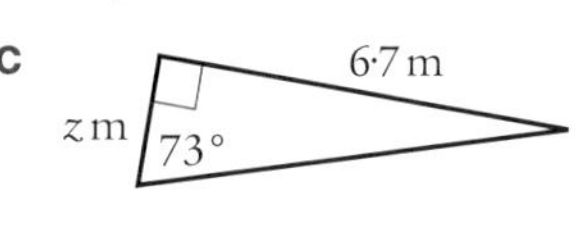

2 For both triangles:

i use the tangent ratio to find the unknown shorter side

ii use Pythagoras' theorem to find the hypotenuse.

a

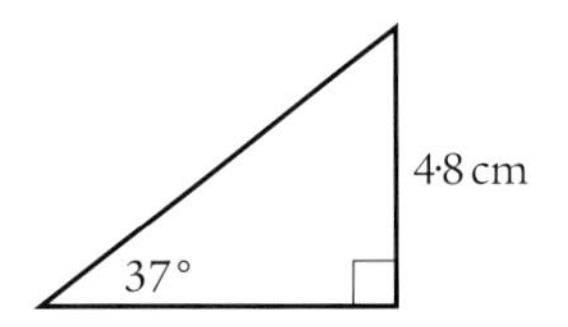

b

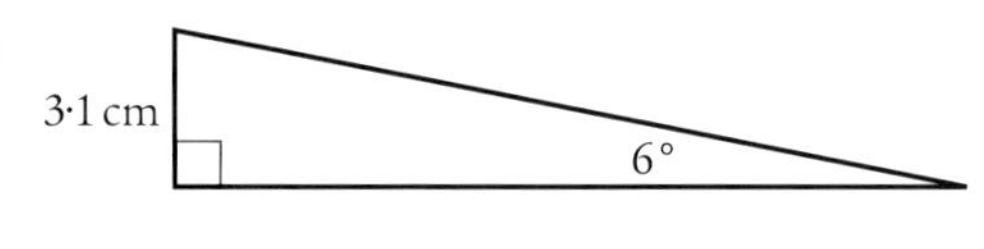

3 Helen and Bryan were watching a Microlite flying by.

Helen measured the angle of elevation of the Microlite as 40°.

The Microlite pilot measured the angle of depression of Bryan as 74° when he was 25 m from Bryan.

In the diagram, D is the point vertically below the pilot.

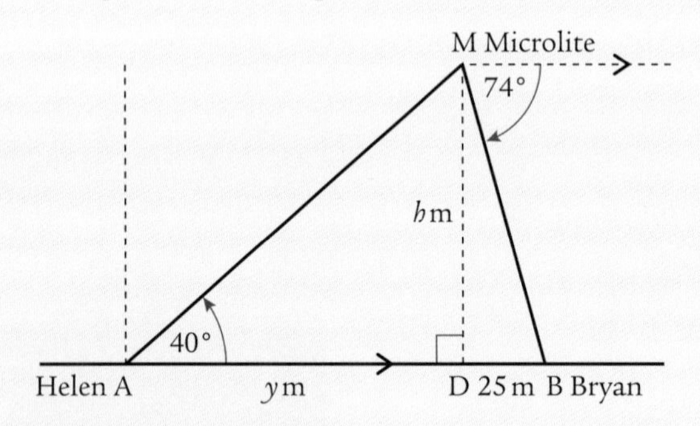

a What is the size of ∠MBD?

b Calculate the height of the Microlite.

c Calculate how far Helen is from: **i** D **ii** Bryan.

4 In a science experiment a ball is rolled down an inclined plane.

The plane is inclined at an angle of 12° to the horizontal.

A camera catches the ball at two different positions on the incline.

A is the point below the ball in the first position and B is the point below it in the second.

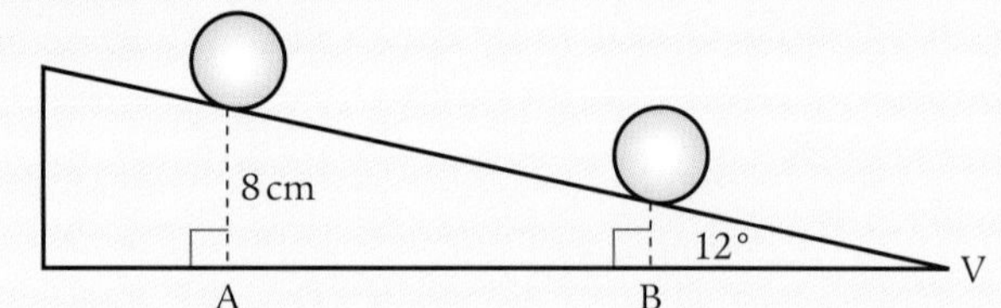

a How far from the bottom of the slope, V, is A?

b The distance between A and B is 10 cm.

What is the height of the ball in the second position?

5 In an experiment with light a partition is set up at right angles to a mirror.

A torch is shone on the mirror and seen by an observer who is 5 metres from the partition.

The ray from the torch hits the mirror at an angle of 20° and, according to the rules of reflection, comes off at the same angle.

P is a point on the partition in line with the observer.

Q is 3 metres behind it and in line with the torch.

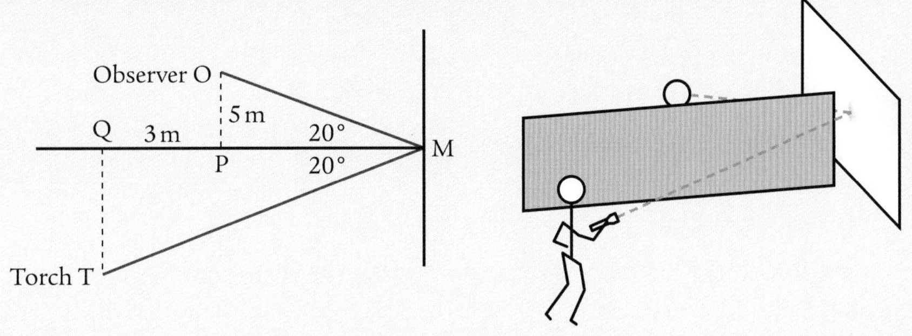

a Calculate the distance PM. **b** How far is the torch from the partition (TQ)?

6 An ornithologist radio-tags a gull to study its movements.

It is released at a point A and is observed flying to a point B, which is on a bearing of 067° from A and 7 km east of A.

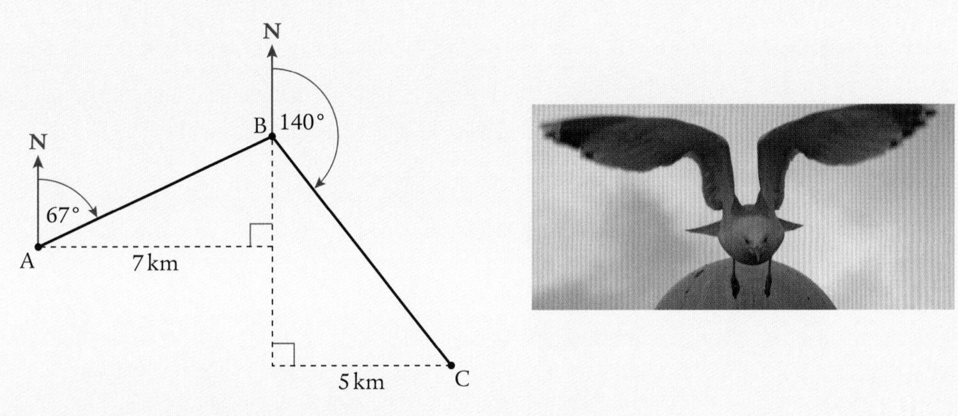

a How far north of A is the point B?

b The gull is then tracked to a point C, which is 140° from B and a further 5 km east of A.

  i How far south of B is C? **ii** How far south of A is C?

7 An astronomer can tell how fast a planet goes round the Sun. So he can precisely date when the radius of the planet's orbit and that of the Earth's are at right angles.

He can easily measure the angle between the Sun and the planet at this time.

a For the planet Venus, this angle, ∠SEV = 35·8°.

The distance between the Sun and the Earth is 1 AU (an astronomical unit). Calculate the distance between the Sun and Venus.

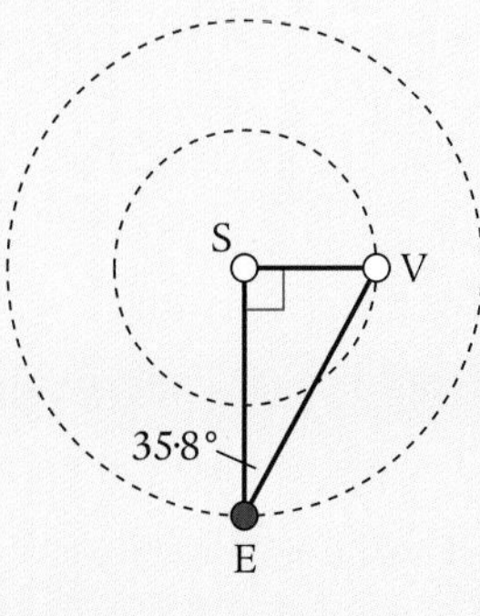

b For Mars, this angle is 56·7°.

Calculate how far Mars is from the Sun in astronomical units.

c For Jupiter, this angle is 79·1°. For Saturn, it is 84·0°.

How much further than Jupiter is Saturn from the Sun?

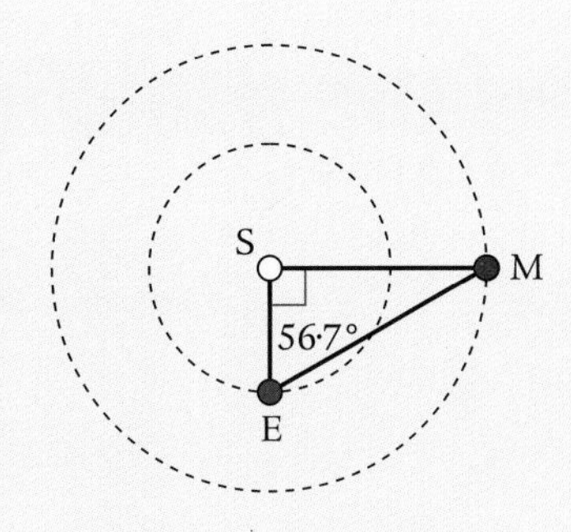

10.2 Finding an angle

When we know the length of the side opposite an angle and the side adjacent to the angle, we can use the tangent tables 'backwards' to find the size of the angle.

What you do on a calculator depends on the model you have.

Most often it is 2nd F [tan] or [$\tan^{-1}$] or [atan]

We will use the middle one. So, for example, $\tan^{-1}(1)$ should return the number 45, because 45° has a tangent of 1.

Example 1

The ship *Ruby* was 3 km west and 2·5 km south of the port of Fairhaven.

What is the bearing of Fairhaven from the *Ruby*?

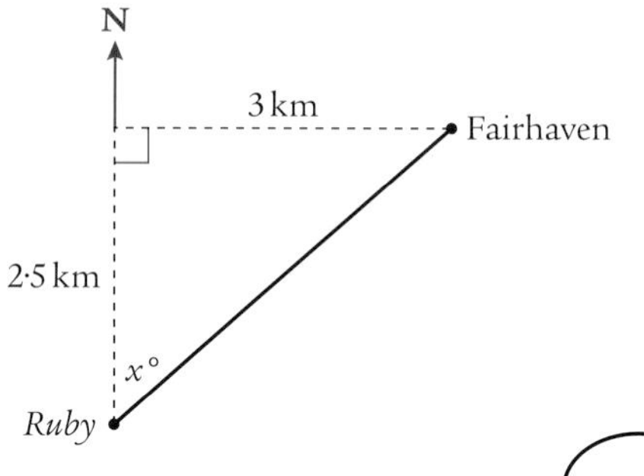

$$\tan x° = \frac{\text{opposite}}{\text{adjacent}} = \frac{3}{2{\cdot}5}$$

$$= 1{\cdot}2$$

$$\Rightarrow \quad x° = \tan^{-1}(1{\cdot}2) = 50{\cdot}2°$$

The three-figure bearing of Fairhaven from the *Ruby* is 050°.

Exercise 10.2A

1 In each case find the size of the marked angle correct to 1 decimal place.

a

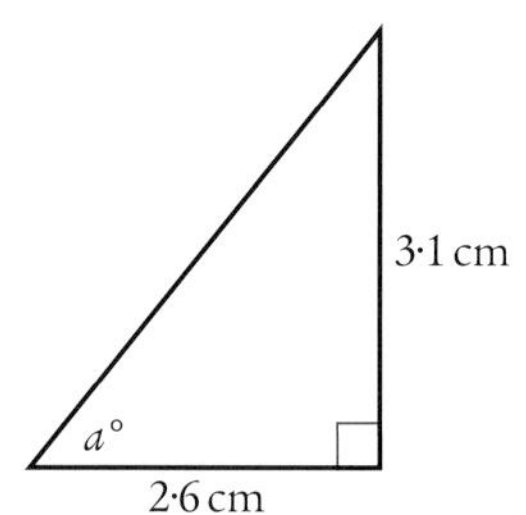

b

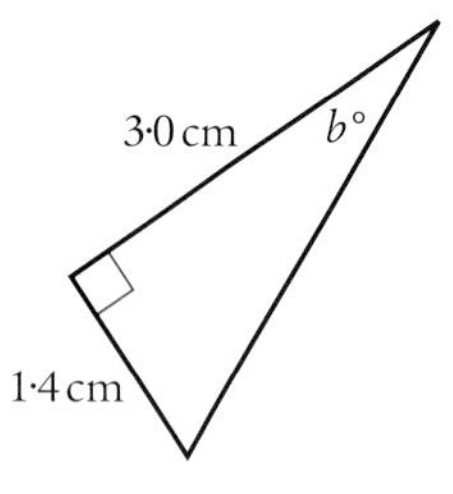

c

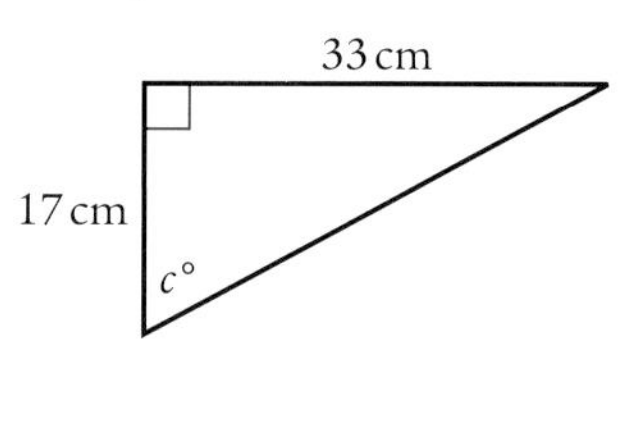

2 Triangle ABC has an altitude BD = 2·0 cm.

D is a point 2·5 cm from A and 4·0 cm from C.

Calculate the size of:

a ∠BAD

b ∠BCD

c ∠ABC.

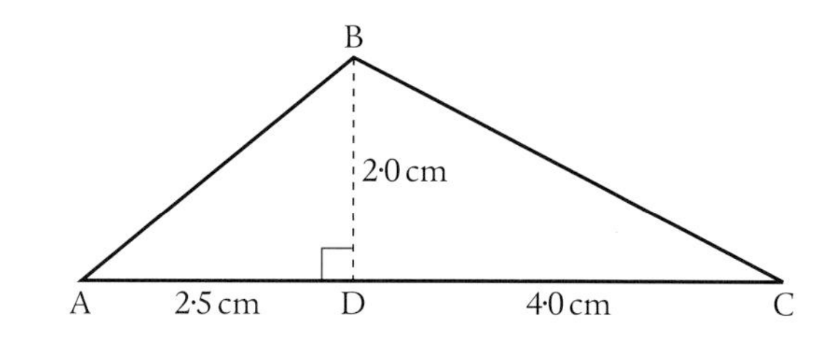

3 We can consider that an isosceles triangle is made from two right-angled triangles placed back-to-back.

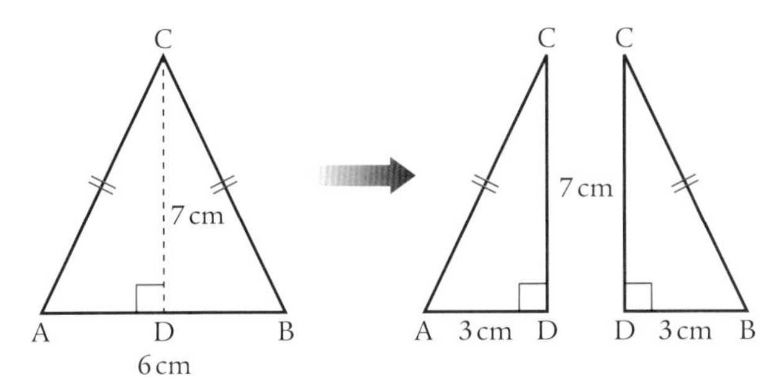

a By considering the right-angled triangle ADC, calculate the size of: **i** ∠CAD **ii** ∠ACD.

b Hence calculate the size of each angle in the isosceles triangle ABC.

c Calculate the size of each angle in these isosceles triangles.

i

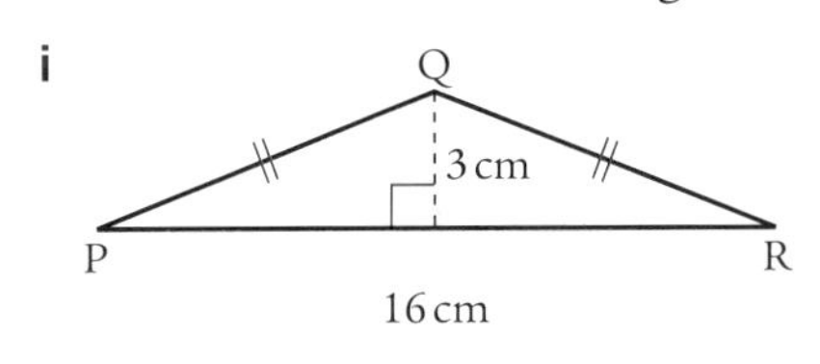

ii

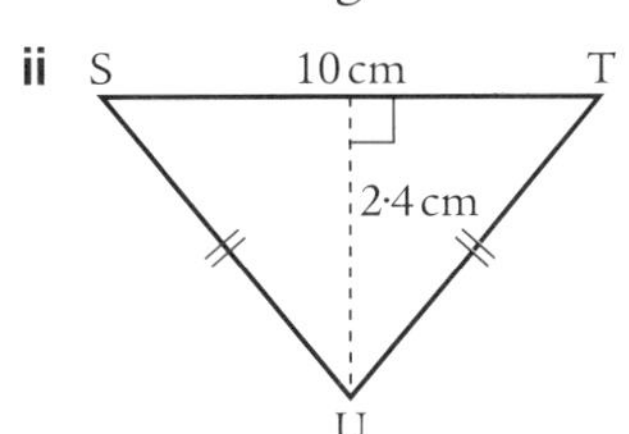

4 Traffic signs warn drivers of unexpectedly steep hills.

They quote the gradient as a **percentage** or a **ratio**.

In each case they are expressing the tangent of the angle.

A gradient of 12% means the tangent is 0·12, one of 1 : 12 means the tangent is $\frac{1}{12}$.

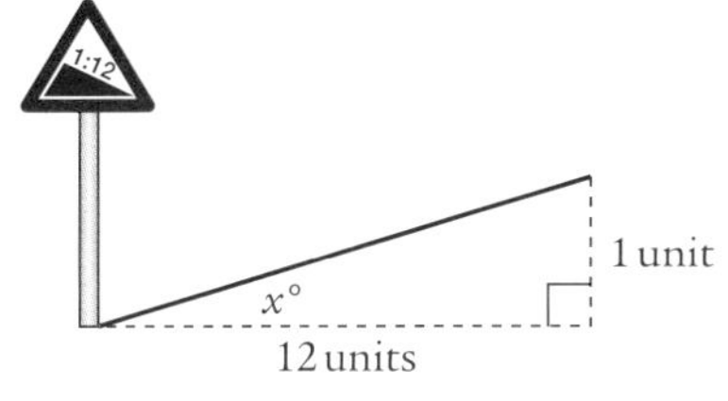

Calculate the angle the road makes with the horizontal when the sign at the bottom of the hill is:

a

b

c

d

5 On average there are 14 deaths and 1200 major injuries a year to workers caused by falls from ladders.

There are standard rules for the safe use of a ladder.

Used on level ground, a ladder should be placed according to the 1-in-4 rule.

a A ladder is set up according to the rule.

What angle does it make with the ground?

b Some people call it the 75° rule. Comment.

c A ladder has to reach 5 metres up a wall.

How far from the wall should the foot of the ladder be?

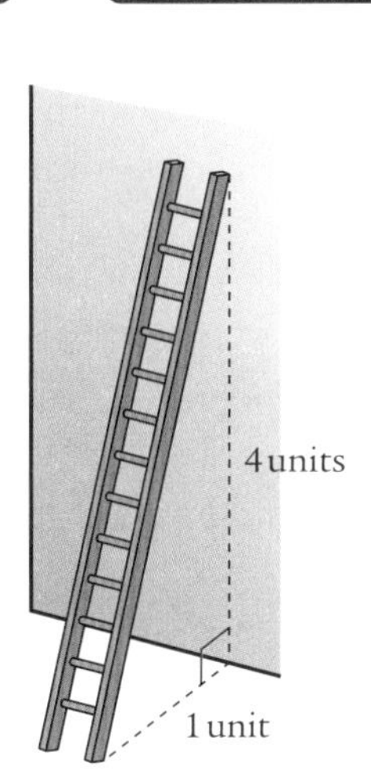

d If the ladder is put up on sloping ground there are two rules:

1. a side slope should be no bigger than 16°
2. a back slope should be no more than 6°.

Which of these are unsafe?

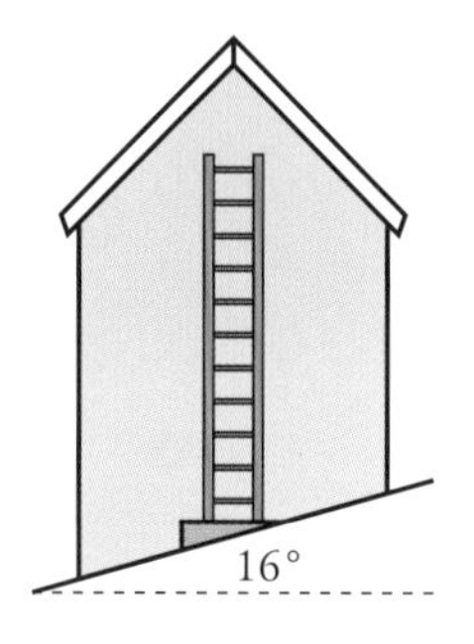

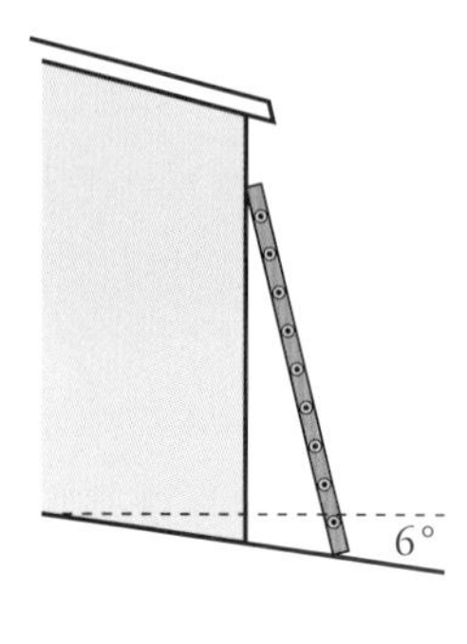

i

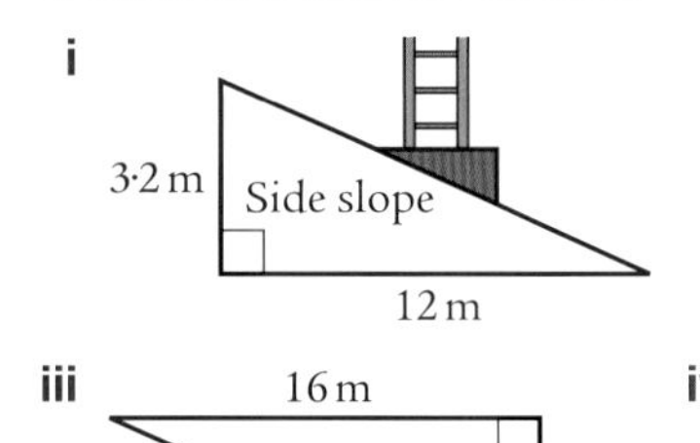

ii

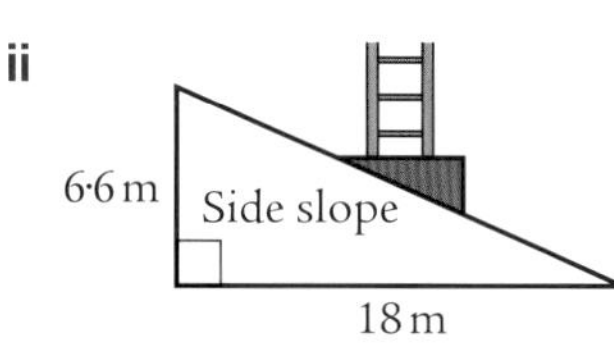

iii 16 m, Back slope, 1·4 m

iv 22 m, Back slope, 3·0 m

6 A steeplejack uses a ladder in the most extreme cases. The 1-in-4 rule just doesn't apply.

The ladder has to be **attached** to the side of the chimney-stack.

Here are some cases where the 4 units are fixed.
We reduce the distance the foot of the ladder is from the wall.

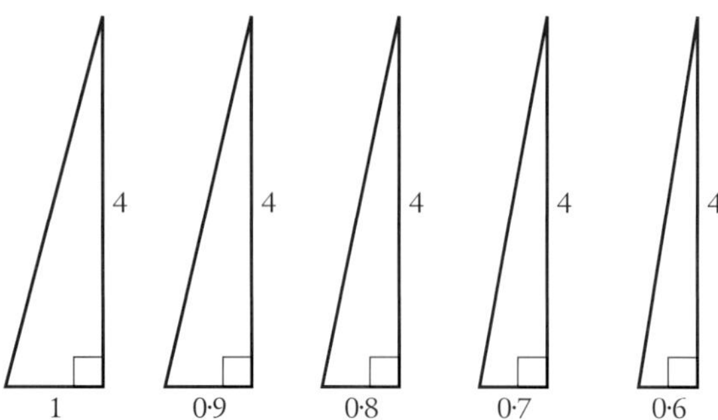

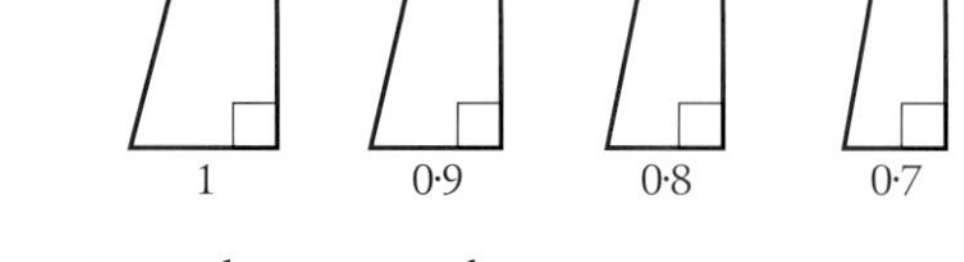

a In each case work out:

i the tangent of the angle the ladder makes with the ground

ii the angle itself.

b When the distance to the wall gets tiny, investigate what happens to:

i the size of the angle

ii the tangent of the angle.

Report on how the calculator deals with the problem encountered.

Example 2

Triangle ABC has a base AB = 9 cm.

This is cut by an altitude at D so that AD = 2 cm.

$\angle CAB = 40°$.

Calculate: **a** the altitude **b** $\angle CBA$.

A sketch gives ...

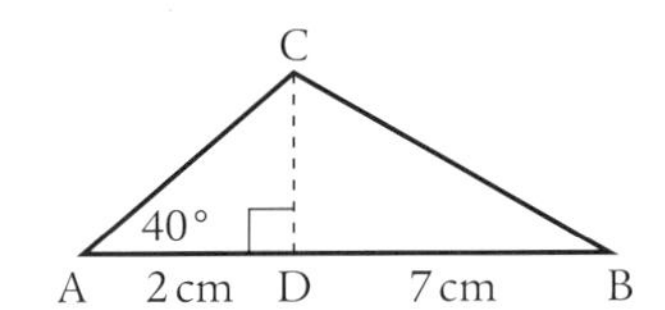

a In triangle ACD:

$$CD = AD \tan 40°$$
$$= 2 \times \tan 40°$$
$$= 1{\cdot}678 \quad \boxed{\tan}$$

Altitude = 1·7 cm (to 1 d.p.)

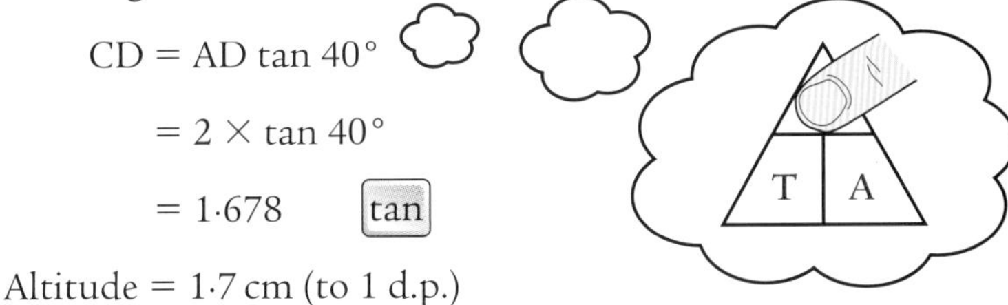

b Now, in triangle BCD:

$$\tan B = \frac{CD}{DB}$$
$$= \frac{1{\cdot}678}{7} \quad \boxed{\tan}$$
$$= 0{\cdot}2397...$$
$$B = \tan^{-1} 0{\cdot}2397 \quad \boxed{\tan^{-1}}$$
$$= 13{\cdot}5$$

$\angle CBA = 13{\cdot}5°$.

Exercise 10.2B

1 Triangle PQR has a base PR = 12 cm.
The altitude from Q cuts PR at S such that PS = 5 cm.
The angle QPR = 72°. Calculate:

a the altitude QS

b the other angles of triangle PQR.

2 A rectangle is 8 cm by 12 cm.

a What angles do the diagonals make with the sides of the rectangle?

b At what acute angle do the diagonals intersect?

3 A kite has diagonals intersecting as shown.
Calculate the size of the four angles of the kite.

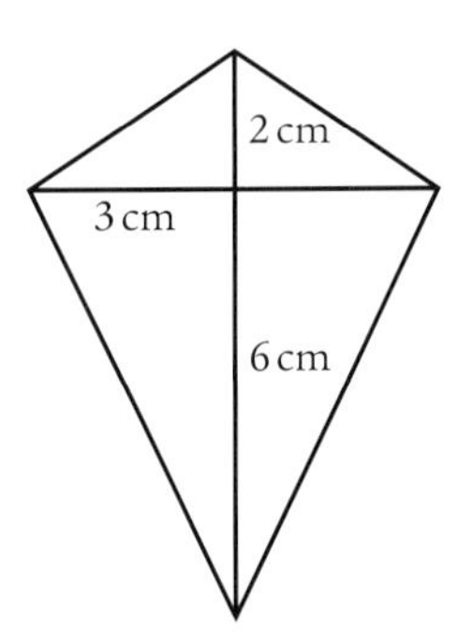

4 When loose material is placed on a flat surface it will form a conical heap.

The picture shows a pile of coal on the dock-side.

The angle marked is known as the **angle of repose** and is the steepest the sides can get before collapsing and starting an avalanche.

Knowing this angle for various materials is very important to engineers.

a Calculate the angle of repose for coal.

b Every child knows that it is easier to make a sandcastle out of wet sand than dry.

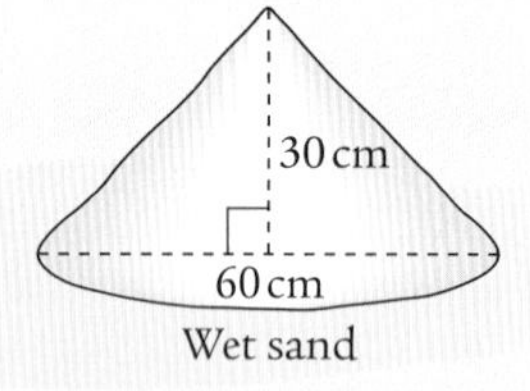

Wet sand

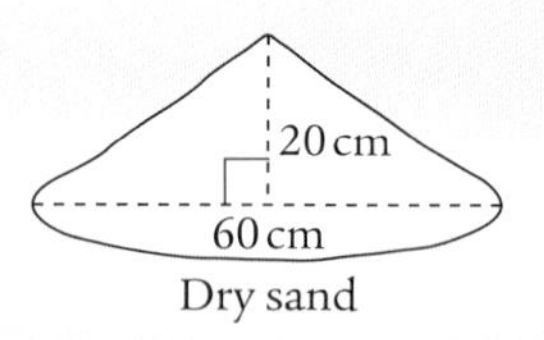

Dry sand

Calculate the angle of repose for each material ... the diagram shows a pile of each sitting at the required angle. **(Treat each cross-section as isosceles.)**

5 A geologist is studying a fault line. It shows up on the surface as a line, but underground it goes off at an angle. The geologist needs to know this angle to calculate forces.

Two metres from the line on the surface, he drills vertically down till he hits the fault, three metres down.

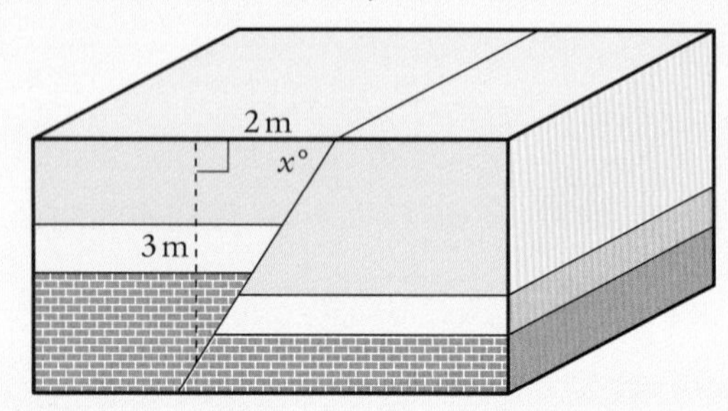

At what angle is the fault running underground?

6 Maps generally work on a grid reference.

The Tay Bridge begins and ends at a roundabout.

The one on the south bank is 1·3 km south and 1·9 km east of the one on the north bank.

a Calculate the angles in the triangle of which the bridge forms the hypotenuse.

b On what bearing is the bridge running as it crosses from north to south?

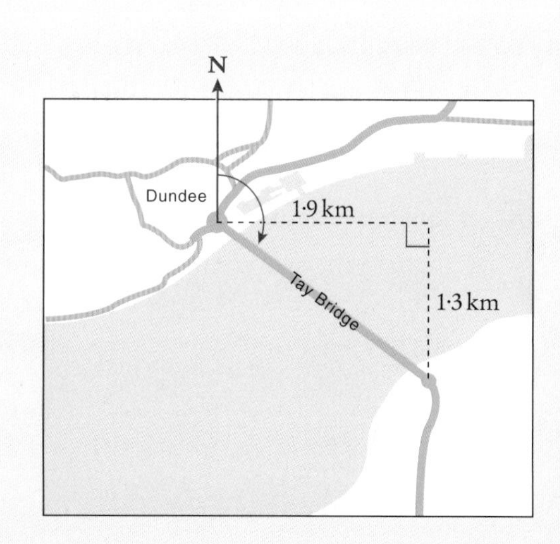

7 For a fly-over at the Jubilee celebrations, the newsreel photographers program the camera to take snaps at fixed positions as the plane goes by.

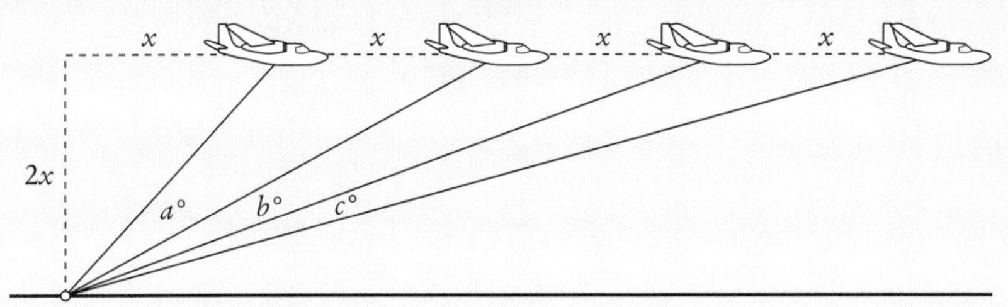

The idea is to take a photo at fixed equal distances in the plane's flight.

Is the **change** that has to be made in the angle of elevation for each position also a fixed amount?

Calculate the values of a, b and c and report your findings.

10.3 The sine of an angle

The tangent ratio is a very useful idea.

However, it isn't very useful for angles close to 90°.

It can't be directly used if the given information doesn't involve the opposite and adjacent sides.

Other ratios were developed for these cases.

One of these is called the sine ratio.

We define it as: $\sin x° = \dfrac{\text{opposite side}}{\text{hypotenuse}}$ (Note 'sine' has been shortened to 'sin')

The memory aid developed for helping with the tangent can be adapted for the sine:

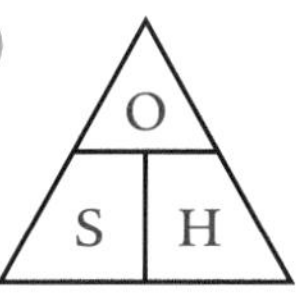

Example 1

A flagpole is 6 metres tall and is held vertical by an 8-metre guy rope attached to the ground.

At what angle does the rope meet the ground?

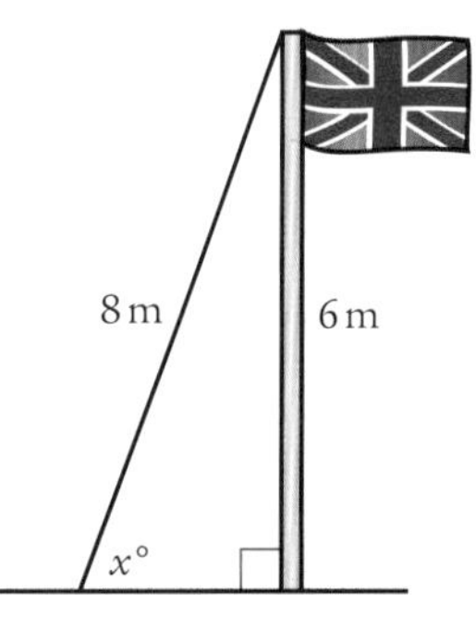

We want the angle marked $x°$.

Opposite it is a 6 m length. The hypotenuse is an 8 m length.

The opposite side and hypotenuse are involved, so use the sine ratio.

We want the angle $x°$.

$$\sin x° = \frac{\text{opposite}}{\text{hypotenuse}} = \frac{6}{8} = 0.75$$

$$\Rightarrow \quad x° = \sin^{-1}(0.75) = 48.6°$$

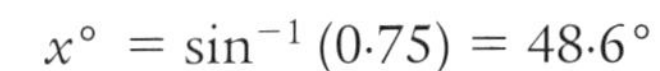

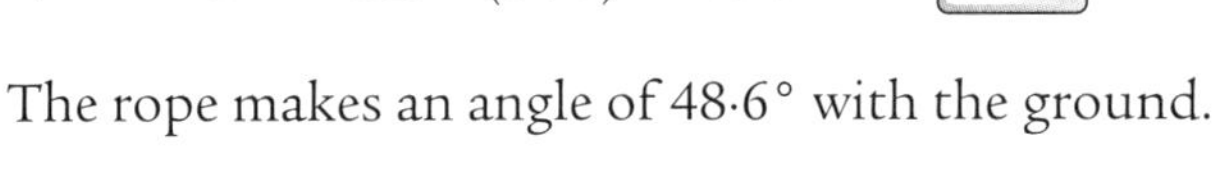

The rope makes an angle of 48.6° with the ground.

Example 2

After an earthquake, a section of the ground collapsed forming a conical hole.

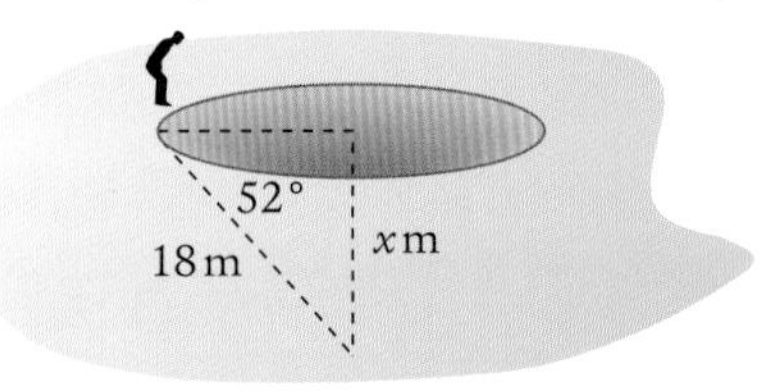

The sloping side of the hole was measured as 18 metres.

The angle it made with the horizontal was 52°.

How deep is the hole?

Once again, the opposite and hypotenuse are involved ... use the sine ratio.

We want the side **opposite** the known angle.

$$\sin A = \frac{\text{opposite}}{\text{hypotenuse}}$$

$$\Rightarrow \quad \sin 52° = \frac{x}{18}$$

$$\Rightarrow \quad x° = 18 \sin 52°$$

$$\Rightarrow \quad x° = 14{\cdot}2$$

The hole is 14·2 m deep (to 1 d.p.).

Example 3

A boy rows across the river from a point A.

The current is strong so that instead of going to B he ends up 100 m down-stream at C.

He has been put off course by 32°.

How far did he actually go in the crossing?

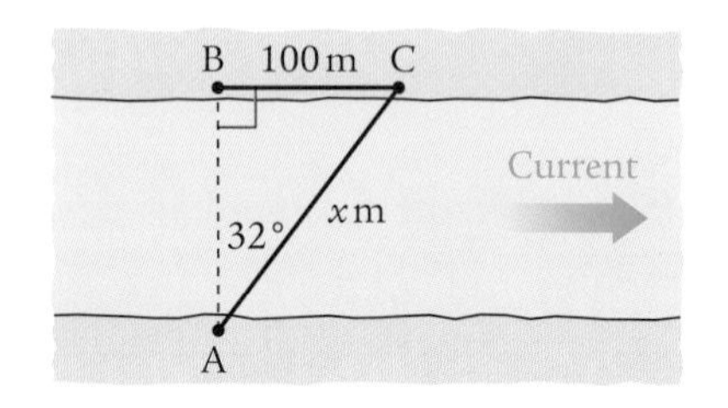

Again, the opposite side and hypotenuse are involved, so use the sine ratio.

We want the **hypotenuse**.

$$\sin A = \frac{\text{opposite}}{\text{hypotenuse}}$$

$$\Rightarrow \quad \sin 32° = \frac{100}{x}$$

$$\Rightarrow \quad x° = \frac{100}{\sin 32°}$$

$$\Rightarrow \quad x° = 188{\cdot}7$$

The boy actually went 188·7 m (to 1 d.p.) in the crossing.

Exercise 10.3A

For Questions 1–4, all lengths are in centimetres.

1 Calculate the size of the side opposite the given angle.

a

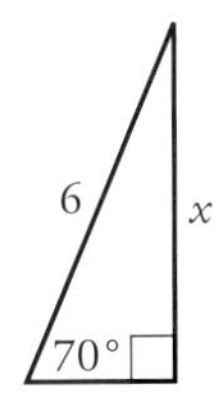

b

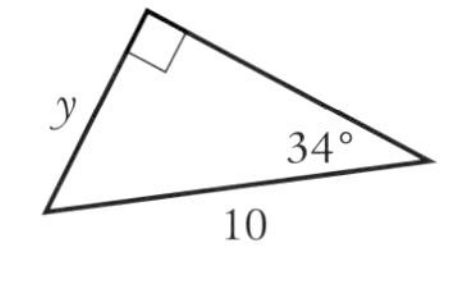

c

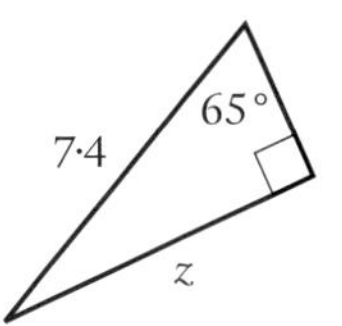

S H

2 Find the length of the hypotenuse for each triangle.

a

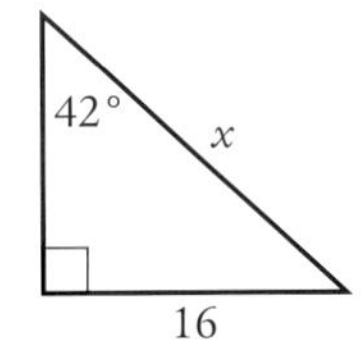

b

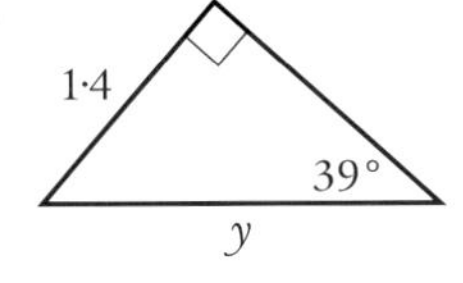

c

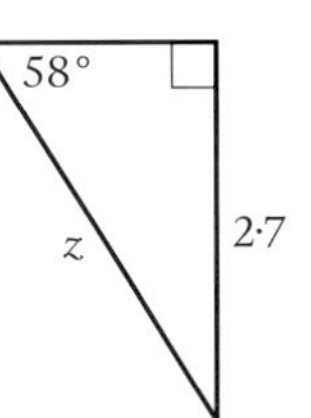

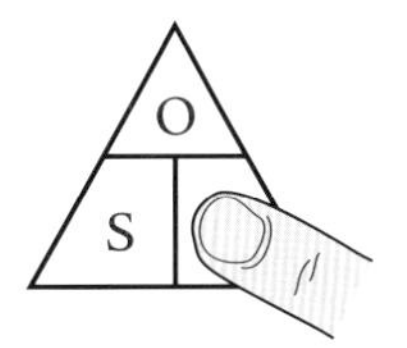

3 Calculate the size of the labelled acute angle in each case.

a

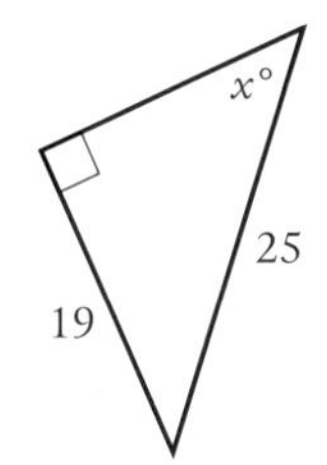

b

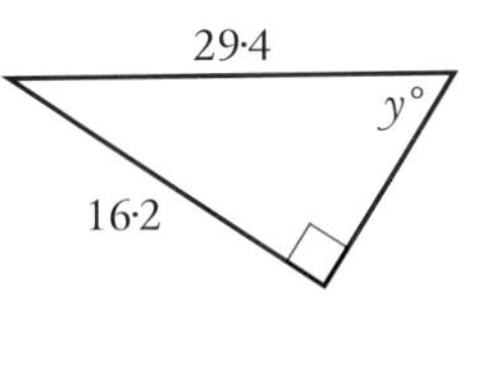

c

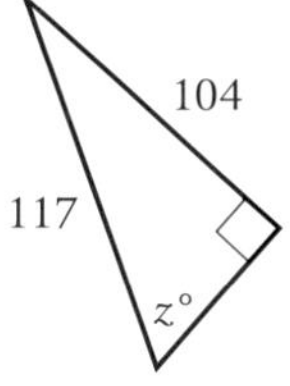

4 Solve the following triangles.

a

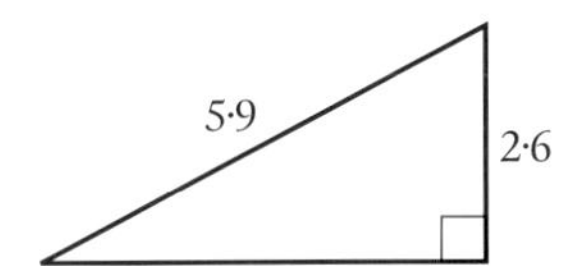

b

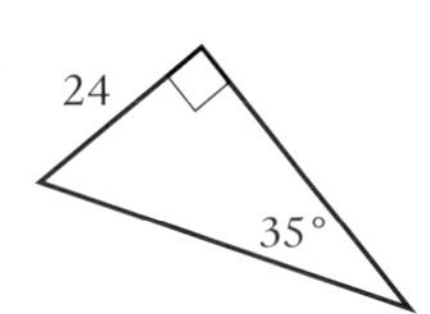

c

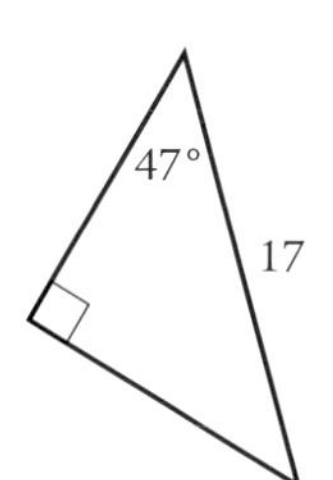

5 The SS *Rebecca* had been sailing north.
It changed direction running on a bearing of 055° for 10·3 km.
How far east of its original course is it?

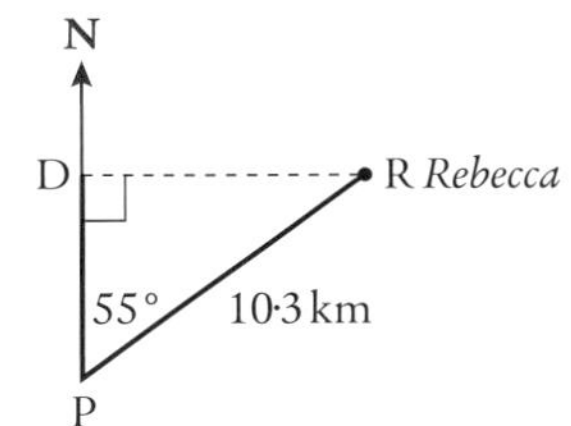

When we say, '**Solve** the triangle', we mean find all the sides and angles you have not been given.

6 Building regulations for staircases state that the steepest you can make stairs is 42°.

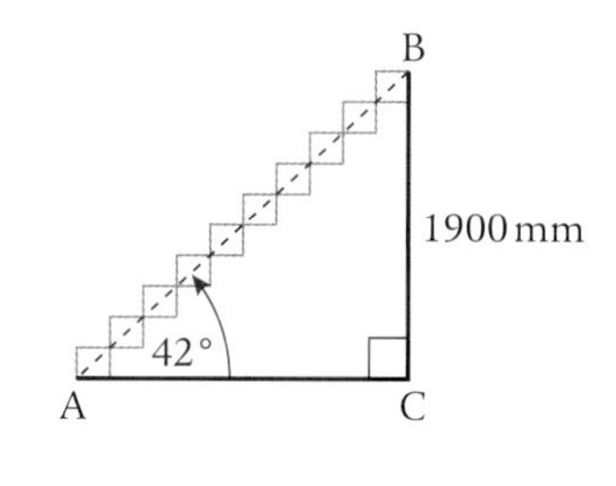

Each step up must be the same, and be between 190 mm and 200 mm.

a An architect designs a staircase pitched at 42° with 10 steps each 190 mm high.

What is the distance AB from the bottom of the stairs to the top?

b In a public building, the angle of the stair has to be 33°.

If each step up is 200 mm and there are 10 steps, what then is the distance AB?

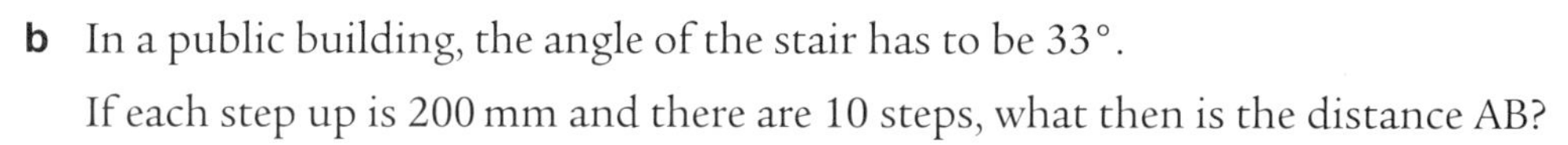

7 In an experiment in Physics, an inclined plane is slowly raised until a block sitting on it starts to slide.

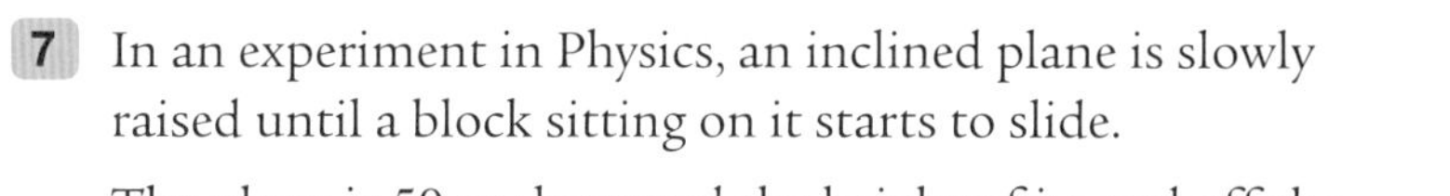

The plane is 50 cm long and the height of its end off the bench, h cm, is measured.

The important feature that is needed is the angle the plane makes with the horizontal, $x°$.

a Material 1 started to slide when $h = 10$ cm.

Calculate the value of x at this point.

b Material 2 didn't slide till the height was 20 cm.

What is the value of x for Material 2?

c One of the materials was a block of ice and the other was a block of wood.

Which material was the block of ice?

8 In the game of snooker, a ball striking the cushion will bounce off at the same angle that it hits it. In the diagram we see a white ball, 8 cm from the cushion, bouncing off the cushion at an angle of 15° and hitting a red ball which is 3 cm from the cushion.

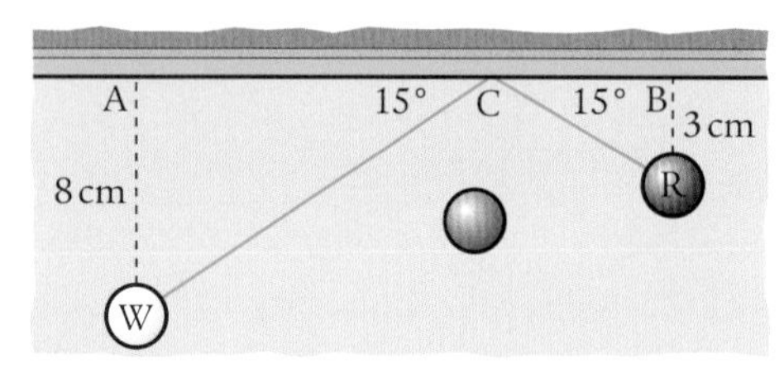

a Calculate the distance **i** WC **ii** CR.

b What is the total distance travelled by the white ball?

Example 4

Triangle ABC has sides AC = 17 cm and BC = 12 cm.

The altitude from C cuts AB at D such that ∠ACD = 55° and ∠BCD = 19°.

Find the length of AB.

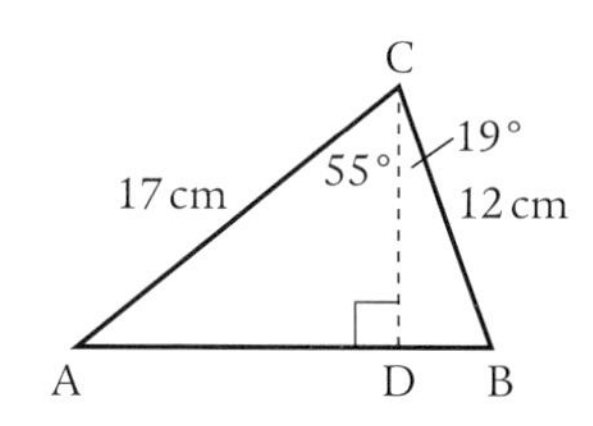

Remember that you only know how to use the sine in a right-angled triangle.

So work in triangles ACD and BCD separately.

In triangle ACD:

$$\sin A = \frac{\text{opposite}}{\text{hypotenuse}}$$

$$\Rightarrow \quad \sin 55° = \frac{AD}{17}$$

$$\Rightarrow \quad AD = 17 \sin 55°$$

$$\Rightarrow \quad AD = 13{\cdot}925...$$

In triangle BCD:

$$\Rightarrow \quad \sin 19° = \frac{BD}{12}$$

$$\Rightarrow \quad BD = 12 \sin 19°$$

$$\Rightarrow \quad BD = 3{\cdot}906...$$

AB = AD + BD = 17·8 cm (to 1 d.p.).

Exercise 10.3B

1 In each example, find the length of the base AB of the triangle.

a

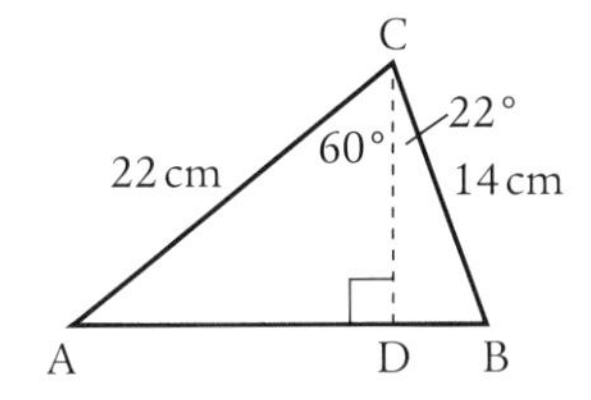

b

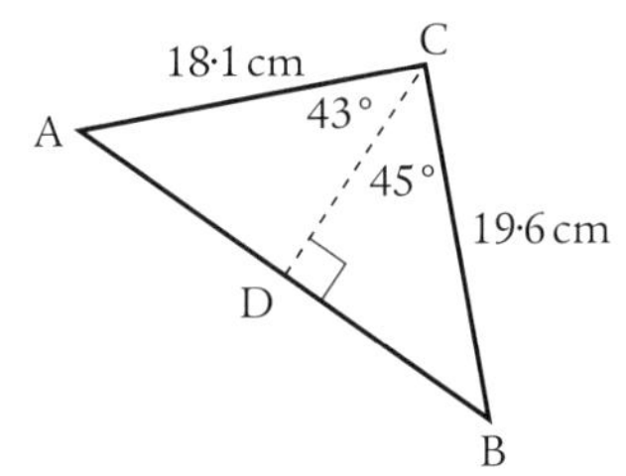

2 The port of Braebank is due east of Ardport.

The *Cumbrian Lass* is sailing due south.

The masters at both ports take a bearing.

From Ardport, the *Cumbrian Lass* is 3 km away on a bearing of 062°.

From Braebank, it is 1 km away on a bearing of 340°.

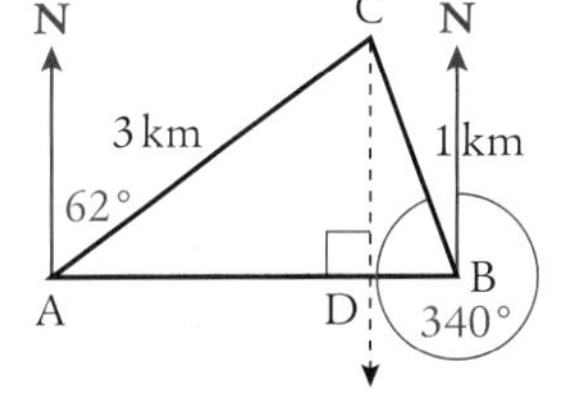

a How far will the *Cumbrian Lass* be from each port at the moment it passes directly between them?

b What is the distance between Ardport and Braebank?

3 A rectangle has a length of 12 cm. Its diagonals are 15 cm long.

a Make a sketch of such a rectangle.

b Find the angle the diagonals make with the sides of the rectangle.

4 **a** The diagonals of a kite, AC and BD, intersect at F.

AF = 4 cm and FC = 8 cm.

∠ADB = 32° and ∠BDC = 58°.

Calculate the perimeter of the kite.

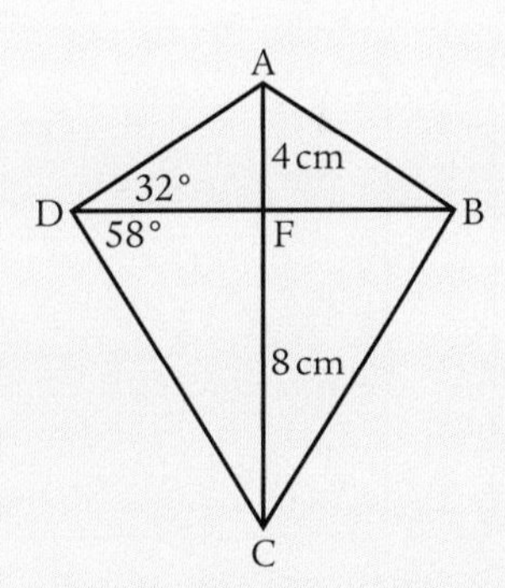

b A V-kite is as shown.

Calculate the size of:

i ∠TPS

ii ∠SPQ.

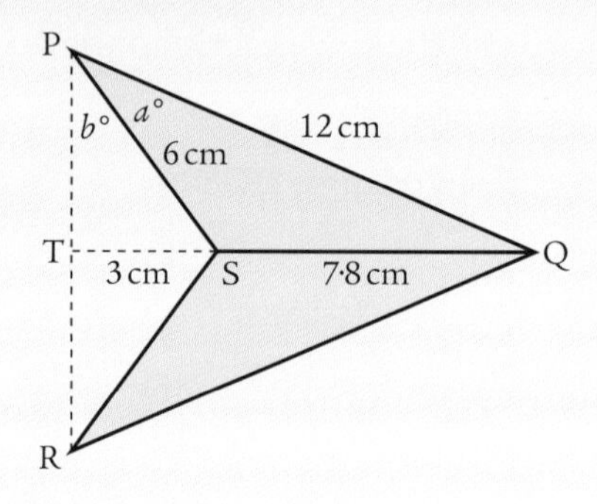

5 From the viewing platform, A, at the top of the Eiffel Tower, I took a photo of its shadow.

The angle of depression to the tip of the shadow, B, was 29°.

Later, from a map, I found the length of the shadow CB to be 429 m.

a What is the size of ∠CAB?

b Calculate the distance between me and the tip of the shadow, AB.

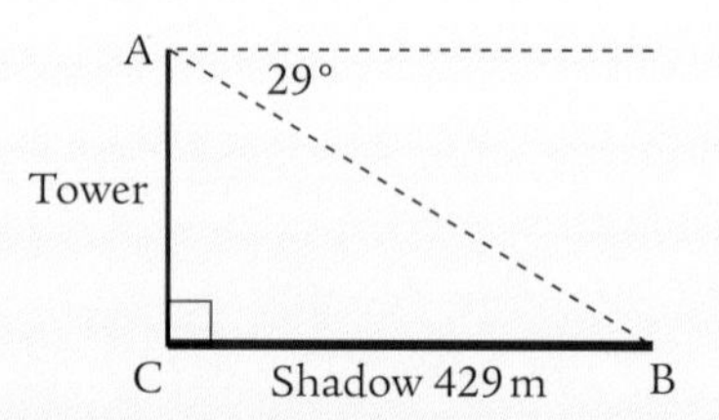

6 If you could **unravel** a helter-skelter, the slide would be the hypotenuse of a right-angled triangle.

The height of the triangle would be the height to the top of the slide.

The angle at the bottom is the slope of the slide.

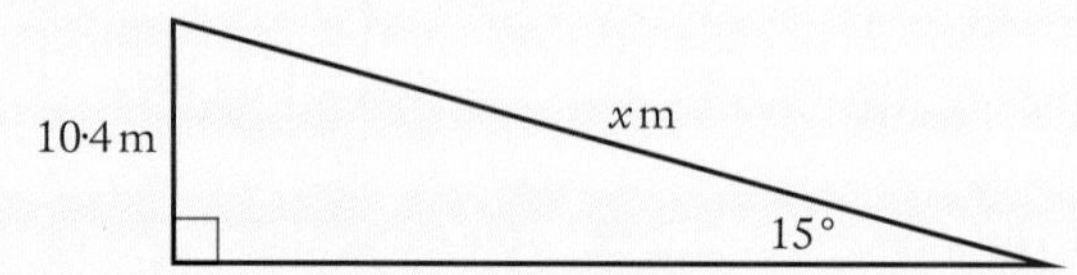

Calculate the length of the slide.

7 A customer buys a flat pack to make a bookcase.

He builds it up on the floor and tries to swing it up into an upright position.

The ceiling is only 2·5 m high. He has bought a kit that is 2·5 m tall and 25 cm deep, so that it fits snuggly into the space available.

However, as he swings it up into position, it jams!

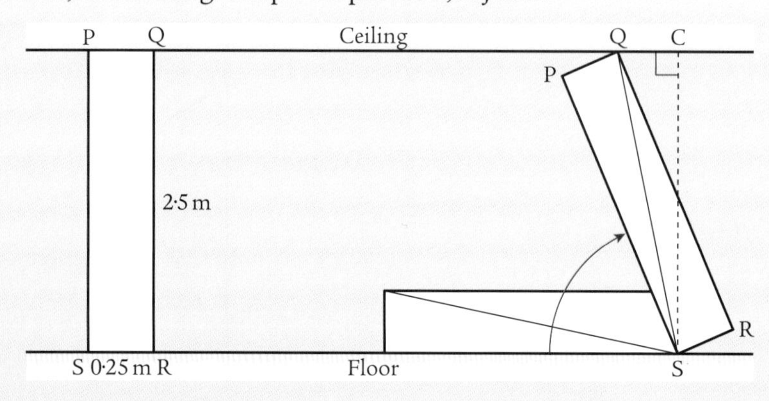

a Use Pythagoras' theorem to find the length of a diagonal of the side of the bookcase.

b Use the tangent ratio to find the size of ∠PSQ.

c Use the sine ratio to find ∠SQC.

d Hence find the angle the unit swings through before it jams.

e Investigate the tallest unit that can be swung up into position in this room, given that it should be 25 cm deep.

10.4 The cosine of an angle

A triangle has three sides.

We have considered problems involving the opposite side and the adjacent side.

We have also considered problems involving the opposite side and the hypotenuse.

The only ratio left to consider is the ratio of the adjacent side to the hypotenuse.

This has been given the name of 'cosine'.

We define it as: $\cos x^\circ = \dfrac{\text{adjacent side}}{\text{hypotenuse}}$ (Note 'cosine' has been shortened to 'cos')

The memory aid developed for helping with the other ratios can be adapted for the cosine:

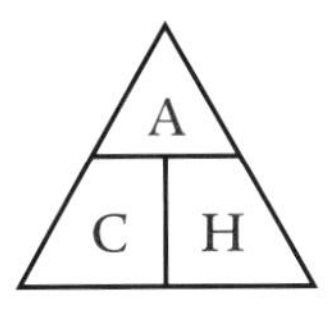

On 15th October 1783 the first manned hot-air balloon flight took place.

The balloon rose to a height of 24 metres and was kept from floating away by a tether of length 26 metres.

What angle did the tether make with the vertical?

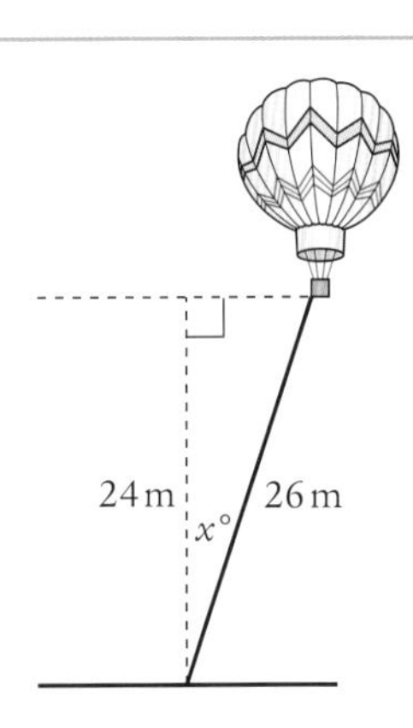

We want the angle marked $x°$.

Adjacent to it is a 24 m length. The hypotenuse is a 26 m length.

The adjacent side and hypotenuse are involved ... use the cosine ratio.

We want the **angle**.

$$\cos x^\circ = \frac{\text{adjacent side}}{\text{hypotenuse}} = \frac{24}{26}$$

$$= 0{\cdot}923$$

$\Rightarrow \quad x^\circ = \cos^{-1}(0{\cdot}923) = 22{\cdot}6^\circ$ cos⁻¹

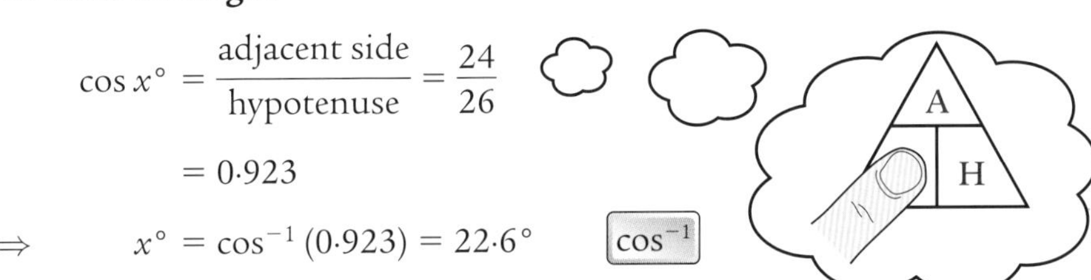

The tether makes an angle of 22·6° with the vertical.

Example 2

A ladder leans against a wall. It is 9 m long.

The recommended safe angle for a ladder is 75° to level ground.

How far should we place the foot of the ladder from the wall?

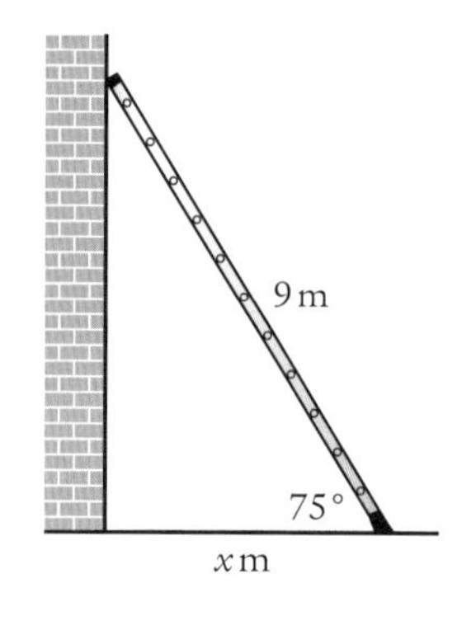

Once again, the adjacent side and hypotenuse are involved, so use the cosine ratio.

We want the side **adjacent** to the known angle.

$$\cos A = \frac{\text{adjacent}}{\text{hypotenuse}}$$

$$\Rightarrow \quad \cos 75° = \frac{x}{9}$$

$$\Rightarrow \quad x° = 9 \cos 75° \qquad \boxed{\cos}$$

$$\Rightarrow \quad x° = 2{\cdot}329...$$

The foot of the ladder should be placed 2·33 m (to 2 d.p.) from the wall.

Example 3

The gangway of a boat makes an angle of 33° with the quayside deck.

It is tethered to the boat and to a point on the quay 10 m from the boat.

How long is the gangway?

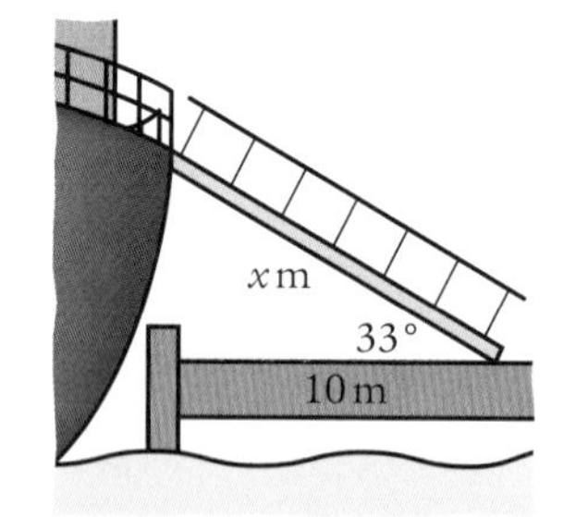

Again, the adjacent and hypotenuse are involved, so use the cosine ratio.

We want the **hypotenuse**.

$$\cos A = \frac{\text{adjacent}}{\text{hypotenuse}}$$

$$\Rightarrow \quad \cos 33° = \frac{10}{x}$$

$$\Rightarrow \quad x° = \frac{10}{\cos 33°} \qquad \boxed{\cos}$$

$$\Rightarrow \quad x° = 11{\cdot}923...$$

The length of the gangway is 11·9 m (to 1 d.p.).

Exercise 10.4A

For Question 1–4, all lengths are in centimetres.

1 Calculate the size of the side adjacent to the given angle.

a

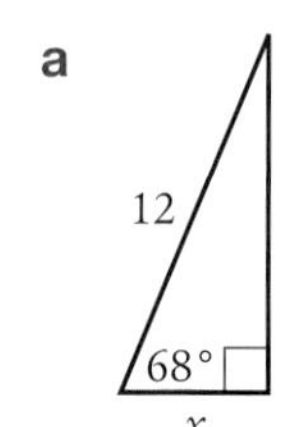

b

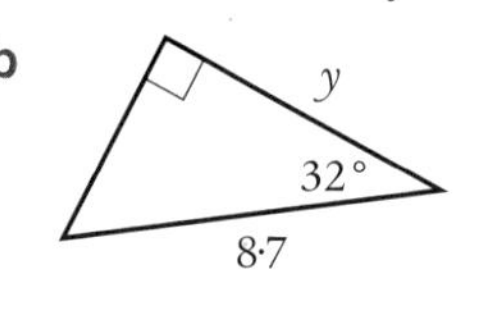

c

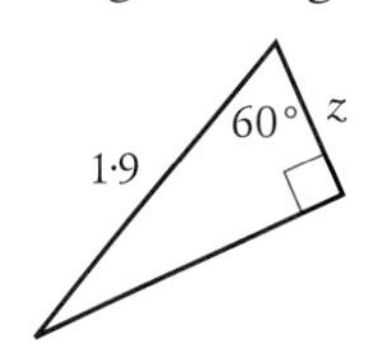

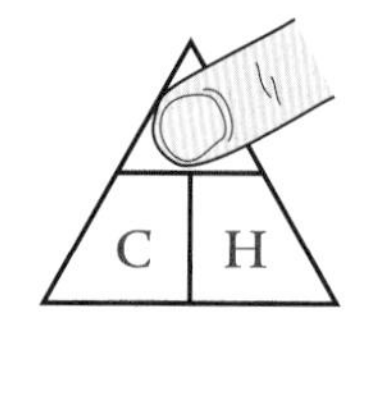

2 Find the length of the hypotenuse for each triangle.

a

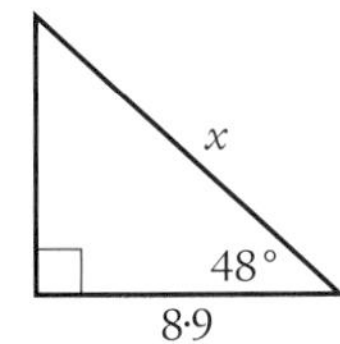

b

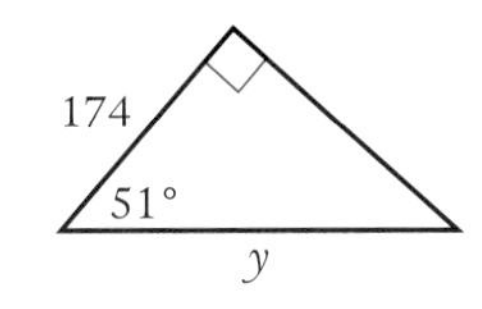

c

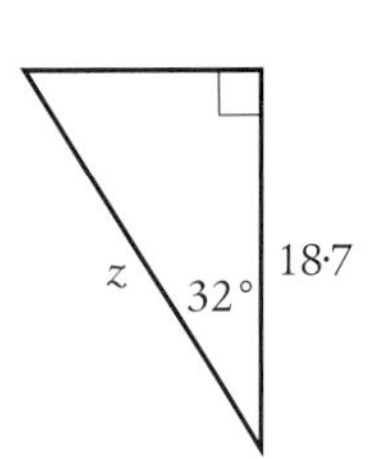

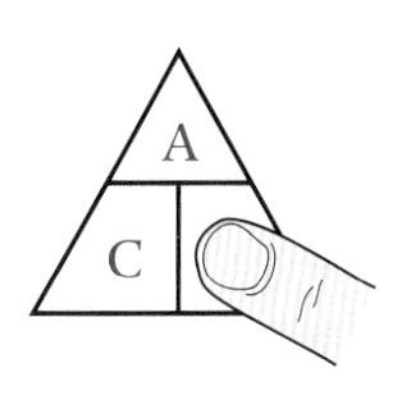

3 Calculate the size of the labelled acute angle in each case.

a

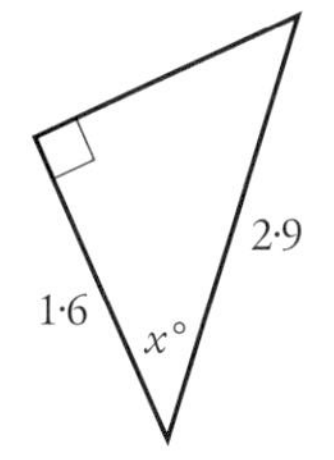

b

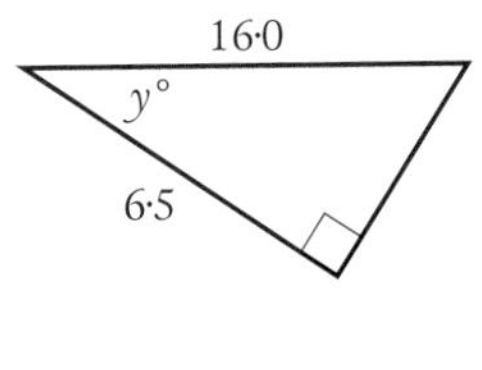

c

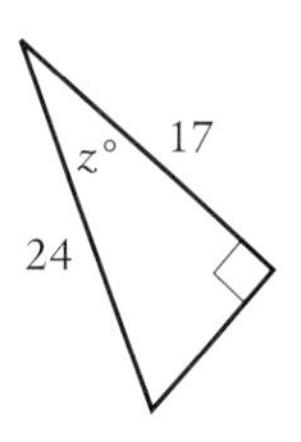

4 Solve the following triangles using the cosine ratio in your first step.

a

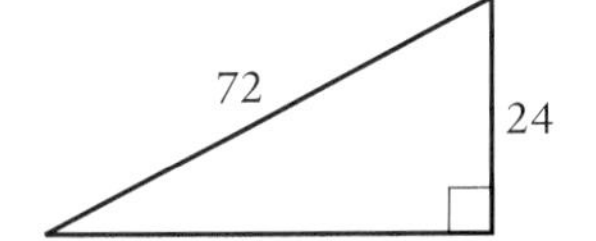

b

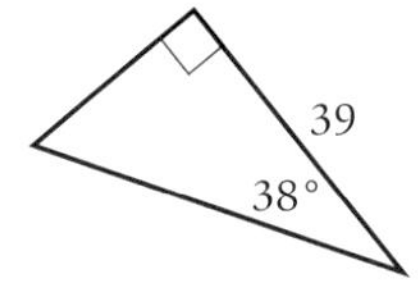

c

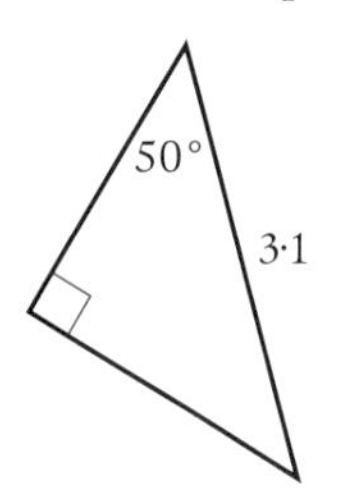

5 The leading edge of a ship's funnel makes an angle of 59° with the deck.

An extension to this edge goes up to a light, L, which is vertically above a point B, 6 m from the front of the funnel.

a Calculate AL, the distance from the foot of the funnel to the light.

b Use Pythagoras' theorem to find the height of the light above the deck.

6 An aircraft maintaining a steady altitude flies 10 km on a bearing of 058°.

a How far north has it travelled during this time?

b It is instructed to alter its bearing to 115° and to maintain this bearing for 15 km.

i How far south will it fly in this leg of its journey?

ii Is it north or south of its original position?

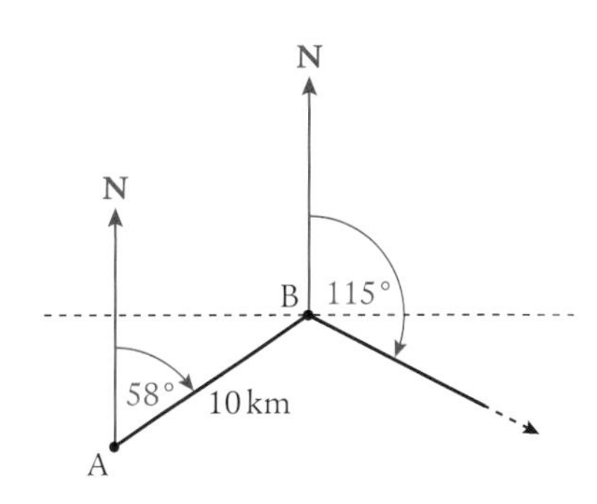

7 At steep hills, road signs indicate the angle the road is to the horizontal.

Earlier we learned that a road sign reading 12% represented an angle of $6{\cdot}8°$.

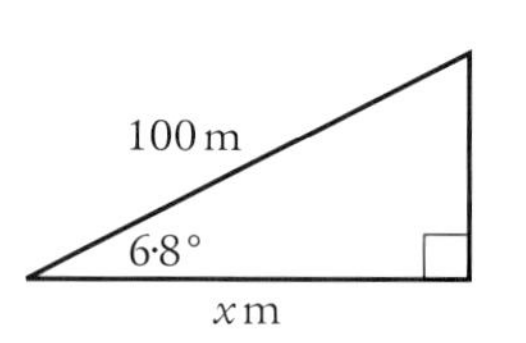

a Calculate for such a hill, how far horizontally you travel when you move 100 m along the road.

b If, instead, the road sign had said 1 : 12, then the road is sloped at an angle of $4{\cdot}8°$.

i If you drive 100 m along this road, what distance horizontally do you go?

ii Is this answer bigger or smaller than the answer to part **a**? Comment.

Exercise 10.4B

1 The rafters of a roof are held rigid by beams and struts.

The struts are at right angles to the rafters and at an angle of 58° to the beam.

The axis of symmetry of the roof is shown.

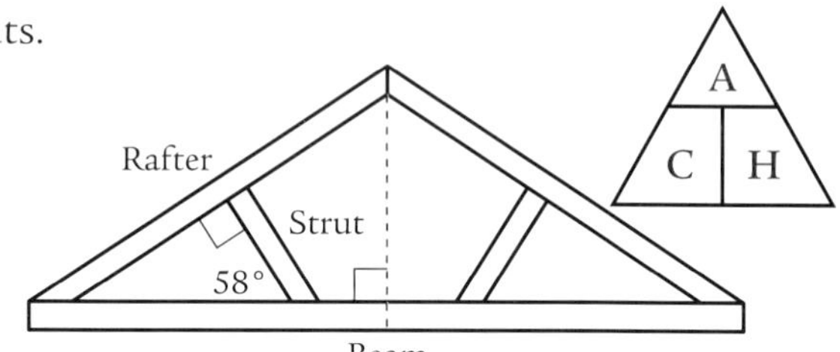

a What is the size of the angle between the beam and the rafter?

b The beam is 10 m long. Use the cosine ratio to calculate the length of the rafter.

c The strut is fixed to the rafter at its midpoint.

Use the cosine ratio to calculate how far along the beam the strut joins it.

2 A rhombus has a perimeter of 20 cm.

Its shorter diagonal has a length of 6 cm.

a Calculate the size of the angles made by the shorter diagonal and the sides of the rhombus.

b Hence find the size of each angle made by the sides and diagonals.

c Calculate the length of the longer diagonal.

3 A boat is dragging at its anchor. The anchor chain is 5 m long.

The boat is 1 m away horizontally from the anchor.

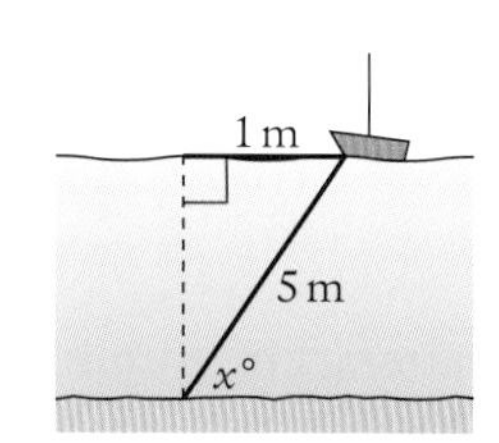

a What angle does the chain make with the sea bed?

b What is the depth of water at this point?

4 The floor of a swimming pool is 25 m long.

The shallow end is 1 m deep and the deep end is 3 m.

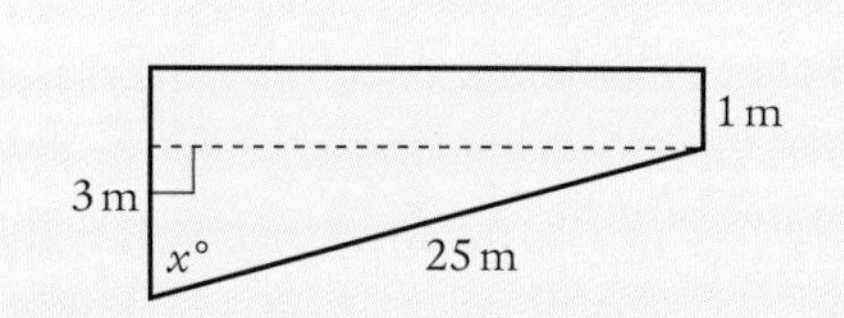

a At what angle is the floor sloping?

b How far do you swim when you do 10 lengths of this pool?

5 In a mining project two vertical shafts were sunk 27 metres apart.

From the bottom of one, a corridor was cut to meet the bottom of the other.

It went down at an angle of 34°.

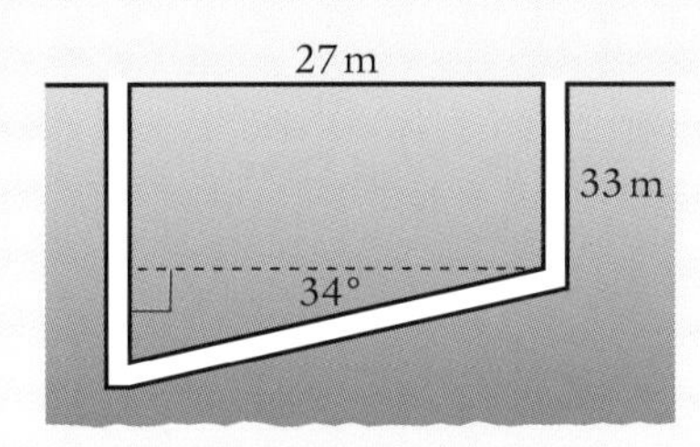

a What was the length of the corridor?

b If the first shaft was 33 m deep, how deep is the second shaft?

6 A circle, centre C, has radii AC and CB of length 10 cm.

The angle between the radii is 60°.

An axis of symmetry is drawn in as a hint.

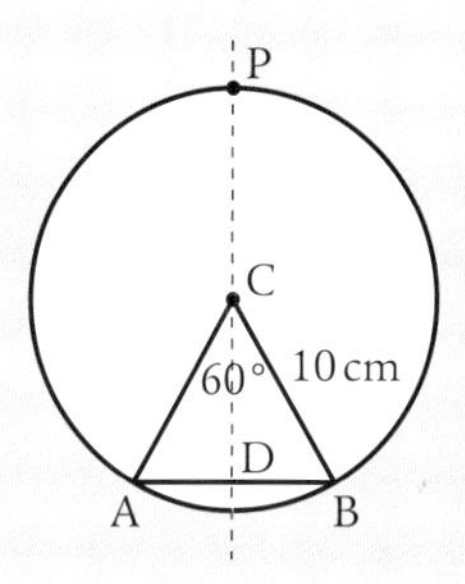

a Calculate the shortest distance between C and the chord AB.

b The cross-section of a small aircraft is circular with a radius of 1·7 m.

The floor can be represented by a chord, as in the diagram above, but the angle ACB is 120°.

Calculate the headroom, DP.

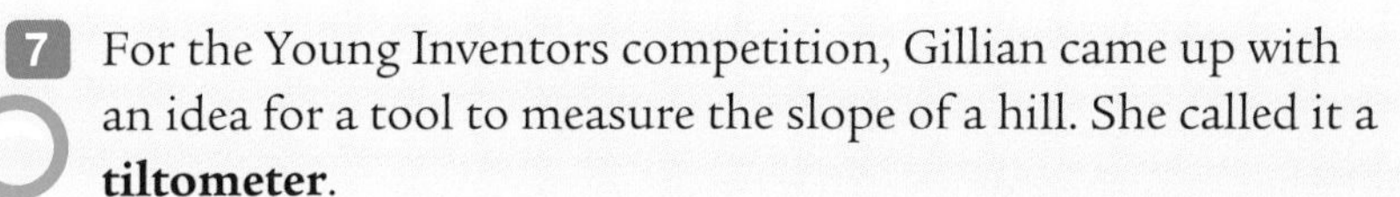

7 For the Young Inventors competition, Gillian came up with an idea for a tool to measure the slope of a hill. She called it a **tiltometer**.

It was a thin see-through cuboid with a square side, half filled with coloured water.

Down two sides of the square the edges were graduated in millimetres.

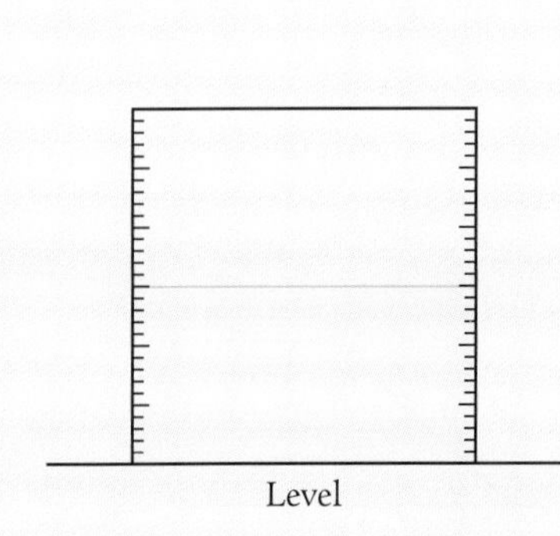

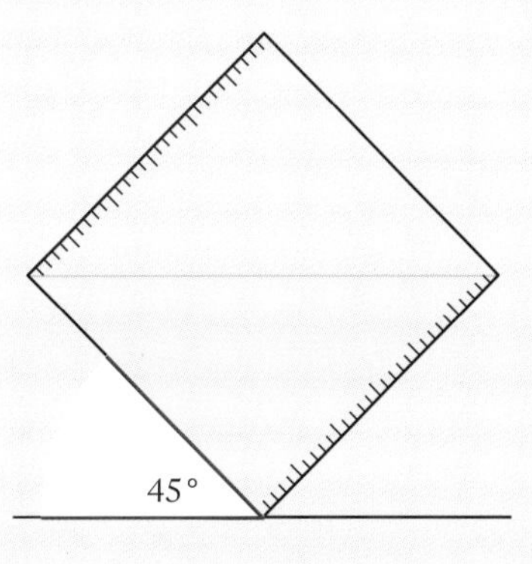

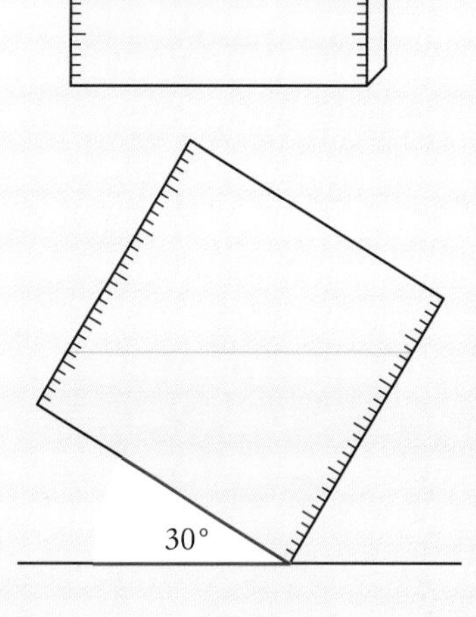

Investigate how the readings on the scale relate to the angle of the slope on which the tiltometer sits.

Bringing it all together

Before we look at preparing for assessment, we should realise that we have to:

- spot which of the three ratios is appropriate
- remember their definitions
- remember, given two parts, what to do with them to find the third part.

Often the mnemonic 'SOHCAHTOA' is used. Note how, taken in groups of three, the letters give the definitions, e.g. SOH means 'Sine is Opposite over Hypotenuse'.

If we put these groups into the triangle memory aids we get a hint at what to do with the parts.

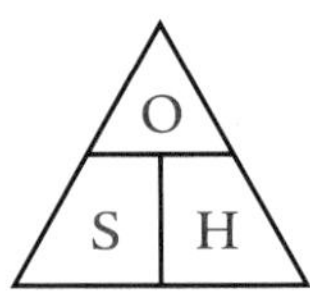

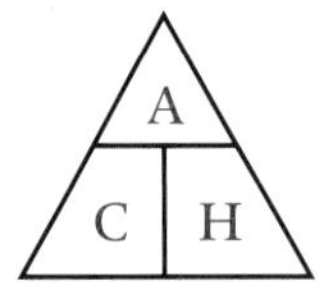

Remember that covering up the part you want gives you a clue as to what to do to find it.

Preparation for assessment

1 Calculate the value of the letter in each diagram.

a

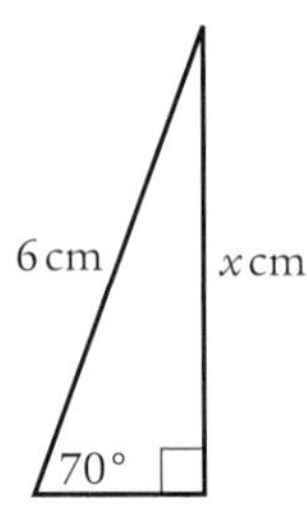

b

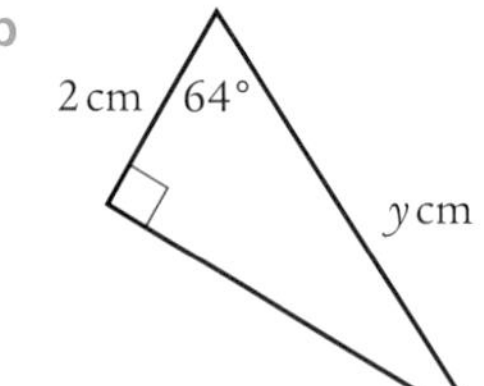

c

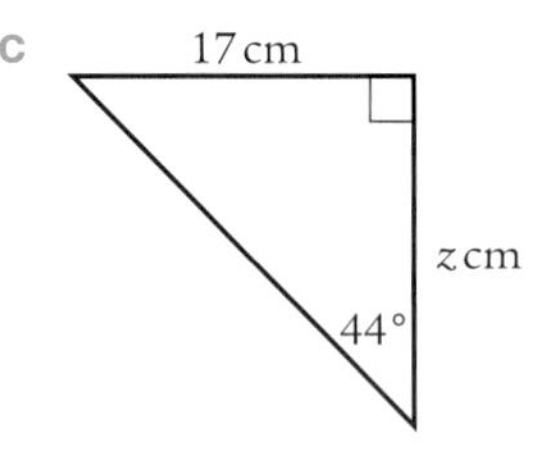

2 Identify a right-angled triangle and then find the value of each letter.

a

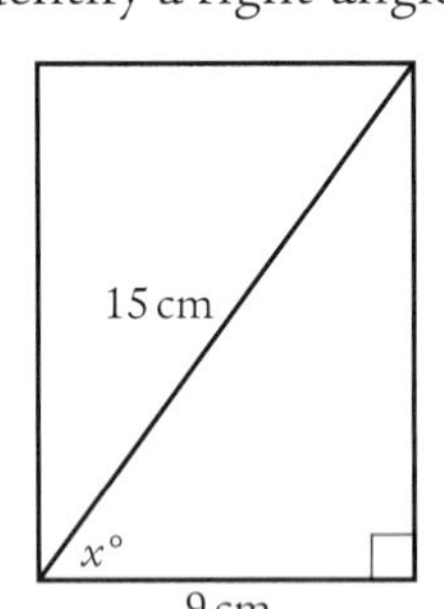

b

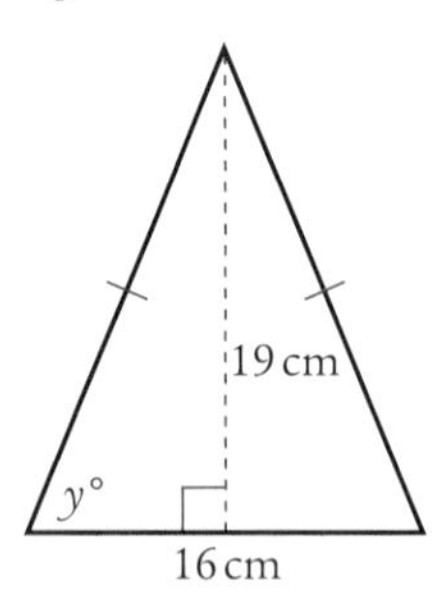

c

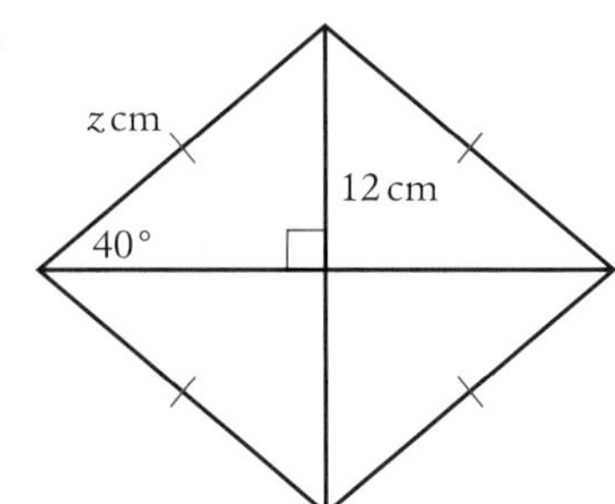

3 Calculate the value of the unknown in each diagram.

a

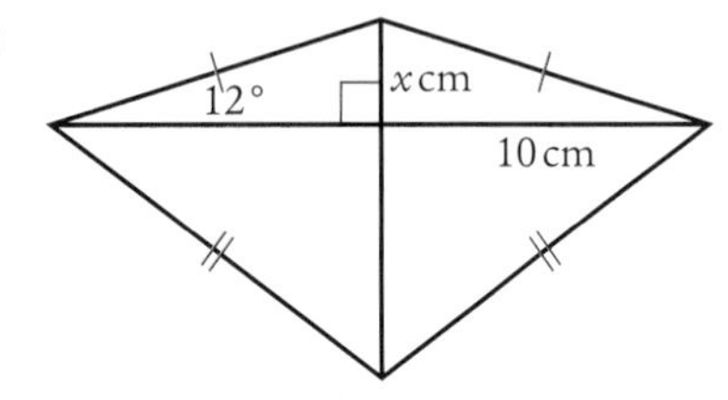

b

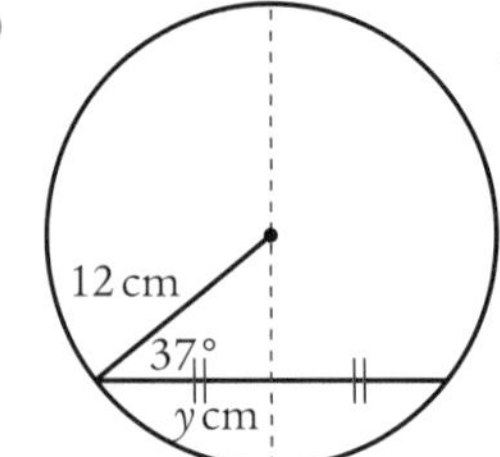

c

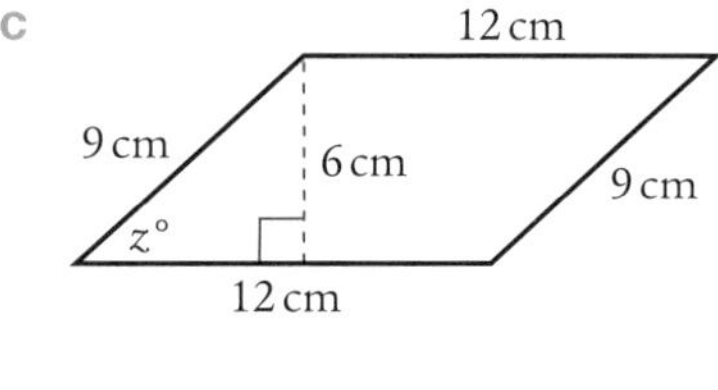

4 An engineering student was examining the Forth Bridge.

He wished to know the angle the top beam made with the horizontal.

He took a photo, placed the picture in a document, drew a line along the beam, and double-clicked.

He was told the line had a height of 1·52 cm and a width of 6·8 cm.

At what angle to the horizontal is the line?

5 A circle has a diameter AB of length 12 cm.

C is a point on the circle.

Chords AC and BC are drawn. BC is 4 cm long.

a What is the size of the angle BCA?

b Calculate the size of the angle BAC?

c Calculate the length of the chord AC.

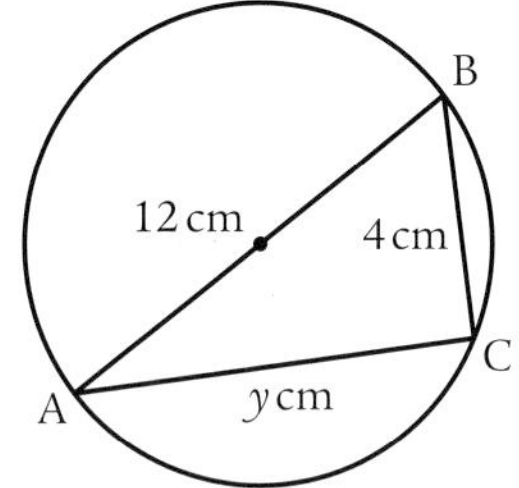
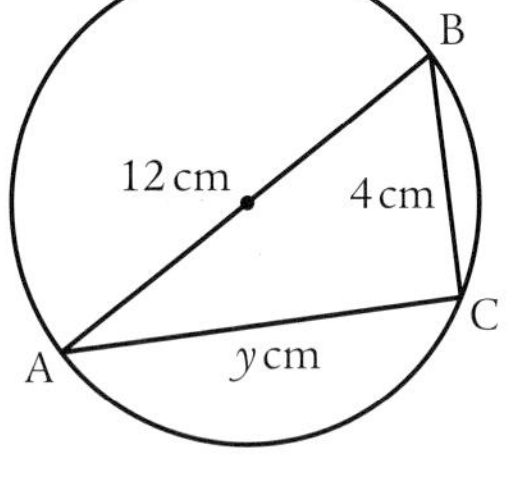

6 A member of the coastguard on top of a 35 m cliff spies a ship drifting towards a reef.

The angle of depression of the ship is measured as 40°.

The angle of depression of the reef is 50°.

a How far is the ship from the base of the cliff?

b How far is the ship from the reef?

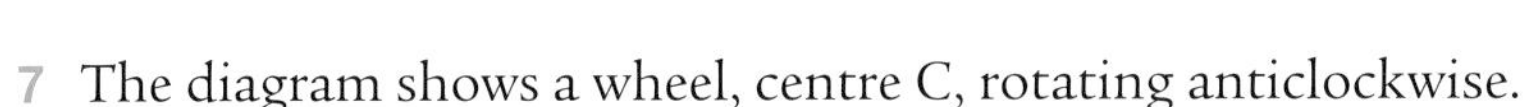

7 The diagram shows a wheel, centre C, rotating anticlockwise.

The point A on the circumference is connected to the point B by a connecting rod.

As the wheel rotates, A goes round in a circle and B is pulled back and forth along the rod CD.

The rod AB is 20 cm long.

When A is at the top of the wheel the angle CAB is 65°.

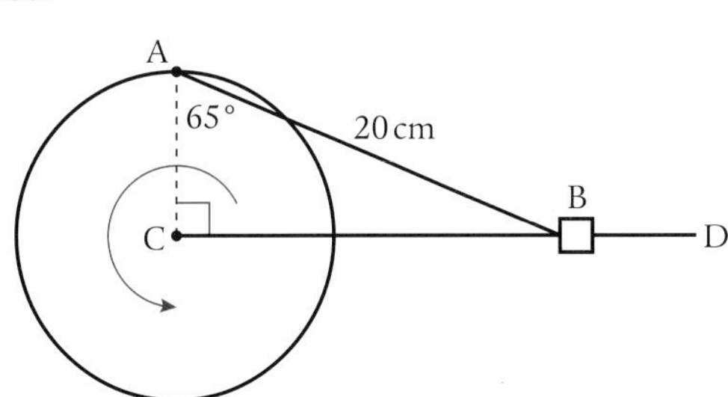

a Calculate the radius of the wheel.

b At another point in the cycle, AB is a tangent to the wheel.

Calculate:

i the angle ABC

ii the distance B is from the centre of the wheel.

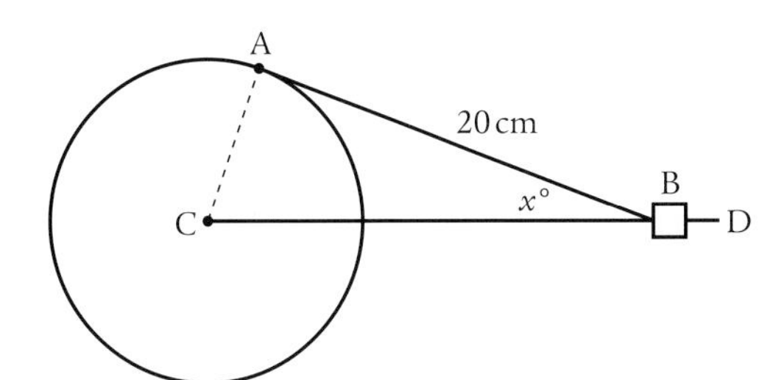

8 Remember the Wallace Monument?

How high above me was the top of the monument?

11 Probability

Before we start...

If you look at the Moon, you will see huge craters that were made when asteroids from outer space slammed into the surface of the Moon.

Scientists track asteroids in case they are heading for Earth.

An asteroid they are presently tracking is about 1·5 km across.

The scientists have calculated it has a 1 in 909 000 chance of colliding with our planet in a few years.

Is a probability of 1 in 909 000 reassuring?

If a collision did happen, what would be the scale of the damage?

Is it a good idea to track asteroids?

The chances of winning the big prize in the lottery are 1 in 13 983 816
... but somebody wins it nearly every week!
Comment!

 What you need to know

1 What is the likelihood of drawing any **one** of these cards from a normal pack of 52 playing cards, if you are only allowed one pick?

2 The graph shows the number of times that four horses have won a race.

a Which of them do you think would be most likely to win the next race they run in?

b What other information would you find useful before making your choice?

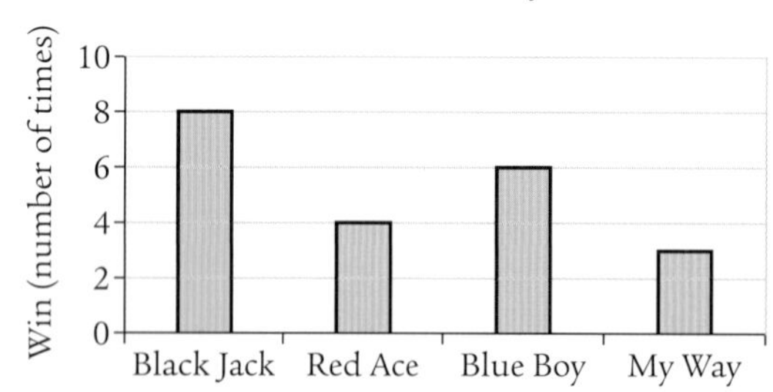

3 A coin is tossed nine times.
The recorded results were H T H H T H H H H.

a Would you expect the next result to be H or T?

b Explain your answer.

4 When two dice are thrown at the same time, the total can be anything from 2 to 12.

a Are you more likely to score a total of 7 or a total of 4?

b Explain your reasoning.

5 The exam results for ten school students who sat maths and physics exams are recorded in the table.

Maths	7	5	3	8	7	8	9	5	6	4	9	6
Physics	8	6	4	4	8	5	10	6	5	7	8	7

A scatter graph to represent this data has been drawn.

a A line of best fit has been added to the graph. Do you agree with it?

b A student scored 7 in physics. Estimate his maths mark.

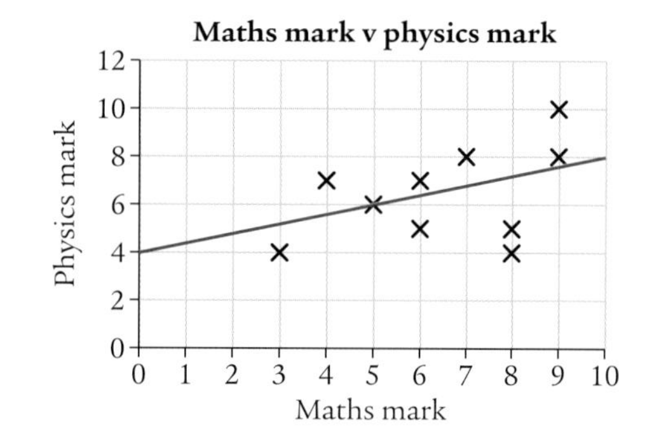

6 The profits, in £M over the last three years, for two companies are shown in the graphs below:

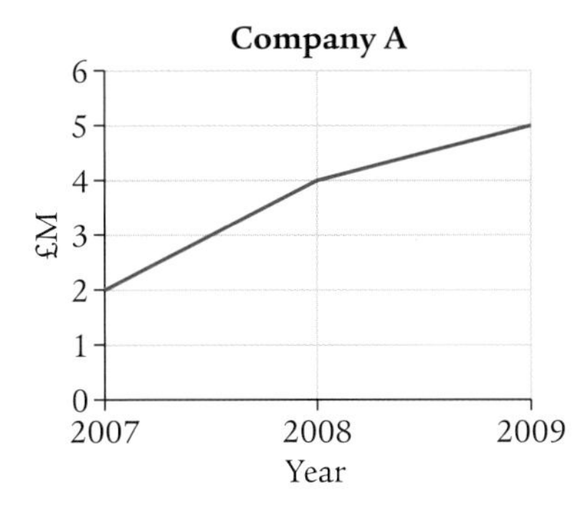

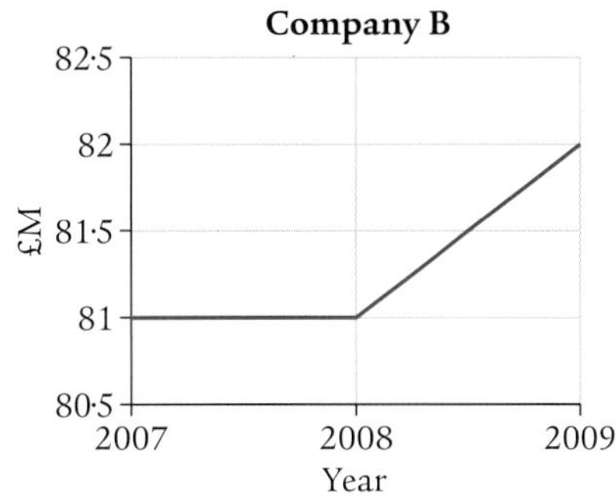

a Which of the two companies might you invest in?

b What other research would you undertake before making your decision?

c How might probability affect your decision?

7 The likelihood of an event can be illustrated on a **probability line**.

One end of the line represents impossibility (0), the other certainty (1).

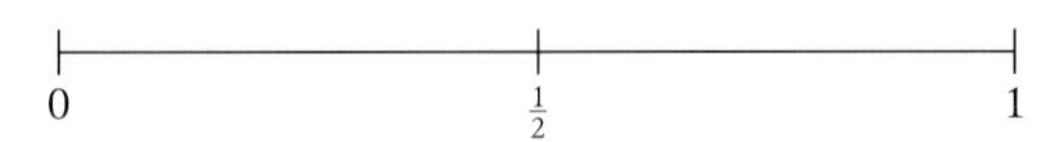

Draw a probability line and indicate on it the likelihood of:

a getting a head when you toss a coin

b drawing a red six from a pack of cards

c being struck by lightning

d you watching TV tonight

e you enjoying mathematics (come on, you know it is closer to 1).

11.1 Simple probability

Probability theory is the mathematics that is used to measure likelihood.

It uses some words in a special way.

An **experiment** or **trial** is an occurrence, like cutting cards, tossing coins, crossing the road.

This experiment will have a list of **possible outcomes**.

- When cutting cards there are 52 possible outcomes.
- When tossing a coin there are 2 possible outcomes, heads or tails.
- When crossing the road there are many possible outcomes including crossing safely, getting knocked down, etc.

An **event** is a set of some of the outcomes ... we call these the **favourable outcomes** of the experiment.

- Wishing a king when cutting cards? ... 4 of the outcomes are favourable.
- Wanting a head when tossing a coin? ... 1 of the outcomes is favourable.
- Getting knocked down when crossing the road? ... how many ways can you get knocked down?

We define the **theoretical probability** of an event, written P(event), by the formula:

$$P(\text{event}) = \frac{\text{number of favourable outcomes}}{\text{number of possible outcomes}}.$$

This only holds when each possible outcome **is equally likely.**

Note: the **favourable outcomes** AND the **not-favourable outcomes** make up **all possible outcomes**

... and so **P(event) + P(not event) = 1**

If the probability of getting a King, P(King) $= \frac{4}{52}$

then the probability of not getting a King, P(not King) $= 1 - \frac{4}{52} = \frac{48}{52}$.

Example 1

Geoff bought a cactus in June 2015. He was told that it would flower sometime in September.

September 2015						
Sun	Mon	Tue	Wed	Thu	Fri	Sat
		1	2	3	4	5
6	7	8	9	10	11	12
13	14	15	16	17	18	19
20	21	22	23	24	25	26
27	28	29	30			

a What is the probability that it will flower at the weekend?

b Calculate:

i P(flowers on Wednesday) **ii** P(doesn't flower on Wednesday).

c His mother's birthday is 27th September.

What are the chances that it will flower after that?
Express your answer as a percentage.

a Number of favourable outcomes = 8 (4 Saturdays and 4 Sundays).

Number of possible outcomes = 30.

P(flowers at weekend) = $\frac{8}{30} = \frac{4}{15} = 0{\cdot}27$ (to 2 s.f.).

b Number of favourable outcomes = 5 (count the Wednesdays).

Number of possible outcomes = 30.

i P(flowers on Wednesday) = $\frac{5}{30} = \frac{1}{6} = 0{\cdot}17$ (to 2 s.f.)

ii P(doesn't flower on Wednesday) = $1 - \frac{1}{6} = \frac{5}{6} = 0{\cdot}83$ (to 2 s.f.).

c Number of favourable outcomes = 3 (28th, 29th or 30th).

Number of possible outcomes = 30.

P(flowers after 27th) = $\frac{3}{30} = \frac{1}{10} = 10\%$.

Example 2

The supermarket 'BigSave' declared that the owners of yellow cars in the car park would receive a £20 token to spend in the supermarket.

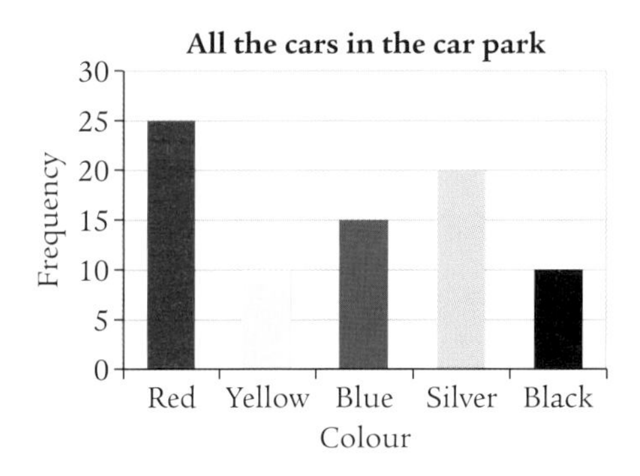

The chart shows the result of a count of the cars and their colours in the car park at that time.

a What is the probability that one of the car owners chosen at random as he goes into the car park will have won a £20 token?

b What is the probability that he won't be a winner?

a Number of favourable outcomes (yellow cars in the car park) = 10.

Total number of outcomes (all cars in the car park) = 25 + 10 + 15 + 20 + 10 = 80.

P(winner) = $\frac{10}{80} = \frac{1}{8} = 0{\cdot}125 = 12{\cdot}5\%$

(Note: any one of the three forms, common fraction, decimal fraction, or percentage, would do unless you're asked for a particular form.)

b P(not winner) = $1 - \frac{1}{8} = \frac{7}{8} = 0{\cdot}875 = 87{\cdot}5\%$.

Exercise 11.1A

1 The first nine prime numbers are written on nine identical blank cards.

They are shuffled and then placed face down.

What is the probability that a card chosen at random will be:

- **a** an even number
- **b** an odd number (not even)
- **c** a number less than 7
- **d** a number greater than 23?

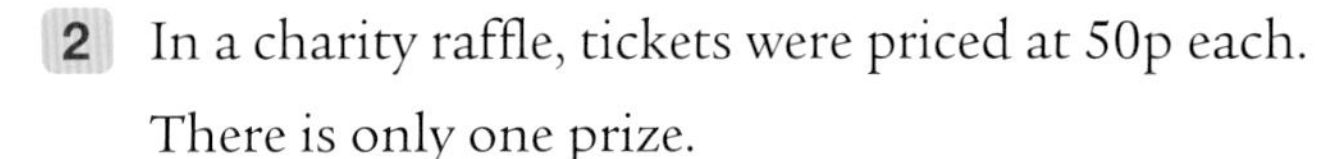

2 In a charity raffle, tickets were priced at 50p each.

There is only one prize.

Anna spent £2 on tickets and Alice spent £4 on tickets.

The total tickets sales came to £120.

- **a** What is the probability that Anna wins?
- **b** What is the probability of Alice being successful?
- **c** What do you notice about the two probabilities in relation to how much they spent?

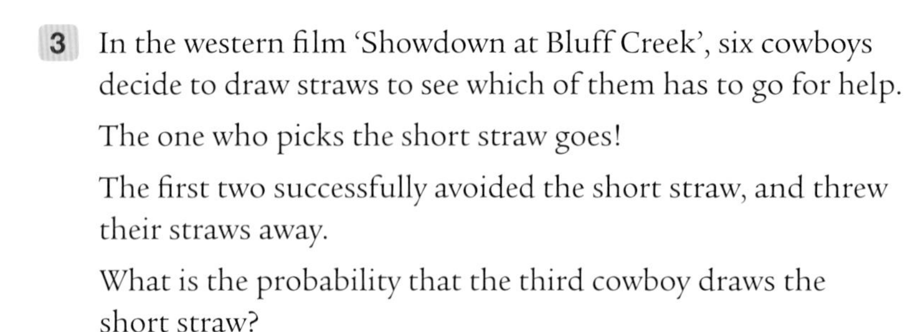

3 In the western film 'Showdown at Bluff Creek', six cowboys decide to draw straws to see which of them has to go for help.

The one who picks the short straw goes!

The first two successfully avoided the short straw, and threw their straws away.

What is the probability that the third cowboy draws the short straw?

4 A group of workers were asked what they earned in a week.

The information is shown in the chart.

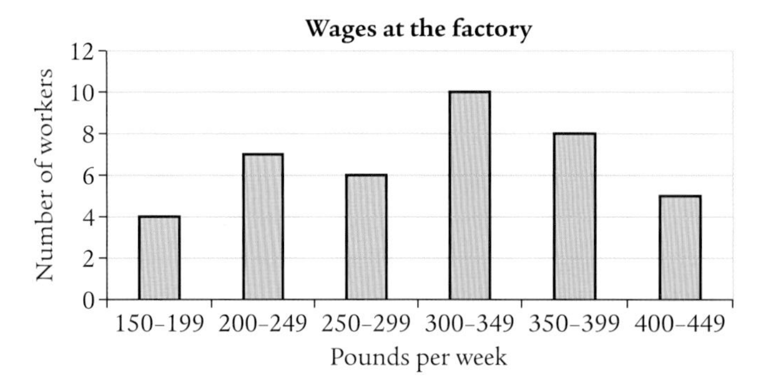

What is the probability that one of the workers, chosen at random, said:

- **a** between £200 and £249
- **b** more than £349?

5 The results of a survey into the percentage of time given to different types of TV programmes are illustrated in the pie chart.

A viewer selects a time of day at random to watch TV.

What is the probability that it will be when:

a an arts programme is showing

b a programme that is **not** sport is showing?

Express the probabilities both as a common fraction and as a percentage.

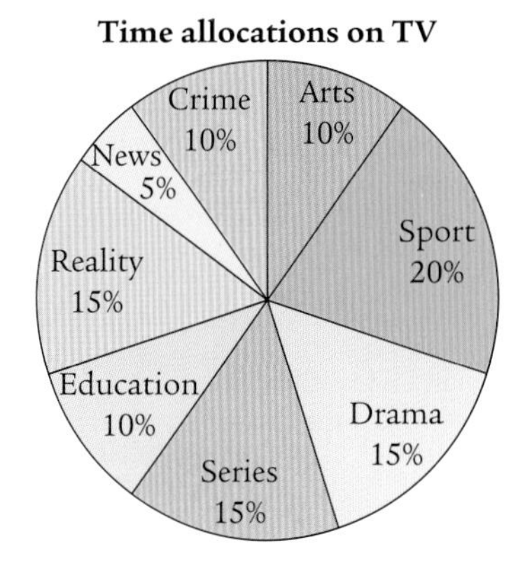

6 The Edinburgh to Glasgow train runs every 15 minutes all day.

What is the probability that if you turned up at any random time, you would need to wait 10 minutes or more for the next train?

(Hint: draw a timeline showing the 15 minutes between trains.)

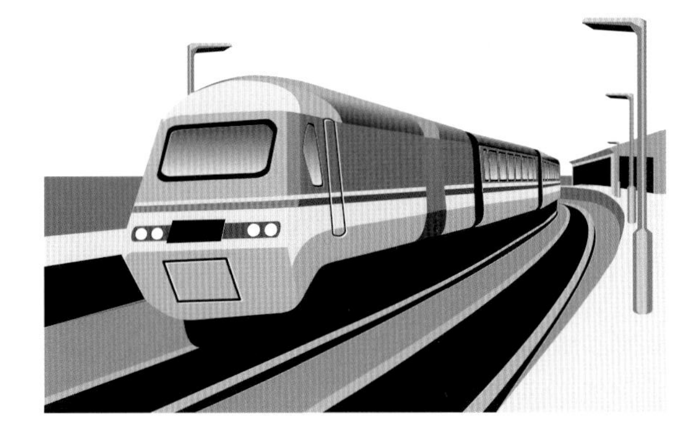

Experimental probability

Sometimes we can't count all the possible outcomes nor establish how many of the outcomes are favourable ... calculating the theoretical probability is beyond us.

However, in order to get an estimate of the likelihood of an event, we often do surveys or experiments and make the assumption that, if we take a big enough sample, the relative frequency of an event can be used as the probability of it happening.

Example 3

On average there are 14 deaths and 1200 major injuries a year to people in the UK caused by falls from ladders. Assume that there are 60 000 000 people in the UK.

Estimate the probability that a person named at random will, next year:

a sustain a serious injury by a fall from a ladder

b die from a fall from a ladder.

a $\text{P(serious injury)} = \frac{1200}{60\,000\,000} = \frac{1}{50\,000} = 0{\cdot}00002.$

b $\text{P(death)} = \frac{14}{60\,000\,000} \approx \frac{1}{4\,285\,714} \approx 0{\cdot}0000002.$

Example 4

Michael used an App on his phone to simulate the roll of a dice.

The frequency with which each number 1 to 6 occurred was recorded.

Score	1	2	3	4	5	6
Frequency	4	6	7	5	2	6

a What is the **theoretical** probability, when rolling a dice, that the score will be:

i 4 ii 5?

b Using the data in the table, what is the **experimental** probability, when using the App, that the score will be:

i 4 ii 5?

c How closely does the App simulate the throw of a dice?

a In theory: i $P(4) = \frac{1}{6}$ ii $P(5) = \frac{1}{6}$.

b In experiment: i $P(4) = \frac{5}{30} = \frac{1}{6}$ ii $P(5) = \frac{2}{30} = \frac{1}{15}$.

c In theory, $P(5) = 0{\cdot}167$ (to 3 d.p.), in experiment (using the App), $P(5) = 0{\cdot}067$.

The App is biased against scores of 5.

Exercise 11.1B

1 The table shows the percentage of pedestrians who die as a result of being struck by cars moving at different speeds.

Car speed (mph)	20	30	40	50
Pedestrian fatalities (%)	5	40	80	99·9

Estimate the probability (giving your answer as a common fraction) that:

a a pedestrian, involved in an accident at 40 mph, dies

b a pedestrian, involved in an accident at 40 mph, lives

c a pedestrian is involved in an accident at 30 mph and survives

d a pedestrian, being hit at 50 mph, survives the accident.

2 The manufacturer of energy-saving light bulbs is confident that 95% of his bulbs will last at least 6000 hours.

a Estimate the probability that a new bulb:

i will last 6000 hours or more

ii will burn out in less than 6000 hours.

b He is confident that 20% of the bulbs will last beyond 15 000 hours.

What is the probability of a bulb not lasting for 15 000 hours?

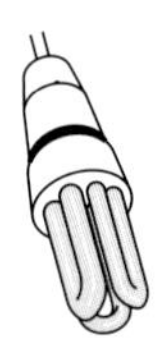

c For every 100 hours beyond 15 000 hours, experiments have shown that another 1% burn out. (There's 19% remaining after 15 100 h, 18% after 15 200 h, etc.)

What is the probability that a bulb will last 16 500 hours?

3 In a round of golf there are 18 holes.

You count how many strokes it takes you to get the ball down each hole.

The bar chart shows **some** of the scores a golfer recorded during her round.

The bars for scores of 4 and 5 are obscured.

She had no holes-in-one.

Golf scores

Frequency: 0 1 2 3 4 5 6 7

Strokes per hole: 1 2 3 4 5 6

a Estimate the probability of this golfer getting a hole-in-3.

b She said that, using the graph, the probability of her getting a hole-in-5 was $\frac{1}{9}$.

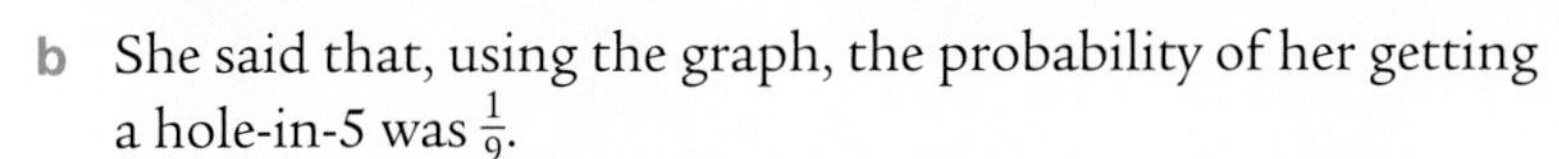

i How many 5s did she have?

ii How many 4s did she have?

c What was the probability that at a hole chosen at random she scored a 4?

4 A family of chemical compounds, known as saturated hydrocarbons, consists of hydrogen and carbon atoms linked in chains.

The simplest is methane with one carbon atom (CH_4).

The next in the family is ethane (C_2H_6) with two carbon atoms linked to six hydrogen atoms.

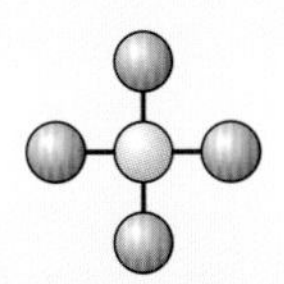

1 carbon atom

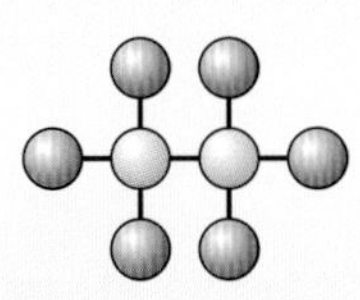

2 carbon atoms

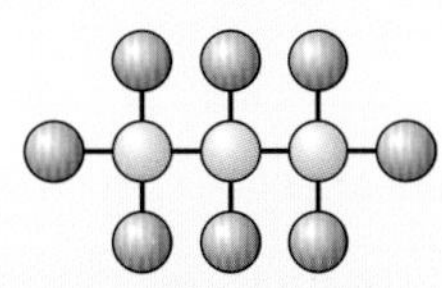

3 carbon atoms

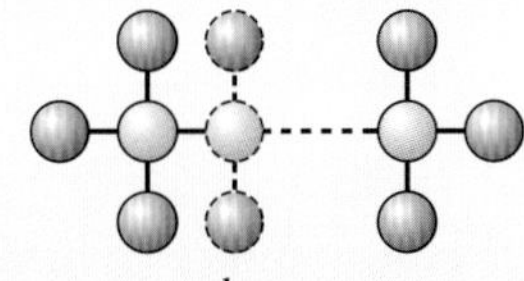

n carbon atoms

a If an atom is targeted at random in an ethane chain, what is the probability it will be a carbon atom?

b Hexane has a chemical formula C_6H_{14}.

What is the probability that an atom chosen at random in a hexane chain will be a hydrogen atom?

c A hydrocarbon chain has n carbon atoms.

i How many hydrogen atoms will it have?

ii What is the probability that an atom chosen at random in this chain will be a carbon atom?

11.2 Linking probabilities

Suppose the probability of getting knocked down while crossing a road at a particularly bad spot was 1 in 100.

Suppose that you manage to successfully cross it 99 times. Does that mean you'll be knocked down the next time you cross?

No. Each crossing is **independent** of the one before it. The outcome of one does not affect the outcome of the other.

Suppose two people were cutting cards and the first person kept the card he cut.

Then the second person's cut is **not** independent of the first. If the first person picked an ace then the second person has one less chance of picking an ace.

In what follows, check that each of the series of events described in the examples are independent.

Example 1

A dice is cast and a coin tossed at the same time.

What is the probability of getting a tail **and** a four?

We need to list all possible outcomes ... and then count the favourable ones.

When two events are considered, it is often useful to make a table:

		Dice					
		1	2	3	4	5	6
Coin	Head (H)	(H, 1)	(H, 2)	(H, 3)	(H, 4)	(H, 5)	(H, 6)
	Tail (T)	(T, 1)	(T, 2)	(T, 3)	(T, 4)	(T, 5)	(T, 6)

There are 12 possible outcomes and only one favourable one, (T, 4).

This means that P(T **and** 4) $= 1 \div 12 = \frac{1}{12}$.

Example 2

Liam says he can toss a coin three times and get three heads.

What is the probability of him managing this?

On the first toss, there are two possible outcomes H and T.

Whatever he gets on the first toss, his second toss could still be H or T. (Independent events!)

Whatever he gets on the first two trials, his third toss could give H or T.

This information can be shown on a tree diagram.

From the diagram, you can see there are eight different outcomes to tossing a coin three times.

Only one is favourable ... H, H, H.

P(H **and** H **and** H) $= \frac{1}{8} = 0{\cdot}125$.

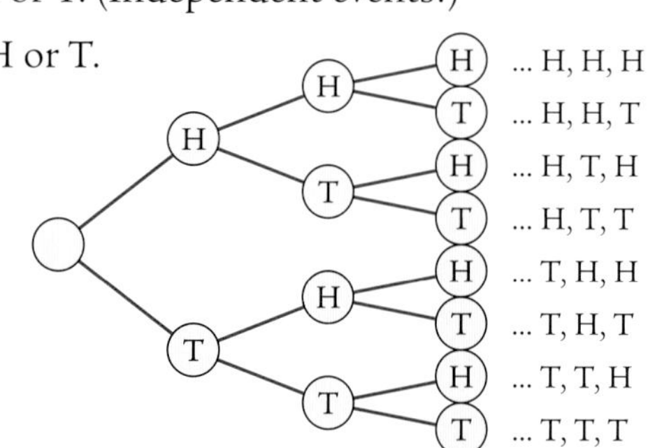

Example 3

A motorist drives through a town centre, which has two sets of traffic lights.

The probability of getting stopped at the first set is $\frac{1}{5}$.

The probability of getting stopped at the second is $\frac{3}{7}$.

a What's the probability of being stopped by both sets, given that the sets of lights work independently?

b Calculate the probability that:

i the motorist won't be stopped by the first set

ii he won't be stopped by the second set

iii he'll make it through the town centre without being stopped.

a Notice that the product of the denominators in the probabilities is $5 \times 7 = 35$.

So imagine 35 trips through the town centre.

$\frac{1}{5}$ of them get stopped at the 1st set $= \frac{1}{5} \times 35 = 7$ ... so 7 possible favourables.

$\frac{3}{7}$ of these get stopped at the 2nd set $= \frac{3}{7} \times 7 = 3 = 3$ favourable outcomes.

From 35 trips we get 3 favourable outcomes: P(stopped at both sets) $= \frac{3}{35}$.

b **i** P(not stopped by 1st set) $= 1 - \frac{1}{5} = \frac{4}{5}$.

ii P(not stopped by 2nd set) $= 1 - \frac{3}{7} = \frac{4}{7}$.

iii Again imagine 35 trips through the town centre.

$\frac{4}{5}$ of them are not stopped at the 1st set $= \frac{4}{5} \times 35 = 28$ ... so 28 possible favourables.

$\frac{4}{7}$ of these are not stopped at the 2nd set $= \frac{4}{7} \times 28 = 16$ favourable outcomes.

From 35 trips we get 16 favourable outcomes: P(not stopped by either set) $= \frac{16}{35}$.

Note: in both parts **a** and **b** the final probability can be calculated directly by multiplying the probabilities together.

This can be summarised by the rule: P(A **and** B) = P(A) × P(B).

Remember that this only holds when the events A and B are independent.

Exercise 11.2A

1 The game of backgammon requires two dice.

The total of the two dice plays an important role in the game.

a Copy and complete the table to show the totals of all possible outcomes.

b From the diagram work out the probability of:

i scoring a total of 7

ii scoring 2

iii getting a double

iv scoring more than 8

v not scoring a total of 6.

		Blue dice					
		1	2	3	4	5	6
Red dice	1						
	2						
	3						
	4						
	5						
	6						

2 A coin is tossed three times.

a What is the probability of getting two heads and a tail occurring in any order?

b What is the probability of getting a head, then a head, then a tail?

c What is making a difference to the answer in parts **a** and **b**?

3 At the local fête, to win a coconut you have to draw a card, from a normal pack, that scores higher than 9 (ace counts high) then cast a die and roll higher than 4.

a What is the probability of drawing a card that scores higher than 9?

b What is the probability of the dice showing more than 4?

c By considering 39 ideal attempts or otherwise, calculate the probability of winning a coconut.

d If it costs £1 to enter, what comment would you make?

(Hint: consider how many coconuts you win and what it costs for 39 ideal attempts.)

4 A Square Go

The first six square numbers are written down on identical blank cards.

They are then shuffled and placed in the lucky dip bag.

A card is picked, the number noted, and replaced.

A second card is picked and the number is added to the first.

		Second pick					
		1	4	9	16	25	36
First pick	1						
	4						
	9						
	16						
	25						
	36						

To win a top prize you must score more than fifty.

a What is the probability of winning a top prize?

b If you don't win the top prize, you can still win an intermediate prize if your score is more than forty.

What is the probability of winning an intermediate prize?

c The stall holder wishes to offer a minor prize, which has a probability of $\frac{1}{6}$.

The poster at the stall will read, 'X or more ... you mustn't score!'.

What number should replace 'X'?

5 At the 'Losing your Marbles' stall there are two rectangular-based tins.

The bases are painted with red shapes on a yellow background.

You drop a marble into the tin, letting it bounce all over before coming to rest.

To score a point the marble must come to rest on a yellow area.

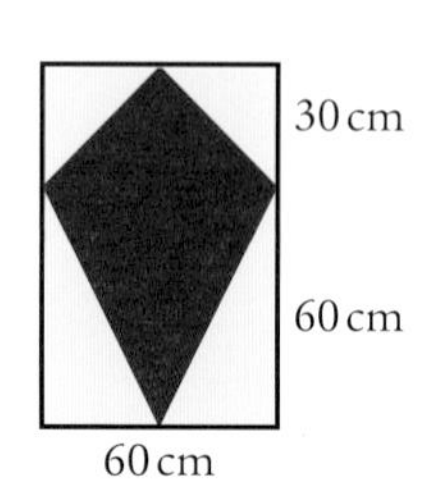

Assume all points on the target have an equal chance of being hit.

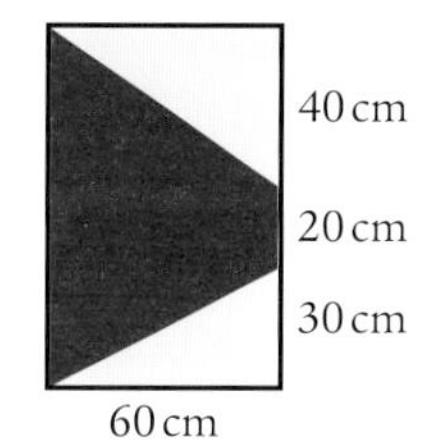

a What is the probability of scoring a point on the kite target? (Hint: consider areas.)

b You can score three points by scoring on the kite with your first marble, and then scoring on the trapezium with a second.

What is the probability of completing this combination?

6 An industrial chemist wants to make a new shade of green that will be used to promote an advertising campaign for 'Summer-fresh Paints'.

There are eight shades of blue, labelled 1 to 8, and five shades of yellow, labelled A to E.

He wants to mix **one of each** to give different shades of green, labelled 1A, 2E, etc.

The company is going to choose one of the shades to call 'Summer-fresh Green'.

a What is the probability that the shade 3E will be chosen?

b What is the probability that Blue 1 will be used in the chosen shade?

c What is the probability that the new shade will be Blue 3 and either Yellow C or Yellow D?

When events are not independent

Consider three cowboys drawing straws.

There are three straws ... one shorter than the rest. Whoever draws the short straw loses.

As each cowboy draws his straw, he keeps it ... which changes the conditions for the next cowboy:

Adam picks from three straws. If he doesn't lose, probability $\frac{2}{3}$, then Ben picks.

Ben picks from two straws. If he doesn't lose, probability $\frac{1}{2}$, then Chuck picks.

Chuck picks from one straw. He is certain to get the short straw.

It is easier to see in a **tree diagram**:

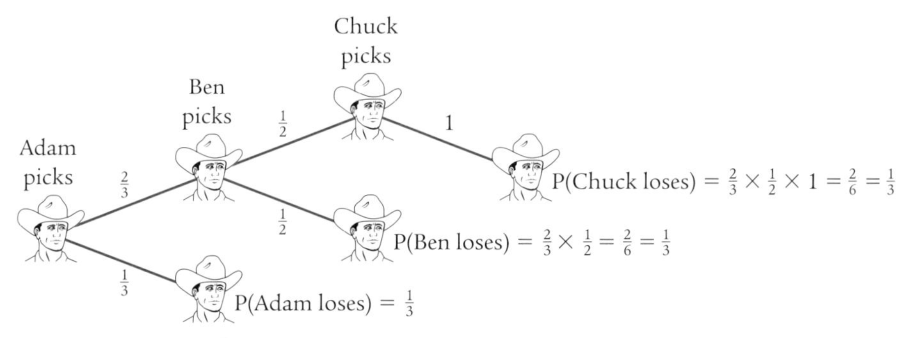

The downward lines indicate getting the short straw ... and the game is over.

The upward lines indicate not getting the short straw ... and the next person picks.

You can find the probability of getting to any point on the tree by multiplying the probabilities marked on each branch of the pathway to that point.

Example 4

To escape his captors, Special Agent James Bend will need to break out of the fortified walls and then swim the shark-infested moat.

Past attempts suggest that only one in eight people manages to get over the walls.

Independently, only 10% survive the moat.

a Calculate the probability of getting over the wall and surviving the sharks.

b Calculate the probability of getting over the wall but not surviving the sharks.

c Comment on the possibility of surviving the sharks but not getting over the wall.

Draw a probability tree:

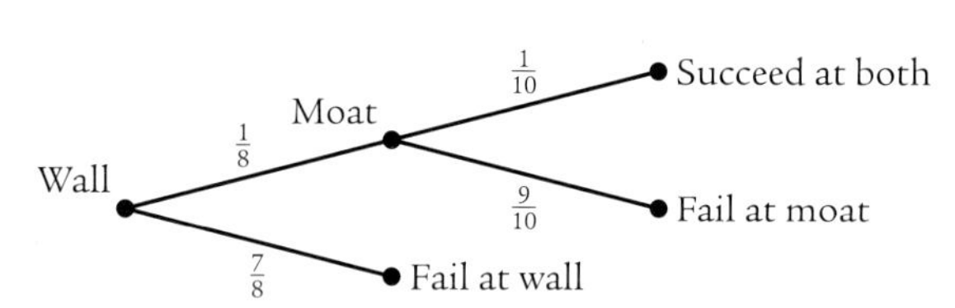

a Probability of surviving both $= \frac{1}{8} \times \frac{1}{10} = \frac{1}{80}$.

b Probability of getting over the wall but not surviving the sharks $= \frac{1}{8} \times \frac{9}{10} = \frac{9}{80}$.

c The situation will never arise ... if he fails at the wall.

Exercise 11.2B

1 For students applying for drama school, only 1 in 165 is accepted.

Three-quarters graduate successfully.

The probability that a registered actor is in work is 1 in 20.

You meet a friend whom you haven't seen for years.
The last time you saw him, he was applying for drama school.

a What's the probability that he graduated and is a working actor?

b What's the probability that he graduated but isn't working at the moment?

c What's the probability that he was accepted but didn't graduate?

2 Four out of five households in the UK have internet access.

Of this group, 60% say they also regularly use mobile devices to access the internet.

What is the probability that a person chosen at random in the UK

a accesses the internet using mobile devices

b has internet access but doesn't use a mobile device

c doesn't have internet access?

3 The ace, 2 and 3 of hearts are shuffled and placed face down on the table.

They are turned over one at a time.

The tree diagram shows all the possible outcomes.

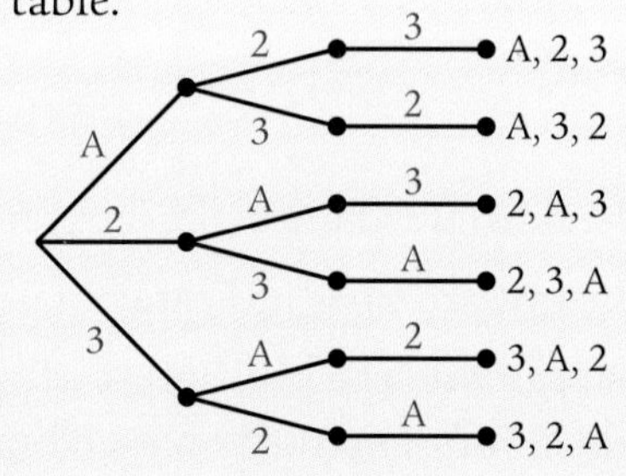

a Make a sketch of the tree and place the corresponding probability against each branch.

b What is the probability that the cards are turned over in the order A, 2, 3?

c What is the probability that the ace is the last card turned over?

d What is the probability that the ace is turned over before the 2?

4 Identical discs are numbered 1, 2, 3 and 4 and then put in a bag.

A gambler claims he can draw 1, then 2, then 3 and then 4, in that order.

When a disc is drawn, it is not replaced.

It has been said that the probability of him achieving this is given by:

$P(1, 2, 3, 4 \text{ in order}) = \frac{1}{4} \times \frac{1}{3} \times \frac{1}{2} \times \frac{1}{1} = \frac{1}{24}$.

a Draw a tree diagram to show that this is the case.

b The ambitious gambler claims he can draw the numbers 1 to 6 in order. What is the probability of him doing this?

c Comment on the effect that adding two extra numbers has on the probability.

d Draw a graph of how the probability of drawing the numbers in order changes as the numbers in the bag changes ... start with 1 only and then 2, 3 and 4.

11.3 What do you expect?

If the probability of a particular outcome of an experiment is $\frac{1}{3}$, we expect this outcome to occur a third of the time.

If we perform the experiment 24 times, then we would expect the particular outcome to occur 8 times.

The number of times a particular outcome is expected to happen is referred to as its **expected frequency ... E(outcome).**

This can be expressed as **E(outcome) = P(outcome)** $\times$ **number of trials.**

Example 1

If a standard pack of cards is cut 26 times, how many times would you expect the following cards to appear?

a An ace. b A diamond. c A black queen.

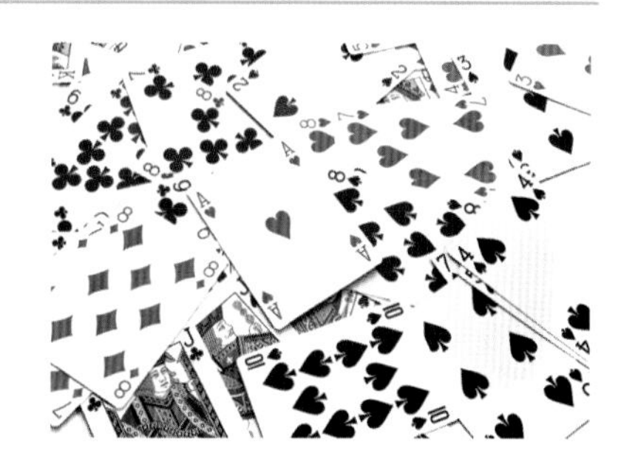

a $P(\text{ace}) = \frac{4}{52} = \frac{1}{13}$.

Number of times pack is cut = 26.

$\Rightarrow \quad E(\text{ace}) = \frac{1}{13} \times 26 = 2$.

b $P(\text{diamond}) = \frac{13}{52} = \frac{1}{4}$.

$\Rightarrow \quad E(\text{diamond}) = \frac{1}{4} \times 26 = 6{\cdot}5$ (either 6 or 7).

c $P(\text{black queen}) = \frac{2}{52} = \frac{1}{26}$.

$\Rightarrow \quad E(\text{black queen}) = \frac{1}{26} \times 26 = 1$.

Example 2

A maker of TVs is confident that 85% of the TVs made will function without fault for at least five years.

a In a batch of 3000 TVs, how many are expected to be working still in five years' time?

b If you buy one of these TVs, what is the probability that it breaks down before the five years is up?

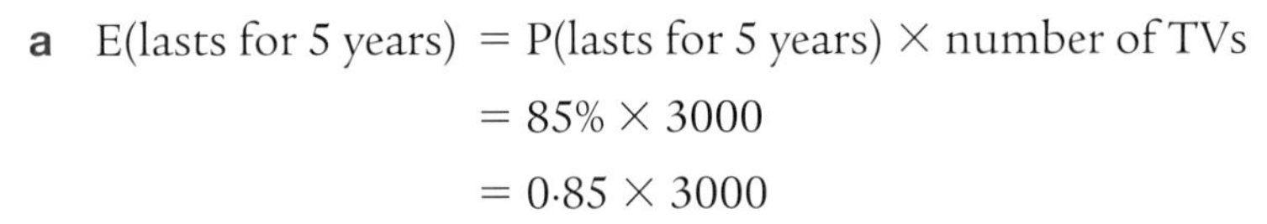

a E(lasts for 5 years) = P(lasts for 5 years) × number of TVs

= 85% × 3000

= 0·85 × 3000

= 2550

It is expected that 2550 TVs will still be working in five years' time.

b P(fails before 5 years) = 1 − P(lasts for 5 years)

= 1 − 0·85

= 0·15

The probability that the TV breaks down before the five years is up is 0·15.

Exercise 11.3A

1 The spinner with numbers 1 to 4 is spun 28 times.

How many 3s would you expect to occur?

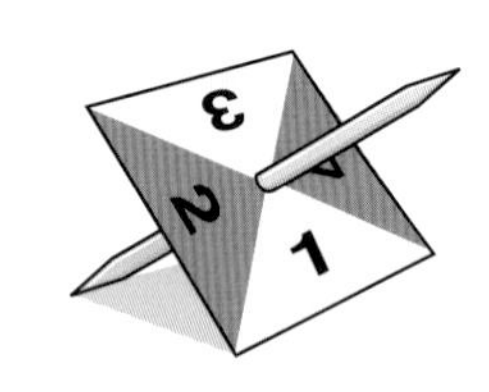

2 A pack of cards is cut 20 times.

How many times would you expect to see a spade result?

3 There are 28 dominoes in a standard box.

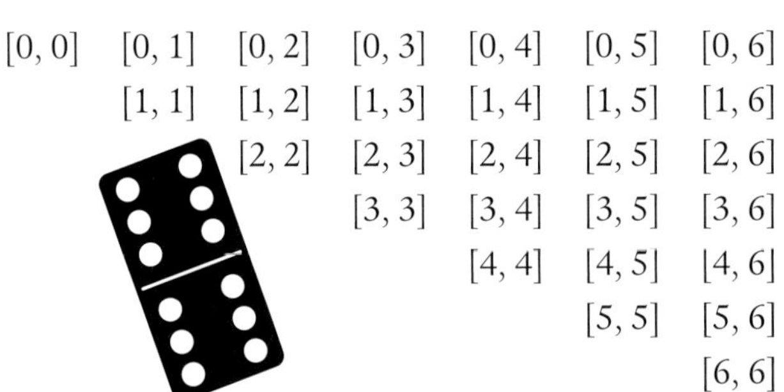

[0, 0]	[0, 1]	[0, 2]	[0, 3]	[0, 4]	[0, 5]	[0, 6]
	[1, 1]	[1, 2]	[1, 3]	[1, 4]	[1, 5]	[1, 6]
		[2, 2]	[2, 3]	[2, 4]	[2, 5]	[2, 6]
			[3, 3]	[3, 4]	[3, 5]	[3, 6]
				[4, 4]	[4, 5]	[4, 6]
					[5, 5]	[5, 6]
						[6, 6]

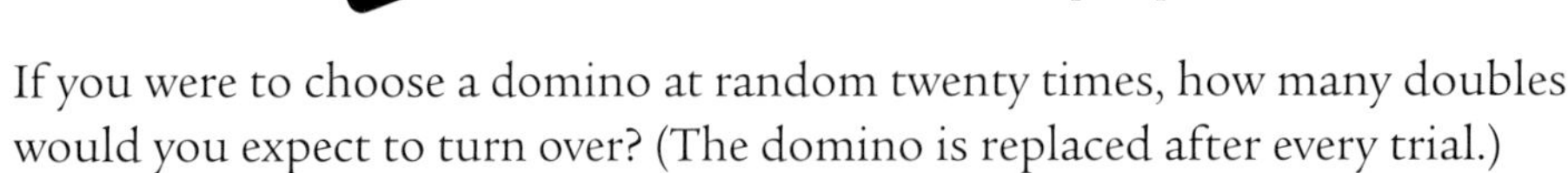

If you were to choose a domino at random twenty times, how many doubles would you expect to turn over? (The domino is replaced after every trial.)

4 Bingo numbers go from 1 to 90.

The bingo hall works out that in a normal month, a game of bingo will be played 360 times.

In how many of the 360 games would you expect the first number called to be a multiple of 8?

B	I	N	G	O
12	25	41	51	63
3	30	37	54	66
7	21	FREE	56	74
1	26	35	50	69
10	17	45	47	64

5 In the casino, it is estimated that a pair of dice will be rolled 9000 times each night at the 'Lucky Seven' tables.

a How many times would they expect to see a total of 7?

b If it costs £10 to enter 'Lucky Seven' and a person who rolls a total of 7 gets £30 back, how much would the casino expect to take in each night from the 'Lucky Seven' tables?

c The casino decides to hold a 'Lucky Eleven' night.

It still expects 9000 plays during the night but this time, 11 is the winning number. It still costs £10 per roll but winners are given back £80.

Do the casino takings go up or down on this night and by how much?

6 The internet is a good source of data.

The following facts and figures were picked from various sites.

- In the UK the risk of death in a road accident in a year is 1 in 16 800.
- The risk of death in a gas related incident in a year is 1 in 1·51 million.
- The risk of being killed by lightning in a year is 1 in 18·7 million.
- The population of the UK is 62 232 000.
- The population of Scotland is 5 222 000.
- The population of England is 51 456 000.

a Estimate the number of people living in the UK who will be killed by lightning.

b Estimate how many more people in England will be killed on the roads than in Scotland.

c How many people in Scotland do you expect to get killed in a year in gas related incidents?

Exercise 11.3B

1 A recent survey of people involved in road traffic accidents in Scotland revealed the probability of fatalities for different groupings. The table below lists the groupings, the probability of being killed and the number of people involved in accidents for each group.

Group	Car	Motorbike	Pedal cycle	Pedestrian
P(of being killed)	0·0127	0·042	0·009	0·023
Number involved	8300	850	800	2053

Estimate how many people died while involved in an accident:

a in a car **b** on a motor bike **c** on a bicycle **d** on foot.

2 One in five of the population of Scotland suffers from hay fever.

Of those who do suffer, only 1 in 20 is **not** allergic to grass pollen.

a In a random group of 450 people, how many would you expect to suffer from hay fever?

b What is the probability that a person chosen at random suffers from hay fever and **is** allergic to grass pollen?

c In a small town with a population of 3800 people, how many would you expect to suffer from hay fever but not be allergic to grass pollen?

3 Recent figures show that of all pregnancies, 1 in 32 will result in twins.

Twins can either be fraternal or identical. Only 12·5% of twins are identical.

In a group of 1000 pregnant women, how many identical twins would you expect to be born? (Use common sense rounding rules.)

4 Climatologists have researched significant storms in the Gulf of Mexico.

- There is a 56% chance of a significant storm upgrading to a hurricane.
- The probability is 1 in 3 that a hurricane upgrades to a major hurricane.

In a ten-year period, 112 significant storms were recorded.

Using the probabilities above estimate:

a how many hurricanes there were in that time

b how many major hurricanes are expected **each year**.

11.4 Scatter graphs

Scatter graphs are ideal for investigating whether one variable might be related to another.

The relation may not be **causal**, i.e. changing one of the variables need not result in the other changing.

If you did a survey in a Primary School, you would find that there was a good relation between the size of shoes the pupil was wearing and their score in a general knowledge quiz ... but it wouldn't mean that if you put people in bigger shoes they would know more.

The relationship is referred to as a **correlation**.

Example 1

The table shows the age and height of ten young people.

Age (years)	14	11	9	13	9	10	15	8	10	13
Height (cm)	165	150	141	160	137	154	174	135	151	166

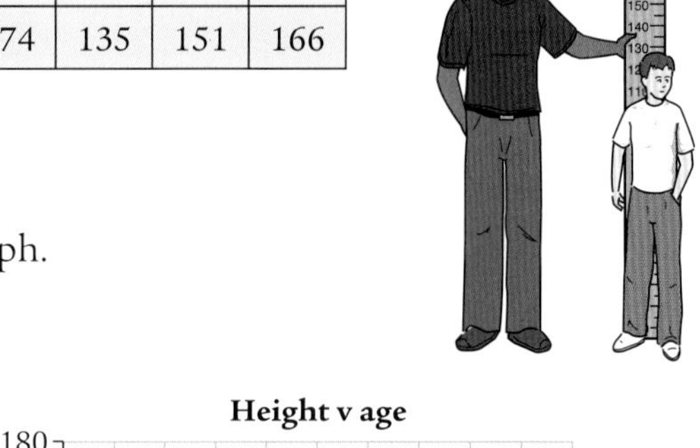

a Show this information on a **scatter graph**.

b Draw a line of best fit.

c Estimate the height of a 12-year-old, from the graph.

d Describe the correlation.

a The scatter graph results in an elongated 'cloud' of points.

b The line of best fit is chosen to run in the general direction of the elongation so that there are about the same number of points above it as below it.

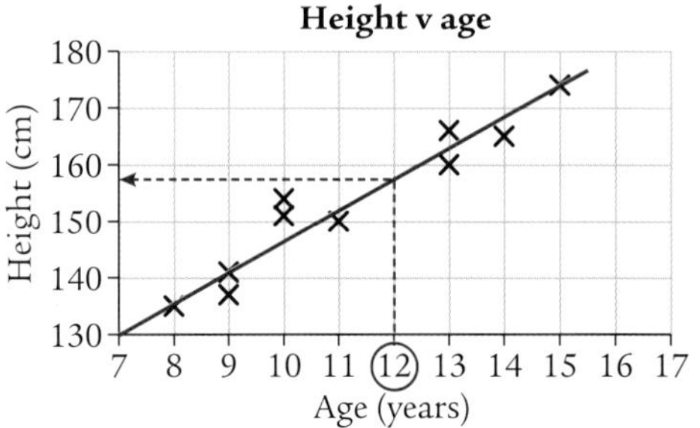

c Treating the line as the indicator of the relationship, it looks like a height of 158 cm corresponds to an age of 12 years.

d The correlation is strong (all the points are close to the line).
The correlation is positive (gradient is positive).
As age increases, height increases.

Example 2

The table shows the number of hours per week ten golfers spent on the practice ground.

It also gives their handicap.

Hours per week	10	2	7	0	8	5	3	12	4	10
Handicap	7	16	8	18	6	9	8	1	16	4

a Show this information on a scatter graph.

b Draw a line of best fit.

c What handicap would you expect a golfer to have if he practised for six hours a week?

d How many hours per week would you expect a 12 handicapper to practise?

e Describe the correlation.

a See graph.

b See graph.

c Six hours of practice correspond to a handicap of 9.

d A handicap of 12 corresponds to four hours of practice.

e The correlation is weak (points are loose and spaced away from the line).
The correlation is negative (gradient is negative).
As the hours of practice go up, the handicap goes down.

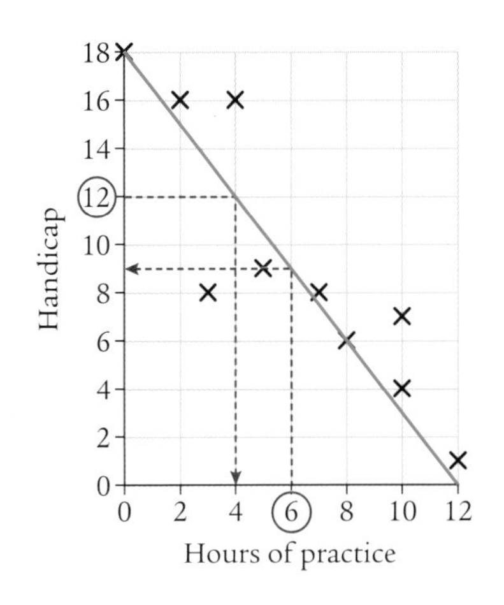

Arnold Palmer, a famous golfer, was once accused of being lucky. His reply was, 'Yes, and the funny thing is, the more I practise, the luckier I get.'

Exercise 11.4A

1 In a school where a number of students did Gaelic, the head teacher chose 12 students at random and compared their test mark in English with their mark in Gaelic.

English mark	23	37	28	26	33	38	29	31	32	29	35	28
Gaelic mark	25	34	26	26	29	35	24	33	37	34	31	26

By showing this information on a scatter graph and drawing a line of best fit:

a estimate the Gaelic mark for a student with an English mark of 30

b describe the correlation.

2 After the last European Football Championships, UEFA did a statistical analysis of the number of shots on goal against the percentage of ball possession the teams had.

Shots on goal	13	9	7	15	11	13	8	12	9	13
Ball possession (%)	55	41	38	62	51	47	41	63	44	54

By showing this information on a scatter graph and drawing a line of best fit:

a estimate how many shots on goal you would expect a team to have, if their ball possession was 57%

b describe the correlation.

3 As part of a fuel efficiency survey, a car manufacturer logged the average speed of journeys (km/h) against the fuel efficiency (km/l), for that journey.

Speed (km/h)	88	93	92	94	91	94	96	89	93	97	88
Efficiency (km/l)	15·8	15·1	14·9	14·8	15	14·9	14·4	15·5	14·8	14·5	15·6

a By showing this information on a scatter graph and drawing a line of best fit:

i estimate what efficiency you would expect at 90 km/h

ii describe the correlation.

b Why do you think the efficiency drops to 14·7 km/l when the speed is 70 km/h?

4 A group of dog owners were asked two questions:

A What is your door number?

B How old is your dog?

Door number	27	48	23	52	37	51	44	21	34	31	12	58
Age of dog (years)	9	3	6	9	5	7	8	9	8	3	6	6

By showing this information on a scatter graph and drawing a line of best fit, answer the following questions.

a Which door number would you expect a four-year-old dog to live at?

b Comment on what you are being asked to do in this question.

5 The table shows the speed of a car and the corresponding distance it took to come to rest when the brakes were applied. The road conditions were variable.

Speed (km/h)	35	63	28	37	46	77	83	61	57	69	71	42
Braking distance (m)	7	25	5	9	14	38	41	23	19	28	29	10

By showing this information on a scatter graph and drawing a line of best fit:

a estimate what braking distance you would expect at 80 km/h

b estimate what speed would you expect to be doing for a braking distance of 33 m

c investigate the stopping distances recommended by the Highway Code

d comment on whether the connection might be causal.

6 During a practice session of snooker, the coach of a young player compared the percentage pot rate and the distance between the cue ball and the target ball.

Distance between balls (m)	0·7	1·3	0·4	1·7	1·1	1·5	0·6	0·8	1·2	0·9
Pot rate (%)	78	64	88	57	72	62	81	77	67	77

By showing this information on a scatter graph and drawing a line of best fit, estimate what percentage pot rate the coach would expect at one metre.

Example 3

So far the line of best fit has been placed using common sense and personal judgement.

This table of values was the result of a survey comparing heights and age.

Age (years)	14	11	9	13	9	10	15	8	10	13
Height (cm)	161	148	141	144	137	154	174	142	151	169

A scatter graph shows that the point (13, 144) is well away from the rest of the cluster of points. This is known as a rogue value. For a clearer picture, rogue values are usually ignored.

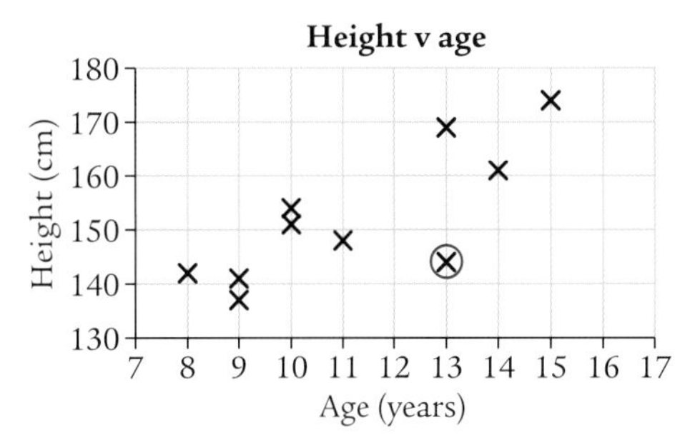

The mean of the five lowest ages is given by:
(8 + 9 + 9 + 10 + 10) ÷ 5 = 9·2.

The mean of the five corresponding heights is:
(142 + 141 + 137 + 154 + 151) ÷ 5 = 145.

The mean of the four largest ages, ignoring the rogue point, is given by:

(11 + 13 + 14 + 15) ÷ 4 = 13·2.

The mean of the four corresponding heights is:

(148 + 169 + 161 + 174) ÷ 4 = 163.

If we remove the rogue value and plot the other nine points, we can plot the two 'average points' (9·2, 145) and (13·2, 163) ... and use them as a guide to draw a straight line.

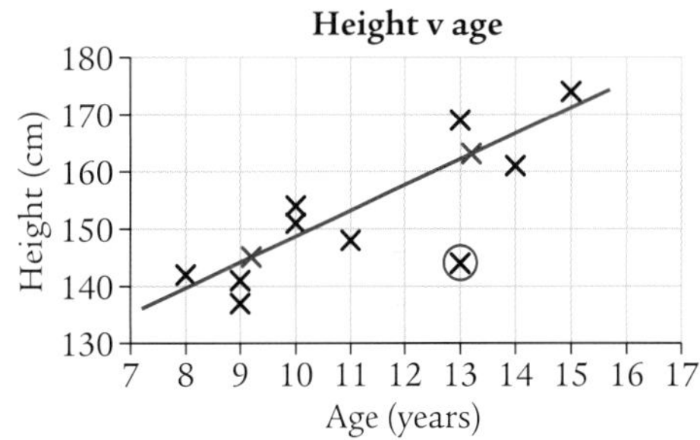

Exercise 11.4B

1 In a science experiment, 50 ml of solution A was mixed with 50 ml of solution B.

A reaction resulted in the mixed solutions turning blue.

The class wanted to see if temperature would influence the time taken for this reaction to occur, so they stored both solutions in baths of water that were at different temperatures.

Temperature (°C)	7	36	24	43	19	14	27	32	39
Reaction time (s)	43	27	30	18	37	35	33	24	22

a By showing this information on a scatter graph and drawing a line of best fit:

i describe the correlation

ii estimate the reaction time when the temperature is 11 degrees.

b Can you explain why temperature influences the reaction time?

2 The students in the physics laboratory did experiments with pendulums.

They changed the mass of the ball, the angle of swing and the length of string.

They were surprised by their results when they found it was only the length of string that affected the time for one complete swing of the pendulum (the period).

Length of string (m)	1·0	1·2	1·4	1·6	1·8	2·0	2·2	2·4	2·6	2·8	3·0
Period (s)	2·1	2·3	2·2	2·6	2·7	2·9	2·9	3·2	3·3	3·5	3·6

a Show this information on a scatter graph and draw the line of best fit.

b In 1851, Léon Foucault, a French physicist, demonstrated the rotation of the Earth on its axis by his newly invented Foucault pendulum.

He suspended a 67-metre wire from the dome of the Panthéon in Paris and attached a 28 kg ball.

Estimate what the period of his pendulum would be.

c A copy of the pendulum can be found in Glasgow's Princes Square.

Investigate how it can be used to tell the time.

3 A Canadian botanist compared the age of trees with their height, to see if predictions could be made about future growth.

Age (years)	17	23	25	19	26	28	19	32	27	20	26	24
Height (m)	6·3	7·1	7·3	5·6	7·2	8·1	6·6	7·1	7·6	6·8	7·7	7·2

a Show this information on a scatter graph.

b Identify the two rogue points, and eliminate them from further calculations.

c Use the lower and upper 'average points' to locate the line of best fit.

d Work out the overall average point and show it on your graph.

4 Students in a physics lab set up two resistors in parallel. To avoid too many variables, the second one always had a resistance of 10 ohms (Ω).

As the first one varied, they measured the total resistance of the parallel circuit.

Resistor 1 (Ω)	5	7	2	4	5	8	3	9	7	6	10	1
Total resistance (Ω)	3·2	4·2	1·7	2·7	3·1	4·6	2·4	4·9	4·4	3·6	4·8	1·0

a Illustrate this information on a scatter graph and draw a line of best fit.

b Mark in the overall 'average point'.

c Your line of best fit should run through the average point. Does it?

11.5 Footnotes

1 Warning on prediction

The line of best fit should really only be drawn between the lowest observed and highest observed x-value. If we estimate a y-value corresponding to an x-value in this region, we can be fairly confident of its value ... such an estimate is called an **interpolation**.

However, if we try to estimate y-values beyond the observed region we can end up with answers that differ wildly from what actually happens. We can't depend on such estimates.

Estimating beyond the observed region is called **extrapolation** and should be avoided in most instances.

With all their data and computers, meteorologists can't really predict the weather accurately beyond two days.

Example 1

The graph illustrates the correlation between the record time for running the mile and the year it was reached.

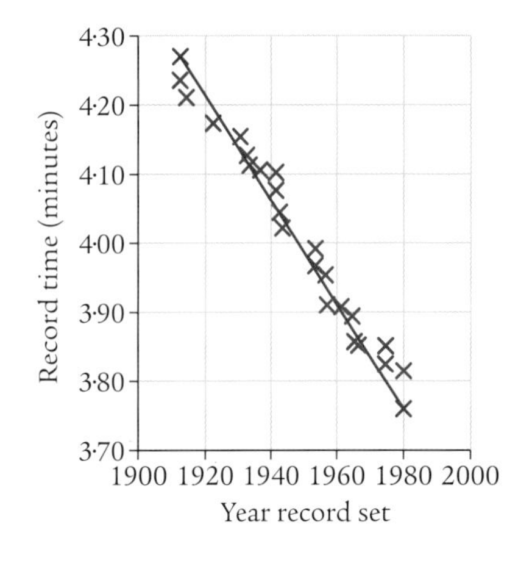

a When does the graph suggest the four-minute mile was attained?

b Is it sensible to use the graph to estimate when:

i the 10-minute mile was broken

ii the one-minute mile might be broken?

a The four-minute mile looks like it was broken in the early 1950s (interpolation).

(Note: the actual year was 1954 by Roger Bannister.)

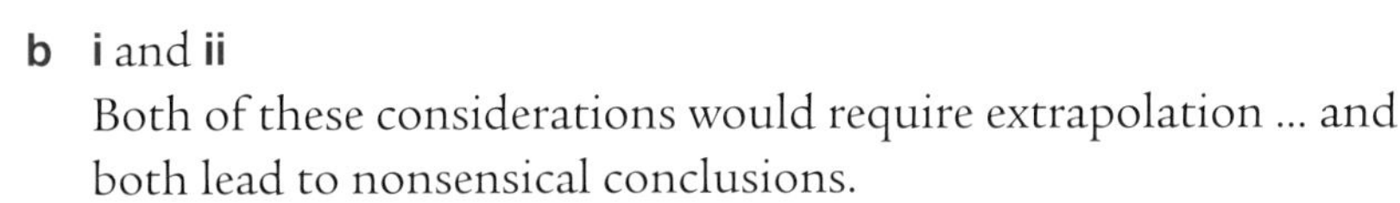

b **i** and **ii**

Both of these considerations would require extrapolation ... and both lead to nonsensical conclusions.

Example 2

The bar graph shows the number of young men living in a small town at the start of the 1900s.

The data was gathered every four years.

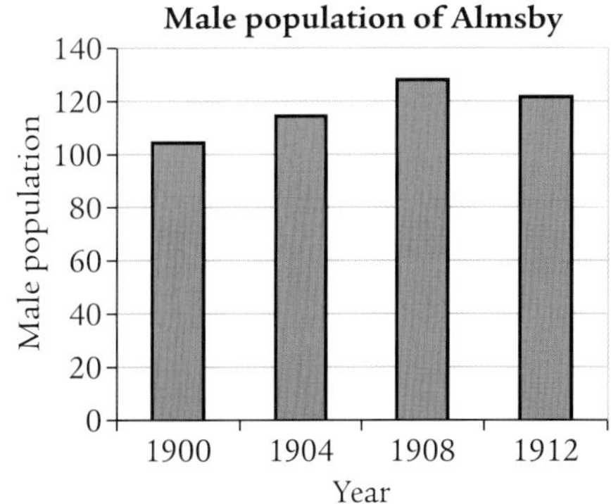

a Estimate the male population for 1916 and 1920.

b The actual figures were 87 and 74. What two historical happenings may have contributed?

a It looks fairly steady at 120 males for each of the years.

b In World War I (1914–1918) millions of young men died.

Between 1918 and 1920 the Spanish flu killed millions all over the world.

2 Consequences

If a coin is tossed to see which team kicks off a game of football, the probability that you call correctly is $\frac{1}{2}$.

If the same coin is tossed to see if someone will win a holiday or lose their car, as a bet, the probability remains $\frac{1}{2}$, but the consequences are hugely different with a major life impact.

The probability of something bad happening may be small, but in decision-making the consequence should always be considered.

Example 3

When deciding if a controlled crossing is necessary, the traffic engineer will look at past statistics, compute the probability of a serious accident and weigh this against the cost of the controlled crossing.

Let's say that at some junction the probability of a fatality is 0·001.

a Does this seem a reasonable risk to take to save money?

b If this junction is crossed 200 000 times a month, what is the expected frequency of a fatality? Comment.

a Probability of 0·001 means that crossing 999 times out of 1000 will be fine.

In other situations, this might be considered an unlikely outcome and not worth guarding against.

b If the junction is crossed 200 000 times a month:

$$\begin{aligned} \text{E(fatality)} &= \text{P(fatality)} \times \text{number of crossings} \\ &= 0{\cdot}001 \times 200\,000 \\ &= 200 \end{aligned}$$

This suggests that if things are not controlled, there will be 200 fatalities a month.

The probability of a particular outcome can be low but if the 'experiment' is conducted many times, the outcome is expected to happen.

3 Making informed choices

People do not like being told what to do. They say that they would far prefer to be given the information that allows them to make an informed choice.

Everyone makes inappropriate choices at times.

It is knowing what the consequences of the choices are and the impact they will have that is important.

Example 4

A holidaymaker needs to be in Edinburgh Airport by 06 30.

She can catch the first train to Edinburgh (£7), which is only in time for the link 80% of the time.

She then uses the link bus (£1) from the station.

Alternatively she can pay £35 for a taxi (which is 100% certain of being on time).

If she misses the flight, the cost of a new ticket will be £250 and there is no guarantee which flight she will be allowed on.

What are her options?

Train and bus fares total is £8.

Probability of missing flight is 0·2 (fairly high).
This means that 1 time in 5 the journey will cost an extra £250.

Taxi £27 more expensive but door-to-door service and 100% certain of being on time.
Major expense for missing the flight.

Imagine five such trips:

by train: expected cost will be $5 \times 8 + 250 = £290$

by taxi: expected cost will be $5 \times 35 = £175$.

In the long run the train is the poorer option by $(290 - 175) \div 5 = £23$ a trip ... and there is also the stress to consider.

What would you do?

Exercise 11.5A

1 The table shows the relationship between the percentage of the population who regularly use a certain social media network and their age group.

Age group	Under 18	18–24	25–34	35–44	45–54	55–64	65+
People using social media network (%)	5	12	34	31	12	4	2

a What is the probability that a person chosen at random from this group will be under 18?

b There are 30 million users in the group, how many will be 65+?

c Describe the relation between age and use of the social media network.

d Why might the under 18 figure be misleading?

2 The world record for the men's long jump is unusual in that there have been extended periods when the record has not been broken. The table shows the year and the record for significant changes to the world record.

Year	1935	1960	1968	1991
Men's long jump record (metres)	8·13	8·21	8·90	8·95

a Could this information be used to predict when the present record will be broken?

b Would you expect the present record to be beaten at the next Olympic Games?

c Research what happened in the 1968 Olympic final and comment on how the geography might have helped.

3 In the lucky draw there are four discs in a bag numbered 1 to 4.

Another bag contains four live scorpions with the numbers 1 to 4 attached.

To win, all you need to do is pull out the number 3 from either bag.

- **a** Which bag would you draw from and why?
- **b** If they changed the scorpion bag to three scorpions with number 3 and one with number 4, which bag would you choose and why?
- **c** What else would you need to consider?

4 A recent publication said that the probability of being injured by fireworks was 1 in 22 000.

- **a** If you knew that the same probability applied to the rope breaking during a bungee jump, what comparisons would you make?
- **b** What would you think if the bungee rope probability went up to 1 in 220 000?
- **c** What probability would you accept before doing a bungee jump?

5 To set the order of play for a board game, it is common that the highest score on the roll of a dice decides who starts.

For a small group of patients waiting for an organ transplant, would rolling a dice to see who got the first available organ be seen as appropriate?
(Base your answer on probability and consequence.)

6 The probability of winning the lottery on any week, where you have purchased one ticket, is about fourteen million to one.

If you were told that the plane you were about to fly in had a fourteen million to one chance of crashing, what comparisons would you make with the lottery probability?

Exercise 11.5B

1 Insurance companies estimate that the probability of a hole-in-one is 1 in 12 500.

At charity golf days it is common to offer a new car for anyone who scores a hole-in-one.

Organisers can insure against the hole-in-one by paying an insurance company about £350. This means that if there is a hole-in-one, the insurers pay for the new car.

- **a** The charity golf day attracts 100 golfers. What is the probability that any one of the golfers gets a hole-in-one?
- **b** Each golfer pays £20. How much would be collected for charity, assuming the insurance is paid?
- **c** The car is worth £12 000. Comment on the consequences of not taking insurance.

2 In 1902 an election was due to be held on the island of Martinique.

The island had a rumbling volcano named Mount Pelée. The ruling Progressive Party was a slight favourite to win the election so they did not want to evacuate the island when the volcano started to become more active.

They concluded that, from previous experience, less than 3% of the 30 000 population would be killed in any major eruption.

a Using their estimate, how many could die in a major eruption?

Unfortunately, Mount Pelée became one of the most violent eruptions in history. Only two people survived the volcanic explosion.

b What was the actual proportion of the population that survived?

c Comment on the decision of the politicians to delay the evacuation.

3 In the UK, 600 000 homes have a 1 in 75 chance of being flooded in any given year.

As flooding becomes more common, insurance companies have increased annual premiums to compensate for the increased number of claims.

For one household in the flood area, the insurance company required a premium of £480 each year, with the understanding that the householder would have to pay the first £2000 of any claim resulting from flood damage. This is called an excess.

a A household had no problems for 19 consecutive years.
How much did they pay for insurance during that time?

b During the twentieth year a major flood resulted in £18 000 of damage.
   i How much did the insurer pay out (remembering the excess)?
   ii Compare this to the payments made by the householder.

c A particularly unlucky house was flooded three times in the first five years of paying the annual premium of £480.
Damages of £8000, £11 000 and £9000 resulted.
How much better off was the householder when compared to an identical case who did not insure?

d Comment on the risk households take when they have no insurance.

3 A car manufacturer sold 300 000 cars of a specific model.

After a while, follow-up research found a major fault in the car that was dangerous if three factors were to happen:

- the car would need to be struck from behind by another vehicle
- the indicator would need to be on and be smashed in the collision
- the petrol tank would need to be below quarter full.

If all three factors were present, there was a good chance a spark could ignite petrol fumes and cause a serious car fire that could kill occupants in the car.

The probability of such a fatal accident occuring in the lifetime of the car is 1 in 50 000. The courts would find the car manufacturer negligent in this instance. They would award damages of about \$8 000 000 each time a fatal fire occurred.

The cost of recalling all cars and rectifying the problem would be \$1500 for each car.

a From a financial point of view, why might the company do nothing?

b From a moral point of view, why is doing nothing a disgraceful decision?

c From a legal perspective, what options does the judge have?

Preparation for assessment

1 To demonstrate probability, a teacher wrote the letters S T A T I S T I C S on ten identical cards, shuffled them and placed them face down.
What would be the probability of:

a choosing the letter S at random

b choosing a letter at random that was not an S

c choosing a letter at random that was neither an S nor a T?

2 How many people, from a group of 30 000, would you expect to have watches that are at least two minutes out, if 96% of wrist watches are within two minutes of the actual time?

3 By researching her family tree, Ann found that three out of five births were female and there was no history of twins. Her two sisters were soon to give birth.
Estimate the probability of Ann having two new nephews in the near future.

4 Last season, the local hockey team won 65% of their games, lost 15% of their games and drew the rest. If these figures were repeated in the coming season, what would be the expected probability of the following sequence of results happening?

a Win, then lose, then draw, in that order.

b Win, then win, then win.

c Two wins and one draw, in any order.

5 When a group of people were asked their weekly take-home pay and how much they saved each week, their replies are shown in the table.

Take-home pay (£)	340	260	270	320	240	300	270	250	290	320	280
Weekly savings (£)	35	20	35	30	10	25	25	15	20	25	30

a Show this information on a scatter graph and drawing a line of best fit.

b How much would you expect someone to save if their weekly take-home pay was £305?

c What percentage of their take-home pay is this?
(Give your answer to the nearest %.)

6 The 'True 2 U' tyre company have a promotion on new tyres.

Buy a set of four for the price of three, if you buy the super R101s at £85 each.

A motorist knows that one of his tyres is right on the legal limit, the tread on two more is 80% used up, and the fourth one is halfway through its expected life. The cheapest tyre that 'True 2 U' has on sale is £38, but it will last about half as long as the R101.

Using probability and cost, list the reasons for and against using the promotion. (Reminder: there are serious legal implications for having illegal tyres on a car.)

7 Remember the asteroids?

If you look at the Moon, you will see huge craters that were made when asteroids from outer space slammed into the surface of the Moon.

Scientists track asteroids in case they are heading for Earth.

An asteroid they are presently tracking is about 1·5 km across.

The scientists have calculated it has a 1 in 909 000 chance of colliding with our planet in a few years.

a Is a probability of 1 in 909 000 reassuring?

b If a collision did happen, what would be the scale of the damage?

c Is it a good idea to track asteroids?

d The chances of winning the big prize in the lottery are 1 in 13 983 816
... but somebody wins it nearly every week!
Comment!

12 Time trials

Before we start...

The Formula 1 Spanish Grand Prix is held in Barcelona in the Circuit de Catalunya.

Circuit de Catalunya

Length of one lap: 4·655 km
Current lap record: Held by Kimi Räikkönen, Ferrari, 2008 ... 1 min 21·670 s
Start line/finish line offset: 0·126 km
Total number of race laps: 66
Total race distance: 307·104 km

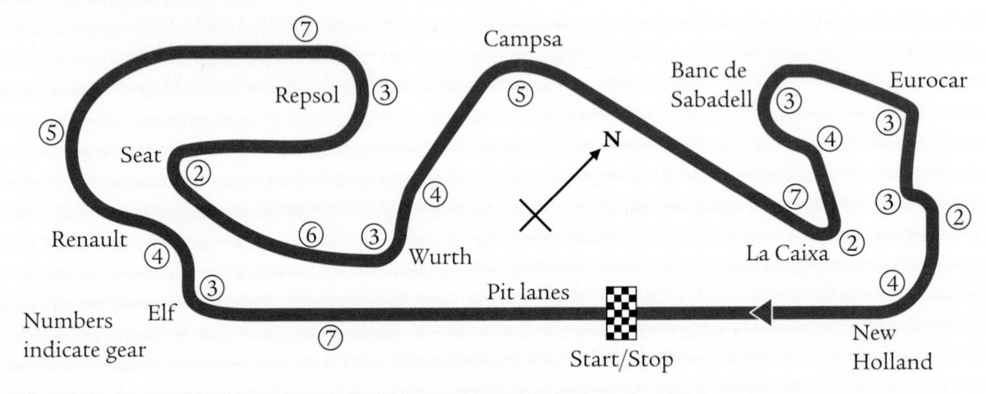

In 2012, the winner was Pastor Maldonado in his Williams car.

The race was completed in 1 hour 39 minutes and 9·145 seconds.

Can we, from the above information, calculate the winner's **average speed** for the race?

What you need to know

1 A Boeing 737 aircraft leaves Lanzarote at 10.55 p.m. and lands in Edinburgh at 3.10 a.m.

Calculate how long the flight took.

2 A storm caused a power cut in the village of Drumness.

It lasted from 17 36 on Tuesday until 11 47 on Wednesday.

How long was the village without electricity?

3 Joe enters a competition to find the Junior Ironman.

It involves running, cycling and swimming, but not in that order.

The graph shows Joe's performance.

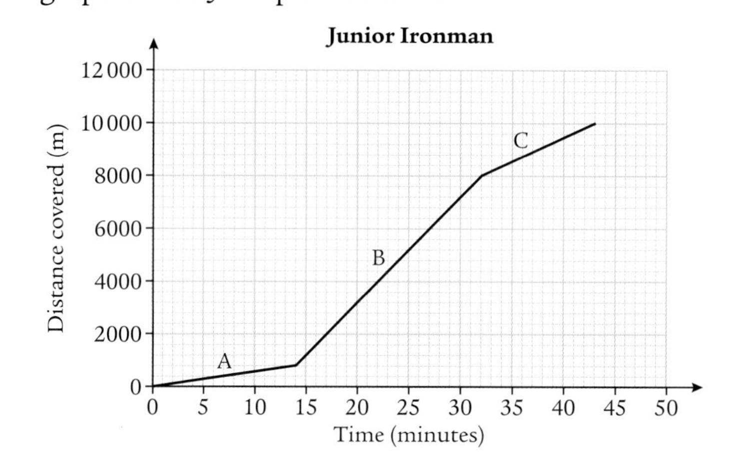

a Which activity is represented on the graph by:

i Section A **ii** Section B **iii** Section C?

b Explain your decisions in part **a**.

c How long did Joe take for each activity?

d What was the distance covered in each activity?

12.1 Rate and speed

A **rate** gives you how one measurement changes with respect to another.

The word 'per' is used to separate the units of the two measurements.

It can be replaced by 'for each'.

A bricklayer lays 240 bricks in 10 minutes.

The rate at which he is working is $\frac{240}{10} = 24$ bricks per minute.

The fastest growing plant is a variety of bamboo.

One grew 852 cm in 8 days.

Its rate of growth was $\frac{852}{8} = 106{\cdot}5$ cm per day.

Speed is a special rate. It is the rate at which **distance** is covered per unit time.

$$\textbf{Average speed} = \frac{\textbf{total distance covered}}{\textbf{total time taken}}$$

This relation between the three measurements, speed, distance and time, is often remembered using the *SDT* triangle:

Cover up the quantity you want and a formula for it will be revealed:

Here we want a formula for speed, S:

$$S = \frac{D}{T}$$

Example 1

Anwar drove at a steady speed for three hours down the motorway from Glasgow to Manchester.

He covered a distance of 186 miles.

a What was his speed?

b Aleesha also drove from Glasgow to Manchester, but didn't drive on the motorway.

She couldn't keep to a steady speed.

In all, she drove 200 miles and it took her 4 hours.

Calculate her **average** speed.

a Anwar drove 186 miles in 3 hours, so in 1 hour he drove:

$\frac{186}{3} = 62.$

His speed was 62 miles per hour (mph).

b Aleesha's **average** speed was $\frac{200}{4} = 50$ mph.

Example 2

Melanie ran the 1500 metres at the school sports in 5 minutes 15 seconds.

Calculate her average speed in metres per second.

(Note the distance must be in **metres**, and the time in **seconds**.)

$$S = \frac{D}{T}$$

$$S = \frac{1500}{315}$$

$S = 4{\cdot}76$ metres per second (m/s) (to 3 s.f.).

(Note that in your science subjects this might be written $4{\cdot}76\ \mathrm{m\,s^{-1}}$.)

Melanie's average speed is 4·76 m/s.

Example 3

A bus leaves Bo'ness at 09 27.

It arrives in the centre of Edinburgh at 10 19.

The distance the bus travels is 15·36 miles.

Calculate the average speed of the bus in mph.

$$S = \frac{D}{T}$$

Time taken = 10 h 19 min − 9 h 27 min = 52 min.

$D = 15{\cdot}36$ miles, $T = 52$ **minutes**,
but for a speed in mph, the distance must be in **miles** and the time in **hours**.

52 minutes = $\frac{52}{60}$ hours = 0·8666... hours

$$S = \frac{15{\cdot}36}{0{\cdot}8666...}$$

$S = 17{\cdot}7$ mph (to 3 s.f.)

The average speed of the bus is 17·7 mph (to 3 s.f.).

Example 4

Andy Murray's tennis serve travelled 17·7 metres in 0·29 seconds.

Calculate the average speed of the serve in:

a metres per second **b** kilometres per hour.

a $S = \frac{D}{T}$

$= \frac{17{\cdot}7}{0{\cdot}29}$

$S = 61{\cdot}0$ metres per second.

b $S = 61{\cdot}0$ metres per second

$= 61 \times 3600$ metres per hour

$= 219\,600$ metres per hour

$= 219{\cdot}6$ kilometres per hour

$= 220$ km/h (to 3 s.f.).

Exercise 12.1A

1 The winners' times in some of the races at the school sports were:

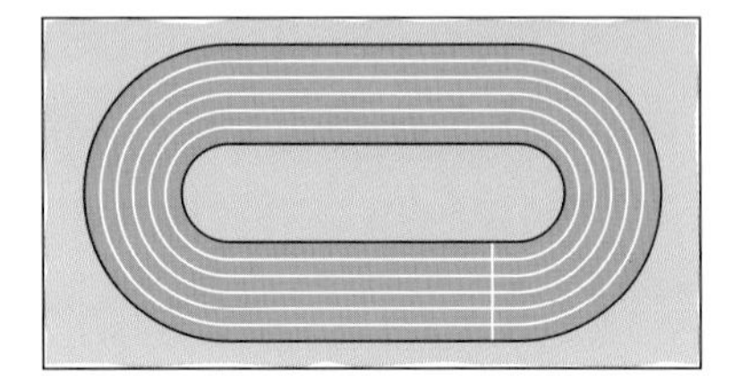

a 100 metres in 13 seconds

b 200 metres in 28 seconds

c 400 metres in 58·4 seconds

d 1500 metres in 4 minutes 52 seconds.

Calculate the speed of each winner in m/s, correct to 1 decimal place.

2 **a** Express these times in hours, giving your answers correct to 2 decimal places where appropriate:

i 5 hours 24 minutes **ii** 10 hours 40 minutes **iii** 1 hour 5 minutes.

b Express these times in minutes:

i 6 minutes 15 seconds **ii** 4 minutes 45 seconds **iii** 8 minutes 20 seconds.

3 Copy and complete the table. Check that your units are correct.

	a	b	c	d	e	f	g
Distance	65 miles	54 feet	831 km	100 m	2800 miles	8 km	7 km
Time	5 hours	3 seconds	6 hours	9·8 s	4 h 30 min	30 min	10 hours
Speed	mph	ft/s	km/h	m/s	mph	km/h	m/h

4 Angela cycled 28 miles in 5 hours.

Betty cycled 22 miles in 4 hours.

a Whose average speed was greater?

b By how much?

5 In 2009, a female cheetah called Sarah covered 100 metres in 6·13 seconds.

Calculate her speed in m/s.

6 The fastest moving animal in the world is the peregrine falcon.

It can dive a distance of 351 metres in 4 seconds.

Calculate the falcon's speed in:

a metres per second

b kilometres per hour.

7 A Japanese 'bullet train' travelled 1380 km in 2 hours 30 minutes.

Calculate its average speed in km/h.

8 The land speed record for a motorcycle was broken in 2010 when a 2600 c.c. Suzuki bike covered a distance of 31·08 miles in 5 minutes.

Calculate the new record speed in mph, to 1 decimal place.

Exercise 12.1B

1 When it is in a hurry, a garden snail can travel a distance of 8 metres in 10 minutes.

Calculate the snail's speed in km/h.

2 **a** At full speed, the *Queen Mary 2* covered a distance of 280 kilometres in five hours.

Calculate its full speed in km/h.

b When sailing at its cruising speed, it travelled 276 km in 5 hours 45 minutes.

Calculate its cruising speed in km/h.

c Calculate the ship's average speed over the two journeys.

(It is **not** 52 km/h! Remember that average speed = **total** distance ÷ **total** time.)

3 **a** The school minibus drove 54 km to Perth in 1 hour 15 minutes.

The return journey took 25 minutes longer.

Calculate the average speed of:

i the journey to Perth ii the return journey iii the round trip.

b The next time the excursion takes place, the teacher assumes it will take the same time.

They have warned parents that the students will be back for 5.30 pm.

When will the bus leave Perth?

4 **a** At the cross-country championships, the winner covered the 8·4 km course in 23 minutes.
Calculate the winner's average speed in km/h.

b The runner who was second came in a minute behind the winner.
What was his average speed?

5 Put these speeds in order, starting with the slowest:
8 m/s 30 km/h 472m/min 0·51 km/min.

6 The London Eye has a diameter of 120 metres.

One revolution of the wheel takes 30 minutes, to the nearest minute.

Calculate the speed of the wheel in:

a m/s

b km/h

c degrees per minute.

7 To make sure people stick to the speed limit, cameras are installed.

Markings are drawn on the road surface.

When a car gets to the markings the camera takes two photographs half a second apart.

On a certain section of road, the speed limit is 50 miles per hour (80 km/h).

- **a** Express this speed in metres per second.
- **b** What is the greatest distance that the driver can travel between photographs to avoid being prosecuted?

8 The *Bonnie Rose* sails from Largs, rounds a buoy and returns to Largs.

Details of its journey are shown in the graph.

Answer the following questions.
(Where appropriate give your answers correct to 3 significant figures.)

- **a** How far is the buoy from Largs?
- **b** How long did it take to reach the buoy?
- **c** Calculate the average speed of the *Bonnie Rose* on the outward journey.
- **d** What was the average speed of its journey back to Largs?
- **e** What was the average speed over the whole journey?

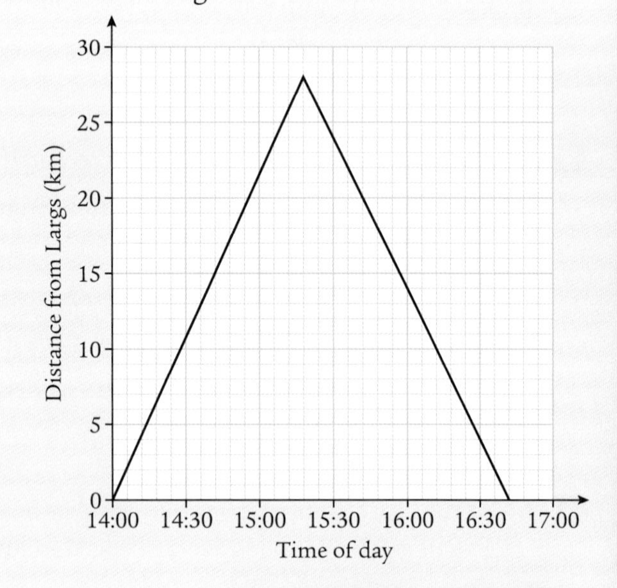

9 On a three-mile stretch of motorway, because of road works, a 50 mph speed limit is imposed.

This is monitored by 'average speed' cameras.

At the start of the stretch a photo is taken of a car and the time of day.

At the end of the stretch another photo is taken of the car and the time.

Using the two photos, the driver's average speed is calculated.

If this is greater than 50 mph then the driver is prosecuted.

After one mile a sign reminds drivers that they are in a monitored stretch.

Accidentally a driver had covered the first mile at 60 mph.

When he saw the sign he slowed down.

At what speed would he have to do the next two miles to cover the three miles at an average speed of 50 mph?

12.2 Distance

Rita plans to run a marathon, a distance of about 26 miles, travelling at a steady speed of 8 mph.

In one hour she would cover a distance of 8 miles.

In three hours she would cover a distance of $8 \times 3 = 24$ miles.

So she estimates she'll do the marathon in just over 3 hours.

Distance = speed × time.

Once again, the *SDT* triangle will help you to remember.

Cover up the quantity you want and a formula for it will be revealed.

Here we want a formula for distance, D:

$D = ST$.

Again we must be careful to be consistent with units ...

Example 1

Calculate the distance travelled by a train in six hours if it goes at a steady speed of 112 km/h.

$$\begin{aligned} D &= ST \\ &= 112 \times 6 \\ &= 672 \text{ km} \end{aligned}$$

The distance travelled by the train is 672 km.

Example 2

A Boeing 737 flew from Glasgow to Tenerife in 4 hours 25 minutes at an average speed of 455 mph.

How far is Tenerife from Glasgow?

$D = ST$

$= 455 \times 4\frac{25}{60}$ — The speed is in miles per hour, so the time must be in hours

$= 455 \times 4{\cdot}4166666...$ — $25 \div 60 = 0{\cdot}4166666...$

$= 2010$ miles (to the nearest whole number).

Tenerife is 2010 miles from Glasgow.

Exercise 12.2A

1 Use the formula $D = S \times T$ to work out the distance travelled in:

- **a** 2 hours, at an average speed of 80 km/h
- **b** 7 hours, at a steady speed of 23 mph
- **c** 6 seconds, at a speed of 9 m/s
- **d** 8 minutes, at 3 metres per minute
- **e** 1 day, at a speed of 12 mph
- **f** 3 minutes, at 2 m/s.

2 **a** Calculate the distance run by each athlete:

Mary	6 m/s, for 1 minute 20 seconds
Peter	9 m/s, for 40 seconds
Fatima	7 m/s, for 1 minute 10 seconds
Scott	4 m/s, for 1 minute 50 seconds.

b List their names in order, starting with the person who ran the furthest.

3 The giant tortoise is said to be the world's slowest reptile.

It managed an average speed of 7·5 cm/s over a 15-minute period.

How far did the tortoise travel in the 15 minutes?

4 Tom drove for three hours at an average speed of 68 mph.

Ann drove for five hours at an average speed of 39 mph.

- **a** Who drove further?
- **b** By how much?

5 **a** Indira calculates that if she cycles from home at an average speed of 24 km/h for three hours, she will reach her grandmother's house.

How far from Indira's house does her grandmother live?

b Indira only manages to average 21 km/h.

How far short of her grandmother's house is she after cycling for three hours?

6 The new Golden Flash motorbike is having a test run.

It runs for 90 minutes at 70 mph, 2 hours 30 minutes at 85 mph and 30 minutes at 120 mph.

What was the total distance covered by the bike?

7 In geography we learn that glaciers' movement depends on their location and the steepness of the mountains that they travel down.

To measure the speed, poles are stuck in the glacier and the distance travelled is measured.

In Greenland, a glacier can cover 20 m per day.

In Antarctica, the Byrd glacier has been measured moving at 2 m per day.

a How far do you expect a pole stuck in the Byrd glacier to move during the month of January?

b How much further do you expect a pole stuck in the Greenland glacier to travel during the same period?

Exercise 12.2B

1 **a** What **decimal** fraction of an hour is:

i 12 minutes　　ii 20 minutes?

b Kelly jogs most lunch-times.

She keeps a note of her times.

She runs each day at an average speed of 11 km/h.

Calculate how far Kelly jogs each day.

Monday	12 minutes
Tuesday	20 minutes
Wednesday	15 minutes
Thursday	30 minutes
Friday	18 minutes

2 Stuart cycled at 56 km/h for 5 minutes.

Susie cycled at 32 km/h for 9 minutes.

a Who cycled further?

b How much further?

3 A train travels at an average speed of 135 km/h from Paris to Madrid. The journey takes 8 hours.

a How far apart are Paris and Madrid?

b The train leaves Paris at 09 30. How far would you expect it to travel by 11 15?

4 A group of hill walkers have arrived at Painters' Point. Their aim is to reach Blackness Castle.

There are two routes to the castle – a short but very hilly route, or a longer but flatter one.

Half of the group take the shorter route and reach the castle in 2 hours and 30 minutes, walking at an average speed of 4 km/h.

The other half take the longer, flatter route and reach the castle in 2 hours 20 minutes, having averaged 5·5 km/h.

Calculate the length of each route.

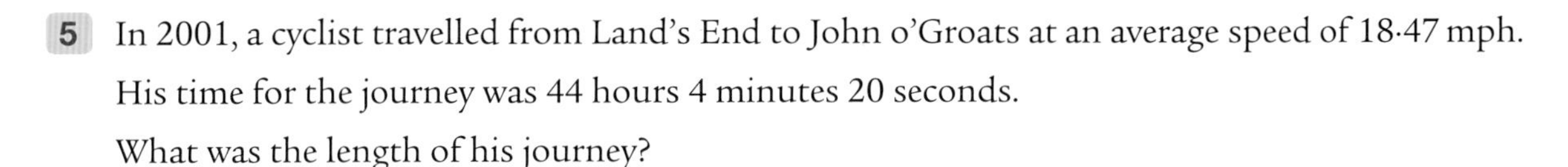

5 In 2001, a cyclist travelled from Land's End to John o'Groats at an average speed of 18·47 mph.

His time for the journey was 44 hours 4 minutes 20 seconds.

What was the length of his journey?

6 Two yachts set off from the same point. The *Sceptre* headed north at a speed of 21 km/h. The *Crescent* headed east at a speed of 25 km/h.

a How far had the *Sceptre* travelled after 20 minutes?

b How far apart are the two yachts after 20 minutes?

c At what rate, in km/h, is the distance between the yachts growing?

7 In geology we learn of the San Andreas Fault that runs through California.

The western side is moving north at an average speed of 35 mm per year.

Of course most of the time it doesn't move, it 'saves up' all the movement for one big shift.

When it does move, an earthquake results.

The last time it really moved was 1906.

At the given speed, how much should the fault have moved by 2014?

8 The planet Earth travels on a path that is roughly circular.

It travels through space at a speed of 1·6 million miles per day.

a How far does it travel in a year (365·25 days)?

b You should know the formula $C = \pi D$. Use it to find the distance of the Earth from the Sun in millions of miles.

9 The people forming the Highway Code decided that a driver will take $\frac{2}{3}$ of a second to react to an emergency.

The distance the car goes in this time is called the 'thinking distance'.

A car is travelling at 30 mph or 13·33 metres per second.

a How far will the car travel while the driver thinks to brake at this speed (to the nearest metre)?

b What is the thinking distance at 60 mph?

c The Highway Code gives all the thinking distances from 20 mph to 70 mph.

Can you reproduce the table?

12.3 Time

Again the *SDT* triangle will help you remember.

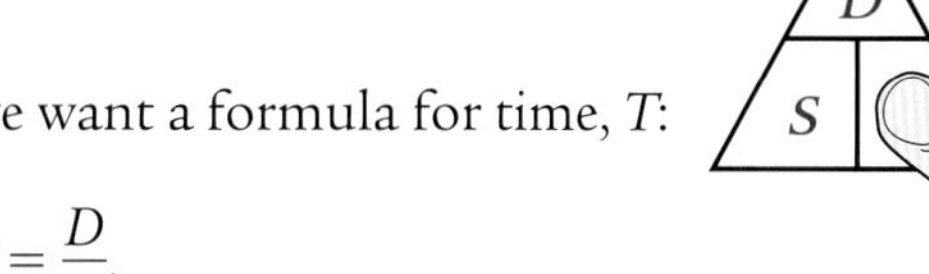

Cover up the quantity you want.

Here we want a formula for time, *T*:

$$T = \frac{D}{S}.$$

Example 1

A lorry driver is travelling from Glasgow to Inverness.

He reckons he should be able to average 50 mph.

The road distance between the two cities is 170 miles.

He left at 2 p.m. He's been asked for his ETA (estimated time of arrival).

When should he arrive at Inverness?

The lorry covers 50 miles each hour.

$$T = \frac{D}{S}$$

$$T = \frac{170}{50} = 3{\cdot}4 \text{ hours.}$$

Now 0·4 of an hour = 0·4 of 60 minutes = $0{\cdot}4 \times 60 = 24$ minutes.

He will take 3 hours 24 minutes.

So his ETA is 5.24 p.m.

Exercise 12.3A

1 How long does it take to travel:

- **a** 150 miles at 50 mph
- **b** 175 miles at 25 mph
- **c** 100 km at 40 km/h
- **d** 44 m at 8 m/s
- **e** 85 km at 20 km/h?

2 Time these journeys.

- **a** A fly flying 48 metres at a speed of 3 m/s.
- **b** A heron flying 320 metres at a speed of 8 m/s.
- **c** A spider crawling 26 centimetres at a speed of 4 cm/s.
- **d** A horse galloping 2000 metres at 13·9 m/s. (Give your answer in minutes.)

3 Naturalists suggest that hedgehogs are developing the habit of not curling up when there is danger. Their theory is that when they cross the road, those that curl up have a greater chance of being run down. Close-circuit cameras make some observations.

- **a** Harry the hedgehog crosses a road at 1·2 metres per minute.

  How long does Harry take to cross a road 6 metres wide?
- **b** Harry makes the return journey at a speed of 90 centimetres per minute.

  How long does it take him this time?

  (Note: 0·666... minutes = $0{\cdot}666 \times 60$ s = 40 s.)

4 Megan is preparing to do a 10-kilometre walk to raise money for charity.

In training she has averaged 7·2 km/h.

At this speed, how long will she take for her walk?

5 The rail journey from Edinburgh to London covers a distance of 450 miles.

The train travels at an average speed of 60 mph.

It leaves Edinburgh at 17 48. When does it arrive in London?

6 Petra runs an 800-metre race at an average speed of 6·5 metres per second.

a At this speed, how long does she take to run the race? (Give your answer in minutes and seconds.)

b The winner's average speed was 6·8 metres per second.

How much quicker was the winner's time than Petra's?

Exercise 12.3B

1 Use the formula $T = \dfrac{D}{S}$ to answer the following.

How long does it take to travel:

a 8 kilometres, at 24 km/h (answer in minutes)
b 15 miles at 75 mph (answer in minutes)
c 3 metres at 18 m/min (answer in seconds)
d 1 kilometre at 8 km/min (answer in seconds)?

2 The land speed record was broken in 2010 when an average speed of 414·3 mph was obtained.

How long at that speed did it take to travel one mile?
(Give your answer in seconds, correct to 3 significant figures.)

3 The winner of the Grand National horse race covered the 4·486 miles of the course at an average speed of 30·59 mph.

a Calculate the winner's time in:

i hours, to 4 significant figures

ii minutes, to 4 significant figures.

b The horse that was second in the race had an average speed of 30·26 mph.
How many seconds slower was it than the winner?

4 Sheila has a meeting in Ayr that begins at 14 30.

She plans to drive the 70 miles there from her home in Edinburgh.

She reckons she can drive there at an average speed of 50 mph.

a What is the latest time at which she should set off to be at the meeting on time?

b Sheila leaves home at 13 00. After an hour she has only travelled 42 miles.

Can Sheila still arrive on time for the meeting if she increases her speed, but keeps within the speed limit of 70 mph?

5 Light travels at 186 000 miles per second.

a How long does it take to reach the Earth from the Sun, 93 000 000 miles away?

b Astronomers have to deal with vast distances. To make this manageable they treat the **light year** as their unit of length. This is defined as the distance light will travel in 365 days.

If we think of the speed of light as 0·186 million miles per second, how many million miles are there in a light year?

6 When you go hill walking you should always tell someone where you are going and give them an estimated time of your return.

To help them work this out, hill walkers make use of **Naismith's rule**.

This says that the walkers should estimate their time on the basis of 1 hour for every 4 km of distance as measured on the map. They should then add 10 minutes for every 100 m gone up or gone down.

a From Corrie to Goat Fell is a distance on the map of 4 km. It is an ascent of 874 m.

Estimate the time of return to Corrie if the walkers set off at 9.15 a.m.

b From Arrochar to Ben Arthur and back is a distance of 8·8 km. It is a climb of 880 m.

Calculate how long the walk should take.

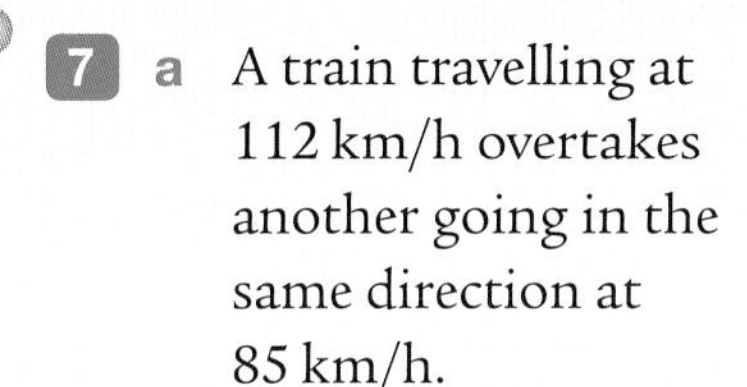

7 a A train travelling at 112 km/h overtakes another going in the same direction at 85 km/h.

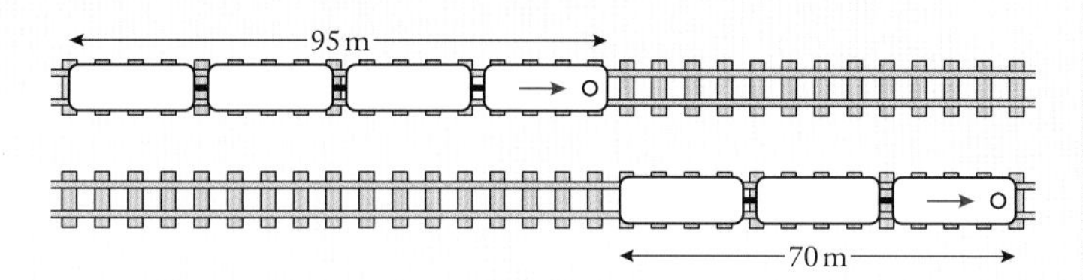

Their lengths are 95 m and 70 m respectively.

How long does it take for the faster train to pass the other completely?

b If the same trains were to pass each other in opposite directions, how long would it take for them to pass?

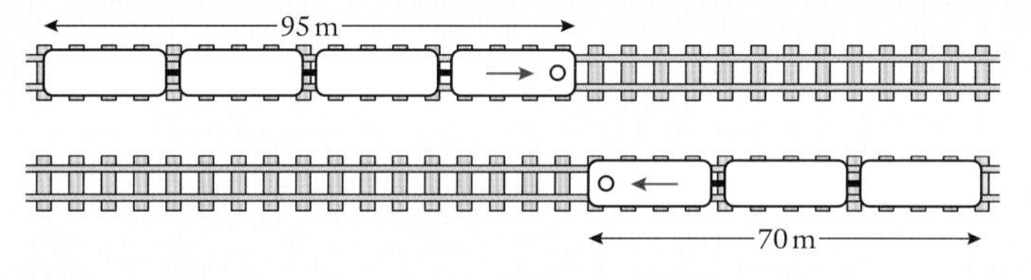

Making sure of speed, distance and time

Given any two, you should be able to find the third.

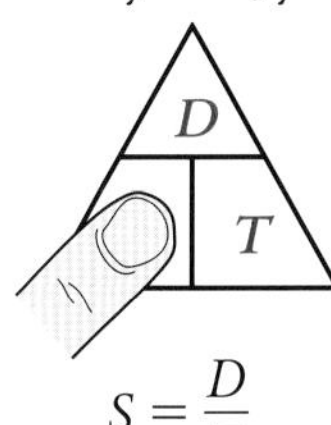

$S = \frac{D}{T}$

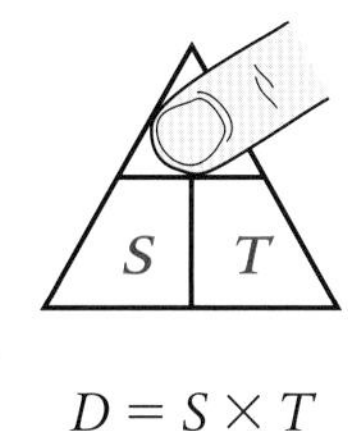

$D = S \times T$

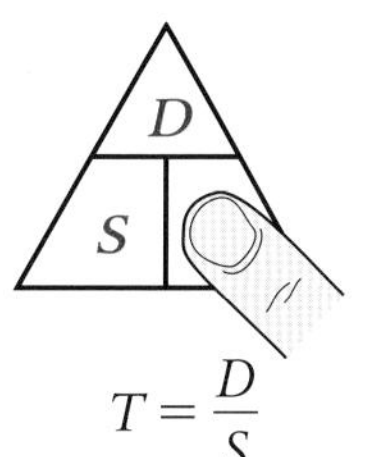

$T = \frac{D}{S}$

These triangles may help when the strategy is yours to pick.

Preparation for assessment

1 Calculate the missing entries in the table.

	a	b	c	d	e	f
Speed	km/h	12 m/s	9 cm/s	mph	30 mph	72 km/h
Distance	243 km	m	3·6 m	24 miles	5 miles	km
Time	9 h	1 min	s	6 min	min	1 h 45 m

2 Dan drove from Glasgow to Edinburgh.

It took him 1 hour 5 minutes.

He averaged a speed of 84 km/h.

a Express 1 hour 5 minutes in hours only, correct to 3 decimal places.

b What was the length of his journey?

3 Ann drove along the motorway for $2\frac{1}{2}$ hours.

She covered a distance of 195 kilometres.

Calculate her average speed.

4 On a rail journey from Aberdeen to Perth the train travelled at 92 km/h for 15 minutes, then at 126 km/h for the remaining 40 minutes of the journey.

How far is it from Aberdeen to Perth?

5 Archie's go-kart can reach a speed of 24 mph.

a At that speed how long would it take him to cover a distance of 60 miles?

b The go-kart broke down after 25 minutes. What distance had it travelled before it broke down?

6 A 150-metre length of road was resurfaced in three days.

a Calculate the average speed of resurfacing in metres per day.

b Each working day is eight hours long. What is the average speed in metres per hour?

7 Tom needs to travel from Perth to Bristol.

He consults the train timetable.

Leave Perth	09 18	12 35	15 40
Arrive Bristol	14 04	17 30	20 16

a The distance by rail is 344 miles.
Calculate the average speed of each train.

b Tom takes the fastest train.
When does it leave Perth?

c Perth is 32 miles from Edinburgh.
Estimate when the train reaches Edinburgh.

8 The graph shows Stuart's trip to his friend Peter's house and back home again.

He went by bus.

a How long did it take Stuart to reach Peter's house?

b How far from Stuart's house does Peter live?

c Calculate the average speed of the bus to Peter's house.

d How long did Stuart stay at his friend's house?

e What was the average speed of the return journey?

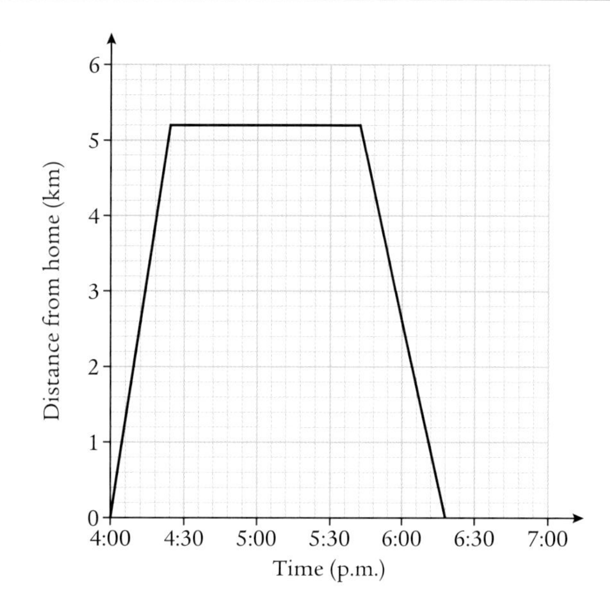

9 A car travels 60 kilometres from A to B at an average speed of 20 km/h and returns at 40 km/h.

Calculate its average speed for the whole journey. Why is it not 30 km/h?

10 a We know 8 kilometres ≈ 5 miles.

So, 80 km/h ≈ 50 mph.

We also know 0 km/h ≈ 0 mph.

Using this information, draw a graph similar to this, which can be used as a ready reckoner to convert speeds in miles per hour to kilometres per hour.

The horizontal axis needs to go to 140 km/h. The vertical axis needs to go to 70 mph.

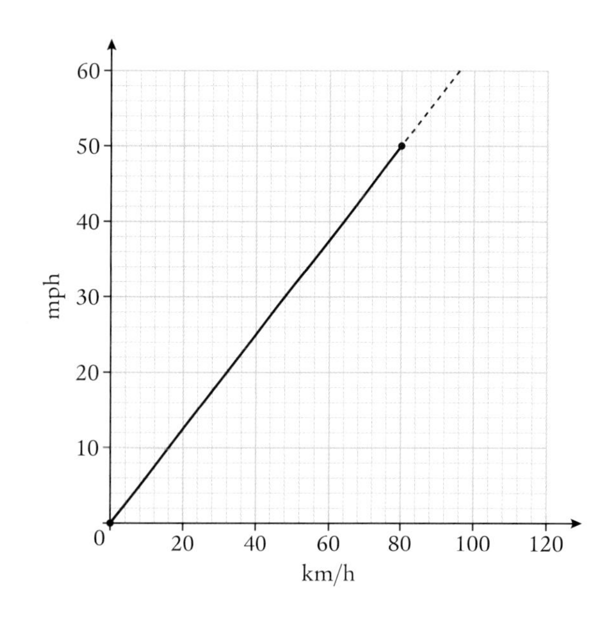

b Use your graph to change these km/h speed limits to mph.

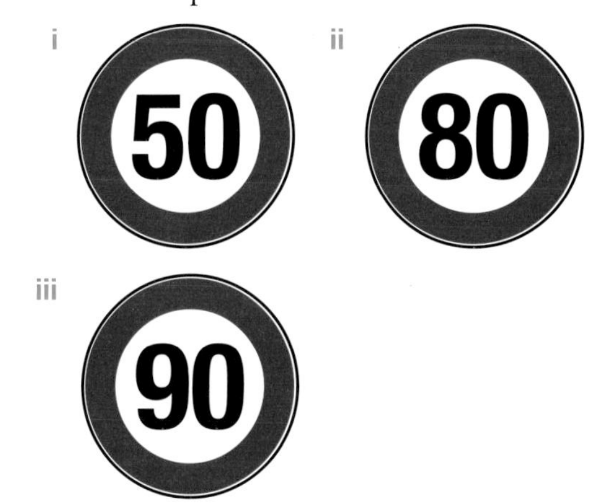

c Use your graph to change these mph speed limits to km/h.

i 30

ii

iii

d Use your ready reckoner to help you explore the thinking distances discussed earlier.

11 In history we learn that the Romans measured distances using 'paces'.

A pace was the distance stepped out by two strides.

A distance of 1000 paces (2000 strides) is *mille pasuus* in Latin ... from which we get our word 'mile'.

Our equivalent to a pace in those days was a 'yard' (1760 yards made a mile) and, smaller than that, three feet made a yard.

We still find these units used and have to understand them when we read history books.

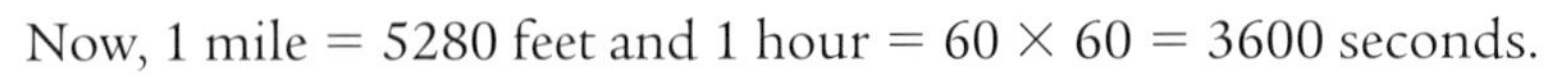

Now, 1 mile = 5280 feet and 1 hour = 60×60 = 3600 seconds.

$\Rightarrow$ 1 mile per hour = 5280 feet per 3600 seconds

$\Rightarrow$ 1 mph = $\frac{22}{15}$ feet per second (ft/s) (Check this by 'cancelling' $5280 \div 3600$)

So 1 mile per hour is the same as $\frac{22}{15}$ feet per second
and 1 foot per second is the same as $\frac{15}{22}$ miles per hour.

a A speed of 12 ft/s = $\left(\frac{15}{22} \times 12\right)$ mph = 8·18 mph.

In the same way, express these speeds in miles per hour.
Where appropriate give your answers correct to 3 significant figures.

i 5 ft/s ii 22 ft/s iii 50 ft/s iv 75 ft/s

b A speed of 25 mph = $\left(\frac{22}{15} \times 25\right)$ ft/s = 36·7 ft/s.

Express these speeds in feet per second.
Where appropriate give your answers correct to 1 decimal place.

i 10 mph ii 30 mph iii 45 mph iv 70 mph

c Investigate where feet and inches might still be used today.

12 Two 'average speed' cameras are situated one mile apart on a motorway.

A speed limit of 70 mph is in force.

A driver passes the first camera travelling at 85 mph.

He reduces his speed to 65 mph a quarter of a mile later and remains at that speed until he passes the second camera.

a Calculate the driver's average speed over the mile.

b i Is he over or under the 70 mph limit?

ii By how much? (Hint: first find the total time it took the driver to cover the mile.)

13 Remember the Formula 1 Spanish Grand Prix?

The race was 307·104 km long.

In 2012, the winner was Pastor Maldonado in his Williams car.

The race was completed in 1 hour 39 minutes and 9·145 seconds.

Calculate the winner's average speed for the race.

Preparation for assessment

No calculator

You will be asked to do a test that requires you to perform calculations without a calculator.

The topics tested will include:

- whole number percentages
- the mean of a data set
- calculating a fraction of a quantity where the numerator is not 1
- adding two decimal numbers and then subtracting from the result
- multiplying a decimal number by a whole number.

Example 1

A shopkeeper had a laptop costing £450. He took 14% off the price during a sale.

What was the sale price?

10% = £45 $\quad \frac{1}{10}$ of £450

1% = £4·50 $\quad \frac{1}{10}$ of 10%

3% = £13·50 $\quad$ 3 times 1%

So 14% = £63·00 $\quad$ Adding 10% + 1% + 3%

So sale price is £450 − £63 = £387.

Example 2

A chef has to convert Fahrenheit temperatures to Celsius for a recipe.

His method is to subtract 32 and then find $\frac{5}{9}$ of the result.

Convert 95 °F into Celsius for him.

95 − 32 = 63

$\frac{1}{9}$ of 63 = 7 $\quad 63 \div 9$

So $\frac{5}{9}$ of 63 = 35 $\quad 5 \times 7$

The temperature 95 °F is the same as 35 °C.

Example 3

Calculate: **a** $25 \times 12 \times 4$ **b** 21×18.

Often calculations can be made easier by rearranging the terms or factorising.

a $25 \times 12 \times 4 = 25 \times 4 \times 12$

$= 100 \times 12$

$= 1200$

b $21 \times 18 = 21 \times 2 \times 9$

$= 42 \times 9$

$= 42 \times 3 \times 3 = 126 \times 3$

$= 378$

Exercise Test A

1 Evaluate:

a $38{\cdot}24 + 54{\cdot}97 - 37{\cdot}54$

b $640{\cdot}1 - 36{\cdot}28$

c $375{\cdot}6 \times 8$

d $6642 \div 9$.

2 Calculate:

a $2 \times 13 \times 50$ **b** $24 \times 25 \times 8$ **c** 43×14 **d** 21×19.

3 **a** Express each number correct to 3 significant figures:

i 6·0451 **ii** 7268 **iii** 45·089.

b Round these to 2 decimal places:

i 14·208 **ii** 7·181 **iii** 0·017.

4 Calculate:

a $\frac{5}{9}$ of 72 **b** $\frac{2}{3}$ of 894 **c** $\frac{3}{5}$ of 625.

5 Calculate the following:

(You may want to remember $50\% = \frac{1}{2}$, $25\% = \frac{1}{4}$, $12\frac{1}{2}\% = \frac{1}{8}$, $20\% = \frac{1}{5}$, $10\% = \frac{1}{10}$, $33\frac{1}{3}\% = \frac{1}{3}$.)

a 5% of £60 **b** 8% of 80 kg **c** 10% of £75

d 20% of 18 m **e** $33\frac{1}{3}\%$ of 366 days **f** 50% of 5·45 seconds

g 75% of 12 g **h** 45% of 220 hours **i** 35% of 40.

6 **a** The price of a pair of trainers rose from £25 to £30.
Calculate the percentage increase in the price.

b A car purchased for £10 000 is now worth only £7500.
Calculate the percentage decrease in its value.

7 Novak typed a 5000-word essay in 1 hour 40 minutes.
Calculate his rate of typing in words per minute.

8 How many hours and minutes are there from:

a 22 17 on Tuesday until 11 05 on Wednesday

b 10.26 p.m. on Thursday until 3.25 a.m. on Friday?

9 **a** Share £84 in the ratio 5 : 2.

b Michele is paid £37·50 for a five-hour shift in the café.
How much should she be paid for an eight-hour shift?

10 **a** Add the measurements 3·25 m and 16·7 m, and take your answer from 20 m.

b Multiply 8·36 by 7.

11 **a** Calculate 9% of £68.

b There were 16 400 spectators at the City versus Rovers football match.
80% of them were City supporters. How many City supporters were there?

12 In a science experiment the room temperature was taken every day of the school week at noon:
18 °C, 24 °C, 21 °C, 19 °C and 23 °C.

Calculate the mean noon temperature for the week.

Calculator allowed

You will also be asked to do a separate test where the use of a calculator is permitted.

You could be asked to:

- solve a linear equation where you might have to simplify it first
- work with area or volume to solve problems
- compare data sets
- create and use a formula
- use the relationship involving speed, distance and time
- use Pythagoras' theorem to solve a problem
- use trigonometry to find a side or angle of a right-angled triangle
- solve a problem that involves coordinates and shape.

Exercise Test B

1 The population of the United Kingdom in 2006 is given in the table:

UK population (2006)					
Total population	Under 16	Adult working and unemployed	Adult retired	Males	Females
	11 946 800	36 092 200	10 791 400	28 618 900	30 211 500

a There are two ways to calculate the total population of the United Kingdom.
What are they?

b Use one of the ways to calculate the total population.

c How many people under 16 would there be if their numbers fell by 10%?

d How many females would there be if their numbers increased by 5%?

e Round each number in the table to the nearest 1000.

f By how many do the females outnumber the males?

2 The table shows the extremes of temperature on each continent.

	Africa	Asia	Europe	N. America	S. America	Oceania
Hottest (°C)	57·8	55	48·0	56·7	48·9	50·7
Coldest (°C)	−23·9	−71·2	−58·1	−63	−32·8	−25·6

a Which continent has recorded the highest temperature?

b Which continent has the greatest difference between its highest and lowest temperatures?

c What is this maximum difference?

d The coldest natural temperature ever recorded on Earth is −89·2 °C in Antarctica in 1983.
How much colder is that than the coldest temperature in Africa?

e The temperature on Uranus has dropped as low as −224 °C.
How much colder is that than the coldest temperature ever recorded on Earth?

3 Sam is looking for the best deal for calls and texts for his mobile phone.
He has to choose from the following five companies:

	Calls per minute	Texts
Com-Com	25p per minute or £5 for up to 150 minutes	10p, or £5 for 100+ texts
Talk-Tel	25p per minute for first 3 minutes then 5p per minute for the rest of the day	14p
Call-Com	25p	12p
Calltex	30p	10p
Com-Call	26p	10p

a In a typical day for Sam he makes 1 hour of calls and sends 12 texts.
Which company offers Sam the best deal?

b Sam's sister Sophie makes on average 10 minutes of calls each day and sends 50 texts.
Which company offers Sophie the best deal?

4 The annual landings and values of the three most common fish into the UK by UK vessels over a five-year period are:

	2006		2007		2008		2009		2010	
	Tonnes	Value	Tonnes	Value	Tonnes	Value	Tonnes	Value	Tonnes	Value
Haddock	38 900	45·3	32 300	39·9	31 900	35·0	34 800	34·2	31 700	36·2
Cod	12 900	20·7	12 800	21·7	9 800	20·3	11 600	20·7	14 700	28·6
Whiting	12 300	9·7	13 100	11·7	11 400	10·7	10 100	9·3	8 900	9·4

(Note: the values are in millions of £s, i.e. 45·3 represents £45 300 000.)

a In which year was the most **i** haddock **ii** cod **iii** whiting landed?

b What was the value of each of the landings in **a**?

c What was the total tonnage of the haddock landed from 2006 to 2010?

d What was the total value of the haddock landed over these years?

e What is the difference in tonnage between the lowest weight of whiting caught and the highest?

f What is the difference in value between the cod landed in 2006 and the cod landed in 2010?

g What is the total value of the three kinds of fish caught in 2010?

h If the total value in **g** were 5% higher, what would it be?

5 **a** Mrs Brown leaves her car in the car park from 2.37 p.m. until 4.26 p.m.

i For how long is the car parked?

ii How much has she to pay?

b Write down a formula for the charge, £C in terms of T, the number of hours paid for.

c Use your formula to calculate the charge on a six-hour stay.

6 Preparing for a cycle race, Rae cycled at 37 mph for 24 minutes, then at 30 mph for 12 minutes. How far did she cycle altogether?

7 A communications satellite orbits the Earth at a height 18 700 miles above the Earth's surface.

A signal travels at the speed of light, 186 000 miles per second.

How long does it take the signal to reach the satellite and travel back to Earth?

8 The distance–time graph shows the journey of two coaches, one travelling from Glasgow to Aberdeen and the other travelling from Aberdeen to Glasgow. Both coaches set off at 11 a.m. Both coaches stopped at services on their journeys.

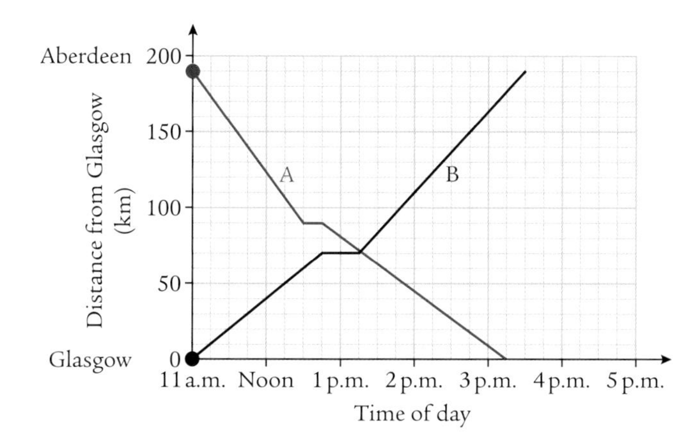

a What is the length of the road journey from Glasgow to Aberdeen?

b How far had each coach travelled before it stopped at services?

c How long had each coach been on the road before it stopped at services?

d Calculate the average speed of each coach on the first part of its journey.

e When did the coaches pass each other?

f How far were the coaches from Aberdeen when they passed each other?

g When did each coach arrive at its final destination?

h What was the average speed of each coach for the second part of its journey?

9 **a** Simplify and solve the equation for x.

$5x + 3 = 2x + 9$

b Here are some clues about a number represented by x:

x is a whole number

x is an even number

x is a triangular number

$x + 5 < 70$

$x - 8 > 20$

Find the number.

10 **a** ABCD is a trapezium.
Find its area.

b Calculate the perimeter of the trapezium, correct to 1 decimal place.

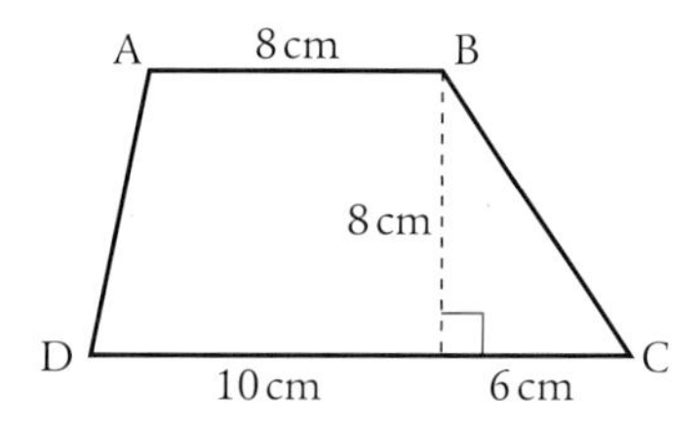

11 **a** On a coordinate diagram plot the points P(2, 1), Q(3, −3) and R(−2, −3).
Find the coordinates of S such that PQRS is a parallelogram.

b Calculate its area.

c **i** Using a scale factor of 2, draw PQ′R′S′, an enlargement of PQRS.

ii Write down the coordinates of Q′, R′ and S′.

d Compare the area of PQ′R′S′ with the area of PQRS.

12 A motorbike travels for two hours at an average speed of s km/h, and then for three hours at an average speed of $(s - 5)$ km/h.
Altogether it travels 430 kilometres.

a Form an equation in s.

b Hence calculate the two speeds.

13 Karen is making patterns using cocktail sticks.

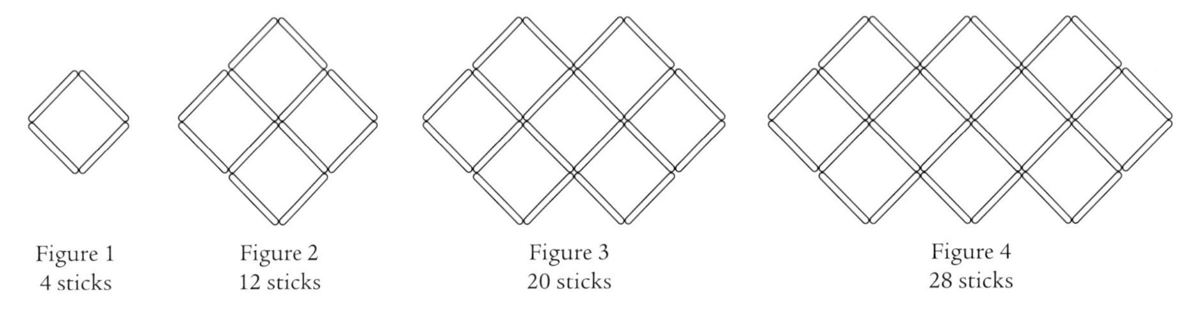

She is building each pattern by doing the same thing each time to the one before it.

a Write down a formula for S, the number of sticks used, in terms of f, the number of the figure.

b How many sticks are needed to make the 6th figure?

c Karen used exactly 100 sticks to make a pattern.
What is the pattern's number?

14 Hampden Park is the football stadium in Glasgow where the Scottish football team plays its matches against other countries.

Its playing surface is a rectangle, with dimensions 116 yards by 75 yards.

The centre circle is 10 yards in diameter.

A free kick has been awarded to Scotland on the halfway line at the edge of the centre circle, A.

The Scotland captain sees that the opposition goalie is not in his goal so he shoots and scores in the centre of the goalmouth, B.

What distance did the ball travel to cross the goal line?

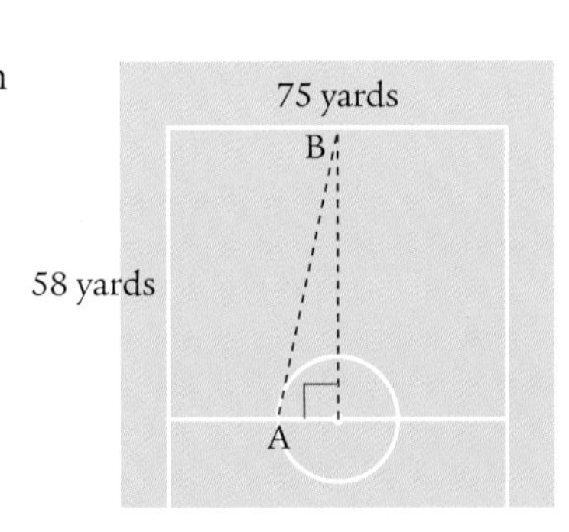

15 A garage is planned with dimensions as shown.

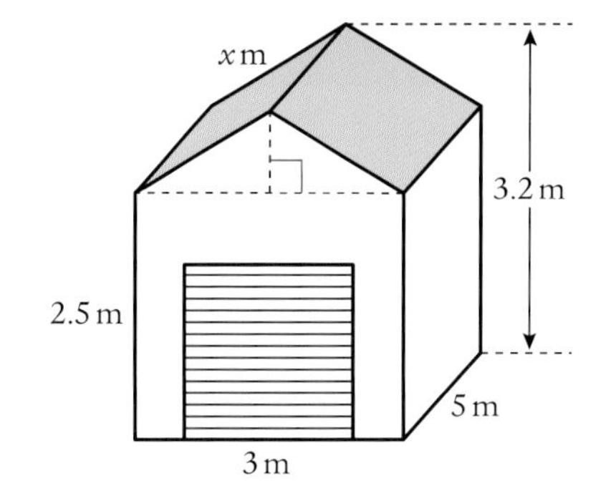

a Calculate the volume of the garage ... the cost of heating will depend on this.

b **i** Calculate the value of x to one decimal place.

ii The roof needs waterproofing. What is the area to be covered?

c At what angle to the horizontal is the roof sloping?

16 The *Rose* (R) is in distress.

It is on a bearing of 075° from the *Daisy* (D).

The *Flying Swan* (F) is on a bearing of 165° from the *Rose* and is 18 km due east of the *Daisy*.

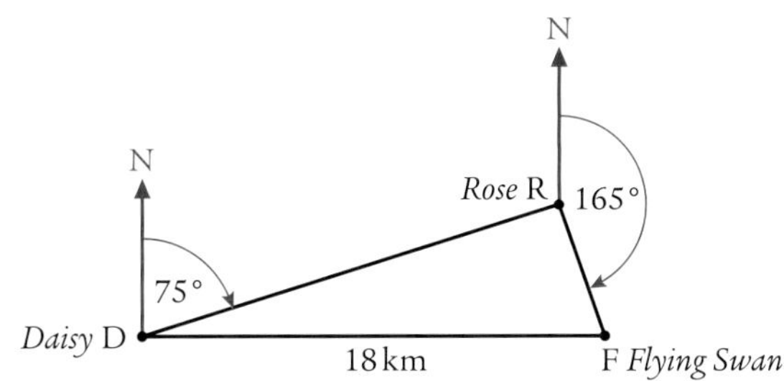

a Calculate the size of ∠DRF.
Explain your answer.

b Calculate the distance *Daisy* and the *Flying Swan* are from the *Rose*.

17 The times of the ten athletes in the 100 metres sprint and the 110 metres hurdles in a decathlon are shown in the table.

100 m sprint (s)	10·5	10·6	10·9	10·2	10·4	10·3	10·7	10·3	10·6	10·4
110 m hurdles (s)	13·6	13·9	14·0	13·4	13·6	13·6	13·8	13·5	13·8	13·7

a Calculate the mean and the range for the 100 metres, to 2 decimal places.

b Calculate the median and the range for the 110 metres hurdles.

c **i** Draw a scatter diagram, and a best fitting straight line.

ii From your line estimate the 110 metres hurdles time for a competitor who runs the 100 metres in 10·8 seconds.

iii Estimate a 100 metres time for a competitor who ran the 110 metres hurdles in 13·45 seconds.

Index

Acknowledgements

The authors and publishers are grateful to the following for providing photographs:

Barry Clarke Photography: 314b;
Crash Media Group/Alamy: 105, 132;
Fotolia: /© **anweber** 63, /© **blende64** 66t, /© **lilufoto** 111, /© **Aleksandr Kulikov** 112, /© **Delphimages** 281, /© **chrisdorney** 282, /© **kuassar** 291, /© **miles5** 300;
iStockphoto: 64t, 64mr, 64bl, 64br, 66b, 74, 123, 129, 241m, 241b, 277, 305, 318b, 306, 323, 311, 314t, 318t;
Mary Evans Picture Library/Alamy: 240t;
Royal Mint Museum: 39;
Visit Scotland: 113

All other photographs were provided by the authors.

Microsoft product screenshots reprinted with permission from Microsoft Corporation.